Social learning towards a sustainable world

Social learning

towards a sustainable world

Principles, perspectives, and praxis

edited by:

Arjen E.J. Wals

Cover art: Annelies Wals

ISBN: 978-90-8686-031-9

First published, 2007
Reprint, 2009

Wageningen Academic Publishers
The Netherlands, 2009

This book is dedicated to my father, Harry Wals, whose love for people and nature inspired not only me but all of those he touched around the world. With his charisma, energy, and youth, he was, without ever using the term himself, a catalyst of social learning.

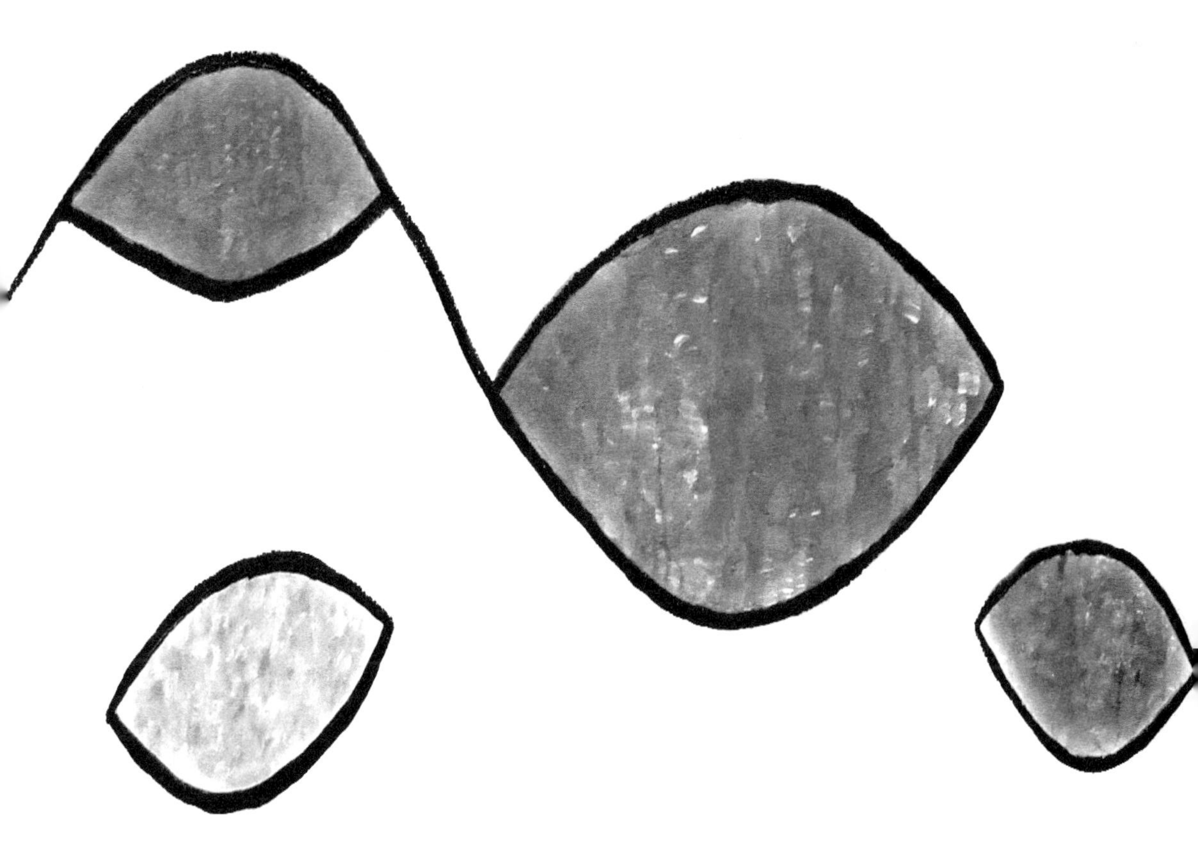

Table of contents

Acknowledgements 11

Foreword 13
Fritjof Capra

Introduction 17
Arjen E. J. Wals and Tore van der Leij

PART I PRINCIPLES 33

Chapter 1: Minding the gap: The role of social learning in linking our stated desire for a more sustainable world to our everyday actions and policies 35
Harold Glasser

Chapter 2: Riding the storm: towards a connective cultural consciousness 63
Stephen Sterling

Chapter 3: The practical value of theory: Conceptualising learning in the pursuit of a sustainable development 83
Anne Loeber, Barbara van Mierlo, John Grin and Cees Leeuwis

Chapter 4: Social learning revisited: Lessons learned from North and South 99
Danny Wildemeersch

Chapter 5: Learning based change for sustainability: Perspectives and pathways 117
Daniella Tilbury

Chapter 6: The critical role of civil society in fostering societal learning for a sustainable world 133
Richard Bawden, Irene Guijt and Jim Woodhill

Chapter 7: From risk to resilience: What role for community greening and civic ecology in cities? 149
Keith G. Tidball and Marianne E. Krasny

Chapter 8: Reaching into the holomovement: A Bohmian perspective on social learning for sustainability 165
David Selby

Chapter 9: Towards sustainability: Five strands of social learning 181
Robert Dyball, Valerie A. Brown and Meg Keen

PART II PERSPECTIVES 195

Chapter 10: Participatory planning in protected areas: Exploring the social-science contribution 197
Joke Vandenabeele and Lieve Goorden

Chapter 11: Social learning amongst social and environmental standard-setting organizations: The case of smallholder certification in the SASA project 209
Rhiannon Pyburn

Chapter 12: Social learning processes and sustainable development: The emergence and transformation of an indigenous land use system in the Andes of Bolivia 229
Stephan Rist, Freddy Delgado and Urs Wiesmann

Chapter 13: From centre of excellence to centre of expertise: Regional centres of expertise on education for sustainable development 245
Zinaida Fadeeva

Chapter 14: Learning about corporate social responsibility from a sustainable development perspective: A Dutch experiment 265
Jacqueline Cramer and Anne Loeber

Chapter 15: Social learning for sustainable development: embracing technical and cultural change as originally inspired by The Natural Step 279
Hilary Bradbury

Chapter 16: Corporate social responsibility: Towards a new dialogue? 297
Peter Lund-Thomsen

Chapter 17: Social learning as action inquiry: Exploring education for sustainable societies 313
Paul Hart

Chapter 18: Social learning and resistance: Towards contingent agency 331
Marcia McKenzie

Chapter 19: Sustainability through vicarious learning: Reframing consumer education 351
Sue McGregor

Chapter 20: Social learning for sustainability in a consumerist society 369
C.S.A. (Kris) van Koppen

PART III PRAXIS 383

Chapter 21: Partnerships between environmentalists and farmers for sustainable development: A case of Kabukuri-numa and the adjacent rice fields in the town of Tajiri in Northern Japan 385
Yoko Mochizuki

Chapter 22: Social learning in the STRAW project 405
Michael K. Stone and Zenobia Barlow

Chapter 23: Social learning in situations of competing claims on water use 419
Janice Jiggins, Niels Röling and Erik van Slobbe

Chapter 24: Exploring learning interactions arising in school-in-community contexts of socio-ecological risk 435
Rob O'Donoghue, Heila Lotz-Sisitka, Robert Asafo-Adjei, Lutho Kota and Nosipho Hanisi

Chapter 25: Professional ignorance and unprofessional experts: Experiences of how small-scale vanilla farmers in Uganda learn to produce for export 449
Paul Kibwika

Chapter 26: Multi-level social learning around local seed in Andean Ecuador 465
Marleen Willemsen, Julio Beingolea Ochoa and Conny Almekinders

Chapter 27: Learning and living with the Earth Charter 483
Michael C. Slaby, Brandon P. Hollingshead and Peter Blaze Corcoran

Epilogue: Creating networks of conversations 497
Arjen E.J. Wals

Afterword 507
Michael W. Apple

About the contributors 509

Index 523

Acknowledgements

In times when academics increasingly feel the pressure to publish in peer reviewed journals that preferably have a 'significant impact factor' and are discouraged from 'wasting' their time on other kinds of publications, it is refreshing to find colleagues who still find the time and have the energy to publish in edited volumes like the one in front of you. When the Dutch Interdepartmental Programme on Learning for Sustainable Development – which has earmarked 'social learning' as a key process in moving towards a more sustainable society – asked me to edit a volume on social learning in the context of sustainability, I had to think twice about it. Would I be able to find the right people? Suppose I would find them, would they be able to make time to write a chapter for an 'old fashioned' book? Would I be able to cover the key perspectives and principles (if they exist at all) on social learning? Would I find some highly reflective cases studies to match some of these principles and perspectives? Would I be able to do all this in less than a year, which is what the Dutch Interdepartmental Programme wanted? Would I be able to do all this and still be able to do the rest of my job in a way that I could keep it? Of course the answer to all these questions is: no, although I do still appear to have a job.

In order to be able to complete a daunting task in a fuzzy field, one needs a supportive social network, people with commitment and energy who are able to create space for themselves and for others to contribute to a project like this, but also people who work closely with you and people in your work environment that can tolerate an increasingly absent-minded, unavailable and, indeed, at times, sloppy colleague. I am very fortunate to have such a network of dedicated and high-energy colleagues around the world, to have people like Tore van der Leij (throughout) and Kate Collins (towards the end) working alongside me, and to have colleagues who not only (still) tolerate me but also lifted me up when I got bogged down.

Also, I am deeply indebted to all the contributing chapter-authors, to Fritjof Capra, author of the foreword, and Michael Apple, author of the afterword, for their enthusiastic response to my request for a contribution. Within months they often found co-authors in their own networks and were able to produce high quality chapters, making my work as an editor so much easier. It is wonderful to have 53 authors from countries spanning six continents, united in one book.

Furthermore, Roel van Raaij of the Dutch Ministry of Agriculture, Nature and Food Quality, deserves a lot of credit for being a driving force behind the Dutch policy-scene in creating support for this book. I also would like to thank the people at Wageningen Academic Publishers for preparing, publishing and promoting this book. I want to highlight Tore's role in the preparation of this manuscript. It was a pleasure to work with one of my former environmental education students on this book. Tore's insights and role of shadow reader of everything submitted have been invaluable throughout. Without his help this book would not have appeared.

Finally I want to thank Anne, Brian and Kendra for putting up with me when I was stressed out or chose to go to the office on Sundays instead of spending the day with them.

Foreword

Fritjof Capra

Creating communities and societies that are ecologically sustainable is the great challenge of our time. What is sustained in a sustainable community is not economic growth, development, market share, or competitive advantage, but the entire web of life on which our long-term survival depends. We do not need to start from zero to design these communities, but can model them on nature's ecosystems, which are sustainable communities of plants, animals, and microorganisms. Since the outstanding characteristic of the biosphere is its inherent ability to sustain life, a sustainable community is one that is designed in such a way that its ways of life, businesses, economy, physical structures, and technologies honor, support, and cooperate with nature's inherent ability to sustain life.

What is the place of learning in sustainable communities? How can such learning be organized and facilitated? What are some underlying principles? These are some of the key questions that this book seeks to address from the perspective of social learning. Both ecological communities and human communities derive their essential properties, and in fact their very existence, from their relationships. Sustainability is not an individual property, but the property of an entire network. The important concept of feedback, which was discovered in cybernetics in the 1940s, is intimately connected with the network pattern. Because of feedback in living networks, these systems are capable of self-regulation and self-organization. A community can learn from its mistakes, because the mistakes travel and come back along these feedback loops. Next time around we can act differently. This means that a community has its own intelligence, its own learning capability. In fact, a living community is always a learning community.

In the last century we have seen an evolution in education from nature conservation to environmental education to education for sustainability. This evolution parallels the shift from objective 'content' knowledge to contextual knowledge that characterizes ecological, or systemic, thinking. Explaining things in terms of their contexts means explaining them in terms of their environment. This shift encourages educators to serve as facilitators and fellow learners alongside students. It encourages a shift from 'transmissive' expert-based teaching and learning to transformative, community-based learning. In the former, content knowledge had to be transferred to citizens of all ages, particularly in the early years. People had to learn 'about' nature in order to better understand and protect it. In time, educators recognized the importance of more experiential learning, linking the development of competencies of the head (cognitive), heart (emotional), and hands (skills). Today, education for sustainability is less a matter of transmitting the content of ecology to citizens, and more about utilizing the principles underlying ecological processes in helping communities and their members respond to the challenge of sustainability in ways appropriate to their situations. These principles not only help people to better understand nature, but help them to better understand themselves and the communities in which they live and work, and to design education for sustainable living based on those principles.

Recent work in the theory of living systems has shown that such systems generally remain in a stable state, but that, every now and then, an open system will encounter a point of instability, in which there is either a breakdown or, more frequently, a breakthrough – the spontaneous emergence of new forms of order. This spontaneous emergence of order at critical points of instability (often referred to simply as 'emergence') is one of the hallmarks of life. It has been recognized as the dynamic origin of development, learning, and evolution.

Flexibility and diversity are key features of a resilient and sustainable system as they help a system cope with disturbances, so that these disturbances can become triggers for learning, for adaptive responses, or even whole system redesign. A diverse community is a resilient community, one that can adapt to changing situations better than homogenous communities, simply because the learning capabilities and the creative powers available to such a community are greater. But diversity offers a strategic advantage for a community only if there is a vibrant network of relationships and if there is a free flow of information through all the links of the network. When the flows are restricted, suspicion and distrust are created, and diversity becomes a hindrance instead of an advantage. If there is fragmentation, if there are subgroups in the network, or individuals who are not really part of the network, then diversity can generate prejudice, friction, or even destructive conflict.

Leadership in a sustainable learning community, to a large extent, consists of continually facilitating the emergence of new structures and incorporating the best of them in the organization's design. This type of systemic leadership is not limited to a single individual, but can be shared, and responsibility then becomes a capacity of the whole.

How does one facilitate emergence? Emergence can be facilitated by creating a learning culture, by encouraging continual questioning and rewarding innovation. In other words, leadership means creating conditions rather than giving directions. Above all, facilitating emergence means building up and nurturing a network of communications with feedback loops. The first step toward this goal might be loosening previously designed structures and thereby creating more flexibility. Another important aspect is creating an emotional climate that is conducive to emergence. This means a climate of warmth, mutual support, and trust, but also a climate of passion with plenty of opportunities for celebration. Finally, we need to realize that not all emergent solutions are viable. Therefore, a culture fostering emergence must include the freedom to make mistakes. In such a culture, experimentation is encouraged, and learning is valued as much as success. One of the main problems, in business as well as in education, is that organizations are still judged according to their designed structures, not according to their emergent structures. But I would hope that the rise of social learning will create more room for collaborative learning, diversity, and systemic thinking, and lead to more attention to emergent structures and to leadership that facilitates emergence.

This book brings together a range of ideas, stories, and discussions about purposeful learning in communities aimed at creating a world that is more sustainable than the one currently in prospect. This learning is called social learning to emphasize the importance of relationships, collaborative learning, and the roles of diversity and flexibility in responding to challenges and disturbances. In the spirit of social learning theory, the contributors to this book do not all agree on the meaning of social learning or on the specific actions by which to create a more sustainable world. In that sense, the book is designed to expand the network of conversations through which our society can confront various perspectives, discover emerging patterns, and apply learning to a variety of emotional and social contexts.

Introduction

Arjen E. J. Wals and Tore van der Leij

> "Let's face it, the universe is messy. It is nonlinear, turbulent, and chaotic. It is dynamic. It spends its time in transient behaviour. On its way to somewhere else, not in mathematically neat equilibrium. It self-organizes and evolves. It creates diversity, not uniformity. That's what makes the world interesting, that's what makes it beautiful, and that's what makes it work" (Donella Meadows 2005, p. 204).

People, communities and even companies both big and small, around the world are becoming aware that our current way of living in the short run, already for many, and in the long run, for many others, is unsustainable. After two decades of talk about sustainability and sustainable development, it appears easier to identify what is *unsustainable* (i.e. ecologically, socially, economically, ethically, culturally and environmentally) than to identify what it is to be sustainable. What is clear by now is that to break deeply entrenched, unsustainable patterns (assumptions, behaviours and values) requires a new kind of thinking inspired and informed by powerful learning processes that simultaneously lead to individual and collaborative action and transformation. David Selby (this volume) even speaks of a need for 'quantum learning.' The nature of sustainability-challenges seems to be such that a routine problem-solving approach falls short, as transitions towards a sustainable world require more than attempts to reduce the world around us into manageable and solvable problems. Instead, such transitions require a more systemic and reflexive way of thinking and acting with the realization that our world is one of continuous change and ever-present uncertainty. Einstein's observation that we cannot solve today's problems with the same kind of thinking that led to these problems in the first place, holds true even more today. This new kind of thinking means that we cannot think about sustainability in terms of problems that are out there to be solved or 'inconvenient truths' that need to be addressed, but to think in terms of challenges to be taken on in the full realization that as soon as we appear to have met the challenge, things will have changed and the horizon will have shifted once again.

After twenty years or so of talk about sustainability and sustainable development, both in theory and in practice, it has become clear that there is no single outlook on what sustainability or sustainable development means. It is also clear that there is not one process alone that will confidently realize its achievement. Determining the meaning of sustainability is a process involving all kinds of stakeholders in many contexts, and people who may not agree with one another. There are different levels of self-determination, responsibility, power and autonomy that people can exercise while engaged in issues or even disputes related to sustainable development. In dealing with conflicts about how to organize, consume and produce in responsible ways, learning does not take place in a vacuum but rather in rich social contexts with innumerable vantage points, interests, values, power positions, beliefs, existential needs, and inequities (Wals and Heymann 2004, Wals and Jickling 2002). The amount of space individuals have for making their own choices, developing possibilities to act and for taking responsibility for their actions, varies tremendously. This volume presents social learning not just as a naturally-occurring phenomenon but also as a way of organizing learning and communities of learners. This is not to suggest that there is some kind of consensus about the meaning of social learning. As Parson and Clark (1995, p. 429, also quoted by Glasser, this volume) write:

> "The term social learning conceals great diversity. That many researchers describe the phenomena they are examining as 'social learning' does not necessarily indicate a common theoretical perspective, disciplinary heritage, or even language. Rather, the contributions employ the language, concepts, and research methods of a half-dozen major disciplines; they focus on individuals, groups, formal organizations, professional communities, or entire societies; they use different definitions of learning, of what it means for learning to be "social," and of theory. The deepest difference is that for some, social learning, means learning by individuals that takes place in social settings and/or is socially conditioned; for others it means learning by social aggregates."

Although the idea of social learning is a bit messy in and by itself, in this book it tends to refer to learning that takes place when divergent interests, norms, values and constructions of reality meet in an environment that is conducive to learning. This learning can take place at multiple levels i.e. at the level of the individual, at the level of a group or organisation or at the level of networks of actors and stakeholders. In their recent book on environmental management Keen *et al.* describe social learning as "... the collective action and reflection that occurs among different individuals and groups as they work to improve the management of human and environmental interrelations (Keen *et al.* 2005, p. 4).

From a social learning perspective, the emergence of sustainability in the context of education can be viewed both as an evolving product and as an engaging process. Hence, sustainability as a social learning *process* is more interesting than sustainability as an expert pre-determined transferable *product* (i.e. as set by a policy, code of behaviour, charter or standard) which is not to say that the people using the term in this book are not pre-occupied by realising the kind of change in people, organisations and societies, that will ultimately lead to a world that is more sustainable than the one currently in prospect. Through facilitated social learning, knowledge, values and action competence can develop in harmony to increase an individual's or a group's possibilities to participate more fully and effectively in the resolution of emerging personal, organisational and/or societal issues. In social learning, the learning goals are, at least in part, internally determined by the community of learners itself.

The point of social learning is perhaps not so much what people should know, do or be able to do, which could be an embodiment of authoritative thinking and prescriptive management, but rather: How do people learn? What do they want to know and learn? How will they be able to recognize, evaluate and potentially transcend social norms, group thinking and personal biases? What knowledge, skills and competencies are needed to cope with new natural, social, political and economic conditions, and to give shape and meaning to their own lives? How can social learning build upon people's own knowledge, skills and, often alternative, ways of looking at the world? How can the dissonance created by introducing new knowledge, alternative values and ways of looking at the world become a stimulating force for learning, creativity and change? How can people become more sensitive to alternative ways of knowing, valuing and doing, and learn from them? How do we create spaces or environments that are conducive to the emergence of social learning? These questions, all addressed in this volume, not only suggest that learning in the context of sustainability is relatively open-ended and transformative, but also that it is rooted in the life-worlds of people and the encounters they have with each other.

This volume represents the first comprehensive attempt to present social learning in the context of education and sustainable development. The book contains three parts: principles, perspectives, and praxis. Part One provides a rationale for the book, as well as a number of key interpretations and principles of social learning in the context of sustainable development. Furthermore, it raises critical issues with regard to social learning as a 'tool' in moving towards a sustainable world. Part Two, contains contributions from a range fields that are challenged by sustainability issues and the need for systemic change. Perspectives offered include those of:

- organizational learning and environmental management within a framework of corporate social responsibility and 'the greening' of business and industry;
- interactive policy-making and multi-stakeholder governance;
- education, learning and educational psychology;
- multiple land-use, indigenous land-use and integrated rural development;
- consumerism and critical consumer education.

In Part Three, Praxis, these multiple perspectives are complemented by a number of reflective case studies of people/organizations/communities using forms of social learning to move towards sustainability. Each case study seeks to provide sufficient richness to allow the reader to relate and mirror the stories to his or her own experiences and thoughts and to stimulate praxis.

We will briefly touch upon all the contributions to the three parts of this book so as to give you a flavour of what is ahead and perhaps help you find a meaningful pathway through this book. You may wish to read this book from front to back, but perhaps you are interested in a particular perspective or a particular context in which social learning is applied. The overview below will hopefully help you determine a sensible reading sequence.

Principles

Part One, focussing on principles of social learning, opens with a chapter offered by Harold Glasser. Glasser focuses on the gap between people's stated, widespread concern for the environment and the future and their generally unsustainable actions, lifestyles, and public policies. He explores the potential of using a broadly conceived social learning framework as a strategy for both better understanding the existence of the disconnection and bridging this gap. According to Glasser social learning usually refers to an interactive, participatory, negotiated approach to, or process for, guiding collective problem solving and decision-making that incorporates innovation diffusion, systems theory and systems learning, soft systems theory, adaptive management, organizational learning, conflict management, multiple and distributed cognition, and Habermas' Theory of Communicative Action. Glasser's approach to, what he calls, active social learning is concerned with both improving the process of decision-making and creating more eco-culturally sustainable outcomes. He does caution us, however, on an a-critical use of social learning when writing: "While many of the ideas and concepts embraced by advocates of social learning have great potential to help facilitate a transition to eco-cultural sustainability, the term currently runs the risk of being perceived as a silver bullet or panacea". In order to increase the potential of social learning in creating a more ecologically sustainable world, Glasser ends his chapter by presenting eight challenges for research. Some of these provoking challenges, although not all of them, are indeed addressed in many of the subsequent chapters.

In Chapter 2 Stephen Sterling echoes the arguments made over recent years for a more ecological or relational cultural worldview particularly in the West and Westernised societies. Such a worldview is needed, Sterling argues, in order to help us 'ride the storm' of increasing environmental breakdown, inequity and social and political instability, and significantly increase our chances of breaking through to a more sustainable world. The chapter examines the nature of an ecological worldview, an ecological epistemology and of ecological consciousness, reviews evidence for their emergence as manifested in cultural change, debate and practices, and looks at implications for change in educational policy and practice. Sterling makes an important distinction between social learning as contingent (emergent and arising) and intentional (learning by design). In the end Sterling is looking for a learning-based breakthrough to a changed worldview which is both collective and connective.

In Chapter 3 Anne Loeber, Barbara van Mierlo, John Grin and Cees Leeuwis propose some principles for social learning in the context of sustainable development. They start with a reflection on sustainable development suggesting that the nature of sustainable development is conducive to situation improvement mechanisms that are social learning based. In their view, sustainable development is a challenging concept for two reasons. First, it is an essentially contestable concept that is claiming, normatively, to offer desirable directions for action, and at the same time demanding practical change. Therefore, it must be elaborated in an action-oriented way, reflecting a contextual balance between what is deemed desirable and what may be made feasible. Secondly, it is an essentially 'revolutionary' concept; its elaboration and implementation imply an opening up of existing routines, rules, values and assumptions embedded in the institutions that have co-evolved with earlier, 'unsustainable' modes of socio-technological development. Therefore, the elaboration of sustainable development into practical options for action must include a reflexive perspective, i.e. a critical scrutiny of things that are usually taken for granted, in such a way that path dependencies are challenged.

Loeber, van Mierlo, Grin and Leeuwis argue that both types of characteristics of the 'sustainable development' concept imply the need for learning and beg the following questions, which they address in their chapter:

- If sustainable development is an essentially contestable concept, can and should learning then lead to agreement- and if so, what does 'agreement' mean?
- If learning for sustainable development is to contribute to action for sustainable development, then how can the relationship between learning and action, that is, between 'thinking' and 'doing', be conceived?
- If reflexivity is involved in action for sustainable development, then learning must include, not bracket, fundamental values, worldviews and identities. How can that be achieved?
- If path dependencies need to be critically questioned, then learning must somehow take into account the relationship between long-term visions of the

future, and short-term action. How can these two time spans be meaningfully coupled in practice?

In Chapter 4 Danny Wildemeersch links social learning to four different dimensions of participatory processes: planning/action, reflection, communication, and negotiation. He conceives each dimension in terms of a basic tension:

- Planning/action: the tension between need and competence;
- Reflection: the tension between reflectivity and reflexivity;
- Communication: the tension between unilateral and multilateral communication;
- Negotiation: the tension between conflict and co-operation.

Wildemeersch suggests that successful social learning requires a balance between these basic tensions. Having been involved in a number of social learning oriented projects, in both Western and non-Western contexts, he concludes that these projects do not have unambiguous positive results. Wildemeersch explores the reasons for the relative successes and failures of social learning experiments. He pays particular attention to the power mechanisms which intervene in participatory learning processes. In doing so he looks at the different contexts (both Western and non-Western) and tries to analyse how these contexts (structural and cultural) influence processes of social learning for sustainable development.

In Chapter 5, Daniella Tilbury explores the links between social learning and sustainability. She attempts to define the characteristics of learning-based change approaches as well as the key principles which underpin these approaches to sustainability. Tilbury explores a variety of pathways to social learning processes which engage stakeholders in a consideration of power, politics and participation for change. In her chapter Tilbury makes the connection between social learning and current developments in the international policy arena with regards to education for sustainable development (ESD) as embodied by organizations such as UNESCO.

Richard Bawden, Irene Guijt and Jim Woodhill maintain that progress in relation to sustainable development, hinges on a social capacity for different sectors and interests in society to be able to constructively engage with each other. They argue in Chapter 6 that this is of critical importance for leadership of civil society organisations (CSOs) and civil society activism. Bawden, Guijt and Woodhill believe that the effectiveness of civil society depends on its capacity to engage individuals and organisations across all sectors in processes of critical reflection and learning. Understanding and being able to work with power differences and related conflicts is in their view central to such learning. The chapter explores the concept of 'a learning society' focusing in particular on the role of civil society in generating more inclusive and dialogic forms of democracy. The authors consciously opt for the term 'societal learning' instead of social learning as it helps move away from a simplistic group-based learning notion and refers directly to the capacity of societies and communities to be more learning-orientated in the way they tackle important issues related to a more sustainable world.

In Chapter 7 Keith Tidball and Marianne Krasny apply resilience theory to urban-socio-ecological systems. They refer to the work of so-called urban community greeners and other civic ecologists who integrate place-based activities with learning from multiple forms of knowledge including that of community members and outsiders, and with civic activism such as advocating for green spaces. Tidball and Krasny show that this kind of learning can build human, social, natural, financial, and physical capital that becomes integrated into constructive, positive feedback loops. In this way, community greeners integrate diversity, self-organization, and learning to create the conditions that spawn resilience in the face of disaster and conflict. They conclude that urban community greening, local biodiversity monitoring, and similar activities are tools that could become part of a larger civic ecology "tool kit" for building urban resilience.

In Chapter 8 David Selby suggests that there is broad agreement amongst sustainability proponents that the world, as we know it, as well as, our place in it, is at risk. There is disagreement, however, about where we presently stand on a continuum between 'redeemability' and 'irredeemability.' Selby also notices a common agreement amongst sustainability proponents that transformation, and hence transformative learning, is vital if we are to achieve sustainability. The problem is that, while we have a goal of transformation through learning, the processes and modalities of learning we employ carry more than a residue of what has fanned the flames of unsustainability. If we are to make a quantum leap towards sustainability, we need quantum learning. David Selby argues that social learning for sustainability calls for learning processes divested of mechanistic influences and residues. A deeply embedded mechanistic worldview lies behind the global mega-crisis while efforts to realize a sustainable world are themselves hampered by our inability to remove residues of mechanism from our sustainability proposals. He suggests that that we are straitjacketed by our failure to see, let alone address, mechanism within our thought processes. To get out of the mechanism trap he proposes to design and support 'dialogical social learning' based upon David Bohm's conception of dialogue. What Selby proposes in the end is, what he calls, quantum learning for sustainability.

In the closing chapter of Part One, Robert Dyball, Valerie Brown and Meg Keen develop five essential strands of social learning. These strands include the need for reflectivity, that is, learning from reflection on ourselves, our knowledge, and our relationship to others and the effect this has on learning processes aimed at achieving sustainability. The second strand recognizes the power of systems thinking in focusing our attention on processes, relationships and interactions, understood as a dynamically changing whole. The diversity of knowledge and values in any sustainability situation demands the third strand, integration, to bring together, and draw strength from, different ways of thinking, different types of knowledge and diverse experience. The fourth strand embraces the inevitable presence of conflict in forging collaboration, not as a force to be avoided, but one to be harnessed in negotiation. This negotiation means working with rules of dialogue that ensure diverse interests can be expressed and taken into account, and power imbalances addressed. The final strand is that of participation, which invites a wide and diverse community into the learning process, thus facilitating the establishment of new and strong partnerships.

The authors suggest that these strands can be woven through the iterative cycles of learning, across various scales of application and between the various partners in the process. Their use can help recognize and avoid the constraints and artificial jurisdictional and disciplinary boundaries that hinder community and institutional collaboration. Creatively applied the strands can open opportunities for creative new approaches to action and learning that support sustainability.

Perspectives

In Part Two a number of perspectives on social learning in the contact of sustainability are introduced. In the opening chapter, Chapter 10, Joke Vandenabeele and Lieve Goorden discuss social learning in the context of participatory planning in protected areas. The image they use to describe the process is an image of Schön: a highland with a view of the swamp. The official who has to facilitate this process is confronted with a choice. Will he/she stay on the safe floor of the cooperation between civil servants? Or will he/she also choose to come down into the swamp of local actors and their particular uses and definitions of the area? According to Schön debates within the swamp of an area is the only way to enhance deeper learning. That is to handle the basic and most important issues in nature conservation. Vandenabeele and Goorden specify three main requirements concerning the quality of participation of local actors in this kind of planning processes: (1) defining content-based arguments for the participation of local actors in the planning process of nature plans, (2) stimulating a reflection on process criteria to enhance participative planning, and (3) making visible the opportunities of process management via the management of the content.

In Chapter 11 Rhiannon Pyburn explores the (social) learning process amongst organizations already working for a more ecologically and socially sustainable world, including; the Fair-trade Labeling Organizations International (FLO), Social Accountability International (SAI), the Sustainable Agriculture Network of the Rainforest Alliance (RA) and the International Federation of Organic Agriculture Movements (IFOAM). Using the Social Accountability in Sustainable Agriculture (SASA) project as a backdrop, she seeks to understand the challenges of multiple-level social learning. The chapter examines the current theoretical literature on social learning in relation to the experience of the SASA project, by analyzing the progression of one project's sub-objective: to address the challenge of smallholder access to certification (fair trade, organic, rainforest alliance and SA8000) in developing countries.

In Chapter 12 Stephan Rist, Freddy Delgado and Urs Wiesmann analyze a typical common property-based Andean land-use as the outcome of long-term social learning processes. Their analysis shows that this land-use system reflects a successive embodiment of ethical principles in humans and nature corresponding to different periods of local history. Furthermore, they observe that embodied ethical principles are emerging from a historically changing interface of local and external forms of knowledge. One of the conclusions the authors draw is that the degree of differentiation among ethical values corresponding to different stages of local history greatly depends on the type of cognitive competence (reflexivity) developed by the members of the communities as part of their lifeworlds. However, their analysis also demonstrates that cognitive competencies can only release their full transformative potentials when they are embedded in a further development of social capital and social as well as emotional competences. Social learning processes are therefore conceived of by the members of the community as addressing the human being in its entire inner and outer dimensions. The findings of the case study are related to the more general debate about main features, dynamics and the enabling or hindering factors of social learning processes in the field of sustainable development.

In Chapter 13 Zinaida Fadeeva presents and analyses the initial experiences of the first group of so-called Regional Centres of Expertise (RCEs) for education for sustainable development (ESD). RCEs are described as networks of formal, non-formal and informal educational organizations aiming to facilitate education for sustainable development in a specific regional and local community. In addition to discussing the critical factors affecting RCE mobilization and the initial outcomes of RCE activities, the chapter attends to the challenges of developing such centres. In particular, Fadeeva discusses the power dynamics of organizations coming together to form an RCE, and the challenges of going beyond the customary actions towards more revolutionary innovations and the complexities of embracing the wide spectrum of ESD. The chapter also highlights the role of 'excellence', 'expertise' and 'knowledge' for learning for sustainable development. The chapter challenges the notion of expertise as fragmented and compartmentalised knowledge and demonstrates how RCEs and a network of RCEs can provide an opportunity for overcoming traditional institutional divides and facilitating social learning.

In Chapter 14 Jacqueline Cramer and Anne Loeber introduce the role of social learning in the context of developing Corporate Social Responsibility (CSR). They do so by zooming in on an interesting Dutch initiative – the Dutch National Initiative for Sustainable Development (NIDO) – to launch a major programme on promoting CSR, entitled 'From financial to sustainable profit'. In their analysis of this initiative Cramer and Loeber found that learning was triggered not only among the company representatives that participated in the programme, but also at the level of the participating companies. The chapter provides some valuable insights when addressing the question of: "How to induce learning processes in the corporate sector that may help further the ambition for a sustainable development?"

One of the conclusions the authors draw is that the advancement of CSR requires learning at all levels: at the level of the individuals in a company, at company level, but also between companies and other external organisations. In the end, Cramer and Loeber state, learning is the key to dealing with corporate social responsibility. Their analysis of the NIDO programme shows how such learning can be triggered, facilitated and 'exploited' to create an impact.

How can we fundamentally change the ways in which we live together – with all living beings and systems – so that future generations not only survive but thrive? This is the key question that Hilary Bradbury addresses in Chapter 15. Bradbury observes that social, cultural and behavioural change does not easily keep pace with technical insight. We know what to do, but seem to be unable to actually do it. With some irony she notes that the sustainability community piles on technical insights without paying much attention to human, behavioural factors that support sustainable change. Bradbury seeks to balance that by describing a learning approach to sustainability that seeks to embrace technical and cultural change for sustainable development. Therefore, this chapter reminds those involved in the work of sustainable development that all change must be implemented by people, as individuals, groups, organizations, or societies.

The important 'people' side of large scale systems change is described using Bradbury's story of The Natural Step – a learning strategy designed to help businesses and other social institutions move toward greater sustainability. The chapter seeks to articulate principles of human systems change highlighting the connections between what sustains people internally and personally with their work on sustaining the external, 'natural' environment. The examples Bradbury uses vividly illustrate that it is not enough to develop a 'right solution' to our sustainability challenges; we must also figure out how our individual and organizational-cultural behaviours can interactively be brought into alignment with sustainable development. The chapter shows that there is a range of ways in which this interplay between the internal and external, or personal and professional can occur and deliver results. Its contribution is therefore to help crystallize a small set of principles from successful, complex change efforts to date.

Like chapters 14 and 15, Peter Lund-Thomsen's Chapter 16 also focuses on the way companies meet the challenges posed by sustainability. Lund-Thomsen discusses the relationship between corporate social responsibility (CSR), development, and social learning within mainstream business settings. He argues that the CSR discourse is mostly defined by business interests and the need for business to frame sustainable development issues in ways that turn questions of social, environmental, and economic justice into technical problems that can be solved through a managerial problem-solving approach. Such an approach means that only financially viable solutions where a 'business case' for CSR can be made receive serious attention from most business organizations. The implication is that issues around conflict, class struggle, and more radical approaches to citizen participation are sidelined in CSR teaching, CSR conferences, and most business practice in developing countries. In the second part, Lund-Thomsen argues that new spaces for social learning about CSR and development need to be opened so that these 'sidelined' issues are brought into the heart of the CSR and development debate. He advocates initiation of a different kind of dialogue where new knowledge, alternative values, and ways of engaging in CSR and development can be introduced to CSR educators, policy-makers, and present as well as future managers. In the end Lund-Thomsen concludes that social learning in relation to CSR is essential if we are to fully appreciate not only the potential but also the limitations associated with CSR in developing countries.

In Chapter 17 Paul Hart directs our attention toward both new learning theory and expanded conceptions of action research as mutually constitutive arguments for education for sustainability. Hart frames action research as a form of social learning. Underpinning notions of action research as a form of social learning is the assumption that knowledge and understanding may be conceptualized beyond formal (i.e. propositional) knowing as socially-situated, practical knowing. By broadening our ideas about learning, Hart suggests, we can explore learning within the context of an individual's participation in socio-cultural practices. Based on his analysis of the school-based Environment and School Initiatives (ENSI) project, Hart suggests that, when viewed as social learning situations, action research groups may help teachers in subtle but profound ways to acknowledge their multiple subjectivities and to name new subject positions as they learn, through accepted social discourses, to see new categories, perhaps blurring boundaries between existing binaries (e.g., cognitive/social perspectives on learning) and create new approaches, as socially and environmentally sustainable educational experiences.

Like Paul Hart, Marcia McKenzie also uses formal education as a backdrop for her chapter. In Chapter 18 she introduces three portraits of resistance, exploring various understandings of agency, activism, and education by combining the voices of students and educators. The three very different educational stances on agency and activism McKenzie that highlights are: 'awareness and inactive caring', 'critique and lifestyle activism', and 'contingency and changing the world'. Discussing these positions in relation to epistemological orientation, program characteristics, issues of class, and other critical factors, the chapter draws on discursive understandings of knowledge to highlight different conceptions of agency and related modes of socio-ecological activism. The effect of these multiple resistances points to the possibility that change results from the interaction of multiple discourses, whether at the individual or societal level. McKenzie suggests that there is always a possible tension between the discourses available and, as a result, the subject's interpretation and use of them. Rather than being free of a discursive constitution, we may work within that constitution, using alternative discourses to resist, modify or re-direct the discourses themselves. Encumbered by constituting discourse, and not at all transparent or independent of power matrices, this alternate notion of reflexivity, McKenzie concludes, becomes a potential tool as educators work to engage students in more reflexive and systemic forms of socio-ecological activism.

In Chapter 19 Sue McGregor addresses social learning in the context of consumer education. She suggests that the 'type' of consumer education that is taught affects the kind of consumer that is 'created' which, in turn, affects sustainability. How can social learning theory inform this issue? McGregor asks. To address this query, the chapter takes a different perspective: How can social learning theory inform the re-conceptualization of consumer education so as to contribute to sustainable consumer empowerment? If consumers cannot learn to find their inner power, their inner voice and potential (become empowered), they will struggle to see themselves as consumer-citizens. They run the risk of continuing in the role of degraded, isolated consumers seen as inputs for the economy and they miss the opportunity to regain their humanity as global citizens. Sustainable development (especially human and social development), McGregor argues, is predicated on people seeing themselves as citizens first and consumers second, often discussed under the rubric of consumer citizenship.

In order to further develop her perspective, McGregor weaves together such notions as consumer accountability and human responsibilities as well as ideas that stem from the new sciences (chaos theory, quantum physics and living systems), including transdisciplinary inquiry and the holomovement principle (see also David Selby's chapter in Part One). Insights from productive, authentic pedagogies and global education pedagogy also shape the chapter. Consumer education is reframed from a social learning perspective so that educators can create a space where alternative, sustainable, reflective thinking can be fostered.

In Chapter 20 Kris van Koppen opens with the observation that consumption lies at the heart of industrialized societies. The 18th and 19th century can be characterised by a revolution in both production and consumption. Nowadays, van Koppen notes, patterns of consumption have a major influence on the institutions, discourses, and practices in society. Van Koppen uses 'consumerism' as a term embodying these patterns. He uses consumerism as a descriptive term that does not per se bear the negative load it often has in environmentalist writing. Rather than arguing against consumerism, van Koppen believes that in social learning for sustainability, the consumerist features of modern society should be taken into full account, using them positively where possible, and resisting them where necessary. Such a strategy begs an open and thorough analysis of the relationships between consumption and social learning. Van Koppen presents a rough sketch of such an analysis, starting from sociological theories of consumerism and then exploring the relationships with social learning and education for sustainability.

Praxis

In Chapter 21 Yoko Mochizuki open up the praxis section of this book with a story of Kabukuri-numa and the adjacent rice fields in the town of Tajiri in northern Japan. Tajiri Town, where the famous wild goose habitat site Kabukuri Marsh is located, takes an innovative approach to community development. With the leadership of the Japanese Association for Wild Geese Protection (JAWGP), a diverse array of local stakeholders – NGOs, farmers, local and national government authorities, researchers – came to be dedicated to managing Kabukuri Marsh to maintain its ecological functions. Overcoming the initial antagonisms between those who called for the protection of wild geese and rice farmers who viewed wild geese primarily as harmful birds, Tajiri Town is aspiring to pursue the preservation of biodiversity (in rice paddies) and sustainable agriculture. This chapter describes processes of social learning for mutually respectful cooperation between 'environmentalists from outside' (who were perceived by local people, especially farmers, as fanatic bird lovers) and 'local people' (who hated birds) and presents a model case of promoting both environmental and economic agendas at the local/regional level.

In Chapter 22 Michael Stone and Zenobia Barlow offer the STRAW project as a reflective case study on social learning for nature restoration. The Students and Teachers Restoring a Watershed (STRAW) project involves 3,000 students yearly in habitat restoration in the San Francisco Bay Area. The program, the product of twelve years of social learning work, includes students, teachers, administrators, ranchers, for-profit businesses, philanthropic foundations, other non-governmental organizations, and governmental agencies. The process by which this disparate collection of stakeholders, with diverse purposes, goals, and values, became a network working together for sustainability can be understood through a variety of ecological and systems principles. The lessons learned are applicable in many settings.

In Chapter 23 Janice Jiggins, Niels Röling and Erik van Slobbe present social learning as a response to the challenge to find more adequate forms of governance of water resources in a European context. Their contribution is based on their involvement in a European six-country study in support of the European Union's Water Framework Directive (WFD). The chapter first presents the WFD as an attempt to manage complex resource dilemmas sustainably, then examines the implications for governance mechanisms, and continues by focussing on the facilitation of social learning as an approach to the coordination of human behaviour that supplements more familiar forms of resource governance. The chapter ends by drawing out the implications for knowledge processes, and by offering some guidelines for social learning.

In Chapter 24, Rob O'Donoghue, Heila Lotz-Sisitka, Robert Asafo-Adjei, Lutho Kota and Nosipho Hanisi report on three cases of teacher researchers working with local communities to mobilise a cultural capital of indigenous ways of knowing in school curriculum contexts. In each case a communal capital of indigenous cultural practice was the starting point for curriculum activities and for deliberative social interactions around livelihoods and lifestyle choices in the African contexts of the research. The purpose of each research project was to explore methodologies involving learners, teachers and community members working in response to the call for more relevant, contextually situated and socially engaged curriculum in a post-apartheid South Africa. Using the three cases, the chapter explores the findings of a National Research Foundation research programme on environmental learning and curriculum. These findings indicate that social change through environment and health education initiatives appears to be more locally generative in social contexts of risk. Here meaningful learning interactions can begin to engage community knowledge and school learning areas (subjects) around questions of sustainable human livelihoods and lifestyle choices.

In Chapter 25, Paul Kibwika illustrates how smallholder farmers in Uganda engage to learn and innovate about a new crop, 'vanilla,' to take advantage of opportunities in international markets without the intervention of research and extension. It highlights the role of interpersonal relationships and intergenerational exchange in social learning for sustainability. He argues that the purported role of extension and research as 'providing information and technologies' stifles social learning for sustainable living. If research and extension are to be relevant in social learning, their functions must be redefined. New areas of focus that Kibwika introduces include brokerage, organisational development, facilitating learning and dialogues, and entrepreneurial development. These require new competences and a reorientation of mindsets – hence, a new breed of professionals.

In Chapter 26 Marleen Willemsen, Julio Beingolea and Conny Almekinders present the process and experiences around the initiation of a seed system conservation project in three Andean provinces of Ecuador. The process they describe aimed to make farmers more aware of the importance of seeds for their agriculture. The awareness-raising was only the beginning of the project and meant to provide a space for farmers to identify desirable actions in regard to their seeds. The involved NGOs and local organisations aimed for a project that dealt with seeds and food security and was to be designed in a participatory way. In the first part of the chapter they elaborate on the issues of use of seeds, genetic erosion and sustainability in the Ecuadorian communities in the Andes and how the authors feel this initiative fits the concept of learning for sustainability. Subsequently, they present the experiences of the actors in the learning process at various levels and cycles. Finally, they reflect on the development of the participation in the learning process over time.

In Chapter 27 Michael C. Slaby, Brandon P. Hollingshead and Peter Blaze Corcoran reflect upon the learning that takes places within the global network of the Earth Charter Youth Initiative (ECYI). The ECYI includes several hundred young leaders from some 40 countries striving to bring alive the values of justice, sustainability and peace as they are formulated in the Earth Charter. The Earth Charter is a statement of ethical principles developed through a large process of worldwide participation by many thousands of stakeholders in meetings and online discourse over eleven years. Within the network of the ECYI, the authors see three levels of social learning that can be discerned: learning processes that focus on the lifestyle of the individual participant, those that take place in small local youth groups, and those that are facilitated through international online communication.

Conclusion

The principles, perspectives and praxes of social learning in the context of sustainability as presented, outlined and discussed in this book, will continue to evolve and new ones will emerge. Both social learning and the search for a more sustainable world continue to draw attention from increasingly overlapping worlds of research, education, policy-making, governance, community organising, and business and industry. If growth in the number of Google-hits is any indication of this increasing attention then consider the following. In the 16 months it took to put this book together the number of hits for 'social learning' grew from just over 400,000 hits to just over 900,000 hits, while the number of hits for 'social learning' and 'sustainability' combined grew from 53,000 hits to 151,000 hits[1]. When randomly opening up some of these pages it becomes immediately clear that there are many ways of conceptualising both social learning and sustainability and that there is a need for a thorough discussion of both terms and there possible relationship. This book addresses that need and invites readers to critically consider social learning as a transitional and transformative process that can help create the kinds of systemic changes needed to meet the challenge of sustainability.

References

Keen, M., Brown, V., and Dyball, R., eds. (2005) *Social Learning in Environmental Management: Towards a Sustainable Future*, London: Earthscan.

Meadows, D. (2005) "Dancing with Systems", in M. Stone and Z. Barlow, eds., *Ecological Literacy: Educating our Children for a Sustainable World*, San Francisco: Sierra Club Books.

Parson, E.A. and Clark, W.C. (1995) "Sustainable Development as Social Learning: Theoretical Perspectives and Practical Challenges for the Design of a Research Program", in L.H. Gunderson, C.S. Holling and S.S. Light, eds., *Barriers and Bridges to the Renewal of Ecosystems and Institutions*, New York: Columbia University Press, pp. 428-460.

Wals, A.E.J. and Heymann, F.V. (2004) "Learning on the edge: exploring the change potential of conflict in social learning for sustainable living", in A. Wenden, ed., *Educating for a Culture of Social and Ecological Peace*. New York: SUNY Press, pp. 123-145.

Wals, A.E.J. and Jickling, B. (2002) "Sustainability in Higher Education from Doublethink and Newspeak to Critical Thinking and Meaningful Learning", *Higher Education Policy,* 15: 121-131.

[1] The first time we googled these terms was in August of 2005 when accepting the challenge to compile this volume. The second time was late November of 2006. The number of Google-hits of course does not necessarily mean that all hits lead to high quality web-sites or that something must be very important. To illustrate this: on this day (November 24, 2006) 'social learning' gets 923,000 hits, 'bull shit' gets 909,000 hits, while 'military intelligence' gets 1,390,000 hits. Unfortunately we failed to type in the latter two terms in August 2005 as well and are therefore unable to establish whether they experienced a similar Google-growth pattern.

Part I
Principles

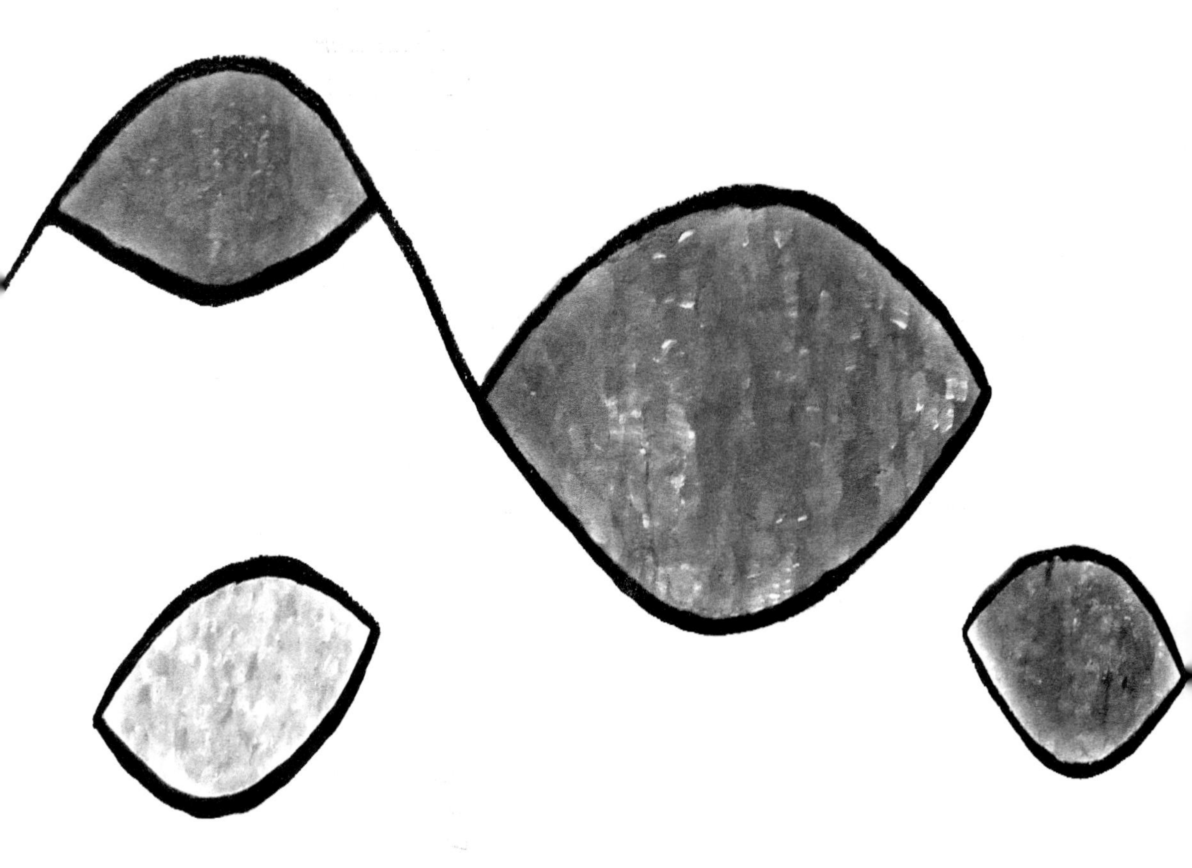

Chapter 1

Minding the gap: the role of social learning in linking our stated desire for a more sustainable world to our everyday actions and policies

Harold Glasser

> There can be few more pressing and critical goals for the future of humankind than to ensure steady improvement in the quality of life for this and future generations, in a way that respects our common heritage – the planet we live on. As people we seek positive change for ourselves, our children, and [our] grandchildren; we must do it in ways that respect the right of all to do so. To do this we must learn constantly – about ourselves, our potential, our limitations, our relationships, our society, our environment, our world. Education for sustainable development is a life-wide and lifelong endeavour which challenges individuals, institutions and societies to view tomorrow as a day that belongs to all of us, or it will not belong to anyone (UNESCO, *Decade of Education for Sustainable Development*).

Introduction

Humans have been both fascinated and tortured by questions regarding our fate and future for at least as long as we have possessed the ability to share our thoughts and document these ruminations. Under the best of circumstances, these musings involve asking a series of questions about the present, past, and future. Where are we? How did we get here? Where do we appear to be heading? Where do we want to go? How do we get there from here?

Many have argued that humankind is currently amidst (and possibly adrift in) an unprecedented transition; one as significant as passage into the Stone Age, the Agricultural Age, or the Industrial Revolution (Speth 2004, Raskin *et al.* 2002, Bossel 1998, Catton 1980). Our fate and future is and always has been intertwined with nature, despite the widespread failure of most humans to act in a manner that reflects a deep understanding of this relationship. And now, for the first time, we have gone full circle, causing the fate and future of nature – and evolution in general – to become entwined with our own (Broswimmer 2002, Wilson 1992, Myers 1979).

The contours of the future we are now forging, however, as always, are yet to be fully determined. Simply restated, the future is emergent and, within limits, plastic. While conscious design is unlikely to afford us the capacity to control the future directly, how we craft our sphere of concern and how effectively we link this to action will likely influence the future in profound ways. A broad spectrum of ostensibly divergent scenarios for the future has been proposed (Hammond 1998, Hawken *et al.* 1982, Catton 1980, Ophuls 1977, Brown 1954). These scenarios range from a perpetuation of the status quo via increasingly authoritarian measures to buoy economic growth under mounting scarcity and inequity; to a barbarized 'Mad Max' future with ecological and social breakdown and consequent population crash; to a radically transformed, more creative, equitable, ecologically and culturally sustainable future.

Notwithstanding the gravity of humankind's overarching predicament, the focus of this chapter is much narrower. My goal is to explore some of the likely requirements and potential stumbling blocks associated with a single strategy for guiding one possible vision of the future – ecocultural sustainability.

I have coined the term 'ecocultural sustainability' to refer to both a state of dynamic equilibrium and a social process that is desirable and ecologically sound[2]. Ecocultural sustainability requires that a society can, at a minimum, continually renew itself and its members by supporting: (1) the flourishing of rich cultural and biological diversity; (2) forms of governance that are just, egalitarian, transparent, and participatory; (3) economies that are sufficient, equitable, accountable, and bioregionally sound; and (4) production and consumption that promotes universalizable lifestyles and keeps its ecocultural wake in-check by both learning from and working with nature and limiting its total life-cycle costs (social, environmental, and financial). Successful implementation of the ecocultural sustainability paradigm rests on both cultivating a form of rationality that integrates reason and emotion and inculcating a balance between the needs of individuals and the imperative of the common good (human and nonhuman). It calls for educational processes and systems that nurture active citizens and open minds by encouraging wonder, creativity, tolerance, cooperation, and collaboration. By propagating the skills to regularly monitor and evaluate the activities of individuals and organizations – to learn from their mistakes and celebrate their successes – it promotes vigorous self-criticism, combats rigidity and apathy, and fosters anticipatory decision-making and adaptive learning. And by cultivating agility to distinguish between needs and wants, meaningful innovation and shear novelty, the sacred and the profane, and maintaining a

[2] The discussion of "ecocultural sustainability" presented here represents a revised and substantially expanded version of a definition I presented earlier (Glasser 2004, p. 134).

balance between specialization and generalization, such societies prepare their individuals, organizations and institutions to counteract maladaptive forces and respond to unforeseen challenges and changes that are beyond their control with hope, joy, imagination, and unruffledness.

The introductory quote from UNESCO touches on four key ideas that undergird a transition to a more ecoculturally sustainable world. These ideas are also consistent with an evolving stream of contemporary thought. First, there is expanding acknowledgment that past and current assumptions, practices, and policies that guide the pursuit of lasting quality of life improvements, in both rich and poor nations alike, require radical and continual reassessment and rethinking[3]. In particular, we must pay much more attention to the relationship between our values and our actions – between the world of our hopes and dreams and the world we are creating with our everyday decisions. Second, there is growing awareness and appreciation that quality of life is composed from an array of multiple, often incommensurable, yet interrelated elements – and that enduring improvements to quality of life are not achievable by individuals in isolation[4]. Furthermore, there is expanding recognition, at least by some, that achieving such improvements rests on paying careful attention to the requirements of the common good (both human and nonhuman)[5]. Third, there is a new level of sophistication and solicitude

[3] Examples of this perspective abound today. They range from Jared Diamond's (2005) assessment of how environmental challenges and poor decision making have figured in the collapse of civilizations throughout the ages, to Jane Jacob's (2004) effort to both illuminate the telltale signs of social decay and suggest strategies for arresting them, to the recent admonitions of the Millenium Ecosystem Assessment Board (2005). They also include the emerging field of 'sustainability science' (Kates *et al.* 2001), which seeks to understand the essential character of nature-society interactions; Ornstein and Ehrlich's (2000 [1989]) study of the mismatch between the character and scale of change in the world that our brains evolved in and the character and scale of change in the world today and the ensuing paradox that salvation can only be generated by awareness and conscious change; and Arne Naess's (Naess 2005, Glasser 2001) characterization of 'deep ecology' and 'shallow ecology' and his corollary effort to promote consistency among our fundamental values, shared assumptions, lifestyles, and concrete actions – particularly as they relate to nature.

[4] See for instance, the United Nations' "Millennium Development Goals" (2005) and their subsequent implementation plan (United Nations 2006).

[5] A broad range of authors argue that lasting improvements to quality of life are tied to a renewed emphasis on community, the "common good", and community self-renewal. Examples include, Daly and Cobb's (1989) effort to redirect the economy towards "community, the environment, and a sustainable future"; Oelschlaeger's (1994) argument for developing community values, rekindling participatory democracy, and eradicating utilitarian individualism as the dominant paradigm of decision choice; Ostrom's (1990) work on collective management of the commons; and Gardner's (1995) work on individual and societal self-renewal.

regarding the linkages between environmental quality and quality of life[6]. Finally, there is a renaissance in the role of, and commitment to, learning as the foundation and primary vehicle for achieving a higher quality of life for all[7].

This emphasis on learning as the locus for creating a more sustainable and desirable world is especially meaningful. The import of this turn toward learning is drawn, only in part, from the fact that the first three ideas are derivative of, or contingent on, effective learning processes. Since the World Commission on Environment and Development's publication of *Our Common Future* (1987), virtually all mainstream discussions regarding the quest for a more sustainable and desirable world have emphasized that lasting improvements to quality of life can only be achieved by stimulating a new era of economic growth. What makes the above discussion on the importance of continually improving quality of life so significant is the conspicuous absence of any mention about the role that economic growth should play. There is an unstated, implicit decoupling of quality of life from standard of living (beyond certain basic requirements). In the introductory quote from UNESCO, *learning*, in some sense, has supplanted economic growth as the metanarrative and vehicle for bringing about a more sustainable and desirable world for all.

This chapter is an exploratory and necessarily preliminary effort to survey the promise – and potential pitfalls – of turning to learning, and social learning in particular, as the foundation and conduit for harnessing the human propensity to contemplate our fate and future. I am not simply concerned with the concept of social learning from the more traditional standpoints of survival and reproductive success (Heyes and Galef 1996), the transmission of culture (Boyd and Richerson 1985), or even the application of particular interpretations of social learning or social learning traditions to problems in psychology and human behavior (Gardner and Stern 1996, Rosenthal and Zimmerman 1978, Bandura 1977), planning and

[6] A wide variety of researchers are attempting to elucidate the connections between quality of life and the state of the environment. Three prominent examples include, Costanza *et al.*'s (1997) work on ecosystems services, Prescott-Allen's (2001) indices of "Human Wellbeing" and "Ecosystem Wellbeing" for 180 nations, and the work of the Millennium Ecosystem Assessment (2005).

[7] This point is echoed by Chapter 36 of *Agenda 21*, which focuses on three programme areas: Reorienting education towards sustainable development; Increasing public awareness; and Promoting training (United Nations 2004). The United States' National Research Council's Board on Sustainable Development goes even further by describing the transition to sustainability as a "process of social learning and adaptive response amid turbulence and surprise" (1999, p. 48). A recent report from the Nordic Council of Ministers (2003) and an anthology from Sweden (Wickenberg *et al.* 2004) demonstrate that education for sustainable development is on some national agendas. Milbraith's (1989) work on humankind's tragic success as a cause for rethinking civilization and the role of learning in envisioning a sustainable society is an important precursor to these efforts.

policy research (Robinson 2003, Friedmann 1987, Friedmann and Abonyi 1976, Heclo 1974), management theory and organizational change (Wegner 1999, McKenzie-Mohr and Smith 1999, Argyris and Schön 1996), human services provision (Goldstein 1981), environmental policy (Fiorino 2001, Webler *et al.* 1995), environmental and resource management (Keen *et al.* 2005, Leeuwis and Pyburn 2002a, Lee 1995), or even sustainable development and sustainability science (Siebenhüner 2004, National Research Council Board on Sustainable Development 1999, Parson and Clark 1995).

I am interested in a more general and, I believe, more fundamental set of nested questions. Is there a common and consistent interpretation of social learning? If not, why? If so, can social learning inspire and foster planned, directed action and behavior that is more consistent with our highest values and aspirations regarding improving quality of life? If so, is this force strong enough to counterbalance the historical tendency toward anthropocentric and ethnocentric approaches that tend to advantage narrow self-interest? In short, does social learning give an edge to anticipatory, holistic, egalitarian, and nonanthropocentric planning processes *and* decisions that favor continual quality of life improvements for all – humans and the biosphere as a whole?

To address these questions, I take a meta-perspective and reflect on the roots of our predicament and the meanings of learning. I touch on the interconnections among learning and information, knowledge, understanding, power, neurobiology, human nature, culture, and values and also consider their relationship to decision-making and action. I step back to consider the individual learning requirements that are necessary to buttress the effective implementation of social learning. Finally, I ask if there is a set of concepts or principles that can be drawn from the various social learning traditions and perspectives – or otherwise identified – to form a coherent social learning for ecocultural sustainability paradigm. In the closing section, I offer a list of 'challenges' that I hope will serve as a tentative outline for a research program for social learning for ecocultural sustainability. I do this with the aspiration of helping to stimulate a larger, collaborative conversation on creating a comprehensive, targeted research program for applying social learning to address the predicament – and promise – of our collective fate and future.

The gap

The following passage identifies a gap between a particular society's ideals and practical reality. Consider what culture and historical period are being portrayed.

> ____ had its deplorable failings and distressing shortcomings; its utopian ideals honored more in the breach than in observance; its 'Sunday preaching and Monday practice'; it yearned for peace, but was constantly at war; it professed such ideals as justice, equity, and compassion, but abounded in injustice, inequality, and oppression; materialistic and shortsighted, it unbalanced the ecology essential to its economy; it suffered from the 'generation gap' between parents and children and between teachers and students; it had its 'drop-outs', 'cop-outs', hippies and perverts; it had 'unisex' devotees, and perhaps even something like a 'mini-maxi' controversy.

The quote is from Samuel Noah Kramer (1981, p. 259-260), the renowned Assyriologist, writing on Sumerian civilization of more than 4,000 years ago. I use this quote to vividly illustrate the timeless nature of the gap between the world of our aspirations, hopes, and dreams *and* the world we create with our policies, practices, and every day actions.

Lest one believe that Sumerian society was unique in being plagued by such a rift – or rather that gaps between a society's values and their practical expression are not widespread – I offer a quote from an Egyptian man contemplating suicide, also from 4,000 years ago (as quoted in Gardner 1995, p. 122). More significant than the mere existence of the gap, its breadth, or this man's awareness of it, perhaps, is his obvious sense of isolation and feeling of paralysis in trying to come to terms with it.

> To whom can I speak today?
> The gentle man has perished
> The violent man has access to everybody.
> To whom can I speak today?
> The iniquity that smites the land
> It has no end.
> To whom can I speak today?
> There are no righteous men
> The earth is surrendered to criminals.

The existence of such a gap is by no means limited to the past. Of particular significance to our contemporary dilemma (and the goal of continuously improving quality of life) is the reference in both quotes to the seduction of material affluence and the corresponding failure to recognize, appreciate, or effectively respond to the predicament of our seemingly interminable quest for ever greater consumption and its potential to undermine the ecological and social basis of our existence.

Robert Prescott-Allen's (2001, p. 13) Human Wellbeing Index (HWI), which integrates countrywide data on health and population, wealth, knowledge, community, and equity in a single, normalized, five-category indicator (good, fair, medium, poor, bad), demonstrates that two-thirds of the world's population live in countries with a bad or poor HWI and less than one-sixth live in countries with a fair or good HWI. Furthermore, the gap between the best and worst is huge (countries in the top 10% have a median that is eight times those in the bottom 10%) and even the top performers (Norway, Denmark, and Finland) have considerable room for improvement. Prescott-Allen's (2001, p. 59) Ecosystem Wellbeing Index EWI, which integrates countrywide data on land, water, air, species and genes, and resource use in a single, normalized, five-category indicator, reveals that no country has a good EWI and that countries with a bad or poor EWI cover almost half of the planet's terrestrial and inland water surface. Moreover, if better monitored, many of the countries with fair or medium ratings would be downgraded (Prescott-Allen 2001). These figures are even more disconcerting when placed in context by the Millennium Ecosystem Assessment (2005, p. 6), which indicates that 60% (15 out of 24) of the ecosystems services that humans depend on for our sustenance are degraded or used unsustainably, that this degradation already causes serious harm to human well-being, and that efforts to increase one ecosystem service frequently result in the degradation of other services.

On the positive side, a broad array of survey data from citizens throughout the world – rich and poor countries alike – demonstrates the existence of sincere, well-intentioned positive environmental attitudes, anxiety about environmental degradation, rudimentary awareness of the environment's role in supporting quality of life, and a stated willingness to trade-off economic development for environmental protection (Coyle 2005, Gruber 2003, Bloom 1995, Kempton *et al.* 1995, Dunlap *et al.* 1993)[8]. In the U.S., where this data has been gathered for over thirty years, these attitudes and concerns have had remarkable staying power (Coyle 2005, Gruber 2003). Furthermore, Prescott-Allen's work demonstrates that increases in human well-being do not necessarily result in greater environmental impact (2001, p. 107). The ways in which human well-being is pursued matter – a high quality of life can be achieved with limited environmental consequences.

What is most surprising or, perhaps, troubling is that while environmental concerns and attitudes are widely supported and long-standing, they have generally not, at least in the U.S., translated into consistent, effective actions and behaviors – voting

[8] For more details regarding public perception of the environment, see Glasser (2004, p. 134-136), which also discusses the widespread non-anthropocentric, non-instrumental expressions of environmental concern and Glasser *et al.* (1994), which discusses the environmental values and concerns of participants in the Intergovernmental Panel on Climate Change (IPCC) process.

habits, purchasing decisions, and lifestyles – for improving environmental quality (Coyle 2005, Gruber 2003, Roper 2002). Similarly, on the international level – except for the widely touted Montreal Protocol – these concerns and attitudes have not generated effective treaties for responding to contemporary, global-scale environmental challenges (Speth 2004). This is 'The Gap' I am speaking of. Simply put, awareness of a problem, accessibility of extensive information on its origins and impacts, and, even, stated concern about it do not guarantee action or imply that, if taken, the action(s) will be appropriate or effective.

The greening of progress

The ideological commitment to sustainable development as continuous improvement in the overall conditions of human life, as discussed in the UNESCO quote, is unavoidably rooted in the notion of progress – at least for those of us in the West. The orthodox view of the idea of progress, which dates back to at least the time of Xenophanes in the late 6th century B.C.E., holds that moral, political, economic, technological, and social betterment are *inevitable* (Nisbet 1980, Edelstein 1967). Such a view of ineluctable, boundless progress became widely adopted in the West during the Enlightenment and continues to be broadly embraced today. This perspective has been justified by – and tied to – humankind's expanding capability to control and manipulate nature (Marx 1996). It is also wrapped up in a conviction that humankind is perfectible (Marx 1996). Yet many of today's interconnected environmental and social problems – over-consumption, poverty, over-harvesting, climate change, stratospheric ozone reduction, over-population, biodiversity loss, pollution, fresh water shortages, invasive species, fisheries collapse, deforestation, over-grazing, erosion, desertification, and salinization – are the unintended, unforeseen (but not necessarily unforeseen or unforeseeable) consequences of a failure to recognize, adequately appreciate, or effectively respond to the reciprocal character of humankind's relationship with nature.

As the *Living Beyond Our Means* statement from the Millennium Ecosystem Assessment's Board (2005, p. 2) points out, "Human activity is putting such strain on the natural foundations of the Earth that the ability of the planet's ecosystems to sustain future generations can no longer be taken for granted". The upshot is that the orthodox view of progress, which has pitted humans against nature, may finally be strained beyond its seams. This idea, however, is not new. A similar argument, based on an early effort to model the relationships among environment, economy, and human population, was made in the original 1972 *Limits to Growth* study (Meadows *et al.*). Paul and Anne Ehrlich (1979) made a related argument, on ostensibly narrower grounds, when they asserted, that human-induced species extinction, like randomly popping rivets from an airplane's wing, was akin to playing Russian Roulette with the fabric of life.

Concerns relating to the planet's overall carrying capacity and the potential of technology to keep pace with changing human needs and expectations are also not limited to the late twentieth century. Over two hundred years ago Malthus (1970 [1798]), possibly underestimating or failing to recognize the potential of agricultural technology to increase crop yields, raised concerns about the limits of agriculture to keep pace with increasing demand from population growth. Jevons (1865) issued a warning about England's ability to maintain its progress and wealth in the face of finite coal reserves and Sears (1988 [1935]) called attention to spreading desertification in the United States due to poor soil conservation practices. More recently, the Post World War II era brought such concerns to a crescendo by spurring a huge growth in literature that began connecting concerns about carrying capacity and resource scarcity to questions about the downside of technology, anxiety regarding effective governance, distress over biodiversity loss, and misgivings about the potential of continuous economic growth to bring the good life for all (Vogt 1948, Osborn 1948, Leopold 1949, Brown 1954).

This solicitude regarding our use of the environment and its role in securing and maintaining a high quality of life, albeit on a more local scale, has ancient antecedents too. Plato (429-347 B.C.E.) was troubled by local climate change caused by deforestation and its effects on agriculture (1989 [1929]). Vitrivius (1st century C.E.), by making an analogy to the neurological problems of lead smelters, called for a ban on the use of lead water pipes (1985 [1934]). And Mencius (372-289 B.C.E.) went through great lengths to argue for sustainable resource management in China:

> If you do not interfere with the busy seasons in the fields, then there will be more grain than people can eat; if you do not allow nets with too fine a mesh to be used in large ponds, then there will be more fish and turtles than they can eat; if hatchets and axes are permitted in the forests on the hills only in the proper seasons, then there will be more timber than they can use (Hughes 1989, p. 19).

These represent only a few of the many rich examples of 'unheeded' foresight that have been gifted to us. While it is clear from the historical record that, at times, actions were taken and regulations were made, the pattern of ecocultural deterioration that often ensued also makes it clear that these efforts were, in the main, unrecognized, inadequately supported, or insufficiently enforced.

I contend that today's sustainability and sustainable development discussions (Kates *et al.* 2005, Robinson 2004, National Research Council Board on Sustainable Development 1999, Bossel 1998, Daly and Cobb 1989, Milbraith 1989) are the contemporary manifestation and integration of these constructive critics of progress' concerns regarding maintaining and improving quality of life. Their key insight is that progress is not inevitable. The variety we get, if in fact we achieve progress, depends on how effectively our institutions, policies, practices, and every day decisions manifest our diverse values and our understanding of how the world works. My conclusion is that while achieving continuous quality of life improvements for all cannot be achieved by abandoning the idea of progress, it also requires a more than superficial departure from the orthodox notion of progress.

The famous American environmentalist, Dave Brower, was fond of saying that he was not "blindly against progress, but against blind progress". This phrase could be a mantra for the less dogmatic, constructive critics of the orthodox notion of progress that I have been describing. Their work suggests that progress is multifaceted and contingent. Progress in one realm need not imply progress in another. In fact, progress in one realm can be inversely related to progress in another. Excessive progress in one realm can even foster a lack of resilience that engenders collapse (Diamond 2005). What's more, past gains can be reversible – and irretrievable, as with lost languages or the skills, traditions, and wisdom that are forfeited when a culture becomes extinct.

I have coined the term 'greening of progress' to refer to the process of modifying the orthodox notion of progress to support a transition to ecocultural sustainability. This revised view of progress incorporates three assumptions. First, progress is an inherently normative idea. The idea of progress cannot be separated from our values and assumptions about human nature (are humans inherently good, bad, both, or neither), technology, economics, what is sacred, and our views about the way the world works. Furthermore, every decision will, almost inevitably, generate tradeoffs. Second, humankind's quality of life is ultimately tied to, and constrained by, our ability to maintain the health and flourishing of nature and the planet's various ecosystem services along with our ability to stay within the planet's biogeophysical carrying capacity. Third, the *rate* and *character* of progress are shaped by our concern for the common good; our ability and proclivities to acquire, process, evaluate, and share information about the current state of affairs (particularly feedback data); the types of decision making processes and criteria we employ; our proficiency at understanding and reflecting our highest concerns in our institutions, policies, and lifestyles; our adeptness at acting in an anticipatory and adaptive fashion (as opposed to a simply reactive one); and our capacity to support individual and institutional self-renewal (Gardner 1995).

In contrast to others (Speth 2004, Raskin *et al.* 2002), I have specifically chosen not to include a formal requirement for radical value change. I have done this because I believe the surveys of the publics' environmental attitudes and concern demonstrate that the underlying values to support such change, while possibly not deep enough or well enough informed by science and a sophisticated understanding of causal relationships (Coyle 2005, Gruber 2003), nevertheless, already exist, are sincere, and are widely embraced. Rather than eliciting a sweeping change in values, the more fundamental and crucial steps may involve better understanding our existing palette of values (and their relative implications for improving quality of life), reprioritizing or realigning our values in relation to this improved understanding, and eliciting greater consistency in their application.

Niels Röling offered a provocative and challenging admonition that alludes to the essential change embodied by my 'greening of progress' perspective when he stated, "Until now man has fought nature. From now on, he will fight his own nature" (translation of French phrase, as quoted in Röling 2002, p. 25). Rather than fight our nature, however, I believe the fundamental challenge is to better understand our nature – and learn how to work with it – to identify levers of change that can help us bring about the change we seek[9]. But what is 'our nature'? What is socially constructed, what is guided by our neurobiology (Damasio 2005), and what is genetically determined (Wilson 1998)[10]? How much leeway do we have in applying learning to adapt to our evolving understanding of the world

[9] For an insightful discussion of the efficacy of different change strategies and how applying particular strategies to problems can prove to be either fabulous or disastrous, see Watzlawick *et al.* (1974). For an application of Watzlawick and his colleagues' theory to the problem of the disconnect between peoples' stated environmental concern and their environmentally destructive behaviors, see Glasser (2004).

[10] For an interesting insight into this crucial and subtle question, consider the perspective of E.O. Wilson (1998, p. 2049), the originator of sociobiology:
"Human nature is not the genes, which prescribe it, or the universals of culture, which are its products. It is rather the epigenetic rules of cognition, the inherited regularities of cognitive development that predispose individuals to perceive reality in certain ways and to create and learn some cultural variants in preference to competing variants."
For another important perspective on this issue, ponder the view of the noted neurologist, Antonio Damasio (2005: p. xx):
"[T]he body, as represented in the brain, may constitute the indispensable frame of reference for the neural processes that we experience as the mind; that our very organism rather than some absolute external reality is used as the ground reference for the constructions we make of the world around us and for the construction of the ever-present sense of subjectivity that is part and parcel of our experiences; that our most refined thoughts and best actions, our greatest joys and deepest sorrows, use the body as a yardstick."
Consider also that it is the environment – as external reality – that gave rise to *homo sapiens sapiens*; the interplay between genetics, the environment, chance, and possibly our own ingenuity (adeptness at learning), in some sense, *crafted* the human brain as we know it today.

and our place in it – and how much does human nature shape our understanding of the world and our place in it? From my greening of progress perspective, I take Röling to mean that environmental management must become much more about managing people – especially the way we learn, form and test our values, and use nature to satisfy our needs and desires – than managing nature, per se (i.e. attempting to control and manipulate soil, forests, marine environments, and ecosystems). I would also modify Röling's insight to incorporate the idea that a greening of progress tradition, or at least a countercurrent, has existed for at least several millennia. But why hasn't this modified view of progress taken hold? The pivotal issue, in my mind, is to clarify the role that learning can play in supporting the greening of progress and in facilitating a transition to ecocultural sustainability.

Individual learning, social learning, and ecocultural sustainability

Given widespread environmental concern and abundant information regarding human induced ecocultural degradation, why does the overall pattern of unsustainability continue to grow? While the information regarding human induced ecocultural degradation is often speculative, uncertain, and contested, I believe the inescapability of human ignorance is not nearly as disconcerting as our dereliction to effectively draw on what we know. If a transition to ecocultural sustainability is ever to take hold, unprecedented individual and collective change must occur. While such change might be catalyzed by some random, external event, my interest here is in the possibility of planned, directed change. Change of this character and scale, however, has no chartered course. My discussion of 'the gap' and the 'greening of progress' demonstrated, that while no society has yet to successfully make such a transition, it is not for lack of interest or effort. Comprehensive, coordinated change – spanning our behavior, practices, policies, institutions, and, perhaps, values – is extremely difficult.

Any planned, directed change by individuals or collectives is built on learning. Using the *Oxford American Dictionary* definition as a rough guide, I define learning more generally as the process of acquiring knowledge, skills, norms, values, or understanding through experience, imitation, observation, modeling, practice, or study; by being taught; or as a result of collaboration. I also note that understanding is interpreted very broadly here to also include intuition, which may be the product of extensive study, spiritual practice, divine inspiration, or even serendipity, rather than conscious reasoning. Contrary to widely held views in social psychology, political science, planning, and management (O'Riordan 1995, Goldstein 1981, Friedmann and Abonyi 1976), I do not believe that learning must necessarily engender behavioral change. Not all learning warrants behavioral change and, sometimes, competing interests, goals, and objectives militate against change. This point has been made vividly clear, for instance, by the pairing of our growing knowledge of, and scientific consensus around, the existence of anthropogenically induced climate change *with* our dismal, individual and collective failure to effectively respond to this knowledge (Speth 2004). It is only through learning, however, that we acquire our values, attitudes, and concerns along with our conception of reality. By acquiring new information (or exploiting existing information), we have the possibility to test these values and concerns against our understanding of reality and, if warranted, we can take measures to rethink our values, realign our behavior and action, or do both. When corrective responses result from anticipatory learning (as opposed to simple adaptation), I refer to them as planned change.

As will become readily apparent, there is currently no common and consistent interpretation of social learning. To give a flavor of the variety of perspectives regarding the meaning of the term social learning, I share below five interpretations:

- Social learning is a higher form of learning occurring in a social context for the purpose of personal and social adaptation (Goldstein 1981, p. 237).
- Social learning is the process by which organisms 'see' their environmental circumstances by intelligence gathering and act with foresight or prepared adjustment. This principle of precautionary but evolutionary adjustment may be a vital one for responding to environmental stress (O'Riordan 1995, p. 4).
- [The] combination of adaptive management and political change is social learning (Kai Lee 1995, p. 228).
- Social learning [is the p]rocesses by which society democratically adapts its core institutions to cope with social and ecological change in ways that will optimize the collective well-being of current and future generations (Woodhill 2002, p. 323).
- 'Social learning' reflects the idea that the shared learning of interdependent stakeholders is a key mechanism for arriving at more desirable futures. With time, the concept of 'social learning' has intertwined with related ideas such as soft systems thinking... and adaptive management... A consistent characteristic of the various approaches is that they advocate an interactive (or participatory) style of problem solving, whereby outside intervention takes the form of facilitation (Leeuwis and Pyburn 2002b, p. 11).

The state of affairs regarding this mélange of views and theoretical approaches has, perhaps, been best captured by Parson and Clark (1995, p. 429):

> The term *social learning* conceals great diversity. That many researchers describe the phenomena they are examining as 'social learning' does not necessarily indicate a common theoretical perspective, disciplinary heritage, or even language. Rather, the contributions employ the language, concepts, and research methods of a half-dozen major disciplines; they focus on individuals, groups, formal organizations, professional communities, or entire societies; they use different definitions of learning, of what it means for learning to be 'social', and of theory. The deepest difference is that for some, *social learning,* means learning by individuals that takes place in social settings and/or is socially conditioned; for others it means learning by social aggregates.

In an effort to shed some light on the distinction between *individual* and *social* learning and offer some clarity and coherence to the situation, I take a more generic, ethological approach. I introduce a broader notion of social learning than typically appears in much of the contemporary literature on social learning and sustainability (Keen *et al.* 2005, Leeuwis and Pyburn 2002a, National Research Council Board on Sustainable Development 1999, Lee 1995). This characterization attempts to offer a unique resolution to Parson and Clark's (1995) concern about the existence of two widely separated views on who can engage in social learning and the character of the process. I view almost all learning by individuals as some form of social learning. The exception is pure trial-and-error learning through direct personal experience, essentially immune from the influence of others (human and nonhuman). This is a time intensive and potentially costly approach for acquiring knowledge, skill, or understanding. Imagine, for instance, learning how to speak without having the benefit of hearing others talk, learning how to decipher written language without having been taught the alphabet, or learning what things are safe or healthy to eat solely through random or even systematic experimentation. While pure trial-and-error learning is demanding and rare, it has clearly been, and continues to be, pivotal to human development. When individuals engage in the process of learning, they more frequently employ observation, imitation, modeling, self-instruction, conversation, and mentoring, among other strategies. All of these strategies, however, rest on some interaction with living beings or, at least, employing the artifacts (e.g. language, tools, books, drawings, videos, music recordings, software, etc.) of living, or once living, beings.

Albert Bandura has argued that modeling, from the standpoint of behavior elicitation, is the most significant form of learning in which individuals engage (1977, p. 22):

> Learning would be exceedingly laborious, not to mention hazardous, if people had to rely solely on the effects of their own actions to inform them what to do. Fortunately, most human behavior is learned observationally through modeling: from observing others one forms an idea of how new behaviors are performed, and on later occasions this coded information serves as a guide for action.

Bandura's social learning theory (1977) explains human behavior in terms of continuous interaction among cognitive, behavioral, and environmental influences. Bandura separated the conditions for successful behavioral modeling into four components:

- Attention – a 'model' behavior in the environment must grab or capture a potential learner's notice.
- Retention – the learner must remember the observed behavior.
- Reproduction – the learner must be able to accurately replicate the observed behavior.
- Motivation – the environment must offer a consequence (reinforcement or punishment) that increases the probability for a learner to demonstrate what they have learned.

While Bandura's social learning theory was developed to explain individual behavior, it can be applied to collectives with great efficacy too.

As long as learning, by individuals or collectives, involves some form of input drawn from others, I characterize it as social learning. The more salient distinction, I find, is differentiating between what I refer to as *passive* social learning and *active* social learning. Passive social learning, by individuals or collectives, rests on the prior learning of others. It does not require inputs in the form of communication or interaction – direct feedback – from other living beings. Passive social learning includes learning that results from reading a newspaper, watching a blacksmith forge a tool, viewing a movie, listening to a radio program, attending a lecture or poetry reading (without questions from the audience), searching the internet, or following a recipe. It also includes observing the practices of, and interactions among, others. Passive social learning has many advantages for cultural evolution over trial-and-error learning because it can lead to the same results at much lower cost in terms of time, effort, and danger. A drawback is that most results must be accepted uncritically – i.e. on trust. Another potential drawback is that it generally requires embracing, actively or at least tacitly, the values and assumptions that are encoded in the transferred knowledge. While the passive social learning process may yield important new insights for the particular individual or individuals involved, it generally has limited applicability for directly spawning substantively new social innovations.

Most learning in our contemporary world, both individual and collective, is passive social learning. Because it relies on the received wisdom of others (frequently experts), passive social learning can be used to readily propagate behaviors that favor narrow interests over the common good. Robert Edgerton (1992) characterizes such social or cultural maladaptation as the maintenance of traditional beliefs, practices, or institutions that: (1) harm people's health or well-being, (2) make it ineffective at coping with environmental demands, or (3) threaten the society's viability. Such maladaptation is ubiquitous today. An example is the orthodox notion of progress, which supports a general belief that environmental problems do not need to be addressed today because new technologies can always be created to cost-effectively address any problems that might subsequently arise. Other related examples include our mindless commitment to economic growth (at seemingly all costs) and our widespread failure to appreciate the many tradeoffs associated with rapidly rising global per capita animal protein consumption. Vested interests and those unwilling to share power, if they can insulate themselves from the effects of maladaptation (assuming they are aware or are concerned about them), generally have a significant interest in perpetuating such behaviors.

Employing Bandura's framework, ecoculturally sustainable behaviors are commonly seen as less appealing, so they fail to grab our attention. The behaviors are frequently unfamiliar so they are less likely to be retained. They are also often more involved or more complex, so they are less likely to be reproduced. Finally, the behaviors are routinely perceived as inconvenient, more expensive, more time consuming, not fun or 'cool', unsafe (as with smaller more fuel efficient vehicles or bicycles), or as activities of the counter culture, so there is little motivation to try them out. The motivation for employing more ecoculturally sustainable behaviors is further diminished for two key reasons. First, a behemoth advertising industry bombards people all over the world with models of people enjoying, or rather basking in unsustainable behaviors, without experiencing any negative side effects or tradeoffs. Second, the negative side effects that do exist are often not readily 'visible' or they are distributed in space and time far away from those causing the impacts.

Maladaptive behaviors, such as corruption; excessive consumption, profligate waste, and exorbitant energy use in the rich countries; and high population growth in many economically disadvantaged countries are widely modeled in the media and in society. It is should be no surprise, as Bandura suggests, that such behaviors are likely to be perpetuated despite widespread information documenting the negative overall consequences of maintaining such behaviours. Simply put, our societal emphasis on passive social learning and our proclivity (by accident or design) for modeling unsustainable, as opposed to sustainable, behaviors severely hampers the possibility of facilitating a transition to ecocultural sustainability.

Active social learning, on the other hand, is built on conscious interaction and communication between at least two living beings. It is inherently dialogical. Active social learning can be broken into three rough categories that are a function of the skills and values of the individuals in the collective and the power relationships that define them. The three categories, which reflect increasing levels of participation by the group members, include:

- Hierarchical – based on predetermined, inflexible relationships between established teachers and learners;
- Non-hierarchical – based on two-way learning, where each participant, as an 'expert' in their own right, shares their knowledge and experience; and
- Co-learning – based on non-hierarchical relationships, collaboration, trust, full participation, and shared exploration.

Hierarchical and non-hierarchical active social learning are widely applied and used with great benefit to expand the penetration of existing knowledge. Co-learning, because of its requirements for team building, complete engagement, 'learning-by-doing' (Dewey 1997 [1938]), and accountability, in addition to supporting the penetration of existing knowledge, supports the generation of new knowledge and novel strategies for addressing real-world problems. Co-learning supports change, positive change in particular, by building capacity in three fundamental areas: critical evaluation of existing knowledge and problems, knowledge generation and penetration, *and* application of this new knowledge to policy, practice, and everyday life.

Active social learning can take place in the context of a conversation, a course employing the Socratic method, dancing with a partner, symphony practice, a community meeting, an open, participatory public review process, and, although less visceral, video conferencing over the internet. Opportunities for cross-fertilization and emergence make it much more effective than passive social learning at creating innovations and widely diffusing novel behaviors. Active social learning, because of the opportunity to directly engage both a broad range of perspectives and the whole human, also has the potential to promote more open, equitable, and competent learning processes. Furthermore, the potential to receive direct feedback from other living beings and gain a palpable 'taste' for the effects of our own unsustainable behaviors offers a powerful motivation for challenging the desirability of the underlying, taken-for-granted assumptions, values, and principles – such as the orthodox notion of progress – that guide our theories-in-use, routinized policies, practices, and individual behaviors. As such, the highest, most diverse and participatory forms of active social learning appear to offer a viable prospect for combating maladaptation. While I run the risk of engaging in a tautology, I believe these highest forms of active social learning can be used with great advantage in our learning environments and decision-making processes to promote a societal shift toward ecocultural sustainability – if they also model the principles of ecocultural sustainability.

An early illustration of this more active form of social learning as co-learning, which relates to building the foundation for ecocultural sustainability, is Lewis Mumford's idea of the 'regional survey' (1970 [1938], p. 381-387). Mumford saw this work as helping to cultivate the cultural base of a progressive civilization. He describes the regional survey as a form of participatory, communal education that utilizes an organic approach to knowledge to help citizens perform and integrate systematic local surveys of soil, geology, history, industry, climate, etc. In Mumford's view this process gives context and meaning to specialized knowledge and thereby forms the vital nucleus of a functional education for political life.

> When the landscape as a whole comes to mean to the community and the individual citizen what the single garden does to the individual lover of flowers, the regional survey will not merely be a mode of assimilating scientific knowledge: it will be a dynamic preparation for further activity (Mumford 1970 [1938], p. 385).

Mumford saw this process, which he credits with helping to create the Boston Metropolitan Park System and the Appalachian Trail, as involving the entire local community, especially schoolchildren. Leaving it to the realm of specialist, expert investigators would make it politically inert. Although Mumford's idea is broader and less instrumental, the idea is akin to today's community sustainability indicators projects.

Active social learning, however, can support widely different levels of engagement and inquiry. It supports multiple loop learning (Argyris and Schön 1996), which can be used to question both existing practices and the values that undergird them, but the depth and character of questioning that the collective chooses to engage in cannot be determined in advance. Similarly, the collective can utilize adaptive learning (Webler *et al.* 1995), but the use of such techniques will be governed by the interest, openness, preparedness, and social dynamic of the collective. Because active social learning can involve diverse players with competing or even conflicting values and interests, I posit that the most successful forms of active social learning will result from non-coercive relationships that rest on building a common language, transparency, tolerance, mutual trust, collaboration, shared interests and, concern for the common good. Such forms of active social learning can employ conflict in a positive way by challenging complacency and encouraging 'out-of-the-box' thinking.

The more active forms of social learning can also facilitate anticipatory responses by examining routinized and inconspicuous practices, such as the creeping escalation of standards for comfort, cleanliness, and convenience (Shove 2003). Examples of activities that benefit from these higher forms of active social learning, include playing in an improvisational jazz band and participating in collaborative, integrated-systems design projects – such as a green building, an organic farm, an ecological design project applying biomimicry, or a green planning initiative, such as those under development in The Netherlands and New Zealand (Johnson 1997). A further benefit of the more active forms of social learning is that their requirement for elevated levels of engagement – especially when diverse constituencies are involved – aids in building critical thinking skills, supports a richer form of rationality that integrates reason and emotion, and promotes contextualization and accountability that are crucial for helping to close gaps between peoples' values and actions.

Two significant potential weaknesses of active social learning come to mind. First, benefits do not accrue automatically from employing the process – active social learning, particularly in its hierarchical forms, can be used with equal ease and effectiveness to support maladaptation (consider efforts to stimulate ethnic conflict by Hitler, the Belgians in Rwanda, and most recently George Bush in the Middle East). I believe realizing the potential of active social learning – as a bulwark against maladaptation and a foundation for the transition to ecocultural sustainability – rests on the collective not only choosing what level process it will employ, but also on the collective making this choice with full awareness of the requirements and demands of these most active, risky, and challenging forms of social learning. A second significant weakness of active social learning is that its success depends on effective capacity building. Success rests at least as much on the preparedness, competence, openness, and maturity of the individuals engaging in it as on the rules that guide particular organizational learning, public participation, or decision-making processes. Furthermore, as wise as the decisions that a group arrives at may be, they are only as good as the potential of the new policies and actions to be successfully modeled and embraced by the society at large. Thus, if a society fails to make the educational infra structure investments to prepare all of its citizens to fully participate in the highest forms of active social learning, it will forever fail to reap its benefits and ecocultural unsustainability will likely fail to be halted.

Toward a social learning for ecocultural sustainability research agenda

I return now to a question posed in the introduction. Does social learning inspire and foster planned, directed action and behavior that favors continual quality of life improvements for all – humans and the biosphere as a whole? This question can now be restated as: Does social learning automatically channel uncoordinated and inharmonious individual actions into collective actions that support and reflect the goals of ecocultural sustainability? In my view, the answer is, 'not automatically'. Perhaps the most important reason is that social learning, as maladaptation, can effectively drive and perpetuate unsustainable behaviors.

In reference to this persistent and troubling issue, I offer a koan from Lao Tzu (1961, p. 145), the sixth century B.C.E. founder of Taoism, which aptly reflects the challenge, promise, and hope of coming to terms with the maladaptive form of social learning:

> To realize that our knowledge is ignorance,
> This is a noble insight.
> To regard our ignorance as knowledge,
> This is mental sickness.
> Only when we are sick of our sickness,
> Shall we cease to be sick.
> The Sage is not sick, being sick of sickness;
> This is the secret of health.

The first crucial step of creating an effective response involves acknowledging, understanding, and appreciating the lure and power of maladaptation. One of the keys to fostering a transition to ecocultural sustainability rests in helping all of society share in this 'secret of health', this wisdom of the Sages.

To paraphrase John Gardner, the great proponent of individual and societal self-renewal, we have before us some breathtaking opportunities disguised as insoluble problems. In an effort to advance the process of turning these ostensibly insoluble problems into breathtaking opportunities, I propose that the social learning for sustainability research community gather together (literally or virtually) to craft a focused collaborative research agenda. In the spirit of trying to help spark this large-scale, collaborative conversation, I offer the following, tentative and unpolished list of eight challenges for review and discussion (the order need not be adhered to rigidly):

1. **Develop a consistent and coherent *working definition* of 'social learning'.**
2. **Initiate a comprehensive, systematic review of existing applications and case studies of 'social learning'.** This component has four main purposes: (1) to document the full range of interpretations of social learning across all disciplines; (2) to document the range of existing applications of social learning; (3) to clarify what aspects of social learning are guided by our neurobiology, genetically determined, guided by our culture, or open to change; and (4) to understand how researchers and practitioners from different disciplines have attempted to funnel uncoordinated and inharmonious individual actions into collective actions that support explicit goals.
3. **Explore the possibility of creating a consistent and coherent *working definition* of 'social learning for sustainability'.**
4. **Identify well-documented, testable social learning 'levers' that have significant potential to help individuals and collectives respond more effectively to situations where they have a vague or general familiarity with a problem – ecocultural unsustainability – but, nevertheless, choose not to respond or respond ineffectively.** Such 'situations' require addressing at least seven issues: (1) having no idea that a potentially serious problem exists; (2) honestly believing that a 'problem' is a not a problem; (3) denying the existence of a problem by simply wishing it away or by ignoring the information (this includes *educated incapacity*, an acquired or learned inability to perceive a problem); (4) accepting the existence of a problem, but perceiving it as easily surmountable; (5) accepting the existence of a problem, but perceiving other problems or issues to take a higher priority; (6) failing to muster adequate support for action; and (7) taking action, but the chosen action proves to be inadequate, mismatched to the problem, or unsuccessful. Two corollary challenges include applying these social learning levers to real-world cases and evaluating their efficacy.
5. **Create well-documented, testable strategies for applying social learning to 'minding the gap'.** Assuming that interest in improving quality of life and concern for the environment are strong and sincere – that people are not hypocritical – it becomes important to identify or create well-documented, testable social learning techniques and instruments to help people to: (1) better understand these values and concerns, (2) put these values and concerns into perspective relative to their other values and concerns (particularly those that are otherwise unstated and taken-for-granted), (3) make the difficult to discern impacts of their actions more conspicuous and glaring, and (4) test how they link their values and concerns to their daily actions and practices. If the outcomes of peoples' actions and practices are widely inconsistent with their highest values and aspirations and if after engaging in this process they see these values as fundamental to their world view, then the real work becomes identifying additional, well-documented and testable social learning strategies

to promote both more consistent individual and public policy decision making for 'minding the gap'. Two corollary challenges include applying these social learning strategies to real-world cases and evaluating their efficacy.

6. **Develop educational strategies to support capacity building for individual learning, so that people are poised to participate in the highest forms of active social learning.** Apply these strategies in the real-world and evaluate their efficacy.
7. **Apply social learning to model strategies for recognizing, understanding, and publicizing maladaptation – and evaluate their efficacy.** Examples might include using the media (internet, movies, videos, etc.) and teach-ins to grab and capture peoples' attention by viscerally highlighting deeply troubling unsustainable behaviors associated with issues such as global climate change, loss of cultural diversity, or the impending water crisis. An important corollary challenge is to also provide strategies for effectively responding to these forms of maladaptation.
8. **Apply social learning to model ecoculturally sustainable behaviors – and evaluate their efficacy.** Examples might include, creating and publicizing a community sustainability indicators project that is directly integrated with policy and practice or creating a new housing project that demonstrates that small, super-energy efficient, green homes are stylish and comfortable as well as cost saving.

As noted earlier, social learning means many things to many people. There is as yet no widely accepted, clear and coherent interpretation of social learning. Social learning may even surpass 'sustainability' and 'sustainable development' for its breadth and diversity of interpretations. It goes without saying that there is also no lucid, well-developed social learning for sustainability paradigm. This, however, is no reason to abandon any of these terms – quite the contrary. A modest degree of vagueness and ambiguity can provide an entry point for all and stimulate a process of clarification, questioning, and conversation that, in the end, may prove far more important than any definitional consensus.

The paradox of social learning is that it can result in our ruination or our renaissance. Our goal is not simply to evade collapse. There is a vital difference between growing ever cleverer and becoming wiser. Steady improvement in quality of life for all rests on developing, and continually renewing, our capacity to bridge the gap between our values and our actions. The secret to making this ostensibly insoluble problem soluble hinges on recognizing that information is not knowledge and knowledge is not understanding. The promise and power of learning for sustainability involves internalizing this distinction *and* learning to appreciate that understanding results from access to information, the capacity to make sense of it, the opportunity to openly debate its significance, the sophistication to draw meaning from it, and the wisdom to put it into context. This is how we build the capacity and conviction – individual and collective – to bring consonance between our highest values and our actions.

While many of the ideas and concepts embraced by advocates of social learning have tremendous potential to facilitate a transition to ecocultural sustainability, the term currently runs the risk of being perceived as a silver bullet or panacea. At its best, active social learning may very well encourage a deeper, more robust understanding of cause and effect, ongoing moral development, and creative, anticipatory problem solving – these benefits, however, are not guaranteed. I have attempted to add some modest clarity and coherence to our understanding of the meanings and potential of social learning and outline some of the challenges before us – but many questions remain unanswered and considerable work and collaboration remains before us.

References

Argyris, C. and D.A. Schön (1996) *Organizational Learning II: Theory, Method, and Practice,* Upper Saddle River, NJ: Prentice Hall.

Bandura, A. (1977) *Social Learning Theory,* Englewood Cliffs, NJ: Prentice Hall.

Bloom, D.E. (1995) "International Public Opinion and the Environment", *Science,* 269**:** 354-358.

Bossel, H. (1998) *Earth at a Crossroads: Paths to a Sustainable Future,* Cambridge: Cambridge University Press.

Boyd, R. and Richerson, P.J. (1985) *Culture and the Evolutionary Process,* Chicago: University of Chicago Press.

Broswimmer, F.J. (2002) *Ecocide: A Short History of the Mass Extinction of Species,* London: Pluto Press.

Brown, H. (1954) *The Challenge of Man's Future,* New York: Viking Press.

Catton, W.R. Jr. (1980) *Overshoot: The Ecological Basis of Revolutionary Change,* Chicago: University of Illinois Press.

Costanza, R., d'Arge, R., de Groot, R., Farber, S., Grasso, M., Hannon, B., Limburg, K., Naeem, S., O'Neill, R.V., Paruelo, J., Raskin, R.G., Sutton, P. and van den Belt, M. (1997) "The Value of the World's Ecosystem Services and Natural Capital", *Nature,* 387: 253-260.

Coyle, K. (2005) *Environmental Literacy in America: What Ten Years of NEETF/Roper Research and Related Studies Say About Environmental Literacy in the U.S.,* Washington, D.C.: National Environmental Education & Training Foundation.

Daly, H.E. and Cobb, J.B. (1989) *For the Common Good: Redirecting the Economy Toward Community, the Environment, and a Sustainable Future,* Boston, Massachusetts: Beacon Press.

Damasio, A. (2005 [1994]) *Descartes' Error: Emotion, Reason, and the Human Brain,* New York: Penguin books.

Dewey, J. (1997 [1938]) *Experience and Education,* New York: Simon and Schuster.

Diamond, J. (2005) *Collapse: How Societies Choose to Fail or Succeed.* New York: Viking.

Dunlap, R.E., Gallup, G.H. Jr. and Gallup, A.M. (1993) "Of Global Concern: Results of the Health of the Planet Survey", *Environment,* 35: 7-15, 33-39.

Edelstein, L. (1967) *The Idea of Progress in Classical Antiquity,* Baltimore, MD: Johns Hopkins University Press.

Edgerton, R.B. (1992) *Sick Societies: Challenging the Myth of Primitive Harmony,* New York: Free Press.

Ehrlich, A.E. and Ehrlich, P.R.(1979) "Ecoscience: The Snail Darter and Us", *Mother Earth News*: 128-129.

Fiorino, D. (2001) "Environmental Policy As Learning: A New View of an Old Landscape", *Public Administration Review,* 613: 322-334.

Friedmann, J. (1987) *Planning in Public Domain: From Knowledge to Action.* Princeton, NJ: Princeton University Press.

Friedmann, J. and Abonyi, G. (1976) "Social Learning: A Model for Policy Research", *Environment and Planning A,* 8: 927-940. Reprinted in H.E. Freeman ed. (1978) *Policy Studies Review Annual 2,* Beverly Hills, CA: Sage Publications, pp. 82-95.

Gardner, G.T. and Stern, P.C.(1996) *Environmental Problems and Human Behavior,* Boston: Allyn and Bacon.

Gardner, J.W. (1995 [1963]) *Self-Renewal: The Individual and the Innovative Society,* New York: W.W. Norton.

Glasser, H. (2001) "Deep Ecology", in Neil J. Smelser and Paul B. Bates, eds., *International Encyclopedia of Social and Behavioral Sciences,* Vol. 6, Oxford: Pergamon, pp. 4041-4045.

Glasser, H. (2004) "Learning Our Way to a Sustainable and Desirable World: Ideas Inspired by Arne Naess and Deep Ecology", in Arjen E.J. Wals and Peter Blaze Corcoran, eds., *Higher Education and the Challenge of Sustainability: Problematics, Promise, and Practice,* Dordrecht: Kluwer, pp. 131-148.

Glasser, H., Craig, P. and Kempton, W. (1994) "Ethics and Values in Environmental Policy: The Said and the UNCED", in J. van der Straaten and J. van den Bergh, eds., *Toward Sustainable Development: Concepts, Methods, and Policy,* Washington, D.C.: Island Press, pp. 80-103.

Goldstein, H. (1981) *Social Learning and Change: A Cognitive Approach to Human Services,* Columbia, SC: University of South Carolina Press.

Gruber, D.L. (2003) *The Grassroots of a Green Revolution: Polling America on the Environment,* Cambridge, MA: MIT Press.

Hammond, A. (1998) *Which World? Scenarios for the 21st Century: Global Destinies, Regional Choices,* Washington, D.C.: Island Press.

Hawken, P., Ogilvy, J. and Schwartz, P. (1982) *Seven Tomorrows: Toward a Voluntary History,* New York: Bantam Books.

Heclo, H. (1974) *Modern Social Politics in Britain and Sweden: From Relief to Income Maintenance,* New Haven: Yale University Press.

Heyes, C., M. and Galef, B.G. Jr., eds. (1996) *Social Learning in Animals: The Roots of Culture,* San Diego: Academic Press.

Hughes, J.D. (1989) "Mencius' Prescriptions for Ancient Chinese Environmental Problems", *Environmental Review,* 13: 15-27.

Jacobs, J. (2004) *Dark Age Ahead,* New York: Vintage.

Jevons, W.S. (1865). *The Coal Question: An Inquiry concerning the Progress of the Nation and the Probable Exhaustion of Our Coal Mines,* New York: Augustus M. Kelley.

Johnson, H.D., with a Foreword by D.R. Brower (1997) *Green Plans: Greenprint for Sustainability,* Lincoln: University of Nebraska Press.

Kates, R.W., Clark, W.C., Corell, R., Hall, J.M., Jaeger, C.C., Lowe, I., McCarthy, J.J., Schellnhuber, H.J., Bolin, B., Dickson, N.M., Faucheux, S., Gallopin, G.S., Grübler, A., Huntley, B., Jäger, J., Jodha, N.S., Kasperson, R.E., Mabogunje, A., Matson, P., Mooney, H., Moore III, B., O'Riordan, T. and Svedlin, U. (2001) "Sustainability Science", *Science,* 292: 641-642.

Kates, R.W., Parris, T.M. and Leiserowitz, A. (2005) "What is Sustainable Development? Goals, Indicators, and Practice", *Environment,* 473: 8-21.

Keen, M., Brown, V. and Dyball, R., eds. (2005) *Social Learning in Environmental Management: Towards a Sustainable Future,* London: Earthscan.

Kempton, W., Boster, J.S. and Hartley, J.A. (1995) *Environmental Values in American Culture,* Cambridge, MA: MIT Press.

Kramer, S.N. (1981 [1956]) *History Begins at Sumer: Thirty-Nine Firsts in Man's Recorded History,* Philadelphia: University of Pennsylvania Press.

Lao Tzu (1961) *Tao Teh Ching,* Boston: Shambhala.

Lee, K.N. (1995) "Deliberately Seeking Sustainability in the Columbia River Basin", in L.H. Gunderson, C.S. Holling and S.S. Light, eds., *Barriers and Bridges to the Renewal of Ecosystems and Institutions,* New York: Columbia University Press, pp. 215-238.

Leeuwis, C. and Pyburn, R., eds. (2002a) *Wheelbarrows Full of Frogs: Social Learning in Rural Resource Management,* Assen, The Netherlands: Koninklijke van Gorcum.

Leeuwis, C. and Pyburn, R. (2002b) "Social Learning for Rural Resource Management", in C. Leeuwis and R. Pyburn eds., *Wheelbarrows Full of Frogs: Social Learning in Rural Resource Management,* Assen, The Netherlands: Koninklijke van Gorcum, pp. 11-21.

Leopold, A. (1949) *A Sand County Almanac and Sketches Here and There,* New York: Oxford University Press.

Malthus, T. 1970 [1798]. An Essay on the Principal of Population. New York: Penguin Books.

Marx, L. (1996) "The Domination of Nature and the Redefinition of Progress", in L. Marx and B. Mazlish eds., *Progress: Fact or Illusion?,* Ann Arbor: University of Michigan Press, pp. 201-218.

McKenzie-Mohr, D. and Smith, W. (1999) *Fostering Sustainable Behavior: An Introduction to Community-Based Social Marketing,* Gabriola Island, British Columbia: New Society Publishers.

Meadows, D.H., Meadows, D.L., Randers, J. and Behrens, W.W. (1972) *The Limits to Growth: A Report for the Club of Rome's Project on The Predicament of Mankind,* New York: New American Library.

Milbraith, L.W. (1989) *Envisioning a Sustainable Society: Learning Our Way Out,* Albany, NY: State University of New York Press.

Millennium Ecosystem Assessment (2005) *Ecosystems and Human Well-Being: Synthesis,* Washington, D.C.: Island Press.

Millennium Ecosystem Assessment Board (2005) *Living Beyond Our Means: Natural Assets and Human Well-Being (Statement from the Board)*, available from http://www.millenniumassessment.org/en/Products.BoardStatement.aspx

Mumford, L. (1970 [1938]) *The Culture of Cities*, New York: Harcourt Brace & Company.

Myers, N. (1979) *The Sinking Ark: A New Look at the Problem of Disappearing Species*, New York: Pergamon.

Naess, A. (2005) *The Selected Works of Arne Naess: Deep Ecology of Wisdom*, Vol. 10, H. Glasser and A. Drengson eds., Dordrecht, The Netherlands: Springer.

National Research Council Board on Sustainable Development (1999) *Our Common Journey: A Transition Toward Sustainability*, Washington, D.C.: National Academy Press.

Nisbet, R.A. (1980) *History of the Idea of Progress*, New York: Basic Books.

Nordic Council of Ministers (2003) *Education for Sustainable Development: Report from the Nordic Council of Ministers' Seminar Held in Karlskrona on 12-13 June 2003*, Copenhagen, Denmark: Nordic Council.

O'Riordan, T., ed. (1995) *Environmental Science for Environmental Management*, Essex, U.K.: Longman.

Oelschlaeger, M. (1994) *Caring for Creation: An Ecumenical Approach to the Environmental Crisis*, New Haven: Yale University Press.

Ophuls, W. (1977) *Ecology and the Politics of Scarcity*, San Francisco: W.H. Freeman.

Ornstein, R. and Ehrlich, P.R. (2000 [1989]) *New World New Mind*, Cambridge, MA: Malor Books.

Osborn, F. (1948) *Our Plundered Planet*, Boston: Little Brown.

Ostrom, E. (1990) *Governing the Commons: The Evolution of Institutions for Collective Action*, Cambridge, U.K.: Cambridge University Press.

Parson, E.A. and Clark, W.C. (1995) "Sustainable Development as Social Learning: Theoretical Perspectives and Practical Challenges for the Design of a Research Program", in L.H. Gunderson, C.S. Holling, and S.S. Light, eds., *Barriers and Bridges to the Renewal of Ecosystems and Institutions*, New York: Columbia University Press, pp. 428-460.

Plato (1989 [1929]) "Critias", in R.G. Bury, trans., *Timaeus, Critias, Cleitophon, Menexenus, Epistles*, Cambridge: Harvard University Press, pp. 255-307.

Prescott-Allen, R. (2001) *The Wellbeing of Nations: A Country-by-Country Index of Quality of Life and the Environment*, Washington, D.C.: Island Press.

Raskin, P., Banuri, T., Gallopin, G., Gutman, P., Hammond, A., Kates, R. and Swart, R. (2002) *Great Transitions: The Promise and the Lure of the Times Ahead*, Boston: Stockholm Environmental Institute and the Global Scenario Institute.

Robinson, J. (2003) "Future Subjunctive: Backcasting as Social Learning", *Futures*, 35: 829-856.

Robinson, J. (2004) "Squaring the Circle? Some Thoughts on the Idea of Sustainable Development", *Ecological Economics*, 48: 369-384.

Röling, N. (2002) "Beyond the Aggregation of Individual Preference", in C. Leeuwis and Rhiannon Pyburn eds., *Wheelbarrows Full of Frogs: Social Learning in Rural Resource Management*, Assen, The Netherlands: Koninklijke van Gorcum, pp. 25-47.

Roper (2002) *Green Gauge 2002: Americans Perspective on Environmental Issues, Yes... But*, New York: Roper ASW and NOP World.

Rosenthal, T.L. and Zimmerman, B.J. (1978) *Social Learning and Cognition,* New York: Academic Press.

Sears, P.B. (1988 [1935]) *Deserts on the March,* Washington, D.C.: Island Press.

Siebenhüner, B. (2004) "Social Learning and Sustainability Science", *International* Journal of Sustainable Development 7 (2): 146-163.

Shove, E. (2003) Comfort, Cleanliness, and Convenience: The Social Organization of Normality, Oxford: Berg.

Speth, J.G. (2004) *Red Sky at Morning: America and the Crisis of the Global Environment – A Citizen's Agenda for Action,* New Haven: Yale University Press.

UNESCO, *Decade of Education for Sustainable Development,* available from: http://portal.unesco.org/education/en/ev.php-URL_ID=23279&URL_DO=DO_TOPIC&URL_SECTION=201.html

United Nations (2004) *Agenda 21: Chapter 36,* available from http://www.un.org/esa/sustdev/documents/agenda21/english/agenda21chapter36.htm

United Nations (2005) *Millennium Development Goals,* available from http://www.un.org/millenniumgoals/

United Nations (2006) *Investing in Development: A Practical Plan to Achieve the Millennium Development Goals,* available from http://www.unmillenniumproject.org/reports/index.htm

Vitruvius (1985 [1934]) *De Architectura: Books VI-X,* Frank Granger, trans., Cambridge: Harvard University Press.

Vogt, W. (1948) *Road to Survival,* New York: William Sloane Associates.

Watzlawick, P., Weakland, J. and Fisch, R. (1974) *Change: Principles of Problem Formation and Problem Resolution,* New York: W.W. Norton.

Webler, T., Kastenholz, H., and Renn, O. (1995) "Public Participation in Impact Assessment: A Social Learning Perspective", Environmental Impact Assessment Review, 15: 443-463.

Wegner, E. (1999) *Communities of Practice: Learning, Meaning, and Identity,* Cambridge: Cambridge University Press.

Wickenberg, P., Axelsson, H., Fritzén, L., Helldén, G., Öhman, J., eds. (2004) *Learning to Change Our World? Swedish Research on Education and Sustainable Development,* Lund: Studentlitteratur.

Wilson, E.O. (1992) *The Diversity of Life,* Cambridge, MA: Harvard University Press.

Wilson, E.O. (1998) "Integrated Science and the Coming Century of the Environment", *Science,* 279: 2048-2049.

Woodhill, J. (2002) "Sustainability, Social Learning, and the Democratic Imperative: Lessons from the Australian Landcare Movement", in C. Leeuwis and R. Pyburn eds., *Wheelbarrows Full of Frogs: Social Learning in Rural Resource Management,* Assen, The Netherlands: Koninklijke van Gorcum, pp. 317-331.

World Commission on Environment and Development (WCED), ed. (1987) *Our Common Future.* Oxford: Oxford University Press.

Chapter 2

Riding the storm: towards a connective cultural consciousness

Stephen Sterling

> "Have we pushed the snooze button so we can sleep awhile longer?" (Lester Brown 2006).

Pressing up against the Earth's limits, we are being confronted with the limits of our thinking: a dawning realisation that the fundamental problem is not primarily 'out there' but 'in here', rooted in the underlying beliefs and worldview of the Western mind. This chapter is about social learning as a cultural shift – that is, the necessity and possibility of a deep change in shared worldview if we are to manifest the transition towards a more liveable and sustainable world whilst workable options remain.

In short, the argument stresses the need for a seismic shift, from the still dominant underpinnings of modernism, through and beyond the inroads of deconstructive postmodernism, and towards a relational, ecological or participative consciousness appropriate to the deeply interconnected world that we have created. This entails a shift of emphasis from relationships largely based on separation, control, manipulation and excessive competition towards those based on participation, appreciation, self-organisation, equity and justice. Increasing numbers of writers are pointing to the emergence and nature of this ecological worldview, predicated on the idea of a co-created or participative reality. In essence, this argument calls for social learning at a cultural, and indeed, global level – a critical self-awareness such that we might become "conscious agents of cultural evolution" in order to create a more sustainable civilization (Gardner, in Brown 2001, p. 206). There is, in Williams' words a need for "relearning on a grand scale", which should be "a core part of learning across society, necessitating a metamorphosis of many of our current education and learning constructs" (Williams 2004, p. 4). Seen from such perspectives as these, 'the learning society' is one that seeks to understand, transcend and re-direct itself through *intentional learning,* – towards what (Raskin *et al.* 2002) and others have called 'the Great Transition'. This scenario is characterised by "a connected and engaged global citizenry" which "advances a new development paradigm that emphasizes the quality of life, human solidarity, and a strong ecological sensibility" (Raskin *et al.* 2002, p.91). In this chapter, I look at the meaning of an ecological worldview, its relation to current dominant worldviews, evidence of its manifestation, the nature of learning that can lead to a more ecological consciousness, and attempt to assess the chances of a learning-based breakthrough to a changed worldview which is both *collective* and *connective.*

Crisis

I start from the assumption, based on numerous reports and indicators that we do indeed face a severe and unprecedented nexus of intractable problems which characterise our age and threaten the quality of life both for human and non-human species. In recent years, there have been an increasing number of high level warnings which state that humanity as a whole has a choice between moving towards more sustainable living patterns or face a scenario of increasing systemic breakdown and possible catastrophe, whether ecological, social, economic, political or some combination (WCED 1987, IUCN, UNEP, WWF 1991, World Resources Institute 2000, Loh 2004, Millennium Ecosystem Assessment 2005, www.millenniumassessment.org, Meadows *et al.* 2005). Such reports have been appearing regularly for some years. In parallel, many environmental educators and activists alike have tended to believe that graphic information about environmental degradation and inequity would be sufficient to cause a shift of thinking commensurate with the scale of the issues.

Amongst some people, a questioning of assumptions is indeed provoked, but for others, maintenance of a deep-seated worldview prevails despite evidence that it may no longer be appropriate to changed conditions. It may be that Chapman's view rings true for many people, who, he suggests:

> ...will not change their mode of thinking or operating within the world until their existing modes are proved beyond doubt, through direct experience, to be failing (Chapman 2002, p. 14).

By contrast, the growing appeals for change in worldview or paradigm themselves demonstrate a significant learning response to changed conditions. Thus, many commentators suggest the root of the 'world problematique' (Peccei 1982), lies in a crisis of perception; of the *way we see the world* (Bateson 1972, Skolimowski 1981, Laszlo 1989, Capra 1996). Accordingly, there are calls for 'a new way of thinking' (Clark 1989, Milbrath 1989, Bohm 1992, Laszlo 1997, Capra 1996, Elgin 1994) or 'reperception' (Harman 1988) which allows us to transcend the limits of thinking that have led to the current global predicament. From this perspective, the challenge of sustainability invokes much more than technical or 'rational' solutions. Laszlo (1997, p. 13) a noted holistic scientist and systems thinker, in a report for the Club of Budapest think-tank, states:

> To live in the third millennium we shall need more than incremental improvements on our current rationality; we shall need new thinking joined with new ways of perceiving and visioning ourselves, others, nature and the world around us.

Similarly, O'Riordan and Voisey (1998, p. 3), writing on the need to achieve what they call 'the sustainability transition', suggest that it "is as much about *new ways of knowing, of being differently human* in a threatened but cooperating world, as it is about management and innovation of procedures and products" (my italics). Such writers follow the logic of Einstein's statement which maintains that problems cannot be solved using the same consciousness or mode of thinking that created them, and that instead we need to change our perception.

> No problem can be solved from the same consciousness that created it. We have to learn to see the world anew (Einstein in Banathy 1995).

Similarly, Bohm (1992, p. 3) adds:

> The reason we don't see the source of our problem is that the means by which we try to solve them are the source.

Mary Clark, in a lengthy work subtitled 'The Search for New Modes of Thinking' argues that it is "the West that is most in need of the 'new modes of thinking' that Einstein demanded" (1989, p. 472) because of the rate of environmental change that the science and technology associated with this worldview has created. This worldview, she maintains, has "grown maladaptive". Similarly, Rich (1994, p. 288) points out the danger of the dominance of this worldview: "the consequences of maladaptation in a single, global culture may entail disaster on a scale unprecedented in human history".

Examination of many writers' descriptions of what the desired 'new modes of thinking' might be, and which might transcend this trap, reveals much use of terms like 'integrative', 'holistic', 'systemic', 'connective', and 'ecological'. Such terms carry meanings which are seen as necessary and logical responses to a thinking culture which the writers perceive as essentially reductionist and dualistic and therefore fragmentary and dis-integrative in its effects.

Before exploring in greater depth the meaning of an ecological worldview, let us now look in more detail at the concept of 'worldview' and the nature of the dominant worldview which the emergent ecological worldview seeks to transcend.

Worldview

> The cultural worldview, or social paradigm, is a story about the way the world works. It is "...the basic way of perceiving, thinking, valuing, and doing associated with a particular vision of reality"(Harman 1988, p. 10).

It is both a projection and reflection of how the world is seen, and is a characteristic of any society from history to the present. In a stable society, the dominant and mainstream story accommodates differences of view and debate within accepted parameters, and on the basis of accepted axioms and assumptions which are often unexamined and unarticulated. It has a *descriptive* aspect, influencing which aspects of and how the world is perceived, and a *normative* and *purposive* aspect which legitimises courses of action. So two components of paradigm can be distinguished, the *eidos* which refers to the cognitive or intellectual paradigm and the *ethos*, which refers to the affective level, values and norms. Further, these give rise to and influence a third dimension, the *praxis*. This term refers to the 'theory in action' and behaviour, both what is done (and not done) and how it is done. Of these three dimensions of paradigm, it is the ethos and perceptual dimension which is often most hidden from people's immediate awareness.

From the anthropologist's external viewpoint, a worldview appears to have the function of maintaining stability and continuity in any given culture or society. Yet this benign influence may become dysfunctional if an 'incoherence' (Bohm, p. 1992) develops between worldview and world. This is the crux of the argument pertaining to the dominant Western worldview. According to Bateson, a critically important figure in the history of systems thinking, our worldview is founded upon an 'epistemological error', a perception of and belief in separateness that makes it so. Bateson (1972, p. 463) states:

> I believe that (the) massive aggregation of threats to man and his ecological systems arises out of errors in our habits of thought at deep and partly unconscious levels.

If Bateson is right, it seems that the dualism of our worldview is a defining characteristic of our individual and collective psyche. Hence, whilst we should surface and examine the deep influences of reductionism, objectivism, materialism, and individualism as interrelating parts of the Western mindset, it appears that dualism is the fundamental key. Arguably, the essence of the modern worldview was (and still is) the perception of 'discontinuities' between subject/object, mind/body, people/nature and other such poles, underlain by a powerful mechanistic metaphor informing our sense of the world. In other words, the Western mind shifted historically from some sense of identity with 'the Other' in pre 1500 worldviews to a profound sense of, as well as intellectual belief in, *separateness.*

Wilber makes an important distinction between 'differentiation' and 'dissociation'. As he suggests, it is one thing to *differentiate* between culture and nature, for example, but quite another to *dissociate* them: 'One of the most prevalent forms of evolutionary pathology occurs when differentiation goes too far into dissociation' (Wilber 1997, p. 73). Yet, dissociation appears to be endemic – one might say systemically endemic – in Western society, worldview, epistemology, language and thought.

Let's now look at the meaning of the ecological worldview which, as suggested above, may be seen as a learning response to the inadequacy of the dominant operational Western epistemology.

Ecological worldview

As the term implies, an ecological worldview and sensibility arises from the intentional identification of *ecology* as an ontological metaphor, to contrast with the underlying Newtonian metaphor of mechanism which informs modernist thought. Within the history of the modernist paradigm, there has always been tension between the dominant *mechanistic* and the alternative *organicist* ways of viewing the world. Hence Capra (1996, p. 17) states:

> The basic tension is one between the parts and the whole. The emphasis on the parts has been called mechanistic, reductionist or atomistic; the emphasis on the whole holistic, organismic, or ecological.

Since the 1960s, Sachs suggests:

> ...ecology has left the biology departments of universities and migrated into every consciousness. The scientific term has turned into a worldview. And as worldview, it carries the promise of reuniting what has been fragmented, of healing what has been torn apart – in short of caring for the whole (Sachs 1999, p. 63).

An ecological worldview then, is essentially a 'living systems' and relational view, wherein everything, including human agency, unavoidably participates in the dynamic condition and future of the whole because everything is part of the whole. Thus, this worldview is variously referred as called 'participative' (Reason and Bradbury 2001) 'co-evolutionary' (Norgaard 1994), or 'living systems' (Elgin 1997). It is sometimes referred to as the 'postmodern ecological worldview' (Zweers

2000) associated with 'revisionary postmodernism' (Griffin 1992), – as opposed to deconstructive postmodernism. It is important to note the difference between the intellectual description of an ecological worldview, and the experience of it. Many writers see it as a profound shift of awareness from detached observer to engaged participant, from objectivity to critical subjectivity.

There is a considerable literature on ecological thought and philosophy. What follows is my interpretation of some of the key ideas emerging from this literature. I have developed a detailed review and argument in my doctoral thesis *Whole Systems Thinking as a Basis for Paradigm Change in Education* (Sterling 2003, p. 200) but for reasons of space, summarise some of the essential ideas and beliefs underpinning the postmodern ecological worldview here.

These include change:

- of *perception* from the prevailing 'I-It'relationship (Buber 1970 [1923]) of objectification and separation between the individual and others, or between the individual and nature, towards dialogic 'I-Thou' relationships, which recognises the 'Other' and that reality is co-created;
- of *assumption* from the separateness of mind and matter, to a panexperientialist view of their co-evolutionary relationship;
- of *conception* of an essentially dead and inert world, to an animate, dynamic and ultimately sacred world to be regarded with reverence;
- of *idea* of separate material 'environment', to a view of our embeddedness in a wider ecology which is both physical and non-material;
- of *focus* from external physical world, to the relation between our inner and outer worlds and the acceptance of multiple realities;
- of *models of order* from dysfunctional hierarchies to healthy holarchies;
- of *disposition* from control to participation;
- of *agency* from outside intervener to co-creator of reality and environment;
- of *belief* in certainty and intervention to recognotion of uncertainty and need for appreciation;
- of *view of evolution* from mechanism to co-evolution; and,
- of *view of knowledge* from a mono-universalism to diversity and contextualism.

The essence of these ideas can be mapped onto a simple but potent model of paradigm, knowing and cognition that I developed elsewhere (Sterling 2003). This model (Figure 2.1) suggests three interrelated areas of human knowing and experience which may be summarised as the domains of Seeing (perception), Knowing (conception) and Doing (design and action).

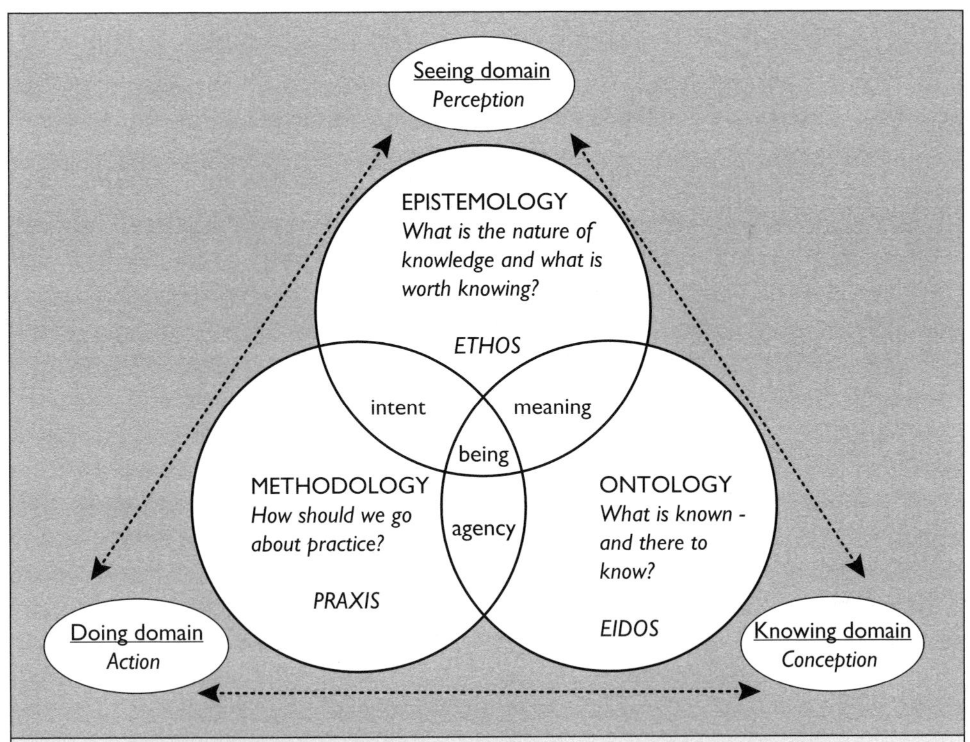

Figure 2.1. Dimensions of knowing and experience.
Epistemology, Ontology, Methodology: dimensions of *Knowing* aspect
Ethos, Eidos, Praxis: dimensions of *Paradigm* aspect
Perception, Conception, Action: dimensions of *Cognition* aspect

I argue that in each area and in the whole, an ecological worldview implies a significant shift or movement:

- in *assumptions* – leading towards greater compassion;
- in *distinctions* – leading towards greater understanding of connectivity; and,
- in *intentions/actions* – leading towards systemic wisdom and action which is more integrative and ecological.

These are summarised in Table 2.1, and explicated further in Table 2.2.

In summary, I argue that in essence the ecological worldview is characterised by the qualities of 'expansion', 'connection' and 'integration' and that these may be seen as a healthy and healing response to the *narrowness of perception*, *disconnective thinking*, and *dis-integrative practice* so often manifested in social

Table 2.1. Shifts in the three domains of knowing associated with an ecological worldview.

Seeing domain **Assumptions**	**Knowing domain** **Distinctions**	**Doing domain** **Intentions / actions**
Extension	Connection	Integration
Re-perception	Re-cognition	Response-ability
Compassion	Understanding	Wisdom

Table 2.2. Towards an ecological paradigm.

Seeing domain – Expansion: a widening of our boundaries of concern to include 'the Other' whether this be, for example, the neighbour or community, distant environments, peoples, and non-human species, or 'the needs of future generations'. A change in the quality of assumptions leading to greater empathy and *re-perception.*
Knowing domain – Connection: the disposition and ability to recognise and understand links and patterns of behaviour and influence between often seemingly disparate factors in all areas of life, to recognise systemic consequences of actions, and to value different insights and ways of knowing brought by others. A change in the quality of our distinctions leading to greater understanding and *re-cognition.*
Doing domain - Integration: a purposeful disposition and capability to seek healthy relationships between parts and wholes, recognising that the whole is greater than the sum of the parts; to seek positive synergies and anticipate the systemic consequence of actions. In systems terms, the concern here is with self-organisation, diversity, systemic coherence, integrity and healthy emergence. A change in the 'quality of our intentions/actions' leading to greater wisdom and *response-ability* (ability to respond).

discourse, policy and practice. This re-perception is manifested for example in the Earth Charter (www.earthcharter.org), which reflects an ecological approach to reorienting human activity to the workings and limits of the biosphere (see Slaby *et al.* 2007).

The changes towards an ecological paradigm outlined here, at individual, group or social level, depend on learning. Let's now turn our attention to this key topic.

Learning

It is often said that learning is a crucial key to a more sustainable future. However, it is obvious that a great deal of learning, both everyday and through formal education, makes no positive difference to a sustainable future, and may indeed make that prospect less rather than more likely. Learning is often promoted as a self-evident good, yet clearly, some learning will enable people to exploit the planet more effectively, or exploit others more fully. The act of learning is itself a neutral process – what matters ethically are the purposes and values behind curriculum and action associated with learning.

An illuminative theory was developed by Bateson from Whitehead and Russell's theory of logical types, and concerns levels of change and learning. Bateson distinguished three orders of learning and change, corresponding with increases in learning capacity, and these have been adopted variously by learning and change theorists, particularly in the field of systemic learning and organizational change, such as Argyris and Schon (1996) (single and double loop learning), and Ison and Russell (2000) (first order and second order change).

A key point is that learning can either serve to keep a system stable, or enable it to change to a new state in relation to its environment. Watzlawick, Weakland and Fisch (1980, p. 50), make the distinction thus: "there are two different types of change: one that occurs within a given system which itself remains unchanged, and one whose occurrence changes the system itself".

While these ideas are often used to describe organisational change, they apply equally to worldview/paradigm change where the worldview is itself seen as a system of thought (Bohm 1992), in relation to individuals, groups or entire societies.

Most learning that goes on within and outside learning institutions *makes no difference at all to individuals' or society's overall paradigm.* This is because, in Bateson's model, it is *first order learning* or basic learning within a consensually accepted framework. This is maintenance learning – adjustments or adaptations are made to keep things stable in the face of change: what Clark calls (1989, p. 236) calls 'change within changelessness'. In most cases, this is not a 'bad thing' but a necessary learning response to ensure stability. Bell and Morse (1999, p. 102) use Maturana and Varela's notion of *autopoesis* in living systems to explain how organisations, or belief systems can act as relatively closed systems in relation to a changing environment. But this becomes a 'bad' thing, or a maladaptation, when first order change is neither appropriate nor an adequate response to significant change in the environment (such as evidence of unsustainability).

By contrast, second order learning, according to Ison and Russell (2000, p. 229):

> ...is change that is so fundamental that the system itself is changed. In order to achieve (this) it is necessary to step outside the usual frame of reference and take a meta-perspective.

The critical point here, is that the urgent challenges of sustainability, require at least second-order social learning – a metacognitive and therefore fundamental questioning and re-ordering of assumptions. Faced with interrelated problems of immense complexity, arguably society is doubly constrained – with most attention and debate focussed within lower order learning levels informed by and nesting within an increasingly dysfunctional worldview. However, the growing debate about alternatives to such trends as globalisation, corporatisation, and growing inequity, in the context of deteriorating and threatened global ecosystems, is perhaps evidence of second order learning amongst some sections of society involving a deep questioning of assumptions.

Beyond this, many commentators call for third order learning or epistemic learning – that is, leading to a complete change of worldview or epistemology. According to the Center for Transformative Learning at OISE at the University of Toronto, transformative learning involves experiencing:

> ...a deep structural shift in the basic premises of thought, feelings and actions. It is a shift of consciousness that dramatically and permanently alters our way of being in the world. Such a shift involves our understanding of ourselves and our self-location: our relationships with other humans and with the natural world (Morrell and O'Connor 2002, p.xvii).

Many commentators see this as involving epistemological and perceptual change, and a transpersonal ethical and participative sensibility. In brief, an expansion of consciousness and a more relational or ecological way of seeing arises, inspiring different sets of values and practices. Whilst a number of commentators interpret Bateson's model in different ways, it affords a powerful insight on the possibilities of and constraints on the 'higher order' learning experience that the crisis of sustainability suggests is necessary. Not least, it indicates that a shift of perception from first order to second order learning, or from second order to third order, often involves resistance for it poses a significant challenge to existing beliefs and ideas, reconstruction of meaning, discomfort and difficulty but also sometimes, excitement.

There is another key point. Hawkins (1991, p. 178) (in the context of organisational change) puts it like this:

> It is not possible fully to understand a level of learning from within that level...we need some people in organisations to be concerned with and involved in Learning III before we can possibly improve Learning II functioning. An organisation needs not only its doers and operatives (Learning I); its strategists and thinkers (Learning II); but also its men and women of wisdom (Learning III).

These are not specialisations but – according to Bateson's theory – different stages of learning capacity in recognising that different people will be at different levels of learning (hence the common term 'higher order learning'). The point is that people who can reach Learning III will influence those working within Learning II and I.

Applied to cultural change, this suggests that any shift in worldview – and certainly any 'acceleration' – requires sufficient members of society to have experienced some form of transformative, epistemic change in order to facilitate and stimulate second order learning amongst greater numbers. Here, it is useful to make a distinction between intentioned learning, or 'learning by design' on the one hand, and reactive learning or 'learning by default' on the other'. As far as possible, it is the task of those who share some form of ecological consciousness to advance intentioned learning in order to build a social intelligence necessary to make the breakthrough to a more sustainable society.

Default learning happens when events impress themselves on the learners' consciousness, by surprise, shock or crisis. Learning by design, by contrast, implies a prior awareness, a willingness and intention to learn in response to a perceived innovation, threat or opportunity. The former is a reactive response, the latter is an anticipative response.

Clearly, crisis can trigger deep learning. Indeed, the unsettled state that crisis provokes can be seen as necessary to generate second or third order learning. Yet if experience of crisis is too far outside normality it can also lead to denial, fear and retrenchment, or a numbing sense of impotence. There is evidence reviewed below, that awareness of mounting crisis and systemic breakdown is indeed causing some deep questioning of assumptions underpinning the Dominant Social Paradigm (Milbrath 1989), yet crisis is an unpredictable if potent agent of change. In addition, we are faced with a time factor: the longer we delay reorientation of our systems towards sustainability, the fewer options we will have available not least as ecosystems degrade further (Bossel 1998, p. 308). Rather we need to 'accelerate the shift' (Gardner 2001, p. 189) to a more sustainable and more peaceable world, with, in all likelihood, less pain than would be caused through waiting for, say, as yet unknown technological solutions.

A recent report suggests that the traditional approach of the environmental movement using a combination of fear and information fails to move or engage with the public (Hounsham 2006). Rather it suggests, the environmental movement needs to connect with people's values, emotions and desires, present positive and realisable visions, and reconcile people's existing *sphere of concern* with their perceived limited *sphere of influence* through facilitating their ability to engage in change. This report confirms the indisputable educational adage, 'start where people are'. The critical challenge, however, lies in reducing the gap between where we are now and where, collectively, we need to get to. A Harvard University international report on sustainability values, for example, points out that whilst 'many requisite value and attitudes are in place...action lags behind' (Leiserowitz *et al.* 2004, p. 37). It seems that social learning agents need to carefully embrace and use the balance between security and challenge in learning situations to facilitate deep questioning, creativity and innovation. There are perhaps several ways forward, mixing elements of critique, visioning, and action as suggested below:

- Using the learning opportunities and openings that the current sense of loss of old certainties present.
- Highlighting paradox and incoherence to invite critical questioning, (for example as regards attempting to realise towards low carbon targets whilst expanding air travel).
- Inviting questions of 'ethical defensibility' in relation to proposals, ideas, policies and actions.
- Working on systemic effects and consequences of proposals, ideas, policies and actions beyond those intended (in advertently).
- Improving feedback signals (information) to indicate connections between everyday decisions and consumption and global issues of poverty, global warming, etc.
- Working by example and participative action on community-led initiatives and innovation.
- Visioning alternative futures and working with ecological design and systemic ideas to develop sustainable systems with multiple gains.
- Networking and sharing experience and learning between those involved in sustainability initiatives.

Change

According to Clark (1989, p. 235) in the last 2500 years, there have been only two "major periods of *conscious* social change, when societies deliberately 'critiqued' themselves and created new worldviews". So, following the example of Athenian and Renaissance societies, it would appear that our own time needs to be a third period of deep reflection and change. Measured against this grand scale, it is possible to see the whole movement of deconstructive postmodernism as an important second-order learning step which has critiqued modernist assumptions and opened the space for change. Whilst most writers about paradigm change follow Kuhn's view of the incommensurability of paradigms, I follow Wilber (1996) in subscribing to an *evolutionary* view of paradigm change whereby new paradigms are 'more adequate' than those they seek to replace. Change, in this view, is characterised by learning and a degree of commensurability and incorporation of previous paradigms as new ones emerge. Hence, I have suggested that we might see ourselves as living within an historic movement from the still dominant modernist paradigm of realism, into the idealist/constructivist position or moment, but at the same time, experiencing the stirrings of an emergent postmodern ecological worldview. The latter incorporates the ecological realism fundamental to much environmentalism, but also fully acknowledges the role of perception and of language emphasised by idealists and constructivists. I have argued this point in detail elsewhere (Sterling 2003) but Table 2.3 summarises the pattern.

The relative influence of root metaphors is roughly illustrated in the Table by whether they are shown in bold, ordinary type or in brackets. Hence, under the postmodern ecological worldview, 'mechanism' and 'text' are subsumed rather than dominant.

Clearly, there is no *simple* shift involved here: many of us hold elements of these paradigms at the same time in our perception or thinking. Heron (1992, p. 251), for example, suggests:

> Today, a significant minority have abandoned the Newtonian-Cartesian belief system in favour of some elaboration of a systems theory worldview. But it may be that they, and certainly the majority of people, still *see* the world in Newtonian-Cartesian terms. It is a big shift for concepts to move from being simply beliefs held in the mind to beliefs that inform and transform the very act of perception.

Perhaps mounting crisis is – and will be – the trigger that hastens significant social learning. The Harvard study mentioned above notes that there are many 'examples of nonlinear, abrupt and accelerated action in response to...powerful, galvanizing

Table 2.3. Mapping fundamental paradigmatic positions: moments, movements and metaphors.

Moments and movements		
First order change	**Second order change**	**Third order change**
Modernism	Postmodernism (deconstructivist)	Postmodernism
Foundationalism	Pragmatism/critical theory	Participativism
Realism	Idealism	Co-evolutionism
Materialism/dualism	Dualism	Panexperientialism
Universalism	Relativism	Relationalism
Objectivism	Subjectivism	Critical subjectivity
Positivism	Constructivism	Participatory knowing
Environmentalism	Ecologism	Whole systems thinking
Hard systems	Soft systems	Whole systems thinking
Root metaphors		
Mechanism	**Text**	**Living systems**
(Organicism)	Mechanism (Organicism)	(Text) (Mechanism)

events' (Leiserowitz *et al.* 2004, p. 37). Lester Brown, long time advocate of deep change towards sustainability, suggests (2006, p.xi):

> There is a mounting tide of public concern about where the world is heading and a growing sense that we need to change course. The rising price of oil and growing competition for this resource are feeding this concern. So, too, are the various manifestations of climate change, such as ice melting and rising sea level. When Hurricane Katrina left in its wake a $200-billion bill – nearly seven times the cost of any previous storm – it sent a message to the entire world.

The awareness that sudden and near-sudden crisis provokes is default learning, and it can lead to disempowerment and despondency. On the other hand, an awareness of incipient crisis combined with foresight can lead to significant 'learning by design'. Such learning may be seen as both preventive and remedial, both anticipative and rooted in current needs. This is evidenced in recent times by growing interest and practice in such diverse yet complementary areas as corporate and social responsibility, industrial ecology, ecological economics, sustainable agriculture, adaptive management, biodiversity and ecological restoration, green chemistry, ecological design and architecture, sustainable construction, local and green purchasing by institutions, renewable energy, ethical investment, local and healthy food, preventive health, environmental justice, ecological taxes, environmental law, education for sustainability, and sustainable communities. At national and international level, whilst resistance and denial undoubtedly remains, there is also clear evidence that concern about global warming and 'peak oil' is beginning to cause some deep re-thinking of norms and assumptions which have underlain public policy for decades. For example, Sweden aims to become 'the world's first oil free economy' by 2020 (Vidal 2006, p. 16). At the same time there is a discernable shift towards self-empowerment in civil society, evidenced, for example by the global 'Make Poverty History' movement launched in 2005. Thus, Rockefeller (2005) mentions 'the growing power and influence of global civil society which is exercised in and through consumer campaigns, shareholder initiatives, and Global Policy Networks all involving thousands of NGOs'. This quote appears in a book on the Earth Charter (Corcoran *et al.* 2005) which itself is giving rise to new movements and initiatives on a global scale. The Harvard study (2004, p. 39)) suggests that sufficient conditions are in place to generate a possible 'take off for sustainable development'.

According to Fritjof Capra (1996, p. 3) the ecological paradigm represents a "new perception of reality" which has "profound implications not only for science and philosophy, but also for business, politics, health care, education, and everyday life." Hence, the ecological worldview has a strong teleological element which urgently challenges the objectivism of modernism, and the relativism of deconstructionism. In essence, it is interested in healthy relationships, in sustainable systems, be they ecological, social, economic, or political, and in exploring and realising such values as creativity, diversity, stability, and resilience in total systems.

However, there is no 'historical necessity' that the ecological worldview will prevail (Zweers 2000). Ecophilosopher Eckersley (1992, p. 52), for example, sees "nothing inevitable" about "a new, ecologically informed cultural transformation". Eisler (1990, p.xx) adds that while a better future is possible:

> "...it by no means follows (as some would have us believe) that we will inevitably move beyond the threat of nuclear or ecological holocaust into a new and better age. In the last analysis, that choice is up to us."

Raskin et al's study of future scenarios suggests that 'the momentum toward an unsustainable future, can be reversed but only with great difficulty.. (and yet)...a planetary transition toward a humane, just and ecological future is possible' (2002, p.95). This double-edged message is echoed by the Earth Council's book *Plan B* (Brown 2006, p. 266):

> One way or another, the decision will be made by our generation. Of that there is little doubt. But it will affect life on earth for all generations to come.

The critical factor is how to assure a sufficiently widespread and adequate learning response. Ballard (2005) suggests that four necessary and interrelated conditions underpin such a response:

- Awareness of what is happening and what is required.
- Agency or ability to find a response that is meaningful.
- Association with other groups and networks.
- Action and reflection.

The task for educators and social learning agents is to facilitate participative and systemic 'critical learning systems' (Bawden 1997) and situations where these conditions can be realised. At heart is the idea of *resilience,* and this a key to the interrelationship between learning and sustainability. Rather than viewing these as distinct fields – as reflected in the terms 'learning about sustainability,' or even 'learning for sustainability' – I have argued for the radical notions of 'learning as sustainability' (Sterling 2003) and 'sustainable education' (Sterling 2001). These emphasize the complementarity and overlap between such educational values as learner autonomy, capacity building and participation on the one hand, and such sustainability values as self-renewal, system health and integrity on the other. In essence, sustainability is about conservation of potential and increasing self-organisation, resilience and adaptive capacity at all nesting levels within social-ecological systems (Bossel 1998, Folke 2003), and learning – reflexive, experiential, experimental, participative, iterative, real-world and action oriented – is intrinsic to this process and challenge.

Conclusion

This chapter has argued that, in order to provide any assurance of a liveable and sustainable future in this century, global society as a whole needs to recognise, embed and enact intentional social learning towards an ecological consciousness and ecological patterns of organisation and human activity. We need, (in Milbrath's words 1989), 'to learn our way through' and with sufficient energy, commitment, humility and positive vision to make a decisive difference at local and global scales within the coming decades.

It is a persuasive yet tricky argument, as it can come across as dictatorial and dogmatic. Where is there space for critique, dissent and pluralism? Certainly, there are those in the ecological movement who advance their argument with a missionary zeal. Given the impending 'storm' and grave issues of risk and urgency, there is a strong case for such passion. Yet sustainable change and sustainable social learning derives from engagement, reflection and self-critique, rather than instruction. Borrowing the ecophilosopher Naess' idea of a 'platform' (Naess 1995), the essential ideas informing an ecological worldview can be broadly shared without prescribing or predetermining ultimate premises, or specific interpretations and actions. These must instead emerge from individuals' and groups' own sense of meaning and from their own learning. In this sense, the relationalism intrinsic to the ecological metaphor and platform gives coherence to and supports plural interpretations and actions appropriate to local cultures and conditions – echoing the ecological principle of diversity in unity. Paradoxically, (adapting a phrase from Ackoff 1999) an ecological worldview yields many different views of the same thing, and the same view of many different things.

I have argued that if we are to 'ride the storm' successfully, we will need to think and act ecologically and with increasing systemic wisdom. This worldview is the product of a diverse but connected social learning movement, particularly over the last few decades. It is also a dynamic and increasingly coherent and cogent basis for further challenge and change in human activity towards sustainability across the globe. Here is an ecological *zeitgeist* in the air towards a consciousness that is both *connective* and *collective*. Whether it manifests sufficiently into a shared cultural base capable of transcending the dominant social paradigm remains to be seen. One way or the other, our destiny hangs on it.

References

Ackoff, R.L. (1999) "Jacket endorsement" in R. Flood, ed., *Rethinking the Fifth Discipline – Learning within the unknowable,* London:Routledge.

Argyris, C. and Schon, D. (1996) *Organisational Learning II,* New York: Addison Wesley.

Ballard, D. (2005) "Using learning processes to promote change for sustainable development", *Action Research,* Vol 3 (2), London: Sage Publications.

Bateson, G. (1972) *Steps to an Ecology of Mind,* San Franscisco: Chandler.

Bawden, R. (1997) "Leadership for Systemic Development" in Centre for Systemic Development, *Resource Manual for Leadership and Change,* University of Western Sydney: Hawkesbury.

Bell, S. and Morse, S. (1999) *Sustainability Indicators – Measuring the Immeasurable?* London: Earthscan.

Bohm, D. (1992) *Thought as a System,* London, Routledge.

Bossel, H. (1998) *Earth at a Crossroads – Paths to a Sustainable Future,* Cambridge: Cambridge University Press.

Brown, L. (2001) *Eco-economy,* London: Earthscan.

Brown, L. (2006) *Plan B 2.0: Rescuing a Planet Under Stress and a Civilization in Trouble,* NY: W.W. Norton & Co.

Buber, M. (1970 [1923]) *I and Thou.* New York: Scribner.

Capra, F. (1996) *The Web of Life,* London: Harper and Collins.

Chapman, J. (2002). *System Failure.* London: Demos.

Clark, M. (1989) *Ariadne's Thread – The Search for New Ways of Thinking,* Basingstoke: Macmillan.

Corcoran, P, Vileda, M. and Roerink, A. (eds) (2005) *The Earth Charter in Action – Towards a Sustainable World,* Amsterdam: KIT Publishers.

Eckersley, R. (1992) *Environmentalism and Critical Theory,* Albany: State University of New York.

Eisler, R. (1990) *The Chalice and the Blade,* London: Pandora/HarperCollins.

Einstein, A., quoted in Banathy, B. (1995) "Developing a Systems View of Education", *Educational Technology,* June 1995.

Elgin, D. (1994) "Building a Sustainable Species-Civilisation – a challenge of culture and consciousness", in McKenzie-Mohr, D. and Marien, M., ed., (1994) *Futures,* Special issue, Visions of Sustainability, vol. 26, no 2, California: Cambridge Millennium Project.

Elgin, D. (1997) *Global Consciousness Change: Indicators of an Emerging Paradigm,* California: Cambridge Millennium Project.

Folke, C. (2003) "Socio-Ecological Resilience and Behavioural Responses", in A. Biel, B. Hansson and M. Martensson, ed., *Individual and Structural Determinants of Environmental Practice,* Aldershot: Ashgate.

Gardner, G. (2001) "Accelerating the Shift to Sustainability" in L. Brown et al, *State of the World,* Worldwatch Institute report on progress towards a sustainable society, London: Earthscan Publications.

Griffin, D.R. (1992) "Introduction to SUNY Series in Constructive Postmodern Thought" in D. Orr, ed., (1992) *Ecological Literacy: Education and the transition to a postmodern world,* Albany: SUNY Press.

Harman, W. (1988) *Global Mind Change: The Promises of the Last Years of the Twentieth Century,* Knowledge Systems Inc., Indianapolis, IN.

Hawkins, P. (1991) "The Spiritual Dimension of the Learning Organisation", *Management Education and Development,* 22(3):172-187.

Heron, J. (1992) *Feeling and Personhood: psychology in another key,* London: Sage. Hounsham, S (2006) *Painting the Town Green,* London: Green-Engage/Transport 2000.

Hounsham, S. (2006) *Painting the town green.* London: Green Engage.

IUCN, UNEP, WWF (1991) *Caring for the Earth – A Strategy for Sustainable Living,* IUCN, UNEP, WWF, Gland.

Ison, R. and Russell, D. (2000) *Agricultural Extension and Rural Development – Breaking out of Traditions, a second-order systems perspective,* Cambridge: Cambridge University Press.

Laszlo, E. (1989) *The Inner Limits of Mankind,* London: One World Publications.

Laszlo, E. (1997) *3rd Millennium – The Challenge and the Vision,* The Club of Budapest, Stroud: Gaia Books.

Leiserowitz, A., Kates, R., and Parris, T. (2004) *Sustainability Values, Attitudes and Behaviours: A Review of Multi-national and Global Trends,* CID Working Paper No. 113, Center for International Development, Harvard University, Cambridge, MA. Electronically retrievable via: http://www.cid.harvard.edu/cidwp/pdf/113.pdf

Loh, J. (2004) WWF, *Living Planet Report 2004,* Gland.

Meadows, D.H., Meadows, D.L., and Randers, J. (2005) *Limits to Growth – the 30 year update,* London: Earthscan.

Milbrath, L. (1989) *Envisioning a Sustainable Society,* Albany: State University of New York Press.

Millennium Ecosystem Assessment (2005) *Ecosystems and Human Well-Being: Synthesis,* Washington, D.C.: Island Press.

Morrell, A., and O'Connor, M. (2002) "Introduction" in O'Sullivan, E., Morrell, A., and O Connor, M. (2002), *Expanding the Boundaries of Transformative Learning,* New York: Palgrave Macmillan.

Naess, A. (1995) "The Deep Ecological Movement – Some Philosophical Aspects", in G. Sessions, ed., *Deep Ecology for the 21st Century,* Boston: Shambala.

Norgaard, R. (1994) *Development Betrayed – The end of progress and a co-evolutionary revisioning of the future,* London: Routledge.

O'Riordan, T., Voisey, H. (1998) *The Politics of Agenda 21 in Europe,* London: Earthscan.

Peccei, A. (1982) *One Hundred Pages for the Future,* London: Futura.

Raskin, P., Banuri, T., Gallopin, G., Gutman, P., Hammond, A., Kates, R. and Swart, R. (2002) *Great Transitions: The Promise and the Lure of the Times Ahead,* Boston: Stockholm Environmental Institute and the Global Scenario Institute.

Reason, P and Bradbury, H. (eds) (2001) *Handbook of Action Research – Participative Practice and Enquiry,* London: Sage Publications.

Rich, B. (1994) *Mortgaging the Earth – The World Bank, Environmental Impoverishment, and the Crisis of Development,* Boston: Beacon Press.

Rockefeller, S. (2005) "The Transition to Sustainability" in P. Corcoran, M. Vileda and A. Roerink, ed., *The Earth Charter in Action – Towards a Sustainable World,* Amsterdam: KIT Publishers.

Sachs, W. (1999) *Planet Dialectics,* London: Zed Books.

Skolimowski, H. (1981) *Ecophilosophy,* Lonson: Marion Boyars.

Slaby, M.C., Hollingshead, B.P. and Corcoran, P.B. (2007) "Learning and living with Earth Charter" in A.E.J. Wals, ed., *Social learning towards a sustainable world – principles, perspectives and praxis,* Wageningen, Wageningen Academic Publishers, pp. 483-495.

Sterling, S. (2001) *Sustainable Education – Re-Visioning Learning and Change,* Schumacher Society Briefing no 6, Dartington: Green Books.

Sterling, S. (2003) "Whole Systems Thinking as a Basis for Paradigm Change in Education – Explorations in the Context of Sustainability", PhD thesis, University of Bath, Sterling, UK. www.bath.ac.uk/cree/sterling.htm

Wilber, K. (1996) *A Brief History of Everything,* Dublin: Gill and Macmillan.

Wilber, K. (1997) *The Eye of Spirit – An integral vision for a world gone slightly mad,* Boston: Shambhala Publications.

Williams, M. (2004) "Preface", in N. Potter et al, ed., *See Change – Learning and education for sustainability, Parliamentary Commissioner for the Environment,* New Zealand: Wellington.

Vidal, J. (2006) "Sweden plans to be world's first oil-free economy", *The Guardian London,* 8 February.

Watzlawick, P., Weakland, J.H., and Fisch, R. (1980) "Change", in M. Lockett and R. Spear, ed., *Organizations as Systems,* Milton Keynes: The Open University Press.

WCED (1987) "Our Common Future", World Commission on Environment and Development, Oxford University Press, Oxford.

World Resources Institute (2000) "World Resources 2000-2001: People and ecosystems, the fraying web of life", UNEP/UNDP/World Bank/World Resources Institute.

Zweers, W. (2000) *Participation with Nature – Outline for an Ecologisation of our Worldview,* Utrecht: International Books.

Chapter 3

The practical value of theory: conceptualising learning in the pursuit of a sustainable development

Anne Loeber, Barbara van Mierlo, John Grin and Cees Leeuwis

Introduction

A book that addresses 'learning and sustainable development' is likely to be picked up with the intention of learning *about* sustainable development. What is it? What does the concept imply in concrete, practical terms? What precisely is the problem that needs to be solved? In the 20-odd years that followed the now famous phrase from the World Commission on Environment and Development (WCED), which described it as a development that "meets the needs of the present without compromising the ability of future generations to meet their own needs" (WCED 1987, p. 8), sustainable development has been defined in numerous ways (Brooks 1992, de la Court 1990), and elaborated in a wide variety of policy measures and business plans. Rather than as an indication of confusion and lack of clarity, in our view this wealth of interpretations is inherently characteristic to the concept, and is one of its major merits. The question of what a sustainable development may entail should *a priori* be answered in plural (cf. Grin 2006). In contrast to earlier environmentalists' interpretations, in which a 'sustainable society' implied a zero-sum trade-off between the economy and the environment, the WCED's description promotes the idea of balancing economic, social and environmental goals. Given that, in addition, each of these dimensions may be understood in various ways, this 'balancing act' is inherently ambiguous: sustainable development may accommodate potentially conflicting values, beliefs and points of view on what is a sensible, desirable and feasible thing to do.

While the concept of sustainable development may be open to multiple interpretations, it is quintessential that any particular specification or interpretation of the concept must involve a substantive elaboration of its meaning in a particular setting or context (Loeber 2004). For its attainment, it is imperative that knowledge on what is 'sustainable' in practice – knowledge on what to do – is indeed acted on as well. To ensure that any particular elaboration is both meaningful to the people whom it concerns as well as practical, *learning* is an essential element of projects and practices that seek to contribute to a sustainable development.

In our view, sustainable development implies a need for learning in three respects. First, it is an essentially *contestable* concept, in the sense that no authoritative, universally valid definition can be formulated. There is no way of determining what is 'really sustainable' other than through processes of collective and contextual deliberation and mutual learning. Secondly, in addition, it is a concept that claims, *normatively*, to offer desirable directions for action. Hence, the learning processes implied in the first characteristic are more than mere 'joint fact finding' exercises, and involve processes of value judgment. From both characteristics, it follows that the sustainable development concept needs to be elaborated in an 'action-oriented' way, in which a balance is found between what is deemed desirable and what may be made feasible, given a particular context. Finding such a balance requires both 'puzzling and powering' (Heclo 1974). In other words, learning for sustainable development is not merely about ideas, but also about the power dimensions involved in the envisioned transformations.

Thirdly, the balancing of feasible and desirable options for action is particularly complex given that sustainable development is a *revolutionary* concept. Its elaboration and implementation may imply system innovation, i.e. an 'opening up' of existing routines, rules, values and assumptions embedded in the institutions that have co-evolved with earlier, 'unsustainable' modes of socio-technological development and that, as a result, tend to reflect and produce the juxtaposition, now considered undesirable, of economic progress and ecological and social balance (Grin 2006). Therefore, the elaboration of sustainable development into practical options for action must include a 'reflexive perspective', that is, a critical scrutiny of things that are usually taken for granted, in such a way that their historically grown self-evidence ('path dependency') is challenged. In our view, this implies a need for learning understood as reflection on the theories, beliefs and assumptions that underlie action.

Practitioners who face the challenge of fostering sustainable development, such as project leaders or facilitators of innovative projects and programmes, in our opinion, are bound to explicitly encourage learning processes. They will be confronted with serious doubts and critical questions as to how to set up projects and programmes that induce learning for a sustainable development in practice. Three of these questions will be discussed in this chapter. The first one arises from the often disappointing effects of awareness-raising campaigns. Learning about the environmental consequences of their behaviour does not necessarily mean that people start behaving in a more environmentally friendly way. Then what is the use of 'changing cognition', and how can the relationship between learning and action, that is, between thinking and doing, be conceived? Since so many environmental and social problems lead to conflicts that require the co-operation of various parties, the second question is whether or not it is possible to learn in groups of heterogeneous actors who hold diverging ideas, and whether it is at all feasible to come to an agreement. And if so, what does 'agreement' mean? The last question addresses the seeming contradiction between 'feasible' and 'revolutionary' change. What is considered to be 'feasible' from the perspective of the people involved might entail mere incremental changes, while more far-reaching ideas for change in the long run, even though desirable, may be considered impossible. How can the ambition to foster fundamental change in the pursuit of a sustainable development nonetheless be kept alive? We will deal with these questions, below drawing upon various bodies of literature on learning. We will do so from the perspective of a practitioner seeking to induce learning processes in the light of the sustainable development ambition.

Conceptualising learning

Learning has been looked at from various disciplines and angles, including cognitive psychology, social psychology, (adult) education studies, management studies, innovation studies, policy science studies, development studies and complex systems thinking. As a result, the concept of learning is used to cover "a wide society of ideas" (Minsky 1988, p. 120). Here we do not attempt to give a full overview of the resulting conceptual richness (for an overview, see Grin and Loeber 2006). Instead, we choose to discuss theories that bear relevance to the perspective on sustainable development outlined above; that is, to dealing with a concept that is essentially normative, contestable and radical. We are especially interested in those perspectives that address action-oriented processes of learning that take place in regular societal contexts rather than in formal educational settings. The aim is to formulate methodological principles in relation to the three questions raised that give guidance to the way innovative projects can be set up.

The learning individual

Let us first consider the learning individual. In theories on learning that focus on the individual, the importance of concrete experience is often emphasised. A central point of reference in this field of so-called experiential learning is Kolb (1984). Drawing on the work of authors such as Lewin, Dewey and Piaget, Kolb developed a model of the 'learning cycle'. In order to learn, Kolb posits, an individual must go through the following stages: experiencing, reflecting, conceptualizing, deciding and acting. Concrete experience in and through action sets the learning process in motion: the experiencing individual observes the effects of his or her actions and reflects on these. Thereupon, s/he conceptualizes the relation between action and effect, and generalises it into theoretical terms. Subsequently, s/he tests the theory by acting accordingly in a subsequent situation.

Not all kinds of experiences lead to learning; learning occurs mainly when there are conflicts between expectations and experiences or between ideas and desires. Such learning may also mean *un*learning or *re*learning, in the sense that people who learn dispose of their old theories and ideas. In this process of relearning, reflection on the experiences is of great importance. Following Kolb's line of reasoning, Leeuwis (2004) argues that for reflection, (proper) feedback on the consequences of one's actions is essential. "The process of stimulating and contributing to learning ... is almost synonymous with organizing and providing good quality feedback" (2004, p. 154). Therefore, in the practice of an innovative project it serves to ensure that there are proper feedback mechanisms. Leeuwis provides some examples from farm management, such as on-farm experiments, visualizing agro-ecological processes that are difficult to observe, comparing farm operations and results, and so on.

Scholars that use Kolb's theory in the field of sustainable development particularly apply the idea of the learning cycle. It offers a concrete framework for developing activities within evolving networks for the different phases of the learning process (see for instance Keen *et al.* 2005). What makes this theory on learning interesting from the perspective of a sustainable development is that it focuses explicitly on the relationship between cognition and action, rather than on the increase of an individual's stock of knowledge.

However, in the light of the sustainable development ambition, this theory has two major limitations. First, it does not problematise the conditions under which learning may be stimulated. The focus is on learning from and through (primarily) individual experience; the theory does not take into consideration the *contextual* aspect, that is, how learning is influenced by social settings. In addition to an individual's personal abilities, the social setting in which learning takes place can be more or less conducive to learning.

A second aspect of Kolb's model that detracts from its practical value for innovative projects is that it overlooks the role of values and interests that influence human action. In the pursuit of a sustainable development, it is imperative to take these into consideration. After all, the concept inherently implies revolutionary change, as we have observed above, and often values and deeply held beliefs are at the core of possible resistance to let go of old habits and common ideas.

One author who did integrate such values and beliefs in a theory on learning is Schön. In his view, cognition cannot be separated from values and beliefs. What is more: nor can cognition and action. According to Schön, actors engage in 'reflection-*in*-action' (1983, p. 54). Central to his work (1983, Schön and Rein 1994) is the idea that people act (and choose their line of action) on the basis of what he calls 'theories-in-use'. A theory-in-use is a mental map of theoretical, normative and empirical considerations that a professional brings to bear on the way s/he tries to solve a problem. Schön conceptualises learning as the process of reviewing such a mental map in the light of 'crises and surprises': unexpected events, and unexpected misfits between the specificities of the problem situation and the theory-in-use, detected through observation and experience. If that is the case, the new and surprising information on the situation (the situation's 'back talk') may lead an individual to change the theories, beliefs and values that underlie his or her actions. If a mere change is made in the definition of the problem encountered or in the way solution strategies are pondered, the reviewing of the underlying notions amounts to so-called first order (or 'single loop') learning. First order learning leaves fundamental notions, preferences and values intact. As a result, this type of learning generally results in incremental changes in an actor's problem-solving strategies. When the fundamental elements in a theory-in-use themselves are the object of reflection, the actor engages in second order ('double loop') learning. This may result in major changes to an actor's strategic choices, objectives and preferences.

The learning processes that Schön describes are of the kind that people constantly engage in every day. Observation and experience provide a continual flow of information through which one can come to reflect on one's goals and actions, and on the way in which these goals and actions relate to each other with regard to the context in which one operates. Schön's explanation of these processes holds two important clues as to the potential role of learning in the pursuit of a sustainable development.

First, by illuminating the relationship between learning and action, that is, between thinking and doing, Schön's work sheds light on the nature of the changes that an innovative project must seek to provoke, namely changes in the theories-in-use that underlie current 'non-sustainable' actions. Often these theories are tacit; they remain implicit and go unnoticed. In order to challenge them, they need to be brought to the surface: people will have to be made aware of their tacit rationalities, and be tempted to reconsider them, for instance by providing 'surprising' feedback or by breaching the routine of daily work. With that, an important answer to the first leading question of this chapter is provided. A second relevant aspect of Schön's insights is that, even though theories-in-use play a role in the actions of various actors in a similar way, they differ in terms of contents depending on professional training and experience, social background, up-bringing and so on. Because of their intrinsic and fundamental divergence, the theories-in-use that people from different professional and cultural backgrounds hold, will influence the possibility for them to learn collectively, a topic to which we will now turn.

Learning in groups and organisations

In Schön's earlier work, the trigger that induces an individual to learn is identified in the relationship between the problem-solving actor and the problem situation encountered. Yet, learning processes in practice take place not only 'in action', but also – as Schön also acknowledges – most notably, in *interaction*, both with others and with the contexts of a problem situation. In the light of the pursuit of a sustainable development, it is particularly useful to conceptualise learning as a social event. After all, the changes implied by the sustainable development concept require joint action by large numbers of actors. There are also more instrumental reasons that make it worthwhile to consider learning as an essentially social practice. Take for instance Kolb's insight that people learn from the feedback they get about the effects of their actions. Even if feedback is improved, as suggested by Leeuwis (2004), long term effects (as well as 'long distance effects') are likely to remain unobserved by the acting individual. As Senge put it: "We learn best from experience but we never directly experience the consequences of many of our most important decisions" (1990, p. 23).

Furthermore, we need others to help us notice not only what we fail to observe because of practical reasons (lack of information due to time lapse or distance), but also because of "what [we] have worked to avoid seeing" (Schön 1983, p. 283). If there is no clear need to reflect fundamentally on the tacit assumptions that underlie common patterns of behaviour, these are often factored out of the discussion. A constant questioning of these would interfere with daily routine. Moreover, people are likely to develop so-called defensive routines (Argyris 1990), which discourage someone from doing so in order to avoid the kind of feelings of uneasiness that occur in confrontations with discussion partners, for instance, the threat of losing face. Such uneasiness may also be experienced when someone is confronted with information that does not match his of her understanding of a situation, such as news of the unforeseen effects of his of her actions. A common response is to avoid such unwelcome information. It is either ignored or dismissed as unimportant or untrue, *unless* others help one to become aware of it and take it into consideration. Notably, second order learning is likely to occur only in those situations where a person is no longer able to 'shut out' dissonant information or when one deliberately wishes to reflect on one's (professional) practices. A setting in which defence mechanisms are dismantled and one is stimulated by others to take into consideration new and possibly counter-intuitive information may therefore encourage and accelerate the learning required for stimulating change towards a more sustainable society. Given that the changes required for a sustainable development are likely to concern the fundamental values that dominate the current way of life, second order learning may be considered imperative.

On the basis of empirical and theoretical work, Grin and Hoppe (1995) and Forester (1999) have outlined the conditions under which second order learning may be induced. They emphasise that an atmosphere of trust and a commitment to reciprocity is essential. Since people are supposed to try and make explicit what usually remains tacit, and thus be encouraged to reflect on what is taken for granted, a rule like reciprocity helps: "I'll let you in on my private considerations if you'll let me in on yours". In addition, second order learning can be triggered by unexpected events with a high impact that encourage groups to scrutinize usual practices, e.g. by natural disasters, such as floods or a threatening regulation. Unexpected events with negative consequences are seen to trigger learning more than positive ones (Grin and Hoppe 1995).

The research done on learning as a social event draws attention not only to the concrete setting in which learning takes place, but also to the relationship between the learning individual and his or her wider surroundings. While authors such as Argyris and Schön (1996) conceive of the relation between an individual and the organisational setting in which s/he operates as subtly interwoven, according to Senge (1990), the organisation itself may be viewed as a learning entity. In his work, Senge explores the prerequisites that organisations need to successfully and effectively adapt to, and anticipate, a changing environment. He identifies five basic competences ('disciplines'), which include the ability of an organisation to help employees elicit their deeply held images and assumptions, so as to open them up to the influence of others, and its capacity to have employees develop a joint understanding of a desirable future that fosters genuine commitment and engagement (rather than mere compliance). The possibility of successfully applying these disciplinary requirements hinges crucially on an organisation's ability to master the 'fifth discipline', that is, the capacity to view and appreciate the organisation as a whole rather than as an accumulation of its constituent parts. This capability is often referred to as 'system thinking'.

It is notably for its focus on systems dynamics, which builds on Forrester's (1968) work, that Senge's ideas have found a receptive audience among people who engage, separately and jointly, in strategies to help realise a sustainable development, e.g. among representatives of corporations that take their social responsibility seriously (Molnar and Mulvihill 2003, Cramer and Loeber 2004). According to Senge, "non-systemic ways of thinking and acting" are at the core of unsustainable practices, and can be tackled by "building [learning] enterprises that operate in greater harmony with larger social and ecological systems" (Senge 2000, p. 1).

It may be the case that through system thinking, organisations can better evolve with, and adapt to, changes in their environment. Yet it remains unclear how the people in (these and other) enterprises may operate together in relative harmony or, better, effective co-operation. After all, as we observed above, actors may differ widely in their most fundamental values, assumptions and beliefs; notions that strongly influence the way they operate. Röling's theory on social learning addresses this question. Röling (2002) posits that stakeholders, indeed each with their own cognition, may develop what he calls 'distributed cognition', through social learning. In interaction, they can learn about each others' experiences, ideas and values and thus have a chance to generate knowledge in which these are more or less incorporated. In that way, stakeholders may work together and engage in collective action or complementary practices, while they may not necessarily share values and aspirations. It is enough if these overlap or are mutually supportive.

Another concept that centres on the idea of complementarity, instead of consensus or complete similarity, has been developed by Grin and van de Graaf (1996). These authors speak of 'congruency' when actors, however heterogeneous in their roles, ideas and values, come to regard some line of action as a meaningful and valuable solution to a problem they experience – however different their problem definitions too. If the solution is compatible with their underlying values and preferences, they may be triggered to act on the insight and, hence, come to act in a congruent manner.

Distributed cognition and congruency are both notions that help us conceptualise the types of agreement we may hope to attain in innovative projects and programmes. Considering that the parties required to provoke an actual change towards a sustainable development may be worlds apart (in the literal and symbolic sense), holding widely diverging interests and outlooks on life, the aim of learning must be to reach some congruency of meaning between these parties, rather than to opt for complete consensus.

This view therefore departs from insights in the literature on consensus building or principled negotiations which hold *that it is essential* for divergent actors to feel that they are mutually dependent in solving a problematic situation. Yet, as we learn from this body of literature, *it is worth aspiring* to develop, among heterogeneous actors, a feeling that they are mutually dependent when solving a problematic situation, in the sense that what they can achieve on their own may be less favourable than what they can achieve together (Susskind *et al.* 1999). As argued by Leeuwis (2000, 2004), enhancement of feelings of interdependence may not only be achieved through learning-based approaches, but also with the help of externally imposed measures such as regulations, fines, deadlines, and so on. If, however, feelings of interdependency remain beyond reach, project managers could foster an alignment of actions by making strategic use of insights in the way in which the achievements of separate actors may enhance and reinforce one another.

Thus, the literature on learning as a social event provides useful and practical answers to the second set of questions that this chapter raised. Still, in order to achieve congruency in action on the basis of a strategic gearing of the actions of many to each other, people must not only be willing but also be *able* to change their practices. An actor's ability to redirect his/her course of action in the light of new preferences critically depends not only on available resources, but also on the institutional arrangements and (juridical, economic, infra-) structural settings that restrict the 'room for manoeuvre' s/he has (or thinks s/he has). This brings us to the nature and importance of system innovation.

Learning and system innovation

It is likely that a sustainable development requires changes in the practices of heterogeneous groups in large networks. Unlike the organisation-based learning situations discussed in the literature presented above (e.g. Senge, Argyris and Schön, Forester 1999, compare Wenger 1998), these groups do not interact in daily life. While members of nested networks may be brought together in temporary, learning-oriented projects, the presence of many others, who do not participate, is likely to be felt via the institutional arrangements and structural settings that restrict the participants' room for manoeuvre. For instance, when people are willing to restrict their use of a car, the effect may be necessarily limited because of the far-off location of shopping malls vis-à-vis residential areas. In this case, the structures resulting from the actions of spatial planners and others that mapped out the location, and the local authorities that approved of the plans present seemingly unchangeable, 'given' conditions to the environmentally-aware consumers that wish to cut back on car miles. To help resolve such 'system imperfections', learning may again play a role.

An important strand in the literature on the relationship between institutions and learning is that of innovation science. Many innovation scientists reason that knowledge and learning are at the heart of modern economies. Lundvall (1992) for instance speaks of a 'learning economy', that is, an economy in which the pace of the creation and destruction of knowledge has become very fast. Although the vocabulary used may suggest differently, the notions that dominate this approach – learning by doing (Arrow 1962), learning by using (Rosenberg 1982) and learning by interacting (Lundvall 1992) – hardly bear any relevance to our present discussion. Of interest, however, is the view on systems developed in that field.

The core notion here is that (national) systems of innovation may either stimulate or slow down processes of learning and innovation. Such a system is "constituted by elements and relationships which interact in the production, diffusion and use of new and economically useful knowledge" (Lundvall 1992, p. 2). Elements of the system of innovation can reinforce each other in promoting processes of learning, or they can block such processes because of system imperfections. In sharp contrast to views in which institutional change is conceived of as merely hindering innovation and lagging behind other – e.g. technical – changes, Lundvall argues that the way in which suppliers and consumers are related, and the institutional set-up within firms, between firms and in policy-making, greatly influences the rate of learning and innovation. Taking the argument a step further, Smits and Kuhlman (2004) argue that the enhancement of innovation (such as for a sustainable development) may therefore be well-served by adding a new type of instrument to the classical repertoire of financial and regulatory instruments: so-called systemic instruments. Systemic instruments are arrangements that create interfaces between actors from different institutional realms; they are supposed to enhance conditions for innovation by, amongst other things, providing a platform for learning and experimenting. Roep *et al.* (2003) have focused on the agency involved in such instruments. They argue that project managers and other central actors in projects for system innovation must influence and facilitate learning processes between actors by strategically connecting unexpected events (what we, following Schön (1983), called 'surprises' above), and changes in action and opportunities for change in routines, rules and assumptions.

While the focus on systems in innovation studies literature is enlightening for understanding the joint influence actors have on institutional arrangements and their accumulated performance, it obscures the relationships between systems dynamics and activities of individual actors or small groups. This is problematic as these relationships lie at the heart of innovation processes. To counter this criticism, recent empirical work has set out to integrate institutional perspectives on innovation with views on learning that address social processes at the micro-level (van Mierlo 2002, Klein Woolthuis *et al.* 2005). In a comparative case-analysis it was found that if deliberate interventions focus on barriers or system imperfections as identified by a project's participants, 'better' learning occurs than in projects in which the activities do not target system imperfections that are perceived as important by the participants. Practitioners hence are advised to ensure a proper match between the intentional activities in a project and the imperfections in the system it intends to address that are flaws or barriers *in the eyes of* the actors involved.

This being said, obviously, an innovative program manager may not wish to 'settle for' resolving merely the imperfections mentioned by a project's participants. First of all, actors may not hold an aligned view on system imperfections. In addition, such an approach might set the standard for the innovations or transformations of the system envisioned undesirably low. As we have seen, people may be inclined to consider the feasibility of change in the light of incremental improvements to the existing situation only, as a result of their 'natural' tendency towards first order learning. Even when second order learning is triggered, the current, relatively stable set of social arrangements and structures in which second order notions have come to develop – and which are by and large guided by 'unsustainable' orientations towards modernization – may remain unchallenged.

In order to develop and endorse more far-reaching notions of change, participants will have to be prompted to consider the 'revolutionary' character of institutional transformations and sustainable development (as we put it in the introductory section). This involves, for instance, a review of existing task divisions, in such a way that new roles and identities are defined which at first glance may seem at odds with the current regimes of justification and reason. For instance, in dealing with intractable problems such as soil dehydration and diffuse pollution, water managers can no longer take societal desires for granted (e.g. keep the land dry at all times and the water at bay by means of stable and high dikes), and control the water accordingly. In the face of these problems, they are asking societal actors to adapt themselves partly to the requirements of the water system (e.g. by accepting occasional flooding of dedicated areas), and making *water quality* managers and other actors co-responsible for *water quantity* management.

This kind of 're-thinking' yet again entails a different type of learning. It requires people to become aware of the fact that the practices considered appropriate in a given setting themselves re-establish and reinforce the rules, conventions and expectations by which they are deemed appropriate, as well as the structures through which they are made at all possible in the first place. Such a theoretically sophisticated understanding of the 'recursiveness' of practices, and the interrelatedness of structure and action (Giddens 1984, Grin 2006) may open up, in practical terms, the 'backdrop of normalcy' against which background a project is given shape, and creates new opportunities for system innovation. In order to enable an understanding of the nature and opportunities of the desired changes, learning may be triggered by 'unpacking' self-evident assumptions. Providing project participants with insights into the social shaping of technology may, for instance, significantly contribute to a clarification of what is considered 'normal' at a certain moment in time (cf. Clausen and Yoshinaka 2004).

Conceptual principles and practical inferences

The literature on learning thus offers several answers to the questions posed at the beginning of this chapter, which challenge managers of innovative projects and other practitioners in the pursuit of a sustainable development. To recapitulate the first question posed (on the relation between cognition and action): while people may often not practise what they preach, it is clear that learning *can* and *does* play a role in sustainable development. Learning, to that end, must not be conceived of as knowledge accumulation on the part of specific groups in society (e.g. citizens, farmers, consumers etc.) in response to the teaching of e.g. governments on how to behave in accordance with a specific interpretation of the concept. By conceptualising learning as the process of reviewing the 'theories-in-use' that actors hold, however, the potential role of stimulating learning in the pursuit of a sustainable development is obvious, as are the practical implications for project management. Innovative projects may provide the settings in which people are helped to explain and scrutinise their tacit theories, beliefs and assumptions and thus helped to radically change their behaviour.

As far as the second question is concerned: given the relationship between cognition and action described here, and the sheer diversity in theories-in-use, we propose to use *learning in social interaction* as the central tenet in the role of projects which aim for a sustainable development. Learning occurs in the course of social practices that entail explicating tacit knowledge such as embodied in technical artefacts. Projects may not only help to explain such knowledge but may also help to align practices – on the basis of a *congruency* of meaning between heterogeneous parties – in a way that sets forth and strengthens dynamics that contribute to a sustainable development.

In order to avoid, thirdly, the potential pitfall that the described relationship between cognition and action entails – the intrinsically conservative, incremental nature of the changes conceived of on the basis of first and even second order learning – additional measures in the design and elaboration of a project will have to be taken. In order to help enlarge the participants' room for manoeuvre and to ensure that the existing institutional setting in which (and against which background) they operate is not taken for granted, another kind of learning is required as well. By stimulating what we may call *system learning*, a project may help actors challenge and redefine the very structures that hinder their progressing aspirations for more sustainable practices.

The main challenge for managers of innovation projects thus is to stimulate learning in social interaction in a way that allows for congruencies in action and system innovations. This conceptual interpretation of the role of learning vis-à-vis sustainable development, based on the theoretical perspectives presented above, offers clues as to the practical set-up of innovative projects:

- Processes of individual learning may be enhanced when, in line with the views of Kolb and other authors, projects explicitly seek to improve feedback mechanisms (e.g. with the aid of visualising techniques).

- Considering the insights from the work of authors like Argyris, Schön and others, second order learning in particular may occur when participants in a project are actively challenged to face and take seriously 'crises and surprises' that are usually ignored in everyday life – e.g. by organising a confrontation between actors with divergent perspectives – in an atmosphere of trust and mutual dependence (reciprocity), and if tacit assumptions, beliefs and values that underlie behaviour are made explicit and are put up for reflection and reconsideration.
- In addition to the reflection on strategies and values mentioned above, following Grin, Leeuwis and other authors, it seems crucial that a project's participants reflect on the role of actors and their relationships in the light of the problem perceived. Therefore, a project could include a 'strategic problem orientation', that is an exploration of the issues at stake, the actor networks around these, and the relationships between the actors and the problem issues perceived. This can provide a basis for working towards alignment, in a process that may well include second order learning towards congruency, enabling joint action towards a sustainable development.
- To endorse the overarching aim of system innovation, a project should furthermore actively seek to facilitate 'system thinking' (as conceptualised by Senge and others, or 'system learning', as it is called here), enabling participants to look at the interrelationships between the structures in which they operate and their own practices in a new light (e.g. concerning time and space: reconsidering the relationship between short-term action and long-term change, between 'here and now' and 'there and then').
- Following Smits, Roep and others, moreover, projects that aim to contribute to a sustainable development could try and function as systemic instruments, that is, to adopt (in their design and elaboration) a systems approach to the problems perceived. If the participants in a project have an aligned view on system imperfections, this may be achieved by taking these as the organisational foci in designing project activities.

In the ways suggested above, projects may adopt a so-called reflexive design, to encourage and facilitate the processes of learning required in the pursuit of a sustainable development. The processes of (system) learning that are thus fostered may be enhanced and strengthened with the aid of methods (such as vision building, backcasting, and reflexive process monitoring) which are being developed in a variety of projects and programmes on sustainable development. In the empirical studies in this volume several of these are presented and described in detail.

References

Argyris, C. (1990) *Overcoming organizational defences – Facilitating organizational learning*, Needham Heights: Allyn and Bacon.

Argyris, C. and Schön, D.A. (1996) *Organizational learning II: theory, method, and practice*. Reading, MA [etc.]: Addison-Wesley.

Arrow, K. (1962) "The economic implications of learning by doing." *Review of economic studies*, 29 (3): 155-173.

Brooks, D.B. (1992) "The Challenge of Sustainability: Is Integrating Environment and Economics Enough?" *Policy Sciences*, 26: 401-408

Clausen, C. and Yoshinaka, Y. (2004) "Socio-technical spaces – a new guide to sociotechnical politics?", in K. Horton and E. Davenport, ed., *Understanding Sociotechnical Action*. Workshop proceedings, Napier University, Edinburgh 3-4 June, pp. 27-30.

Cramer, J. and Loeber, A. (2004) "Governance through learning: making corporate social responsibility in Dutch industry effective from a sustainable development perspective." *Journal of environmental policy and planning*, 6 (3/4): 1-17.

De la Court, T. (1990) *Beyond Brundtland. Green development in the 1990s*, New York/ London and New Jersey: New Horizons Press/ Zed Books Ltd.

Forester, J.F. (1999) *The deliberative practitioner: encouraging participatory planning processes*, London: MIT Press.

Forrester, J.W. (1968) *Principles of systems*. Cambridge, Mass.: Wright-Allen Press.

Giddens, A. (1984) *The constitution of society: outline of the theory of structuration*. Berkeley: University of California Press.

Grin, J. (2006) "Reflexive modernization as a governance issue – or: designing and shaping *Re*-structuration", in J.-P. Voß, D. Bauknecht, R. Kemp, eds., *Reflexive Governance for Sustainable Development*. Cheltenham: Edward Elgar.

Grin, J. and Hoppe, R. (1995) "Toward a comparative framework for learning from experiences with interactive technology assessment." *Industrial & Environmental Crisis Quarterly*, 9 (1): 99-120.

Grin, J. and van de Graaf, H. (1996) "Technology Assessment as learning." *Science, Technology and Human Values*, 20 (1): 72-99.

Grin, J. and Loeber, A. (2006) "Theories of Policy Learning: Agency, Structure and Change", in F. Fischer, G. Miller, and M. Sidney (eds.) *Handbook of Public Policy Analysis: Theory, Politics, and Methods*. London: Taylor and Francis. pp. 201-219.

Heclo, H. (1974) *Social policy in Britain and Sweden*, New Have, CT: Yale University Press.

Keen, M., Brown, V.A. and Dyball, R. (2005) *Social learning in environmental management. Towards a sustainable future*, London: Earthscan.

Klein Woolthuis, R., van Mierlo, B., Leeuwis, C. and Smits, R. (2005). *Tussen actor en systeem. Een theoretische en empirische verkenning van leerprocessen en de rol van NIDO als systeeminstrument*. Wageningen, Wageningen University, Communication and Innovation Studies.

Kolb, D.A. (1984) *Experiential learning: experience as the source of learning and development*, Englewood Cliffs: Prentice-Hall.

Leeuwis, C. (2000) "Re-conceptualizing participation for sustainable rural development. Towards a negotiation approach." *Development and Change,* 31 (5): 931-959.

Leeuwis, C. (with contributions by A. van den Ban) (2004) *Communication for rural innovation: rethinking agricultural extension,* Oxford [etc.]: Blackwell Science.

Loeber, A. (2004) "Practical wisdom in the risk society. Methods and practice of interpretive analysis on questions of sustainable development", Ph.D. thesis, Universiteit van Amsterdam, Amsterdam.

Lundvall, B.A.(1992) *National systems of innovation. Towards a theory of innovation and interactive learning,* London/ New York: Pinter.

Minsky, M. (1988) *The society of the mind,* New York: Simon and Schuster.

Molnar, E. and Mulvihill, P.R. (2003) "Sustainability-focused Organizational Learning: Recent Experiences and New Challenges", *Journal of Environmental Planning and Management,* 46 (2): 167–176.

Roep, D., van der Ploeg, J.D. and Wiskerke, J.S.C. (2003) "Managing technical-institutional design processes: some strategic lessons from environmental co-operatives in the Netherlands", *Netherlands Journal of Agrarian Studies,* 51 (1-2): 195-217.

Röling, N. (2002) "Beyond the aggregation of individual preferences. Moving from multiple to distributed cognition in resource dilemmas", in Leeuwis, C. and R. Pyburn, eds., *Wheelbarrows full of frogs: social learning in rural resource management: international research and reflections,* Assen: Koninklijke Van Gorcum, pp. 25-47.

Rosenberg, N. (1982). *Inside the black box. Technology and economics,* Cambridge a.o.: Cambridge University Press.

Schön, D.A. (1983) *The Reflective Practitioner: How professionals think in action,* New York: Basic Books.

Schön, D.A. and Rein, M. (1994) *Frame Reflection. Toward the resolution of Intractable Policy Controversies,* New York: Basic Books.

Senge, P.M. (1990) *The fifth discipline. The art and practice of the learning organization,* New York: Doubleday.

Senge P.M. (2000) "Building the SoL Sustainability Consortium. Emerging Applications of System Dynamics regarding Language, Leadership, and Decision Making.", paper prepared for the 2000 International Conference of the System Dynamics Society, Bergen, Norway, July 30.

Smits, R. and Kuhlmann, S. (2004) "The rise of systemic instruments in innovation policy", *International Journal Foresight and Innovation Policy,* 1 (1/2): 4-32.

Susskind, L., McKearnan, S. and Thomas-Larmer, J. (1999) *The consensus building handbook. A comprehensive guide to reaching agreement,* London/ New Delhi: Sage Publications.

Van Mierlo, B.C. (2002) *Kiem van maatschappelijke verandering: verspreiding van zonnecelsystemen in de woningbouw met behulp van pilotprojecten [The seed of change in society. Diffusion of solar cell systems in housing by means of pilot projects],* Ph.D. thesis, Amsterdam: Aksant.

World Commission on Environment and Development (WCED) (1987) *Our Common Future,* New York and Oxford: Oxford University Press.

Wenger, E. (1998) *Communities of Practice. Learning, Meaning, and Identity,* Cambridge: Cambridge University Press.

Chapter 4

Social learning revisited: lessons learned from North and South

Danny Wildemeersch

Some ten years ago, we developed a concept of 'social learning' which should enable researchers and practitioners to better understand the nature of the learning processes taking place in groups, communities, networks or other social systems engaged in trying to solve social problems. A long commitment to research and practice in the field of adult and continuing education, and more recently to comparative and intercultural education, made us search for theoretical concepts which would help us understand processes of social transformation as 'learning processes'. Various learning theories which had been developed before with respect to non-formal settings mainly focused on the transformation processes taking place within individuals. We were convinced it would also be relevant to develop a frame of reference which would help us to understand better the collective dimensions of these transformation processes. We thought so, because we observed an increasing interest in engaging groups and communities as vehicles of social change. Ten years after, the time has come for an evaluation. We have applied our theory of 'social learning' as an interpretive framework to understand processes of change in various settings such as project groups in university settings (Wildemeersch 1999), community action groups (Van Rhede 1997), public debate on environmental issues (Vandenabeele and Wildemeersch 1998; Janssens and Wildemeersch 2003), policy planning (Van Duffel *et al.* 2001), and multi-party negotiations in 'third world' settings related to water management projects (Dang 2003) and nature conservation (De Greve 2004)[11]. In our reflection on this research, we will limit ourselves, for reasons of comparability, mainly to projects which focus on environmental issues both in the North and the South. We will in the first place present a short reconstruction of our basic ideas about social learning. We will then raise some theoretical questions about issues of power. In the course of our research activities we have observed an important inhibiting and facilitating impact of power processes on the learning processes. Yet, until now, we have failed to conceptualise these dynamics very well. We will explore to what extent a Foucauldian perspective on power recently elaborated in the context

[11] Other research which has taken place from this perspective, which has been reported on only in Dutch, is not mentioned here.

of 'governmentality'[12] studies may be inspiring. The choice of the cases we will present below (two from the North and two from the South) is directed by the expectation of the differential influences of power, of the differences in scale, and of the differences in socio-political contexts.

The origins of social learning

Initially we defined social learning as the 'learning taking place in groups, communities, networks and social systems that operate in new, unexpected, uncertain and unpredictable circumstances; it is directed at the solution of unexpected context problems and it is characterised by an optimal use of the problem-solving capacity which is available within this group or community' (Wildemeersch 1995, p. 33). The learning within these systems is basically experiential and can therefore be characterised as learning by doing. Experiential learning had in the past been conceptualised mainly with reference to individuals. Our challenge now was to conceive of a kind of experiential learning taking place within groups or systems and to make clear how these groups or systems learn. In view of this, we identified four different activities taking place in groups involved in processes of collective problem solving: action, reflection, communication and negotiation (see Figure 4.1). We related the learning to these four activities, and then spoke of four dimensions of social learning. In each dimension we identified two opposite poles which create a tension. The social learning can be described as the increased capacity of the social system to manage these tensions. The four dimensions and the opposite poles are:

Action

Social learning is linked to processes of social action (e.g. developing a policy plan, organising multi-party negotiations, engaging in participatory processes, establishing a task force or a study group, etc.); the action is triggered both by a particular 'need' (need motivation) and a set of 'competences' (competence motivation) which are present in the social system involved.

[12] 'Governmentality' is a neologism contracting both the notion of governing/governance and the notion of mentality; it refers to the governance of the mentalities which today is, according to the researchers of this Foucauldian school, increasingly replacing old forms of coercive governance (see Rose 1999).

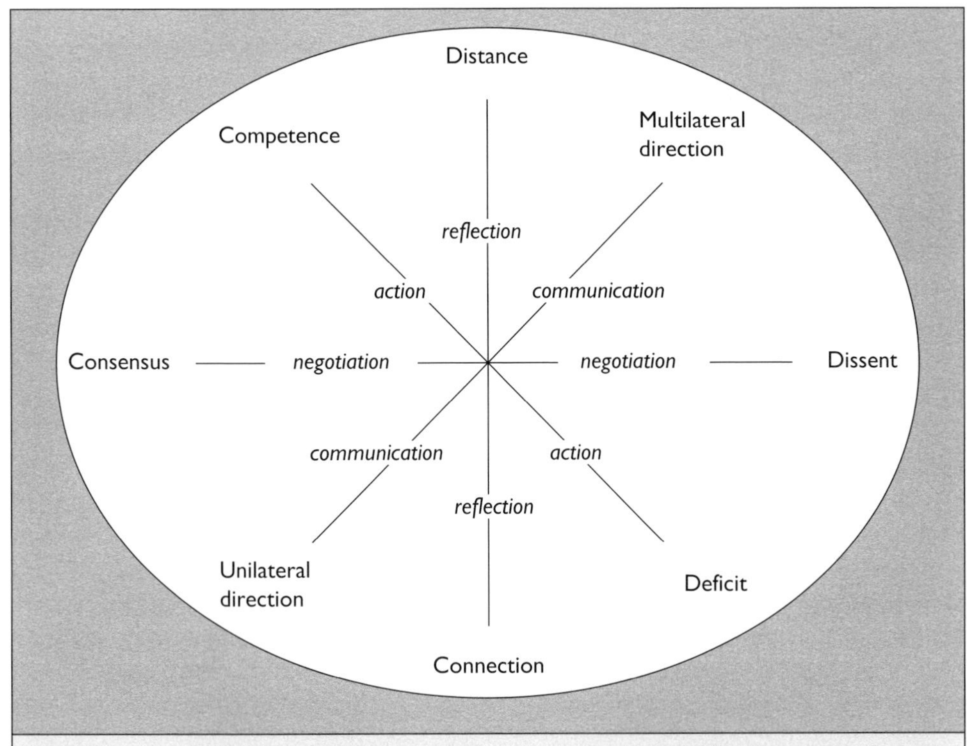

Figure 4.1. Collective problem solving activities and dimensions of social learning.

Reflection

Social learning triggers processes of reflection inside and outside the social system; the reflection dimension balances between 'distance' and 'connection'; taking a distance may lead to questioning of the self-evident aspects of the issue at stake; simultaneously, learning goes together with a process of (dis)identification with particular people, norms or values expressed through symbols, narratives, rituals, etc.; social learning is about finding a balance between these 'rational' and 'emotional' aspects of reflection.

Communication

Learning is inevitably linked to (supported or inhibited by) various communication processes taking place inside and outside the social system; these communication processes can either be 'unilateral' (e.g. inspired by a dominant voice) or 'multilateral' (e.g. inspired by different voices).

Negotiation[13]

Learning is related to processes of negotiation which result from differences of interest represented inside and outside the social system involved; the management of these differences can be consensus-orientated or dissent-orientated, or a combination of both. The creative tension between consensus and dissent can trigger learning within the system.

The actual learning takes place when the social system which is engaged in the process of action somehow manages to find a creative balance along the four axes or dimensions. The balances will be different for every social system, depending on the composition of the system (large or small, homogenous or heterogeneous), its internal and external challenges (high pressure or low pressure), its history (young or old system), the particular context in which it operates (vertical or horizontal), the available competences, its relative openness vis-à-vis the outside world, etc... Therefore, the balance is the result of careful management with regard to these elements which are moreover not stable but in flux as a consequence of the transformation which the system undergoes in time. This management can be organised by agents that operate within the social system or by external agents who operate as formal facilitators[14]. We also emphasised that social learning processes are never value-neutral, as they are related to issues that matter and that therefore often trigger processes of power both inside and outside the group, the network or the community. Yet, as mentioned before, we did not theorise the issue of power very clearly.

Social learning and new forms of governance

While doing research on social learning, and simultaneously being involved in discussions and readings on new concepts of governance, we began to consider these processes of participatory planning[15] in a different way. Before, we had interpreted such planning mainly as possible processes of emancipation and enlightenment, because we had learned to frame them as attempts to redistribute power, resources, and (learning) opportunities. Due to the readings of authors engaged in so-called governmentality studies (Rose 1999, Dean 1999), we now began to see these participatory planning processes as expressions of new forms

[13] Initially we called this dimension 'cooperation'. Later we transformed it into 'negotiation' because the latter notion reflects better the potentially discordant character of the learning process.

[14] We initially distinguished four positions for these agents of change (facilitator, obstructionist, core-actor and go-between); yet, we did not do very much with these distinctions afterwards.

[15] Participatory planning and communicative planning are used in more or less the same sense in this chapter.

of governance which are not necessarily emancipatory but which create, next to opportunities, new forms of dependency[16]. Under the conditions of 'advanced liberalism' (Rose 1999), characteristic for the welfare democracies in the North, traditional forms of coercion are increasingly replaced by new technologies of persuasion and normalisation. These new orientations build upon the ambition of modern citizens who want to be regarded as 'free subjects' able to choose from various products, services, lifestyles, policy orientations, both in the private and public sphere. 'Advanced liberal forms of government thus rest, in new ways, upon the activation of the powers of the citizen. Citizenship is no longer primarily realised in a relationship with the state, or in a single 'public sphere', but in a variety of private, corporate and quasi-public practices from working to shopping. The citizen as consumer is to become an active agent in the regulation of professional expertise; the citizen as prudent is to become an active agent in the provision of security; the citizen as employee is to become an active agent in the regeneration of industry and much more' (Rose 1999, p. 166).

Participatory planning could be considered as one of these new technologies that seemingly create opportunities of (collective) self-improvement and free choice for the participants involved. Therefore, it could be viewed as belonging to the regime of 'governmentality', a power regime which considers people as sovereign citizens of the state, as opposed to the regime of 'sovereignty', which treats people as subjects of a sovereign or of sovereign institutions (Henman and Dean 2004). Under conditions of 'sovereignty', people are forced by violence or by the threat of violence to comply with the wishes of the authorities. Governmentality, on the contrary, creates a power base which does not simply operate from above, but which functions as 'the conduct of conduct'. This horizontal power base is supportive in transforming the behaviour of people with the help of complex assemblages of persons, forms of knowledge, technical procedures and modes of judgement and sanctions. Initiatives of participatory planning could be considered as one such complex assemblage, organising the participation of people with the help of surveys, steering committees, public hearings, civil servants-as-facilitators, etc. These practices are fairly new and complex. Therefore, participants or stakeholders have to engage in 'social learning processes' teaching them how to operate in these conditions of self-direction. This learning is not the result of conventional teaching, which would be the expression of a regime of sovereignty. Instead, governmentality requires self-organised processes of 'learning by doing'. In line with this, social learning could be conceived as a way to manage the complexities, tensions and

[16] This interpretation of participatory practices has been studied by colleagues in our Leuven Department (see: Masschelein and Quaghebeur 2005, Quaghebeur *et al.* 2004, Tessier *et al.* 2004). Other expressions of discomfort about participatory planning are to be found in Cook and Kothari (2004).

contradictions of the new conditions of governance which are currently being developed in various contexts of policymaking and management in advanced liberal societies.

Experiences from the North

After our presentation of the basic dimensions of social learning and an initial attempt to frame social learning in the context of governmentality practices, we will now go deeper into two examples of research on social learning conducted in Flanders (Belgium) between 1995 and 2003. This should result in a more in-depth understanding of the strengths and limitations of social learning with regard to the context in which the process takes place. It will also help us to (re-)frame our understanding of the power dynamics interfering with processes of social learning.

Nature development planning in Flanders (Belgium)

With the help of case study research methodology, we studied the development of 'Communal Nature Development Plans' organised in two separate communes in Flanders in the second half of the nineties (Vandenabeele and Wildemeersch 1997, 1998). This type of planning is an example of the new policy practices which are being stimulated nowadays at the local and the regional level. Here also, the framework of social learning was relevant to interpreting the participatory processes. In our case study, we analysed in two communes, two contrasting cases of small-scale experiments concerning nature development planning: a so-called successful case and a so-called unsuccessful case. It helped us to understand the reasons for success and failure. We defined 'success' both in terms of realising a positive learning and participatory process and of achieving a nature development plan guaranteeing a balanced improvement of the ecological, economic and social conditions, in which tensions and conflicts could be balanced in a productive way.

The following findings explained the differences between the two regions.

- The success of the second case probably had a lot to do with the action orientation of the project. Different stakeholders engaged in a concrete project of sustainable farming. This process mobilised the competencies of the different stakeholders which helped them to appreciate each other's position, point of view and commitment. In the unsuccessful case, the collaboration remained limited to extensive debates.
- A second factor is related to reflexivity. In the second case, the learning process was accompanied by a critical reflection on both the conservationist and the farming point of view. This balancing process, including the critical questioning

of self-evident assumptions, was facilitated by participants in the planning process who, because of their biographical background, operated as brokers between the various communities.

- In the successful case a genuine process of multilateral communication helped contextualise the particular perspectives and broaden the learning opportunities. Multilateral communication presupposes the commitment to engage in a process without precise outcomes. The participants in the second case were prepared to take that risk. In the first case, particular positions were constantly reconfirmed and prevented the communication from opening up and including new voices and perspectives.
- A last factor explaining the successes and failures is the negotiation of consensus and dissent. In the successful region, the debates among the stakeholders were conducted under the supervision of a very competent chairperson of the nature council. He managed to engage the various stakeholders in a reflection on the different definitions of the situation and to prevent an immediate juxtaposition of interests. Conflict was balanced by attempts to arrive at consensus. In the first region these competences were not present to the same extent. Dissent was omnipresent and no attempt was made to find common ground.

On the whole, we came to the conclusion that the social learning process is a vulnerable activity full of tensions and risks. It is an open-ended learning situation without a guarantee of success in advance. The case study taught us that social learning is unsuccessful when no balance is found along the four axes as displayed in the figure, and when extreme orientations are emphasised (e.g. extreme conflict orientation or unilateral communication patterns). We also observed how the search for balance is threatened by power games which come from inside and outside the system. Taking control of these power mechanisms is an important prerequisite for the success of the process.

Forest extension in Flanders (Belgium)

During the period 2001-2003 we engaged in a research project on forest extension in Flanders, together with experts on forestation (Wildemeersch and Lust 2003). The research was related to a policy initiative by the Flemish minister for the environment by which forest land in the territory was to be extended by 10,000 hectares. The measure was fairly controversial. The environmental movement in Flanders was very much in favour of it. The farmers' organisation was opposed to it because they expected a considerable piece of their farmland to be sacrificed to forest land. We did case study research in various locations in Flanders. The main question was how a participatory process could be linked most adequately to a technical process of so-called localisation studies. A 'localisation study' is a planning approach aimed at deciding what territory in a particular area is best suited for the establishment of a new forest, taking into consideration ecological, economical and social aspects. We will focus here on the Roeselare case where we engaged in an action research project aimed at clarifying the dynamics of participation and social learning in the context of a highly controversial policy debate.

Roeselare is a city of about 80,000 inhabitants. Economically, it combines industrial activity with farming activity. Of the total territory of 6,000 hectares, 2,700 hectares consist of 'open' space. The Flemish regional authorities planned to give 200 hectares of the open space a new destination as forest area. In the localisation study, executed by an external agency and commissioned by the Flemish government, three different scenarios were suggested for the establishment of a city forest. The researchers established a so-called 'city forest dialogue' involving various stakeholders from the local community (farmers organisations, environmental organisations, workers organisations, employers, politicians, etc.) aimed at enhancing a 'social learning' process on the level of the local community. The dialogue took place at four meetings of the dialogue group (some 20 representatives) over a period of five months, in combination with other information and discussion activities organised by the municipal authorities and some of the stakeholders. The debates within the dialogue group were very intense and contradictory and reflected major tensions in society at large concerning the organisation of the open space. The conflict between farmers' interests and environmental interests came especially to the fore.

The social learning dynamics were very strongly influenced by the external conditions. In the space of five months, very little change came about in the initial positions and insights of the different stakeholders. The environment-agriculture opposition was so strong that it blocked opportunities to search for a balance along the four dimensions of social learning. The attempt to extend the *planning activity* with a participatory process created a lot of insecurity and unpredictability. The exercise aimed at creating a more stable support base for the establishment of a city forest was not really successful. The need of the various stakeholders to protect their own interest was so dominant that the competencies, which were definitely available within the dialogue group, could never be mobilised to come to a consensus concerning a redefinition of the land use on the communal level. This does not mean that no *reflection* took place in the context of the dialogue group. However, most of the reflection was instrumental to the defence of the interests of the different stakeholders. Hence, most of the *communication* was unilateral. Very rarely were attempts to arrive at decision making based on multilateral communication patterns taken seriously. External interest groups, operating as gatekeepers to the decision making, directed the communication patterns within the dialogue group. Terms of mutual understanding, which developed on certain occasions, were fragile and could never be sustained for very long. As a consequence, the *negotiations* were overwhelmingly discordant. At the end of the dialogue, some kind of compromise was attained, which reflected more or less the power relations within and outside the group. This case teaches us first and foremost that social learning never takes place in a vacuum. When it is related to issues 'that really matter', it will to a certain extent be messy and unpredictable. The dynamics on the small scale are inevitably affected by (political) processes at work on a larger scale. When these large-scale dynamics are overwhelmingly discordant, small-scale social learning will have a limited effect, no matter how good the facilitation.

Experiences from the South

During the last five years, we have also engaged in research related to the social learning framework in a country in the South, namely Vietnam. We[17] have established relationships with research centres in the natural sciences (geology) and in the social sciences (anthropology) in Hanoi. The collaboration is aimed at training researchers in the context of research projects that try to contribute to the development of poor areas, mainly populated by ethnic minorities, in the Northern parts of Vietnam. One research project (VIBEKAP: Vietnamese Belgian Karst Project) on 'Sustainable Water and Land Management and Social Learning' has lasted for five years and finished in the course of 2004. Another research project called LLINC (Limestone Landscape Improvement and Nature Conservation) was started in 2002 and will continue untill 2006. Partners in this project are the Institute of Ethnology (now Anthropology) and the NGO Fauna and Flora International (FFI). Also in this context, we have tried to frame some of the research and development activities in terms of social learning. It is relevant here to present some of the insights and contrast them with findings in the research projects in Flanders.

Developing water supply systems in the commune of Bon Phang

In order to give the VIBEKAP project, which studied water and land management processes in the province of Son La, a practical relevance as well, two micro projects were developed in the commune of Bon Phang, more particularly in the villages Noong O and Nam Tien (Tessier *et al.* 2004). Noong O is known as a black Thai village. This means that most of the inhabitants belong to the Thai ethnic minority. The village totals 62 households and 330 inhabitants. The Nam

[17] Researchers from various disciplines of the natural and social sciences belonging to different Flemish universities collaborated in interdisciplinary research projects on water management financed by the Flemish Interuniversity Council (VLIR: Vlaamse Interuniversitaire Raad).

Tien village is a Khin village, which is the largest ethnic group in Vietnam. This village has 128 households and 650 inhabitants. The projects took place with the support of UNICEF and in collaboration with the local Vietnamese authorities and institutions, including the Water Management Board of each village and a newly established Forest Protection Board. One year after the establishment of new water supply systems in the two villages, a student of ours (Dang 2003) analysed in retrospect the process of social learning which took place there. In the two villages a participatory planning process was started, aimed at including the local inhabitants and the village leaders in a process of decision making on the provision of individual households with water.

The micro projects were quite successful in different ways. In both villages an individualised water supply system, including an operational management system, was established and sustained during the following year. There are several indications that in both villages the commitment during the participation process has extended since then and has established a sense of responsibility which makes the villagers care for the individual and collective system. So, we may carefully conclude that in both cases a social learning process has taken place. Dang (2003) has analysed this process of social learning, paying special attention to the processes of communication. She arrives at interesting observations and conclusions which shine a new light on the dynamics of social learning.

- A first observation relates to the *action dimension*. The social learning seems successful because the participants were able to link their competences to the needs concerning water supply present in the village. The needs were real and acknowledged by most of the villagers. This became clear when all households expressed their willingness to pay a financial contribution. The technical, social, environmental and management competences were mobilised in a positive and adequate way during and after the process of establishing the water supply system. Villagers actively engaged in the planning and the construction of the water supply systems. Most probably, the success of the action was also due to the intensive and creative facilitation of the participation process organised by the researchers. Here the mobilisation of external resources proved to be necessary, due to a lack of experience among the village leaders in organising this kind of participatory activity. And finally, also the input of material resources by external agencies was an important element in creating the conditions of success for the experiment.
- A second observation relates to the *reflection dimension*. The research has not been able to go very deep into that aspect. Elaborate, reflexive interviewing and observation would have been necessary to achieve this. It has not been possible, due to language restrictions and to a lack of experience on behalf of the villagers to engage in reflective interviewing. Nevertheless, we can

observe that reflection processes mostly came about in relation to the more instrumental aspects of the projects: information sessions took place about the contract, the price of the water, the technicalities of the water distribution, the functioning of the water management board, the differences between public and private systems of water supply, etc. In this way, the participants learned to understand how the new water supply system would function and what would be their role in it.

- A third observation relates to the *negotiation* processes. The projects encountered different opinions and sometimes conflicts within the village about the water price, the functioning of the Water Management Board, the representation of the villagers in different boards, the questions of leadership, etc. Apparently, balances were found between these different opinions. Eventually, the two villages succeeded in managing consensus and dissent, which helped them to establish a sustainable water supply system.

One last remark about the *communication* process. It is important, because it helps us understand that in the context of the two villages, communication patterns do not function the way we expect them to from our own more Western point of view. Dang (2003) paid special attention to this aspect, because the researchers were somewhat insecure about the distinction between 'unilateral' and 'multilateral' communication patterns and the extent to which such frames of analysis really applied to the cultural context of ethnic minorities in Vietnam. One of the major observations is that multilateral communication, as we conceive it, is not very prominent (although there were differences between the two villages). Contrary to what the researchers expected, village meetings often function as moments of transfer of information from the village leaders to the families. Dang made various other observations which confirmed that the villagers did not really claim what we would call 'ownership' of the project. The project wanted the villagers to think and decide for themselves on how to manage the water supply system, and to act as if it was their right to decide for themselves. However, just as the villagers enacted their dependency on the village leaders, the village leaders acted as if they were dependent on the researchers who initiated the project.

With respect to this case, we tend to conclude that a particular social learning process has taken place within this project: planning and action was successful, people learned to reflect on how to give a new water supply system a place in their lives, consensus was found among different positions and interests. Yet, all this was mainly directed by unilateral communication patterns which did not reflect a sense of 'ownership' either of the system or of the process. We would expect that a lack of ownership would imply a lack of responsibility for the water supply system. This was not confirmed by the observations. On the contrary, the system has held up much better than other experiments where no participation has taken place.

Social learning and nature conservation: the Ngoc Son – Ngo Luong case

The last case which we will describe in this chapter is, in contrast with the previous one, a large-scale initiative in the North of Vietnam. It is about the negotiations and preliminary studies which were started in 2003 and in which the LLINC project played an active role, both as a research centre documenting the decision-making process and as an active participant/stakeholder in the negotiations taking place on the provincial level. We base our observations again on the research of one of our students (De Greve 2004) who studied every single document which was available and who interviewed the major stakeholders in the process. As such, he reconstructed the process from a perspective of co-management and social learning. We will limit ourselves here to the aspects of social learning.

This case is about the establishment of a nature conservation area in the Pu Luong – Cuc Phuong limestone range, situated in the north-west of Vietnam, approximately 150 kilometres south-west of the capital Hanoi. The range stretches 90 kilometres from Cuc Phuong National Park in the south-east up to the two ridges of the Pu Luong Nature Reserve in the north-west and covers approximately 170,000 hectares. The area is characterised as Karst landscape which is an irregular limestone region with underground streams, caverns and potholes. Karst[18] landscapes have a high ecological and cultural value. To further protect the nature in this area a discussion has started with three different provincial authorities involved to develop protection measures for the area connecting the nature reserve and the national park, namely the Ngoc Son – Ngo Long area. The actions towards conservation have been initiated by the following partners: the Forest Protection Department (FDP) within the Ministry of Agriculture and Rural Development of Vietnam (MARD) and the NGO mentioned earlier, Fauna and Flora International (FFI). The project is funded by the World Bank. Other players in the game are FIPI, the Forest Inventory and Planning Institute of Vietnam, specialised in study activities concerning (de)forestation, and the LLINC project also mentioned earlier. The negotiations between FFI and the Vietnamese authorities started back in 1999, with research exploring the conservation possibilities and a ZOPP workshop (Objectives Oriented Project Planning) attended by 40 participants at about the same time. At the end of 1999 an agreement was reached between the Vietnamese authorities and FFI creating the base for the further coordination of nature conservation activities in the area, with the help of World Bank funding.

[18] A Karst region is a limestone region marked by sinks and interspersed with abrupt ridges, irregular protuberant rocks, caverns and underground streams (Webster Third New International Dictionary).

The researcher has analysed the activities of preparing the decisions on the establishment of a nature reserve or another type of protected area (the final decision about the mode of conservation has not yet been taken) from both a co-management perspective and a social learning perspective. We now go deeper into the findings, thereby combining some of the findings of the two perspectives. The framework used again, relates to the four dimensions of social learning distinguished in this chapter.

- In this case too, the action process was mainly a planning process in which the above mentioned stakeholders were involved over a period of four years. This planning process eventually resulted in a feasibility study which was presented to the political decision makers. The planning process was very linear and straightforward. It did not follow a trajectory and time-schedule clearly defined in advance. It was interrupted on several occasions and took longer than initially expected. It was mainly conducted on a fairly abstract level of decision making predominantly involving authorities and expert groups and not particularly the local communities. This brings De Greve to the conclusion that on the level of these local communities, the balance between the needs and the competences seems to be distorted. Apart from some workshops organised in the margins of the planning process, there were hardly any local people involved in the process. At the planning level, it is obviously the experts who have 'learned' most from the process. The social learning on a wider scale is of a limited scope. De Greve suggests that involving the local communities in the planning process would have been possible when concentrating more closely on the day-to-day concerns of the people in relation to concrete rules, regulations and survival strategies, rather than on abstract discussions on options like 'nature reserve' and 'national park'.
- The reflection process is probably mainly characterised by single loop learning processes. The decision about the way the nature in the area should be protected seems to have been taken by the authorities at a fairly early stage of the process. A nature reserve was the preferred way, while alternative ways of conservation that are currently developed in other parts of the world – and where a better marriage between conservation and development objectives is aimed at – was not really an issue for the authorities or the Vietnamese experts. It was mainly the representatives of the Western institutions that tried to broaden the scope, thereby questioning the basic assumptions of the authorities. As such, they tried to deepen the social learning process into a process of double loop learning. A consequence of this limited scope is that the reflection activities mainly remain focused on the means, the instruments, the effectiveness and the efficiency of the decision-making process and not on creating space for basic disagreement or questioning of the rules.

- Similar to the observations in the previous Vietnamese case (the Bon Phang commune) we observe *communication* patterns which, from the point of view of the researchers, could be defined as unilateral or bilateral rather than multilateral. There were few occasions when different stakeholders were brought together in direct face-to-face interactions. Exceptions were the ZOPP workshop and the workshop where the feasibility study was presented and discussed. Western observers would consider these kinds of workshops as examples of more multilateral communication patterns. Apparently, during the co-management process preparing the final decision, communication seemed to follow the pathways of bilateral interaction (see also Maertens *et al.* 2004). LLINC for instance applied the strategy of approaching the various stakeholders separately and negotiating with them on a bilateral basis. In this way it reproduced the tradition in Vietnam of establishing confidential and dynamic communication patterns next to the official formal and static communication patterns. On other occasions unilateral communication strategies were applied. The research tries to understand this unilateral approach with reference to the way the Vietnamese tend to operate in power relations. Within South East Asian cultures power inequality is not very often disputed publicly. However, in informal networks and through personal and bilateral contacts, the power hierarchy is managed in more subtle ways.
- This brings us to the final observations about the social learning process, related to *negotiation.* There is not much to add to this dimension. It has already been mentioned in previous parts that dissent is not openly expressed in the management of differential interests. Expression of dissent has to happen in a subtle way. Interestingly, De Greve observes that the LLINC project has managed to do that while functioning as a go-between among the different stakeholders, thereby opening the negotiations to unexpected perspectives, viewpoints and interests. According to him, the feasibility study which was very strongly advocated by this research group, can be considered as "an action directed towards increasing the number of perspectives available in the process, while at the same time maintaining a minimal amount of homogeneity between the different perspectives and supporting the creation of 'new', overarching, integrating perspectives" (De Greve 2004, p. 121).

Conclusions

During the last ten years, we have been engaged in quite a variety of research projects linked to processes of social learning taking place in various contexts. These projects have generated a wealth of data which we have tried to interconnect in a systematic way in this chapter. It is important to realise that the concept of social learning is a construction which helps to make sense of social transformation in a particular way. There is no clear indication that a framing of processes of social change in terms of 'social learning' will directly help to improve these processes. What we have first and foremost tried to make clear is that it is relevant to consider social transformation as a social learning process. We hope that the conclusions below will invite some more researchers and (hopefully also) practitioners to look differently at the processes they engage in.

Similarities in social learning in North and South

- We have found that social learning is a relevant framework for looking at processes of social change and that the four dimensions, including the tensions, are interesting analytical tools for making sense of learning in complex settings of collective action.
- We have experienced that, in almost all cases, social learning is a vulnerable activity which can be greatly influenced by the context in which it takes place.
- In particular, when contexts are turbulent and discordant, there is a great chance that these characteristics will affect the inner dynamics of social learning within the systems involved. Attempts to neutralise these external dynamics may be counter-productive.
- In many of the researched processes the social action consisted of planning initiatives. This probably has to do with the fact that our societies are increasingly becoming planning societies, which experience the need to shape and control the future. We have noticed that these planning activities are often fairly abstract and alien to citizens. This restricts the social learning to a privilege of the planning elite that understands the language and the procedures. This elite enjoys engaging with the abstractness and long-term perspectives associated with planning.
- A more concrete action orientation may improve the chances of involving rank-and-file citizens in the participatory planning activities. Small-scale actions that are precisely defined and limited in time, and which are linked to concrete needs, have the potential to motivate and mobilise and to link the learning to wider issues on a more abstract level.
- In participatory planning approaches, the reflective dimension of social learning is often achieved in an instrumental way. The planning constraints limit the scope to 'single loop' learning processes, meaning that the knowledge and the competences gained from the experience are instrumental to the optimisation of the decision-making process. Apparently, 'double loop learning', which creates opportunities for more fundamental questioning of the assumptions, only comes about in 'protected zones' where the participants involved are somewhat liberated from the time and context constraints and pressures.
- Also, multilateral communication is hard to achieve. Unilateral communication is the dominant communication pattern in many participatory planning activities. Being open to unexpected perspectives is rare and, just like double

loop learning, only emerges in conditions which are (made) free of dominating external pressures.
- Adequate facilitation can certainly help to create balances of social learning within the system and to keep the external pressures at some distance. Yet, it cannot completely neutralise the external dynamics, unless it reduces the learning process to a sterilized and irrelevant activity.

Differences in social learning between North and South

- A sense of 'ownership' is not necessarily a major condition for people in the Vietnamese projects to participate and to learn from their participatory experience. Other motivational factors such as sociality, solidarity, obedience to authority and conventionalism seem to play a more important role in the commitment of the participants.
- Consequently, the learning which comes about will rather reflect a top-down expert-layperson relationship, or a more 'paternalistic' relationship reproducing predefined, closed answers to particular problems. In Western settings we encountered more situations where the relationship between the participant and the facilitator was horizontal, whereby the knowledge of the latter could be openly contested, and whereby insights were gained from active self-directed research processes on behalf of all stakeholders. Participants who had not learned to behave like this would feel very uncomfortable in those situations where truth and competence are continually scrutinised by all stakeholders involved.
- This brings us to the final conclusion about power relationships. Apparently, power seems to operate both in a similar and in a different way in the contexts of the North and the South that we researched. Both in the North and in the South, we encountered informal circuits of power operating next to and influencing formal circuits of power, and vice versa. However, in Flanders, interests and power issues are negotiated more openly with less respect for hierarchy and tradition. Therefore, direct multilateral settings of negotiation are more common in the Flemish cases than in the Vietnamese cases. We can probably conclude that power in the Vietnamese cases operates under the regime of 'sovereignty' characterised by more direct and coercive control. Features of social learning are: a unilateral transmission of knowledge, competences and value orientations, vertical relations between leaders and participants and closed processes. In the Flemish cases, power rather operates under the regime of 'governmentality' creating opportunities for participants to engage in more open-ended planning and learning processes, including more horizontal relationships, and to experience more autonomy in setting the learning and planning agenda. However, here too, these governmental practices are restricted when there is strong opposition from both sides.

- The contrast between projects in the North and the South makes us aware of how attempts to introduce the perspective of social learning in planning activities related to initiatives of sustainable development are received in various ways in these different contexts. Simultaneously, this contrast also raises questions on how concepts and practices developed in the North, such as participatory planning and social learning, find their way into policy planning activities in the South. It is important to realise that the social technologies which are introduced in such projects are not just neutral technologies. They are inspired by new approaches to governance, which are currently being developed in the North, in the context of advanced liberal societies. It is therefore relevant to reflect critically about the possible hidden agenda related to development interventions. Development workers and researchers from the North operating in projects in the South may, in spite of their often high moral principles of solidarity and compassion, introduce new forms of social relations reflecting conditions of advanced liberalism. This issue has not been elaborated on very deeply in this chapter. However, it is something that needs more attention in the future.

Acknowledgements

Various researchers, both from the University of Leuven and the University of Nijmegen have been involved in the development and application of this concept in the context of different research projects. The main contributors have been Joke Vandenabeele, Marc Jans, Theo Jansen, Christin Jansen and Katrijn Van Duffel.

References

Cooke, B. and Kothari, U. eds (2004) *Participation: the new tyranny?* London: Zed.

Dean, M. (1999) *Governmentality. Power and Rule in Modern Society*, London: Sage

De Greve, H. (2004) "Multi-stakeholder Negotiations: A Co-Management & Social Learning Perspective on the Ngoc Son – Ngo Luong case, North-Vietnam", unpublished master thesis, Department of Education, K.U. Leuven, Belgium.

Dung, D.T. (2003) "Communication and Social Learning Related to Water and Forest Protection Management", unpublished master thesis, Department of Education, K.U. Leuven, Belgium.

Henman, P. and Mitchell, D. (2004) "The governmental powers of welfare e-administration", Paper prepared for the Australian Electronic Governance Conference. Centre for Public Policy, University of Melbourne, Melbourne Victoria, 14th-15th of April, www.public-policy.unimelb.edu.au/egovernance/papers/15_henman.pdf.

Janssens, C. and Wildemeersch, D. (2003) "Social Learning and Interactive Policy Processes. The case of City Forest Planning in Flanders". In: D. Wildemeersch and V. Stroobants, eds., *Connections: Active Citizenship and Multiple Identities.* K.U. Leuven, Belgium: Department of Education, pp. 141-156.

Maertens, A., Tessier, O., Giang, H.L., Nguyen, H.H., Ta, H.D., Quang, T.O., Wildemeersch, D., Masschelein, J. and Bouwen, R. (2004) "Offering a Virtual Round Table – Some reflections on a Multi-Stakeholder Collaborative Learning Exercise in the Context of the Establishment of the Ngoc Son Ngo Luong Nature Reserve – Hoa Bin Province Vietnam", in O. Batelaan *et al. Trans Karst 2004. International Transdisciplinary Conference on Development Conservation of Karst Regions*, Conference Proceedings, Ha Noi, 13-18 September, pp. 147-154.

Masschelein, J. and Quaghebeur, K. (2005) "Participation for better or for worse?" *Journal of Philosophy of Education*, 39 (I): 51-66.

Quaghebeur, K., Masschelein, J. and Nguyen Hoai, H. (2004) "Paradox of Participation: Giving or Taking Part", *Journal of Community & Applied Social Psychology*, 14: 154-165

Rose, N. (1999) *Powers of Freedom: Reframing Political Thought*, Cambridge: Cambridge University Press

Tessier, O., Van Keer, K., Quaghebeur, K. Masschelein, J., Nguyen Hoai, H., Ho Ly, G. and Wildemeersch, D. (2004) "Giving or imposing the opportunity to participate?" In: A. Neef, ed., *Participatory approaches for sustainable land use in Southeast Asia.* Bangkok: White Lotus Publishers.

Vandenabeele, J. and Wildemeersch, D. (1997) "Sociaal leren met het oog op maatschappelijke verantwoording", *Pedagogisch Tijdschrift* 22 (1/2): 5-22

Vandenabeele, J. and Wildemeersch, D. (1998) "Learning for sustainable development. Examining life world transformation among farmers", in D. Wildemeersch, M. Finger and T. Jansen, eds., *Adult Education and Social Responsibility. Reconciling the Irreconcilible?*, Frankfurt: Peter Lang, pp. 117-133.

Van Duffel, K., Janssens, C. and Wildemeersch, D. (2001) "Policy Planning as Social Learning", in M. Schemmann and M. Bron, eds., *Adult Education and Democratic Citizenship IV*, Krakow: Impuls Publisher.

Van Rhede, A. (1997) *Natuurlijk leren balanceren! Een onderzoek naar organisatie, strategie en leren bij lokale natuur- en milieugroepen,* Faculteit Beleidswetenschappen, Vakgroep Milieu, Natuur en Landschap, Katholieke Universiteit Nijmegen (the Netherlands).

Wildemeersch, D. (1995) *Een verantwoorde uitweg leren (Learning a responsible way out),* Inaugural Lecture, Katholieke Universiteit Nijmegen (the Netherlands).

Wildemeersch, D. (1999) "Paradoxes of Social Learning. Towards a model for project-oriented groupwork", in J. Hojgaard and H. Olesen, eds., *Project Studies. A late modern university reform ?* Roskilde: Roskilde University Press, pp. 38-57.

Wildemeersch D. and Lust, N. eds (2003) *Draagvlakverbreding bij de planning, implementatie en het beheer van stadsrandbossen in Vlaanderen,* Department of Education, K.U. Leuven (Belgium), Vereniging voor Bosbouw in Vlaanderen, Gent University (Belgium).

Chapter 5

Learning based change for sustainability: perspectives and pathways

Daniella Tilbury

Change for sustainability

No country is sustainable or has come close to becoming sustainable and after years of experience in implementing sustainability initiatives there is still no generic recipe for success. As Robert Prescott-Allen reminds us "making progress towards (sustainability) is like going to a country we have never been to before... We do not know what the destinations will be like, we cannot tell how to get there" (2001, p. 2). Given this reality, the people around the globe has come to recognize that sustainability is essentially an on-going social learning process that actively involves stakeholders in creating their vision, acting and reviewing changes (Tilbury and Cooke 2005). This realisation also explains why learning in the context of sustainability is understood as a reflective process rather than as a message or level which must be achieved. There are no templates to be followed or lists to be adhered to. Instead, 'learning through doing' is now seen as vital to help us grow in understanding sustainability, human motivations and visions which provide the key to social change (UNESCO 2002).

The official documents associated with UN Decade in Education for Sustainable Development (UNESCO 2004, 2005, UNESCO – Asia Pacific Regional Bureau for Education 2005) acknowledge that this type of social learning can manifest itself in a variety of forms. It can range from the formal capacity building or training of individuals which often occurs within a college or higher education programme to informal but structured processes in the community or organization which use action learning, reflection and change to improve the effectiveness of a strategy, program or action plan for sustainability.

Learning based change helps challenge established structures and empowers individuals and groups to enable change towards sustainability. Professional development, group cohesion and cultural change are often outcomes also associated with this process.

This chapter explores the links between sustainability and social learning. It attempts to define the characteristics of learning based change approaches as well as the key components which underpin these approaches to sustainability. It explores a variety of pathways to this social learning process which engages stakeholders in a consideration of power, participation and possibilities for change.

Sustainability and social learning[19]

The sustainability literature has clarified that the major problems cannot be solved from our current way of living and will require a shift from traditional ways of thinking and acting upon environmental and socio-economic problems (Milbrath 1996, Environment Canada 2004, SustainUs 2005, Eckersley 1998, Doppelt 2003, UNESCO 2002, 2005). Sustainability will require social learning at a grand scale and although there are no templates for change, there are suggestions that we begin by challenging our existing practice. It has been suggested that environmentalism in the past has been a movement against things – for example stopping pollution and other harmful activities – while the sustainability approach aims to do things differently in the first place, instead of just cleaning up the symptoms of underlying problems (PCE NZ 2004). The sustainability approach is also associated with futures thinking and a move away from a 'doom and gloom' approach which aim to frighten people into action (Tilbury 1995, Tilbury and Cooke 2005).

In essence, three key concepts underpin the notion of social learning for sustainability:

- the need to *challenge the mental models* which have driven communities to unsustainable development. This involves questioning and reflecting upon our actions and developing a much deeper understanding of our social dispositions so that we can re-think and re-design our activities;
- the need for *new learning approaches* which help us explore sustainability and build skills that enable change, such as mentoring, facilitation, participative inquiry, action learning and action research; and,
- the need for utilising pluralism and diversity in joint explorations of more sustainable futures.

Few grasp the fundamental paradigm shift that is required to achieve change for sustainability. Changes to the way we think and learn are needed. For example, sustainability involves more than just understanding how society, environment and economic systems are linked – which is the aspect of sustainability which is more immediately associated with the concept. Sustainability is often graphically

[19] This section refers to the term 'Learning Based Change'. In this context, the term can be used interchangeably with 'Social Learning Based Change' since it focuses on broader social learning rather than curriculum or formal education structures.

represented by three overlapping circles each labelled 'social', 'environmental' and 'economic' representing various dimensions of issues (for an example see Bhandari and Abe 2003, p. 16). Although sustainability does promote holistic thinking this graphical representation is a simplification of what sustainability is really about – which is more about transforming current systems than about merely linking them. Sustainability is about challenging our mental models, policies and practices not just about accommodating dimensions into current work or finding common ground between related programs.

Many groups have struggled to see this difference and have simply changed the label they use to describe their work rather than challenge their thinking and practice. More and more we are seeing the word 'sustainability' being added to the titles of programs, project, activities, departments or units – however, few have actually been redesigned to address new social learning approaches. Many of those who have struggled to understand this difference often refer to the concept of sustainability as 'vague; or tend to interpret the word 'sustainability' literally (Tilbury and Cooke 2005). They are not familiar with the literature or thinking that underpins this concept or recognize the sustainability movement which represents a particular intention – envisioning and negotiating change rather than 'sustaining' the status quo.

The more radical interpretation of sustainability supports the use of learning approaches as ways of exploring the sustainability agenda. These approaches enable people to reflect on their experiences, learn how to make change and move forward. The concepts are not new to the organisational change literature which recognizes that change which is collaborative and context specific (such as that sought by sustainability) must involve learning. It is for this reason that an organisation aligned with sustainability is often defined as a 'learning organisation' (see Senge 1990, 1999, Connor and Dovers 2002).

What type of change?

A great deal of the work which takes place under the label of education or learning for sustainability seeks to engage people in action, for example, consumer action or volunteer conservation action. Learning based change for sustainability takes this a step further, helping learners develop the skills to influence change within a system, organization or wider society. It engages the learner in identifying relationships which can embed change as opposed to single actions which may not challenge root causes. It seeks structural and institutional change rather than focusing on individual change or using end-of-pipe approaches. Systemic change underpins learning based change for sustainability which encourages changes to be mindful of the whole system so that longer-term positive change is more likely to come about. This approach involves the study of how change happens in particular contexts and to consider people's assumptions and strategies for change. By looking at the world in a more holistic way, more systemic changes in our lives and in our society can occur through a 'redesign' of many of our current systems and established ways of living along sustainability principles.

In addition, learning based change encourages education processes which question the thinking and assumptions behind our actions rather than judge our actions. Certain problems can be encountered if particular actions are criticized or demonized without providing an opportunity for people to question why this is the case or without providing alternative and practical solutions. For example, some educators have seen limited value in children coming home from school to lecture their parents about the negative impacts on the environment of using their car. Parent's options may be limited due to socio-economic factors or lack of alternatives e.g. public transport. In any case, being told what not to do is likely to yield unsustainable change. Learning for sustainability focuses on encouraging people to think on why certain decisions are being taken and what the real alternatives available to them are.

The goal is to get to the root of the issues. Traditionally, while citizens have been active in the alleviation of environmental problems they have not addressed issues of sustainability at source. Learning based change for sustainability challenges educators to think beyond raising awareness and go beyond involving learners merely in one-off activities such as cleaning-up or planting of trees. They encourage learners to develop critical and systemic thinking skills, enabling them to get to the core of the issues. This reflects the major shift in thinking from environmentalism to sustainability.

Dealing with the issues at source is an important aspect of this new approach. Critical and systemic thinking enable this by assisting people to identify the root of the issues and to work actively towards trying to address these.

The learning based change approach to sustainability challenges the role of the educators and seeks to break down the traditional teacher-student hierarchy in a classroom as well as the sending out of key messages to target audiences in community education. Learning based change for sustainability encourages collaborative learning environments which do not merely impart knowledge but build capacity of the learner. Negotiation, evaluation and action are essential parts of this process. Approaches such as facilitation and mentoring, redefine the role of the change agent (i.e. the teacher, the policy-maker, NGO-representative, etc.) and encourage learning to be driven by the learner. They challenge traditional power; politics and participation relationships associated with teaching and provide compatible reflective learning and capacity building processes. As Hamú (2004) points out few educators or change agents are trained or experienced in these new learning approaches.

Approaches to learning based change

In essence, learning based change is seen as a process which can motivate and engage people in creating sustainable futures. It is interpreted not only as a process which builds competence but also a change strategy which will assist people and organizations to move towards sustainability. There are a variety of pathways which enable learning based change:

- **Mentoring** provides individuals and groups, who are grappling with sustainability with support, advice and understanding so they can engage with this concept. The process allows people to critically examine opportunities for change within their home community or workplace.
- **Facilitation** encourages learning to be driven by the learner. It equips the learner with the necessary skills and knowledge to take action and participate in change and decision-making. It develops the ability of people to 'critically' reflect on their existing practice and identify the change necessary. The process encourages people to engage in open dialogue and eliminates inequitable power hierarchies by, for instance, not relying solely on expert knowledge.
- **Participative inquiry** is the engagement with and deep exploration of sustainability questions which stimulate new ideas for further interrogation and action. Participatory inquiry offers a new way of understanding and engaging with the community and/or organizational change. It requires participants to collectively strive to understand a question that is important to them by freely examining their existing ideas and practices.
- **Action learning** is a process designed to build capacity using a form of reflection and assessment. The ultimate goal is the improvement of practice. The process involves the participant's developing an action plan, implementing the plan and reflecting on what they have learnt from this. A facilitator assists the participants to develop their plan and learn from their experiences.
- **Action research** is a research method that pursues action (change) and research (understanding) at the same time, through a cyclical process of planning, action, observation and reflection. It aims not just to improve but to innovate practice. Action research provides a valuable process for exploring ways in which sustainability is relevant to the participants' workplaces and or lifestyles. It views change as the desired outcome and involves participants in investigating their own practice. The competence building occurs within a specific context and issue to be addressed. It differs from those processes, often labelled 'capacity building', which consists of training through the dissemination of information of materials.

All these approaches are based on informed collaborative but structured processes which use learning, reflection and change to improve the effectiveness of an organization, strategy, program or action plan. They all engage the learner in exploring notions of participation, power and possibilities of change. Ultimately, they also develop professional skills and learn how to work individually and in a group to achieve change.

Stakeholders are using these approaches to mainstream sustainability approaches within education, training and capacity building in the community. Others are using it to develop Local Agenda 21 plans or Corporate Social Responsibility (CSR) policies or even Education for Sustainability strategies.

Learning based change in practice

'Learning is a process that influences the way people think, feel and act. We learn through experiences throughout our entire lives. Learning happens consciously and subconsciously. We often learn by interacting with people and the environment' (PCE NZ 2004, p. 1).

Learning based change for sustainability is often most effectively used within the Local Agenda 21 context and other planning initiatives where participants are seen as learners as well as contributors (Tilbury *et al.* 2005). It occurs when members of a local community come together to plan for a better quality of life within their local area. In order to move towards a better future, resolve existing conflicts and develop realistic action plans, those involved learn, reflect and negotiate visions for their community. Skilled facilitators provide informal structured learning opportunities during meetings and create a culture of participation, engagement and ownership necessary for implementing sustainability at the local level.

Learning based change for sustainability underpins the organizational change for sustainability literature. Some corporate and government agencies are using the approach as a way of advancing the sustainability agenda internally. They are using: (a) envisioning to align the entire organization with sustainability principles (Tilbury and Wortman 2004); (b) critical and systemic thinking to identify the difference between cause and effect, and understanding the root causes of unsustainability (Doppelt 2003); (c) participation in decision-making as a way of motivating stakeholders to engage in changes for sustainability (Government of Victoria 2004); and, (d) developing partnerships to stimulate dialogue and assist with the implementation of sustainability strategies and action plans (Tilbury and Cooke 2005).

Doppelt's text on 'Leading Change for Sustainability' (2003) supports the learning based change approach in business, government and civil society. It provides insights into the components of this approach as well as case studies which document its value and impact. The book contrasts case studies which have used learning based change approaches with those which have worked with traditional models of training and change – arguing that long term change for sustainability is attainable if this emerging approach is used.

An IUCN publication 'Engaging People in Sustainability' (Tilbury and Wortman 2004) also documents case studies from around the globe, from executive education to volunteer conservation programmes, which use a learning based approach to bring about change for sustainability. Although a variety of terms may used to describe these examples, common components underpin these case studies (see Key Components). The book documents how learning based change motivates, equips and involves both individuals and institutions in reflecting on how they currently live and work. The process assists them in making informed decisions and creating ways to work towards a more sustainable world.

Key components of learning based change for sustainability

Learning based change seeks to implement systemic change within the community, institutions, government and industry through a process which is underpinned by the following key components:

Systemic thinking

Systemic thinking is a way of thinking based upon a critical understanding of how complex systems such as environmental and social systems function by considering the whole rather than the sum of the parts. Systemic thinking offers a better way to understand and manage complex situations as it emphasizes holistic, integrative approaches which take into account the relationship between system components. Systemic thinking works towards long-term solutions that are vital to addressing issues of sustainability. It is a critical component of learning based change for sustainability as it assists people to understand the systems they are attempting to change.

Traditionally we have come to understand things by taking them apart, deconstructing and breaking down components into smaller parts. Sterling (2004) argues that in a complex and every changing world there is a strong argument that analytical thinking is not enough and that it might indeed be increasing our problems. Systemic thinking offers a better way to understand and manage situations marked by complexity. It can replace the old ways of thinking challenging fragmented thought with its emphasis on integrated and adaptive management.

Systemic thinking challenges the current tendency to segregate thought. It encourages us to see connections between things and how 'this' relates to 'that' or recognize that there might be implications to our actions which were not foreseen (Sterling 2004). 'Joined-up thinking', 'integrative thinking', 'relational thinking' and 'holism' are words often used to describe systemic thinking. Systemic thinking recognises that we are sometimes blinded by our current ways of thinking which do not often recognize the importance of connections or linking thinking.

Envisioning

Envisioning is a process that engages people in conceiving and capturing a vision of their ideal future. Envisioning, also known as 'futures thinking', helps people to discover their possible and preferred futures and to uncover the beliefs and assumptions that underlie these visions. It helps learners establish a link between their long term goals and their immediate actions. It also helps contextualize socio-environmental contexts within one's own ambitions and attempting to resolve differences in expectations. Envisioning offers direction and provides impetus for action by harnessing people's deep aspirations which motivates what people do in the present. It contrasts with the doomsday projections of the future which disempower people by their negative images.

Today's media are dominated by stories of poverty, environmental degradation, species extinction, corruption and terrorism. While such issues require urgent attention, basic knowledge about them does not lead us to a clear path to action, nor does it motivate participation in their solution. Rather, such all encompassing negativity often leads to feelings of powerlessness, apathy, guilt and disillusionment, clouding the path towards real solutions.

Many current educational practices are focused on trying to problem-solve their way out of unsustainable development rather than on creating alternative futures (Hicks and Holden 1995). In addition, some traditional education programs and resource have offered a particular view of the future which is not questioned in any way or do not encourage people to engage n change.

Sustainability facilitators and educators have been exploring futures thinking and envisioning tools as a way of helping people, schools, communities and organisations to see 'sustainability' not as a vague concept but something that is directly relevant to their lives. Key questions relating to this process include:

- Q. What assumptions underpin my vision?
- Q. What has influenced or informed my/someone else's vision
- Q. How and why might others not agree with my vision?
- Q. What are the implications of this vision for life, work and everyday choices and actions?

Critical thinking and reflection

Critical thinking and reflection challenges us to examine the way we interpret the world and how our knowledge and opinions are shaped by those around us. Critical thinking leads us to a deeper understanding of the range of community interests and the influence of media and advertising in our lives. It helps identify power relationships within the community and question the cultural assumptions which influence our choices.

Critical thinking is triggered by a questioning process which helps people uncover assumptions, challenge assumed knowledge and question current thinking. This questioning might take place through dialogue in a workshop, during a meeting, through role-playing exercises, or through constructing visual maps.

Throughout the course of a day, people constantly absorb information by reading newspapers, listening to radio, watching television and browsing the internet. They frequently interact through conversations with family, friends, social groups and work colleagues or school peers. They are targeted by companies seeking to sell their products or services. All of these sources influence how people perceive the world and what is considered to be of value in everyday living.

Moreover, these sources present a particular viewpoint, or have bias. Media interests shape the news. Corporations influence government regulation. National interests and priorities reflect cultural perspectives. Through advertising, companies encourage people to consume or link products to feeling of self-worth and status. Friends, family and co-workers also influence as they can lead to 'group thinking' where many simply adopt the opinions and views of those around them – sometimes subconsciously. Critical reflective thinking empowers the individual to identify these influences in their thoughts and actions and to clarify for themselves whether they are making the appropriate choices (Tilbury and Wortman 2004).

Ultimately, sustainability depends on fundamental changes in lifestyles and the choices people make day-to-day. These changes can come about by critical questioning of our current dispositions as well as the social assumptions and practices which threaten our quality of life.

Partnerships for change

Working towards sustainability will require transformation of social structures. These and other challenges of sustainability are daunting and so many are finding networks and partnerships as a vehicle for sharing responsibilities and learning how to address issues. Over the past ten years, many voluntary, multi-stakeholders initiatives, partnerships between government, NGO and business have begun to take root, demonstrating that they are a motivating force for change towards sustainability.

Because they are non-hierarchical, partnerships can be a strong innovative force in transforming institutions such as within the formal education sector and reorienting them towards sustainability (Henderson and Tilbury 2004)[20]. Cross-sectoral partnerships among local, regional and national groups can add value to local initiatives by helping change larger institutional frameworks while maintaining local relevance.

References to networks and partnerships have featured regularly in many pronouncements and international commitments on sustainability which reflect the prominent role they have played in discussions ever since 'Agenda 21'. It was at the Rio Summit (UNESCO 1992) where partnerships were identified as a critical component of sustainability. The Summit promoted an 'action-oriented' formulation of sustainability partnerships. Since then, there has been an increasing recognition that partnerships which share learning experiences can accelerate the process of change towards sustainability (UNESCO 2002). The World Summit on Sustainable Development reinforced this view, ending with a call for greater global partnerships for sustainability.

Today, over 290 voluntary and self organising partnerships between government NGOs and the private sector have been registered with the United Nations (Tilbury and Cooke 2005). These partnerships cut across several themes relevant to sustainability, from health to consumption and poverty alleviation. Many focus on the benefit of capacity building or technology transfer while others seek to affect change in institutional frameworks. Partnerships are also at the core of the Implementation Plan of the UN Decade in Education for Sustainable Development (2005-2014). They are encouraged as a key component of programs across the

[20] Non-hierarchical partnerships are providing greater scope for stakeholders to take ownership and commit to institutional change. However, there have been examples of hierarchical relationships were a stakeholder has provided democratic leadership and enabled participating groups with the space to engage and commit to sustainability outcomes e.g. Australian Sustainable Schools Initiatives (AUSSI). See www.deh.gov.au

spectrum, from formal education to community-based projects and also from international networks down to regions within a country. It argues that planners and managers can increase the effectiveness of their programs by including a range of stakeholders in their design and management. Partners should include not only those with a diverse range of interests and perspectives on sustainability, but those from various levels – local to regional to national and international.

Partnerships for change provide both formal and informal opportunities for learning. Learning can take place during a meeting or through structured exchanges which allow reflection, development of understanding and questioning of mental models. Partnerships also strengthen ownership and commitment to sustainability actions.

Participation

Participation aligned with sustainability goes beyond mere consultation processes to involve people in joint analysis, planning and control of local decisions. In its 'truest' form it can be self-initiated and directed with participants having full control of the process, decision and outcomes.

The word 'participation' is very commonly used in learning for sustainability policies and programs. Participation can take many forms that involve stakeholders in varying degrees, ranging from consultation and consensus building to decision-making, risk sharing and partnerships. Some describe these different levels of participation on a continuum ranging from manipulation of passive participation to an increasingly shared process and finally, to full stakeholder engagement in, and ownership of, decisions and outcomes. When used in the learning for sustainability context participation is linked to notions of decision-making for sustainability rather than merely consultation or active engagement.

Participation in and for sustainability is an important way of recognizing the value and relevance of 'local' or 'context-specific' knowledge. If properly undertaken, this knowledge becomes part of the decision-making process and weighed up with knowledge from other sources. Solutions are developed relevant to each community or stakeholder group. Rather than relying on outside specialists or managers, participation can engage more stakeholders in becoming part of the process of self-governance and decision-making. Successful participation for sustainability involves a wide range of stakeholders and provides opportunities to build a shared vision, a greater sense of unified purpose and community identity.

Through participation, people can build skills to take control of both the decision-making process and responsibility for outcomes. This greater control leads to greater motivation to participate in actions, whether they are community projects, political action, democratic decision-making or community leadership roles.

Genuine participation in the learning experience is essential to build people's abilities and empower learners to take action for change towards sustainability. Through participation learners are at the centre of the active participatory experience with learning, facilitation and decision-making in the hands of the learners themselves. In learning based change approaches, the community leader, mentor or educator is not considered to be the 'expert' by instead is a facilitator dedicated to helping learners to rethink and take decisions and actions aligned with sustainability. This process of participation is more likely to lead to permanent changes as compared to participation in one-off events. Building skills for participation gives people the opportunity to actively participate, build knowledge and develop leadership skills that contribute to action. It challenges the power bases in our society which have led us to unsustainable development.

Documented experiences suggest that networks and partnerships are helping participants to:

- create synergy in their work to maximize opportunities for all involved;
- combine resources, talents and foster long-term relationships to encourage mutual benefit and development;
- reflect on the values and missions and can create a space to create shared visions as well as develop new ideas and strategies;
- motivate action for the future as they provide a forum for mutual support and encouragement; where successes can be celebrated;
- build expertise and capacity which can help to secure financial and technical support from funding sources. As individual partners may be specialized in one area, they might lack the staff or financial ability to commit to long-term change to sustainability. By combining resources and financial assets and pooling technical skills with others they can develop the broad and long-term ideas and strategies for change;
- break hierarchies and challenge traditional power structures; and,
- help to challenge mental models by bringing together individuals and groups with different perspectives and from different levels – when learning together shifts in perspectives and more long-term change is likely(adapted from UNEP DTIE 2005).

In reality, the achievement of successful partnership outcomes – based on common objectives, clearly defined deliverables, where ownership is shared among all partners – has been identified as a major challenge (IISD 2005).

The ten year challenge...

Framed from a social learning perspective, sustainability is seen as an emerging and reflective process which challenges traditional notions of education and sees learning as the cornerstone of social change. Learning based change differs significantly from processes which have been traditionally described as education. With the conceptual pathway to sustainability not clear, the quest for sustainability requires new learning as well as different forms of knowledge to involve people across the social sectors in learning for change. However, setting the pathways for this form of social learning will not be a simple task.

A ten year challenge has begun in 2005 under the banner of the UN Decade in Education for Sustainable Development. The Decade brings with it the momentum required to promote and support learning based approaches across the social sectors.

The Decade, under the leadership of UNESCO, seeks to give an enhanced profile to education and learning in the achieving sustainability and provide opportunities for developing a vision of sustainability (UNESCO 2005). UNESCO defines the vision for the Decade as one where everyone has the opportunity to benefit from quality education and learn for a sustainable future and contribute to social change. The next ten years may provide platform for this new learning to take centre stage as a strategy for achieving social change for sustainability.

References

Bhandari, B. and Abe, O. (2003) (Eds.) *Education for Sustainable Development in Nepal: Views and Visions*, Japan: International Institute for Global Environmental Strategies.

Connor, R.D. and Dovers, S.R. (2002) "Institutional change and learning for sustainable development", Working Paper 2002/1, Centre for Resource and Environmental Studies.

Doppelt, B. (2003) *Leading Change Towards Sustainability: A Change Management Guide for Business, Government and Civil Society*. Sheffield, England: Greenleaf Publishing Ltd.

Eckersely, R. (ed.) (1998) *Measuring progress: Is life getting better?* Melbourne: CSIRO Publishing.

Environment Canada (2004) *Sustainable Development Strategy 2004-2006*, available from http://www.ec.gc.ca/sd-dd_consult/.

Government of Victoria, Department of Sustainability and Environment (2004) *Effective Community Engagement*, Melbourne, Victoria.

Hamú, D. (2004) "Forward", in D. Tilbury and D. Wortman, *Engaging People in Sustainability*, Gland, Switzerland and Cambridge, UK: Commission on Education and Communication, IUCN.

Henderson, K. and Tilbury, D. (2004) *Whole-school Approaches to Sustainability: An International Review of Whole-school Sustainability Programs*, Canberra: Australian Government Department of the Environment and Heritage and Australian Research Institute in Education for Sustainability.

Hicks, D. and Holden, C. (1995) *Visions of the Future: Why we need to teach for tomorrow*, Staffordshire, England: Trentham Books.

International Institute for Sustainable Development (IISD) (2005) *Partnerships*, available from http://www.iisd.org/networks/partnerships.asp

Milbrath, L. W. (1996) *Learning to Think Environmentally: While there is Still Time*, Albany, NY: State University of New York Press.

Parliamentary Commissioner for the Environment (PCE), New Zealand (NZ) (2004) *See Change: Learning and Education for Sustainability*, Wellington, New Zealand: Parliamentary Commissioner for the Environment.

Prescott-Allen, R. (2001) *The Wellbeing of Nations: A Country-by-Country Index of Quality of Life and the Environment*, Covelo, CA: IDRC/Island Press.

Senge, P. (1990) *The fifth discipline: the art and practice of the learning organization*, New York: Doubleday/Currency.

Senge, P. Kleiner, A., Roberts, C., Roth, G., Ross, R. and Smith, B. (1999) *The Dance of Change: The Challenges to Sustaining Momentum in Learning Organizations*. NY: Doubleday/Currency.

Sterling, S. (2004) "Higher Education, Sustainability and the Role of Systemic Learning", in Peter Blaze Corcoran and Arjen EJ. Wals, ed., *Higher Education and the Challenge of Sustainability: Contestation, Critique, Practice and Promise*, Dordrecht: Kluwer Academic Publishers.

SustainUs 2005, *Joined up thinking – can it really be done?* available from http://www.sustainus.com/modules.php?name=News&file=article&sid=84.

Tilbury, D. (1995) "Environmental Education for Sustainability: Defining the new focus of environmental education in the 1990s", *Environmental Education Research* 1(2): 195-210.

Tilbury, D. and Cooke, K. (2005) *A National Review of Environmental Education and its Contribution to Sustainability in Australia: Frameworks for Sustainability*, Canberra: Australian Government Department of the Environment and Heritage and Australian Research Institute in Education for Sustainability.

Tilbury, D. and Wortman, D. (2004) *Engaging People in Sustainability*, Gland, Switzerland and Cambridge, UK: Commission on Education and Communication, IUCN.

Tilbury, D., Coleman, V. Jones, A. and MacMaster, K. (2005) *A National Review of Environmental Education and its Contribution to Sustainability in Australia: Community Education*, Canberra: Australian Government Department of the Environment and Heritage and Australian Research Institute in Education for Sustainability.

United Nations Educational Scientific and Cultural Organisation (UNESCO) (1992) *United Nations Conference on Environment and Development: Agenda 21*, available from: http://www.un.org/esa/sustdev/documents/agenda21/index.htm.

United Nations Educational Scientific and Cultural Organisation (UNESCO) (2002) *Education for Sustainability, From Rio to Johannesburg: Lessons Learnt from a Decade of Commitment*, report presented at the Johannesburg World Summit for Sustainable Development, Paris.

United Nations Educational Scientific and Cultural Organisation (UNESCO) (2004) *United Nations Decade of Education for Sustainable Development 2005-2014: Draft International Implementation Scheme*, Paris.

United Nations Educational Scientific and Cultural Organisation (UNESCO) (2005) *Report by the Director-General on the United Nations Decade of Education for Sustainable Development: International Implementation Scheme and UNESCO's Contribution to the Implementation of the Decade*, Paris.

United Nations Educational Scientific and Cultural Organisation (UNESCO) and Asia and Pacific Regional Bureau for Education (2005) "Working Paper: Asia-Pacific Regional Strategy for Education for Sustainable Development", Working Paper, UNESCO, Bangkok.

United National Environment Program (UNEP) Division of Technology, Information and Economics (DTIE) (2005) *Business Awards for Sustainable Development*, available from http://www.uneptie.org/Outreach/business/award.htm.

Chapter 6

The critical role of civil society in fostering societal learning for a sustainable world

Richard Bawden, Irene Guijt and Jim Woodhill

> "The themes of the future that are now on everyone's lips have not originated from the foresightedness of the rulers or from the struggle in parliament – and certainly not from the cathedrals of power in business, science and the state. They have been put on the social agenda against the concentrated resistance of this institutionalised ignorance by entangled, moralising groups and splinter groups fighting each other over the proper way, split and plagued by doubt. Sub-politics has won a most improbable victory" (Beck 1994, p. 19).

Introduction

What Beck refers to above as 'sub-politics' is the activism of civil society. Democratically we live in a paradoxical era. Much of the world claims to be democratic and the ideals of democratic governance are central to much international rhetoric. Yet a closer look at what counts for democracy is disturbing and very far from a system that enables citizens at large to engage in debate, discourse and judgement about the issues of our time. Indeed, the majority of citizens and particularly the poor, marginalised and vulnerable are very far removed from influencing the decisions that affect their lives. Those that do engage shout from the sidelines. Also paradoxically, many politicians feel that they are responding to, or even trapped by, what they perceive as public opinion. Humanity is caught, it seems, in systems of governance that radically constrain rather than unleash our collective learning capacities. We have an urgent need to reinvent our systems of governance and ideas about democracy in order to generate learning societies in which the immense challenges of sustainable development can be tackled with moral insight, collective intelligence and wisdom about humanity's future.

In today's world, certainly in the west and increasingly elsewhere, there is a dynamic of power between the state (government), the economic sphere and civil society that has enormous consequences for the types of changes that are possible and the manner in which social, political and economic change can be brought about. No sphere is all powerful and all power is relational, yet each has the power to at least partially subvert actions of the other spheres to which they are opposed. Progress, particularly in relation to sustainable development, hinges on a social capacity for different sectors and interests to be able to constructively engage with each other. This is of critical importance for leadership of civil society organisations (CSOs) and civil society activism. Of course there will always be a place and a need for radical confrontational activism, but increasingly, we believe that the effectiveness of civil society will hinge on its capacity to engage individuals and organisations across all sectors in processes of critical reflection and learning. Understanding and being able to work with power differences and related conflicts is in our view central to such learning. However, CSOs and their leaders are often locked in a paradigm of adversarial politics that blinds them to a critical questioning of their own assumptions and strategies. This risks making them ineffective in engaging the private and government sectors in the sort of dialogue and action necessary for deep institutional change.

In this chapter we will explore the concept of what a learning society means, focusing in particular on the role of civil society in generating more inclusive and dialogical forms of democracy. To ground our argument, we draw from a study (Guijt 2005) of civil society participation (CSP) undertaken for four Dutch development NGOs who fund the work of CSOs that work with extremely marginalised groups. In drawing on this study, we link the ideas of a learning society to practical strategies for CSOs so they can better understand power dynamics within various domains of democratic participation.

The civil society initiatives involved in the study show how power-conscious and injustice-challenging work can lead to societal learning in ways that improve the lives of marginalised groups and those living in poverty – and begin to change the relationship between citizens and rulers. We consciously opt for the term 'societal learning' instead of social learning as it helps move away from a simplistic group-based learning notion and refers directly to the capacity of societies and communities to be more learning-oriented in the way they tackle important issues related to a more sustainable world (also see 'The Societal Learning Response' below).

The examples of civil society initiatives from the study include the Butterfly Peace Garden in Sri Lanka (Perera and Walters 2005), which receives children in mixed (ethnic and religious) groups to help them overcome their traumatic experiences. The healing and reconciliation effects also extend to the wider community by bringing in the parents and teachers to share what the children do with art, play and counselling. A societal alternative – co-existence in spite of divisive conflicts – is being learned. Another example is from CALDH in Guatemala (Gish *et al.* 2005), a human rights NGO, that has engaged the youth in electoral and human rights monitoring. It has given the youth the self-confidence to interact with 'important people' and made the authorities see them as people with capacities. In Colombia, the work of the NGO Conciudadania has helped women "to become protagonists in their own right with self understanding and capacity to resist very adverse and painful circumstances... building a peace culture which challenges patriarchy and promotes non violence" (Pearce and Vela 2005). A final example is from Uganda, where CSOs have been actively training citizens to monitor state budgets, by comparing actual expenditure with promised resource allocations – and challenging those involved where discrepancies are ascertained (Mukasa *et al.* 2005). This is fostering the growth of critical consciousness, from village to national levels, in concrete ways that challenge politicians and bureaucrats to live up to new standards of governance. Although these examples are from the South, the challenges for seeking resolution of challenges to sustainable development are equally pertinent in the North – particularly in this age of political pathos and self-centred consumerism at all costs.

For those interested in the relationships between sustainability, democracy, societal learning and civil society, this chapter offers three threads of analysis. First, we look at the challenges and conceptualisation of sustainability and development and why these require investment in capacities for learning. Second, our notion of societal learning is explained, emphasising the importance of politicized reflection. Third, we discuss the domains of civil society participation in which learning is needed and how CSOs can be strategic through conscious reflection on the power dynamics in which they are embedded and which they wish to influence.

Sustainability and development

There are still those who would argue that the idea of talking about sustainability and about development in the same breath is tantamount to ecological blasphemy and a key indicator of a technical discourse that will inevitably lead to the decline of human civilization and our very planet. They view development as the prime enemy of sustainability. Notwithstanding the undoubted destructive impacts of some modernist approaches to and techno-economic practices of development on bio-physical and socio-cultural environments, we argue that that the quest for a more sustainable world, in fact, hinges on development – yet of a profoundly different kind. For us, sustainability provides a moral and intellectual focus for social as well as individual cognitive development. We extend this into an endorsement of Milbraith's (1989) call for the need for us humans "to learn our way out" of our present situation and towards a more sustainable world. This is fundamentally a call for the collective development of the more refined, more relevant, and more complex epistemic capabilities that can be achieved by individuals in appropriate 'learning situations' (Perry 1968). And this then is the essential challenge for, and advantage of so-called societal learning, which we wish to portray as a critical process of engagement of 'multi-stakeholder' constituencies that have issues of common concern about the way the world and its people should be treated.

Thus, the contestability of the concept of sustainability and of its derivatives such as sustainable worlds, sustainable development and so on, is a practical strength rather than a weakness. It gives rise, as one commentator has eloquently suggested, to an 'agenda of good questions' about how we should live our lives (Davison 2001) and particularly in this paper, of critical reflections on how we, as citizens of the world 'see' that world around us as the basis for what we do in, and to it. Hence, we must reflect critically on such key matters as the boundaries we construct around the issues with which we choose to engage and, perhaps most essentially, of the moral and intellectual frameworks and processes which we adopt as our ways of dealing with them. It is crucial that these reflections are not mere exercises in awareness or consciousness-raising, but extend to motivations, capacities, and opportunities for active and responsible responses. Engagement with the issues and with each other collectively in the search for consensual judgments about what needs to be done is vital in their resolution. From this perspective, inclusive participation in this quest for public judgment is not only essential, it is a basic human right (DFID 2000) which if denied for whatever reason, is a denial of social justice. And herein lies the primary significance of power relationships between individuals, within communities, between individuals, communities and social institutions, and so on. There are many different manifestations of the distorting influence of power, with none perhaps more insidious in its impact than the power of knowledge itself and the particular ways by which it is known (Habermas 1984).

All of this dictates the need for us to develop the intellectual and moral capabilities to deal with these responsibilities, as well as be empowered and enabled, by relevant institutions, to have the social, cultural, and political freedom to express them. Such a systemic perspective, on what Sen (1999) refers to as 'capabilities' for development as freedom, also emphasises the vital importance of 'power' as a further key focus of criticality and a central aspect of 'capability development'. We will return to the question of power below.

The issues we face in the quest for a sustainable world are innately complex and dynamic and our approaches for dealing with this clearly inadequate. What has been done in the name of modernizing improvements in the past have all too often come to have unintended consequences of such a scale that they constitute a whole new generation of problematic issues that actually reflect severe inadequacies in the modernization process. We have reached an age of what some refer to as 'reflexive modernization' where the central challenge is no longer mastery over nature or release from traditional constraints, but over the logic and industrial practices of modernization itself (Beck 1992). The institutions of modernity have to confront risks that are the side effects of their own activities as manifested by such phenomena as ecological devastation, global warming, nuclear contamination, social and cultural dislocation and even insurgencies, pollution effects on public health, the emergence of pandemics, economic and political collapse, and so on. The direction and goals of what was considered progress has been characterized by the instrumental nature of techno-economic rationality. Over time the techno-economic decisions that really impact on society have come to rest predominantly with scientists, bankers and corporate managers – and not with either the citizenry on the one hand or with elected governments on the other.

Paradoxically, as parliamentary democracy slowly spreads across the globe, so the distance grows between the world of expert policy making and the world of public opinion, and the result includes a continuing decline in the quality of public participation in their own affairs and the erosion of self-governance. As Yankelovich (1991) observes, over the years the nature of the relationship between experts and public "has grown adversarial rather than mutually supportive," and this has led to the development of a very significant gap that separates the public from the experts. Meanwhile, even as individuals within society have come to rely on knowledge and information generated by specialists, which they routinely interpret and act upon as they live their everyday lives (Giddens 1979), institutions that assist in the process of public judgment remain very thin on the ground. Furthermore, experts do not always agree on matters of mutual interest, which often leaves citizens in the position of not knowing whom to believe. This circumstance is exacerbated when experts are hired by advocates who support a particular position and consequently, 'to engage' all too often means 'to invite conflict' between and among contesting groups. Herein lies the significance of the adoption of a critical stance on engagement. Hence it is not simply a matter of participation, but one of critical reflection and dialogues and acts that challenge roles, power structures, taboos, etc.

As our opening quote illustrates, the structural transformation of modernity, driven by risk, does not erupt as a class revolution as Marx had predicted, but instead, it 'creeps in through the back door on cat's paws' (Beck 1994). Sooner or later, the escalating risks of modernity start to become unacceptable to the polity, and indeed a problem for the techno-economic sphere itself. As the risks mount, a dialectic of control starts to operate that is increasingly marked by newly critical questions about 'matters to hand' as well as about the assumptions and beliefs that frame them. Reflexive modernization comes to resemble a paradigmatic or 'epistemic' revolution. Science is no longer accepted as the sole source of knowledge for action for modern society, at the same time that religious beliefs become pluralistic and thus unacceptable as a basis for focussed political decisions. It is this situation that also dictates that we develop the intellectual and moral capabilities to deal with our new realities and responsibilities, as well as be granted the social, cultural, and political freedom to express them. As Sen (1999) argues, such freedom is a primary and paramount condition if we are to choose a life that we *each* value. And it is these circumstances that propel societies towards a reformulation of the roles that dialogue, discourse and societal learning should play in shaping the future – a democratic imperative for restructuring core institutions and for critically re-appraising how we come to learn to make collective judgments that are of common concern to us – in all of our diversity.

While sustainability and sustainable development remain contestable constructs (Davison 2001), they are central aspects of emerging approaches to what we might refer to as the 'post-productionist' paradigms of late modern development that embrace the critical reflexiveness outlined above. Foundational to all of these is an emphasis on 'participation', the processes of learning and of discourse, and of communication in its broadest sense. As the authors of a report from the Department for International Development in the UK submit, "rights will become real only as citizens are engaged in the decisions and processes which affect their lives" (DFID 2000).

The required responses – that still need to develop – are themselves complex in nature, which demands an integration of different human characteristics and capabilities including moral judgment with intellectual reason, passion with logic, reflection with action, and of theory with practice. These sobering observations lead us to ask if we have the capacities, motivations, and/or opportunities to create less inadequate, more appropriate responses. As Dietz and his colleagues have suggested, devising ways "to sustain the earth's ability to support diverse life, including a reasonable quality of life for humans, involves making tough decisions under uncertainty, complexity, and substantial biophysical constraints as well as conflicting human values and interests" (Dietz *et al.*2003).

Our own response to this question takes the form of the hypothesis that the primary focus for any act of development towards a more sustainable world ought to be the moral and intellectual developments of all of the actors who should be involved in that act. In other words, the central theme of development of sustainability should be the critical learning capabilities of multi-stakeholder constituencies who are facing complex, problematic matters of common concern. This inevitably means engaging civil society in problematising societal injustice, challenging those hindering the claiming of rights, reinfusing democracy with responsiveness, and so forth.

The societal learning response

As will be evident from the above, one of the key features of modern society is that it must now learn how to respond to the often negative consequences of its own actions (Beck 1994) and also how to overcome the epistemic constraints that currently-held assumptions impose on 'seeing' and 'doing' things differently. This implies that modern societies need to learn more quickly, more effectively, and much more critically than societies in the past that faced slower and less globally interconnected social and natural changes – and even then faced dramatic disturbances to the extent of being wiped out (Diamond 2005). This, in turn, requires citizenries willing to participate actively in democratic deliberations, and capable of learning collectively, with and from each other. The citizens of any such society who are unable to innovate collectively in response to changing environments or in attempts to influence the manner by which those environments evolve, run the risk, at the very least of some form of collective crisis, and at worst of the annihilation of their society. The study of Dutch NGO support for civil society participation (Guijt 2005) illustrate ample cases of citizens taking up this very challenge, to create spaces for their sub-cultures, to challenge the core values on which their societies are based, such as conflict, oppression and corruption, and to dare to envision alternatives and make these reality.

David Korten was one of the first to highlight the connection between participation and what he called social learning in development, linking this to the shift from production to people-centred development. In this regard, he argued for the relevance of engagement practices that feature forms of self-organization that 'highlight the role of the individual in the decision process' while also calling for the 'application of human values in decision-making'. He went on to draw particular attention to the fact that the knowledge-building processes of people-centred development are perforce based on social learning concepts and methods (Korten 1984). Following this lead, many others, over the past two decades or so, have adopted as a common theme, "the *full* participation of people in the processes of *learning* about their needs and opportunities, and in the *action* required to address them" (Anon. 2004, emphasis in the original). However, as the mounting critiques of participatory development argue (cf. Hickey and Mohan 2004), this has not always been informed by political astuteness or by clear learning principles or strategies.

All social change requires learning of some form, but the questions here are how societal-wide learning processes can be made more critical, self-reflexive, transformative and thus, more effective. The first step to be taken in this latter context is to increase understanding of the process of learning itself, and how the 'quality of learning' and of learning capabilities can be developed.

We strongly identify with the notion, promoted by writers such as John Dewey, Kurt Lewin, and David Kolb, of learning as the basic process by which we humans make sense of our experiences of the ever-changing world and which serves as the basis for the adaptations we seek and enact. As other people are invariably involved in one way or another in the process of each individual's learning, all learning from this 'experiential perspective' can be considered to be social in its nature. An explicit adoption of the term 'societal (social) learning' however, serves to emphasize the idea of bringing together different stakeholders (actors) who have an interest in a particular problem situation and engaging them in processes of dialogue and collective 'sense-making for action'. This form of 'social learning', as Cornwall and Guijt (2004) declare, entails more than simply group-based learning, but rather "bringing together a range of unlikely comrades in multi-stakeholder processes of fact-finding, negotiation, planning, reassessing, and refocusing". It might be more useful then to refer to 'societal learning' rather than social learning as an overarching concept.

Societal learning can be defined as the process by which communities, stakeholder groups or societies learn how to innovate and to adapt in response to changing social and environmental conditions. Societal learning actively engages different groups, communities, and multi-stakeholder constituencies in society in a communicative process of understanding problematic situations, inter-personal conflicts and social dilemmas and paradoxes, and of creating strategies for improvement. It is also manifestly a self-reflexive, self-critical process that has been seen to comprise three different levels of 'cognitive processing' (Kitchener 1983).

Kitchener distinguishes between cognition, meta-cognition, and epistemic cognition as a hierarchical sequence through which individuals monitor how they conduct their own basic cognitive tasks. In essence, we learn to deal with 'matters to hand' of everyday concern (learning level one or cognition), we learn how to deal with how we deal with such matters to hand (learning level two or meta cognition), and we learn about the nature of knowledge and of the impact of the way we come to know anything (learning level three or epistemic cognition). Kitchener viewed all three levels as developmental. In essence, as many other studies have reaffirmed, our mental frameworks or intellectual and moral perspectives, develop in a manner that increasingly accommodates complexity (West 2004). While the focus of the vast majority of this work has been on the individual, there are clear implications here for the collective cognitive development in societal learning situations too.

Domains of civil society participation for societal learning

It is, of course, one thing to mount the intellectual argument for societal learning and quite another to bring about the institutional changes required in current governance and democratic systems that enable a free flow of ideas, open dialogues, and new partnerships between unlikely allies.

Societal learning, as we see it, represents a paradigm from which to engage in transforming current forms of governance and democracy, which have more often than not proved themselves manifestly inept in coping with the pluralistic challenges of sustainability. Improving the way societies learn challenges us to think about the role of civil society, the way media works, the type of education we receive, and the relationship between science and society and the real meaning of corporate social responsibility. This involves understanding the limitations of existing institutions and mechanisms of governance and experimenting with multi-layered, learning-oriented and participatory forms of governance. And this, in turn, requires tools for reflection by civil society – to which we turn toward the end of this section.

What are the ways in which civil society can foster greater participation and learning in our systems of governance? Civil society organisations are often associated either with those radical organisations that confront society with a need for change or charity organisations that 'give' to the poor and needy. Our argument is that across all forms of civil society action there is scope for more participatory and societal learning-oriented action. To illustrate this we shall briefly introduce six domains of civil society participation, within each of which CSOs can be expected to play certain roles that lead to specific achievements at the level of individual citizens (Guijt 2005). We argue that societies need 'learning' to occur at all levels if citizens are to engage actively in the (re)construction of more sustainable futures. Central in these six domains is the embeddedness of 'societal learning', which we will illustrate.

The first domain refers to 'citizenship strengthening', with the emphasis of societal learning being on helping people to understand their rights and be able to constructively and effectively engage in claim-making, collective action, governance and political processes. This type of learning is very foundational, on which other forms build. It is essential in contexts where democracy is new for most citizens (and participatory democracy a very distant dream indeed), the state has yet to construct the most basic of required structures and daily life poses such a drain on energy that more structurally challenging political engagement by people is a low priority (cf. Buchy and Curtis' study on Guinea 2005). A second equally foundational domain is that of 'trust/dignity/culture/identity', where people learn – through interactions – to have mutually respectful social relationships and engender trust in others based on positive experiences. Learning in this case focuses more on developing and 'living' an alternative model of human interaction than is locally dominant.

Civil society also needs to be active in 'CSO governance, programming, monitoring, and accountability', a third domain. Societal learning focuses both on organisational learning in terms of strategising but also on learning how to enact democracy within the organisations that profess to represent citizens. This requires CSOs to learn about being responsive to the rights, values, aspirations, interests and priority needs of their constituencies. A fourth domain is that of 'citizen participation in local development and service delivery'. Much CSO activity of this type focuses on building local capacities to critically question existing inadequate services, have visions about better alternatives and then design, implement and monitor these.

The fifth domain is one that receives much attention – 'citizen/CSO participation in advocacy and structural change', in which citizens call society at large or particular power holders to debate about unjust policies or implementation. This requires both internal learning about best strategies but also acts as a lever for broader societal learning about the issues that are the focus of advocacy efforts. The final domain is that of 'citizen participation in economic life', one in which many CSOs and citizens are undergoing a steep learning curve as it concerns the relatively new forms of market engagement by poor, vulnerable people on their terms and for their needs, and making the concept of pro-poor economic growth a reality. Here again, learning is internal on strategies, negotiation capacities and indeed, visions for an economic alternative – but also learning in society at large about this alternative model.

In practice, of course, a single CSO is often active within several complementary domains, while each domain is populated by the efforts of many CSOs, whether or not in deliberate collaboration.

Strategising for learning around power dynamics

Together, the six domains of civil society participation described above constitute a quite comprehensive agenda for societal learning. How though, can CSOs be strategic in their pursuit of a more sustainable world that hinges on understanding power differentials and challenging these? Knowledge about the power relations can help CSOs to be more strategic about the contribution they seek to make to societal learning within the domains in which they are active.

One tool for reflection is the so-called 'power cube' which encourages a perspective in which all these forms of civil society engagement are valued and, if discussed with CSOs can help them consciously to shape integrated strategic action. The 'power cube' formed the methodological core of the study of Dutch NGO support for civil society participation (Guijt 2005). This 'spaces, place, power' framework (Cornwall 2002, Gaventa 2003) offers ways to examine participatory action in development and changes in power relations by and/or on behalf of poor and marginalised people. As Gaventa says (2005, p. 3): "Despite the widespread rhetorical acceptance of participation, rights and deepened forms of civil society engagement, it is clear that simply creating new institutional arrangements will not make them real and will not necessarily result in greater inclusion or pro-poor policy change. Rather, much will depend on the nature of the power relations which surround and imbue these new, potentially more democratic, spaces."

The 'power cube framework' enabled a deep questioning of 'civil society participation' in terms of three analytical dimensions: space, place and dynamics of power (see Figure 6.1). This framework understands power "in relation to how spaces for engagement are created, the levels of power (from local to global), as well as different forms of power across them" (ibid, p. 2). By using this lens on citizen action, there is potential to assess the possibilities of transformative action by citizens and how to enlarge these.

In the framework, the dimension of *'places'* refers to the levels on which participatory action is focused or where it occurs:

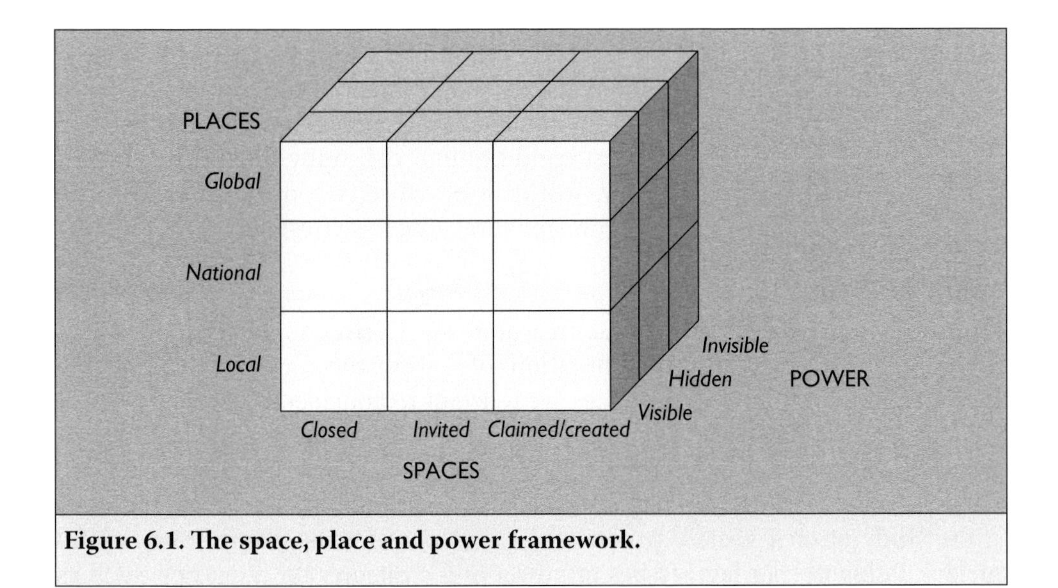

Figure 6.1. The space, place and power framework.

- Local – e.g. household, district, sub-county or municipal level fora and councils.
- National – e.g. national alliances and fora, consultations, parliament.
- International – e.g. global alliances, global governance institutions.

'Spaces' are understood to be spaces of engagement filled by power of varying kinds, visible and invisible, including knowledge and discourse. Thus, a 'space' is an arena, process or mechanism within which people communicate about issues, share information, make decisions and take actions, or in which civil society (people and organisations) seeks to influence decisions which affect daily life:

- Closed spaces – official or unofficial spaces to which only certain people or interest groups are invited, and others are excluded.
- Invited spaces – formal or informal spaces in which powerful officials invite people or organisations to be consulted or to make their views known.
- Claimed spaces – formal or informal spaces created by those who seek to have greater power and influence.

Power dynamics are played out within 'spaces' in each 'place' in various ways, with participatory activities relating to different aspects of empowerment. VeneKlasen and Miller's (2002) distinguish three ways in which 'participation' can affect power relations:

- Visible power – the ability to influence formal decision-making processes, with power as 'agency' openly held and used by people and interest groups and empowerment being the having a voice and influence in formal processes.
- Hidden power – setting the agenda behind the scene, exclusion of others, mobilisation of bias and interests to shape agenda and outcomes, with empowerment being the ability to influence what appears on the agenda.
- Invisible power – deeper social conditioning, culturally embedded norms, effects of knowledge, ideology, worldviews, what is considered within the realm of the possible, with empowerment relating to self-esteem, power within, changes in cultural norms.

The 3x3x3 framework is a heuristic that does not restrict analysis to only three levels, three spaces and three manifestations of power. Each context will determine in particular, which levels and spaces are relevant to consider. Use of the cube in Colombia (Pearce and Vela 2005) identified eight relevant levels and five different spaces for engagement.

In the study of civil society participation, the power cube proved very useful for "discussing deeper level issues of power and strategies for advocacy, such as choosing when and how to engage in different spaces" (Mukasa *et al.* 2005). Its main value was to stimulate more critical reflection about relationships, power shifts, and strategies of the CSOs in relation to the societal challenges they were addressing – be it retroviral drugs for the poor, budget monitoring, legal support to Muslim women in refugee camps and women plantation workers, children's rights in rural schools, the right to be registered at birth, breaking the power of narco-traffickers, and the myriad other issues that threaten the sustainability of their societies. It helped to understand relationships of power between key stakeholder groups in each context and per issue, showing the dynamic reality of civil society action as CSOs move across spaces and places to make the most of opportunities that can further their goals. Without such reflection on where CSOs are positioned vis-à-vis other actors in a range of political arenas, the risks of creative and brave but insignificant action becomes very real.

Conclusion

The social sciences are littered with numerous theories of social change. Some emphasize the role of free will and human reason, while other argue that change is largely determined by power dynamics and existing social structures. Our line of argument, in line with the work of Giddens, sees all these dynamics as present in any social situation. However, we see sufficient evidence to believe that societies have the capacity to use free will and human reason to learn about the dynamics of power and change what might at first seem unassailable and unsustainable structures of domination. Examples of CSO action can be found of what we mean by societal learning.

We also feel that much more conscious work is needed to develop this capacity with the civil society organisations on which we have focused in this chapter. While Beck's 'themes of the future' may be on everyone's lips, the changes needed to deal with this agenda depend on a different order of engagement by the "entangled, moralising groups and splinter groups" to which he refers than has been the case to date. There remains an enormous challenge to develop further the capabilities of civil society to engage with issues from a societal learning perspective and bring about change at the scale on which sustainability clearly depends.

References

Anon. (2004) "Editorial statement", *Participatory Learning and Action,* 50.

Beck, U. (1992) *Risk society: Towards a new society,* London: Sage Publications.

Beck, U. (1994) "The re-invention of politics: Towards a theory of reflexive modernization", in U. Beck, A. Giddens and S. Lash, eds., *Reflexive Modernization: Politics, Tradition and Aesthetics in the Modern Social Order,* Stanford: Stanford University Press.

Buchy, M. and Curtis, M.Y. (2005) Guinea Country Report on the Programme Evaluation "Assessing Civil Society Participation as supported In-Country by Cordaid, Hivos, Novib, and Plan Netherlands". MFP Breed Network, The Netherlands.

Cornwall A. (2002) "Making spaces, changing places: Situating participation in development", IDS Working Paper No.170, IDS, Brighton.

Cornwall, A. and Guijt, I. (2004) "Shifting Perceptions, Changing Practices in PRA: From Infinite Innovation to the Quest for Quality", *Participatory Learning and Action,* 50:160–167.

Davison, A. (2001) *Technology and the Contested Meanings of Sustainability,* New York: State University of New York Press.

Department for International Development (DFID) (2000) *Realising Human Rights for Poor People: Strategies for Achieving the International Development Targets,* London: DFID.

Diamond, J. (2005) *Collapse. How Societies Choose to Fail or Survive,* London: Allen Lane.

Dietz, T., Ostrom, E. and Stern, P.C (2003) "The Struggle to Govern the Commons", *Science,* 302:1907–1912.

Gaventa, J. (2005) "Reflections on the Uses of the 'Power Cube" Approach for Analyzing the Spaces, Places and Dynamics Of Civil Society Participation and Engagement. MFP Breed Netwerk, the Netherlands.

Gaventa, J. (2003) "Towards participatory local governance: Assessing the transformative possibilities" paper presented at IDPM conference, *Participation: From Tyranny to Transformation? Exploring New Approaches to Participation in Development,* Open University and University of Manchester, February 2003.

Giddens, A. (1979) *Central Problems in Social Theory,* London: Macmillan.

Gish, D., Navarro, Z., Pearce, J. and Pettit, J. (2005) Guatemala Country Report on the Programme Evaluation "Assessing Civil Society Participation as supported In-Country by Cordaid, Hivos, Novib, and Plan Netherlands". MFP Breed Network, The Netherlands.

Guijt, I. (2005) Synthesis Report of Dutch CFA Programme Evaluation. "Assessing Civil Society Participation as Supported In-Country by Cordaid, Hivos, Novib and Plan Netherlands 1999-2004". MFP Breed Netwerk, the Netherlands.

Habermas, J. (1984). *The Theory of Communicative Action* in Thomas McCarthy, trans., *Vol. 1. Reason and rationalization; vol 2. Lifeworld and systems: A critique of functionalist reason,* Boston: Beacon Press.

Hickey, S. and Mohan, G. (2004) *Participation: from tyranny to transformation? Exploring new approaches to participation in development,* London: Zed Books.

Kitchener, K.S. (1983) "Cognition, Meta-cognition, and Epistemic cognition: A Three Level Model of Cognitive Processing", *Human Development,* 26:222–232.

Korten, D.C. (1984) "People-Centered Development: Toward a Framework", in D.C. Korten and R. Klaus, eds. *Contributions Toward Theory and Planning Frameworks,* West Hartford CT: Kumarian Press.

Milbraith, L.W. (1989) *Envisioning a Sustainable Society: Learning Our Way Out,* New York: State University of New York Press.

Mukasa, G., Pettit, J. and Woodhill, J. (2005) Uganda Country Report on the Programme Evaluation "Assessing Civil Society Participation as supported In-Country by Cordaid, Hivos, Novib, and Plan Netherlands". MFP Breed Network, The Netherlands.

Pearce, J. and Vela, G. (2005) Colombia Country Report on the Programme Evaluation "Assessing Civil Society Participation as supported In-Country by Cordaid, Hivos, Novib, and Plan Netherlands". MFP Breed Network, The Netherlands.

Perera, S. and Walters, H. (2005) Sri Lanka Country Report on the Programme Evaluation "Assessing Civil Society Participation as supported In-Country by Cordaid, Hivos, Novib, and Plan Netherlands". MFP Breed Network, The Netherlands.

Perry, W.G. (1968) *Forms of Intellectual and Ethical Development in the College Years*, New York: Holt, Rinehart and Winston.

Sen, A. (1999) *Development as Freedom*, New York: Anchor Books.

VeneKlasen, L. and Miller, V. (2002) *A New Weave of People, Power and Politics: the Action Guide for Advocacy and Citizen Participation*, World Neighbors: Oklahoma City.

West, E.J. (2004) "Perry's Legacy: Models of Epistemological Development", *Journal of Adult Development*, 11:61–70.

Yankelovich, D. (1991) *Coming to Public Judgment: Making Democracy Work in a Complex World*, Syracuse NY: Syracuse University Press.

Chapter 7

From risk to resilience: what role for community greening and civic ecology in cities?

Keith G. Tidball and Marianne E. Krasny

Introduction

One of the greatest risks following a natural disaster or conflict in cities is the ensuing social chaos or breakdown of order. Failed cities, such as parts of New Orleans following Hurricane Katrina and Baghdad following war in Iraq, can be viewed as socio-ecological systems that, as a result of disaster or conflict coupled with lack of resilience, have "collapsed into a qualitatively different state that is controlled by a different set of processes" (Resilience Alliance 2006). Communities lacking resilience are at high risk of shifting into a qualitatively different, often undesirable state when disaster strikes. Restoring a community to its previous state can be complex, expensive, and sometimes even impossible. Thus, developing tools, strategies, and policies to build resilience before disaster strikes is essential.

The Resilience Alliance has led the way in developing a broadly interdisciplinary research agenda that integrates the ecological and social sciences , along with complex systems thinking to help understand the conditions that create resilience in socio-ecological systems. Through consideration of diverse forms of knowledge, participatory approaches, and adaptive management, in addition to systems thinking, the Resilience Alliance integrates multiple social learning 'strands' (Dyball *et al.* 2007). Although the resilience work has not focused on cities, its approach is consistent with a call by the Urban Security group at the U.S. Los Alamos National Laboratory for "an approach (to studying urban ecosystems) that integrates physical processes, economic and social factors, and nonlinear feedback across a broad range of scales and disparate process phenomena" (Urban Security 1999).

Social-ecological systems exhibit three characteristics related to resilience: (1) the amount of change the system can undergo and still retain the same controls on function and structure, (2) the degree to which the system is capable of self-organization, and (3) the ability to build and increase the capacity for learning and adaptation (Resilience Alliance 2006). Diversity is fundamental to retaining functional and structural controls in the face of disturbance. Biological diversity provides functional redundancy, so that if one species declines (e.g. a nitrogen-fixing species), other species providing the same ecosystem services will continue to function Similarly, when diverse groups of stakeholders, including resource users from different ethnic or religious groups, scientists, community members with local knowledge, NGOs, and government officials, share the management of a resource, decision-making may be better informed, stakeholders may be more invested in and supportive of the decisions, and more options exist for testing and evaluating policies.

Self-organization refers to the emergence of macro-scale patterns from smaller-scale rules, such as the emergence of ecosystem patterns related to nutrient cycling or plant size distributions as a result of evolution acting at the species level (Levin 2005), or the development of a market economy in laissez-faire political systems. Participation of local residents in managing their own resources also may be viewed as a form of self-organization and can lead to adaptive learning and eventually greater resilience (Olsson *et al.* 2004). For example, following a hurricane on the island of Montserrat, local people involved in rebuilding undertook development projects, such as building a community center and implementing new farming practices, which were not directly related to disaster recovery but were integral to longer-term resilience strategies (Vale and Campanella 2005). In another example, refugees living in camps in Somalia and Kenya learned new methods of growing food, which they took back to their communities following resettlement (Smit and Bailkey 2006).

The Montserrat and African cases provide examples of positive feedback loops, which are also critical to resilience theory. People acquired skills and new knowledge, and applied them to enhancing community development, food security, and the local environment. This, in turn, should create a system that is more resilient in the face of a new disturbance or disaster. One challenge for the development community is how to foster local leadership and action leading to positive feedback loops that lead to resilience. This is in contrast to some interventions that result in destructive, positive feedback loops, such as when following a conflict lack of meaningful employment opportunities for men leads to violence, which in turn leads to destruction of infrastructure and even fewer employment opportunities.

Building resilience through nurturing diversity, self-organization, adaptive learning, and constructive positive feedback loops is consistent with calls for a shift in disaster relief thinking from identifying what is missing in a crisis (needs, hazards, vulnerabilities) to identifying the strengths, skills, and resources that are already in place within communities (IFRC 2004). Such thinking parallels recent calls for asset-based approaches in international development, which emphasize building on existing natural, social, human, financial, and physical capital. However, tools and policies that are consistent with asset-based approaches to building resilience in cities are sorely lacking.

In this chapter, we argue that urban community greening and other 'civic ecology' approaches that integrate natural, human, social, financial, and physical capital in cities, and that encompass diversity, self-organization, and adaptive learning and management leading to positive feedback loops, have the potential to reduce risk from disaster in cities through helping communities to develop resilience before a disaster, and to demonstrate resilience after disaster strikes. We realize that an emphasis on community greening may be counterintuitive, given that many urban residents have unmet fundamental needs including sanitation, personal safety, and land tenure. However, we contend that some individuals and communities take it upon themselves to improve their environment even under the most difficult conditions, and that such action not only is part of resilience but should be incorporated into asset-based development and educational schemes.

In making our argument, we build on and add to existing literature on resilience and draw on our own experience with urban community greening. First, we apply resilience theory to *urban* socio-ecological systems, an important gap in a body of literature focusing largely on aquatic, agricultural, and marine systems. Second, we expand on Vale and Campanella's (2005) comparative analysis of resilience narratives from cities experiencing disasters, which focuses largely on the built rather than the *natural environment*, and on efforts led by government, the private sector, and outside NGOs, as opposed to *community-based* initiatives to build resilience . Perhaps more important, we propose an asset- and community-based *tool*, urban community greening, which can serve as the focus of future adaptive co-management, social learning, and research into resilience in cities. We show how urban community greening builds multiple forms of capital in ways that are distinctly different from other types of greening, and that contribute to diversity, self-organization, and adaptive learning and thus provide the conditions necessary for resilience in socio-ecological systems. Finally, we integrate resilience theory and urban community greening to propose a new 'civic ecology' framework in which to view urban community greening and other socio-ecological, participatory, asset-based approaches to building resilience in cities.

Urban community greening

Community-based efforts to create green spaces in cities, such as community and living memorial gardens and community forestry, are distinct from other types of greening, including green political movements or more formal 'pedigreed' landscapes such as city parks and botanic gardens (Hough 2004). An example of urban community greening comes from Soweto township near Johannesburg,

South Africa, where local residents, many of them immigrants from more rural areas, have taken it upon themselves to reclaim a hill that was overgrown and the scene of rampant sectarian violence during apartheid. Today the Soweto Mountain of Hope (Lindow 2004) is a vibrant garden and outdoor 'community center' incorporating protest sculptures, a women's kitchen and meeting circle, dance and drumming classes and concerts, and huts reflecting the building styles of diverse ethnic groups in South Africa. The Soweto Mountain of Hope also acts as a memorial to victims of AIDS; the garden is along a major thoroughfare leading to a large cemetery and a number of garden plots are planted in the shape of AIDS ribbons. Given Johannesburg's high crime rate and its designation by some as a city at risk of 'failing' (Norton 2003), the Soweto Mountain of Hope is an example of community-based resilience under conditions that commonly follow disaster or conflict. It also provides a test case for how such community-based efforts might enhance resilience in the face of future conflict.

Similar to what occurred in Soweto, the community garden movement in North America can be viewed as a community-based response to urban crime and decay. As city dwellers in New York and elsewhere experienced rising violence and abandonment by politicians in the 1970s, they refused to accept that they and their neighborhoods were the "troubling by-products of urban growth and decay...problems to be solved by politicians, city planners, and environmental professionals" (Anderson 2004). Instead, they took it upon themselves to transform crime- and trash-ridden vacant lots into urban landscapes that represented a new kind of nature incorporating ecological and cultural value. We contend that the active engagement of these community members, many of whom were low-income minorities and immigrants, helped to build stronger, more resilient neighborhoods prior to disaster, and that their efforts would be revisited following disaster. For example, after 9/11, many community gardens became living memorial gardens, whose purpose was to create an outlet for grief and a unifying, community-building demonstration of solidarity and support, all of which can contribute to resilience .

Thus, as opposed to more formal city parks, urban community greening refers to the leadership and active participation of city residents who take it upon themselves to build healthier sustainable communities through planning and caring for 'socio-ecological spaces' and the associated flora, fauna, and structures. Urban community greening encompasses community gardens where city dwellers share a gardening space, often by dividing it into individual family plots and common areas such as benches and casitas; memorial gardens created spontaneously by community members following disaster and conflict; trough gardens where individuals plant in troughs located throughout a city; gardening and tree planting along green areas created by transportation corridors such as railroads and highways; as well as sacred groves of trees and other forms of community forestry. It also encompasses urban agriculture (Smit and Nasr 1999), although the emphasis is more on building individual and social resilience than on food production per se. We contend that whereas greening in general enhances mental, physical, and community health, urban community greening builds natural, human, social, financial, and physical capital in unique ways with important implications for building resilience prior to and following a disaster or conflict.

Building resilience

Numerous studies have shown that the ability to see or experience green space can reduce domestic violence, quicken healing times and reduce stress, improve physical health, and bring about cognitive and psychological benefits for children and adults (Sullivan and Kuo 1996, Ulrich 1984, Hartig *et al.* 1991, Kaplan and Kaplan 1989, Taylor *et al.* 1998, Wells 2000). In addition to building human capital, green areas in apartment complexes have been demonstrated to build social capital through fostering a sense of safety and reducing crime rates in cities (Kuo and Sullivan 2001, Kuo *et al.* 1998). Furthermore, throughout the last century and continuing today, gardening also has been a means for soldiers and victims of war to fight back for their own mental well-being, as well as for the disenfranchised to become involved in acts of defiance. Gardens themselves represent resilience in that they "resist not only environmental difficulty but also social, psychological, political, or economic conditions" (Helphand 2006).

We can expect urban community greening at a minimum to foster the same sorts of resilience-building human and social capital as other types of green space. More important, urban community greening has been demonstrated to build additional forms of capital that relate directly to the diversity, self-organization, and adaptive learning characteristics of resilient societies.

Diversity and the ability to maintain function and structure in the face of change

In densely populated cities, community greening contributes to landscape heterogeneity, adding multiple, small-scale, distributed patches to the green spaces created by formal parks. Furthermore, urban community gardens are sites of biological diversity generally reflecting the cultural and ethnic diversity of the surrounding community. For example, in Sacramento, California, Mien refugee gardeners grow Asian varieties of squashes, eggplants, and beans; in New York City, Latin American gardeners plant alache, epazote, and papalo; and in Grahamstown, South Africa, community gardeners grow a diversity of 'imifino' or wild, edible greens. Whereas the biologically diversity found in community greening generally is not native, it potentially could foster ecological resilience, such as when planting little used varieties reduces risks from insect and disease. Furthermore, the genetic, species, and landscape diversity associated with small-scale agriculture gains importance when cities are viewed as socio-ecological systems. For example, the diversity of fresh produce gathered from community and school gardens in South Africa is seen as playing an important role in helping HIV/AIDS affected individuals maintain healthy immune systems, and thus contributes to individual resilience.

Community greening may also foster human diversity. In North America, South Africa, and Bosnia-Herzegovina, internally-displaced individuals and immigrants representing a diversity of ethnic groups can be found working together in community gardens. Furthermore, community gardens tend to be meeting places for people of all ages and sometimes from a range of economic status.

A question arises as to whether 'human' diversity, such as that represented in urban community greening, is critical to resilience, and if so, what types of diversity are important (e.g. ethnic, views about natural resources management, gender, age). Certainly, one can point to resilient cities in which cultural diversity was not a factor, including Tangshan China following the 1976 earthquake, Gernica following Franco's collusion with the Germans to bomb this Basque stronghold, and Tokyo following earthquakes, fires, and war (Vale and Campanella 2005). In these cases, either strong governments or private industry played a major role in rebuilding, often with the express purpose of setting a political agenda (such as demonstrating a more open economy following the death of Mao in China, or destroying Basque culture in Guernica). On the other hand, new immigrants have been instrumental in rebuilding North American cities after disaster, including Irish and German immigrants following the 1835 fire in New York City (Page, 2005), and Latin American immigrants following civil unrest in the 1990s in Los Angeles (Fulton, 2005) . And efforts to foster participatory natural resources management are built on the assumption that engaging diverse stakeholders in decision-making creates a larger portfolio of more equitable and better-informed land management policies. Research addressing the differences in past rebuilding efforts, and in the ability to rebuild following future disasters, among cities varying in the degree to which they incorporate a diversity of stakeholder perspectives and cultures could help shed light on the question of the importance of human diversity in resilience in urban systems.

Some development and disaster-relief efforts specifically use community greening to nurture diversity, reconciliation, and recovery among ethnic groups that have been engaged in war and conflict. For example, Jews and Palestinians plant trees together in Israel and Palestine to promote the human contact they believe is necessary for achieving peace (Serotta, no date). Through an American Friends Service Committee sponsored project in Bosnia-Herzegovina, people from different ethnic groups, including war veterans and widows, work side by side to grow food for themselves and their families (AFSC 2006).

Active participation and the capacity for self-organization

In community gardens and other community green spaces we have visited across North America, we hear the stories of individuals, often refugees, who have experienced serious trauma as a result of disaster, war, or civil strife, and who while perhaps unable to find or hold a job, are welcomed into a community garden where they are able to plant seeds, water, remove weeds, and otherwise work with the land to create food and beauty while regaining emotional stability. Similarly, in South Africa, poor township women engaged in gardening were able to find solace following domestic violence, gained greater control over their household food security and consumption, and experienced a greater sense of stability in coming to new, often transient communities (Slater 2001).

These examples of 'gardens as horticultural therapy' (Worden *et al.* 2004) demonstrate how community greening creates human capital, and we have seen in the section on diversity above how community greening fosters natural capital. Community greening also creates financial and physical capital, which, along with human and natural capital, leads to social capital. For example, in South Africa, community gardens often are designed as a means for unemployed community members to produce food and earn money, and North American gardens produce fresh food that is not otherwise available to families and elderly neighbors, and that is sometimes sold to create income for gardeners. Furthermore, through bringing in high-quality soil, constructing roof-top and other water collection systems, and building 'casitas' or sheds for social activities and cooking, gardeners contribute to the physical capital in cities. Community gardens also become a safe space where youth and adult neighbors come to socialize, participate in cultural events (e.g. concerts, harvest celebrations), relax, learn about gardening, exercise, and enjoy nature (Armstrong 2000, Hynes 1996, Patel 1991, Rees 1997, Saldivar-Tanaka and Krasny 2004, Schmelzkopf 1995). Unlike many other development efforts, which create a sense of dependency, through engaging community members in producing things of value, community greening can create independence and self-reliance (Gutman 1987) .

Because community gardens generally engage participants in multiple forms of communal activity and community action, they can serve as active training grounds for civic participation (Westphal 2003). For example, in many cities, community greeners organize to secure and defend a right to use land that more powerful city government and business interests would like to develop commercially. They also actively plan and manage what is grown and the activities that are allowed to occur at these sites. Such planning often entails working with people from diverse backgrounds to solve problems, such as how to sanction gardeners who do not follow rules about pesticides and weeding, or how to work with the city to provide a water system or more effective police protection. Through these activities, community greeners gain multiple competencies, ranging from how to grow food and proper nutrition to how to work in multicultural groups to advocate with city government (Hynes 1996, Pinderhughes 2001). They also create social networks, the ability to take an active role in controlling violence and other aspects of community life, and a sense of self-efficacy and empowerment (Slater 2001, Westphal 2003).

In most cases, community greeners themselves initiate the myriad of activities that occur in community green spaces, which in turn lead to increased human, social, and other forms of capital and enhanced food security. Viewed as a socio-ecological system, community gardens nurture constructive, positive feedback loops and are self-organizing, i.e. new system-level patterns emerge from the interactions of people and plants within the system, and these changes in the larger community in turn create greater opportunities for individual community members.

Capacity for learning and adaptation

In social systems, institutions and networks that foster learning and store knowledge and experience, create flexibility in problem solving, and balance power among interest groups play an important role in adaptive capacity (Berkes *et al.* 2000, Roling and Wagemakers 1998, Scheffer *et al.* 2000). Given that individuals engaged in urban community greening work, organize, and learn together, and often gain a sense of empowerment and self-efficacy that leads to action and advocacy, community greening can be viewed as an institution or network that contributes to social learning related to community development and food security.

Two scenarios we have observed in New York City provide examples of the role of community gardening and related community-supported agriculture (CSA) and farmers' markets in social learning. Brook Park Community Garden in the Bronx is the focus of multiple activities in the neighborhood. It includes vegetable plots and memorial flower plantings to commemorate victims of 9/11. A wealth of youth education activities occur on the site and an asphalt area that has not yet been converted to green space serves as a site for dance lessons. Canoes along the border fence attest to the garden's participation in a larger advocacy campaign to restore the nearby East River. At specified times each week, community members reflecting the ethnic diversity of the surrounding neighborhood drop by to pick up farm produce that is brought in from a rural CSA farm. The diversity of people and activities present in the garden provides a rich opportunity for sharing and learning. The garden itself can be viewed as an 'experiment' in managing for food security and community development in cities (BPCG 2006).

Compared to the richness of activities, structures, and land uses in Brook Park Community Garden, the farmers' market next to the fence surrounding the former site of the World Trade Centers, consisting of four long tables on a concrete walkway, may appear sterile at first glance. But viewed as a community initiative to bring back activity and life to the disaster site consisting of rubble, imposing signs extolling the recovery efforts, and grandiose plans for a new monument, the 'ground zero' farmers' market takes on new significance. The individuals who were engaged in the farmer's market prior to 9/11 watched the falling towers; today they see the market as the first step in creating the ecological, social, and cultural diversity needed to bring back their community .

The American Community Gardening Association (2006) provides a network for learning from these and the thousands of other community greening programs across North America, but often greening efforts in poor communities do not have the resources to participate in its activities. In Africa, Asia, and Latin America, we can find numerous examples of urban and community agriculture involving multiple NGO and community partners, and the Resource Centres on Urban Agricultural and Food Security provides a network for learning from these efforts (RUAF 2006, Smith and Bailkey 2006). A need exists for greater networking to further leverage the social learning potential of these and the many other community greening initiatives internationally.

What's missing? Civic ecology, adaptive co-management, and social learning

Thus far, we have argued that urban community greening, through creating human, social, and other forms of capital, plays an important role in fostering diverse, self-organizing, and adaptive communities, i.e. communities that one would expect to demonstrate resilience in the face of disaster. We also have provided examples from Bosnia-Herzegovina, the Middle East, and New York City where community greening was used as an intervention strategy specifically to promote resilience following conflict or disaster. Other examples of the use of greening as an intervention following disaster include using raised beds to grow traditional foods in mobile home parks following Hurricane Katrina, and community agriculture projects implemented at refugee camps to address environmental, economic, and psychological damage following the 2005 tsunami in Sri Lanka, and after fighting in Somalia. Interestingly, through participating in agricultural training programs in camps, refugees may take home new and more varied agricultural techniques than they had before displacement, and thus foster adaptive learning more broadly (RUAF 2006).

What then remains to be done? We contend that the next step is for policy makers and researchers to work to formally integrate urban community greening into adaptive co-management strategies for building communities that are resilient prior to disaster, and able to recover after disaster . As part of this adaptive co-management strategy, we should seek to mobilize the cooperation and 'spontaneous leadership' that emerge through urban community greening to build networks that will participate in management and research decisions. Our recommendations build on the work of Weinstein and Tidball (2007), who suggest that policy makers, NGOs, and international agencies should seek to shape the environment by creating an enabling environment for development and growth, security, peace, stability, and societal healing through leveraging existing local skills, infrastructure and markets.

To guide these efforts, we propose an approach that builds on four factors identified as critical to natural resource management during periods of change and reorganization: (1) learning to live with change and uncertainty; (2) nurturing diversity for resilience; (3) combining different types of knowledge for learning; and (4) creating opportunity for self-organization towards social-ecological sustainability (Folke 2002). Our approach also expands on our ongoing work using community gardens as sites for community and youth sustainability education in cities, through which we have developed a program that integrates multi-cultural and intergenerational understanding, learning from community members and scientists, and civic action (Krasny *et al.* 2006). Combining these perspectives, we propose 'civic ecology' as an approach to natural resources management, education and empowerment, and community development. Civic ecology seeks to help people to organize, learn, and act in ways that increase their capacity to withstand, and where appropriate to grow from, change and uncertainty, through nurturing cultural and ecological diversity, through creating opportunities for civic participation or self-organization, and through fostering learning from different types of knowledge. In the context of this discussion, the ultimate goal of civic ecology is to build social-ecological resilience prior to and following disaster or conflict in cities . Note that education is an integral component of civic ecology, and that the type of learning that occurs through civic ecology education (Krasny and Tidball, 2006; Table 7.1, Figure 7.1) has many parallels to a definition of social learning that integrates negotiation, reflexivity, participation, and systems thinking as strategies to incorporate ecological complexity and the diverse experiences and knowledge of multiple stakeholders in addressing management issues (Dyball *et al.*, 2007).

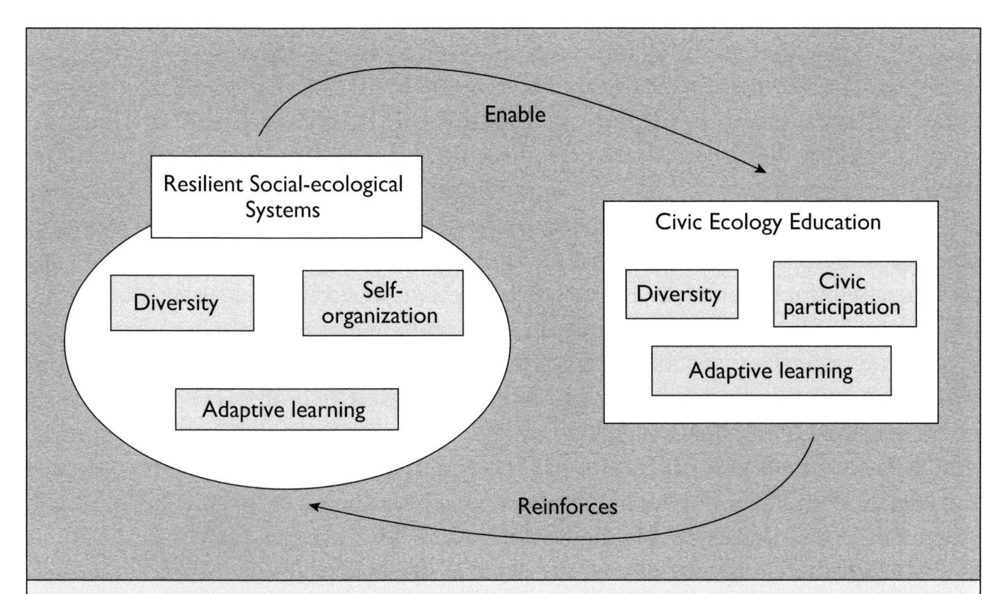

Figure 7.1. Civic Ecology Education draws from and helps people learn about the assets of social-ecological systems, including diversity, self-organization, and adaptive learning. The education efforts in turn foster diversity, participation, and adaptive learning, and thus reinforce existing resilience in social-ecological systems. See also Table 7.1.

Table 7.1. Resilience Attributes (Walker et al. 2001) and Civic Ecology Education.

Resilience attributes	**Civic ecology**
Diversity	Programs incorporate diverse forms of knowledge (western scientific and traditional)
	Youth and adult participants represent diverse cultures
	Programs take place at sites with biodiversity
Self-organization	Activities include local actions and advocacy to improve the community and environment
Adaptive learning	Within one program, educators and youth learn how to conduct better programs and how to improve the environment
	Across programs, educators share what they have learned

South Africa provides some good examples of multiple civic ecology approaches being incorporated into government and foreign donor policy, in particular through programs of the South African National Biodiversity Institute. For example, the Cape Flats Nature initiative employs urban township residents in leading biodiversity monitoring and management efforts, with the goal of preserving native plant communities and promoting ecologically- and socially-conscious tourism (Davis 2005). Another example is the Greening the Nation Programme, which seeks to create jobs, alleviate poverty, and build human capacity through engaging people in creating indigenous species nurseries and gardens at schools, street tree planting, greening of cemeteries, and other greening-related work (SANBI 2006). Similarly, through a joint Columbia University-UNESCO (2006) effort in Cape Town, a team of foreign and local specialists drawn from government and civil society are collaborating to create an urban biosphere reserve as a tool for socially inclusive and environmentally friendly forms of urban management. Although none of these projects is specifically described as building resilience, their integrated social equity and environmental objectives would indicate their potential in building a society able to bounce back from ongoing violence and conflict .

Conclusion

We have used a social-ecological systems framework to help understand the potential of urban community greening and other civic ecology approaches in building resilience and thus reducing risk in the face of disaster and conflict . Urban community greeners and other civic ecologists integrate place-based activities, such as planting community gardens or monitoring local biodiversity, with learning from multiple forms of knowledge including that of community members and outsiders, and with civic activism such as advocating for green spaces, for financial security, and for reduction of crime and violence. In so doing, they build human, social, natural, financial, and physical capital that becomes integrated into constructive, positive feedback loops. In this way, community greeners integrate diversity, self-organization, and learning to create the conditions that spawn resilience in the face of disaster and conflict.

Urban community greening, local biodiversity monitoring, and similar activities are tools that could become part of a larger civic ecology 'tool kit' for building urban resilience. Should relief and development NGOs, governments, international agencies, the scientific community, and community greeners work together to foster, implement, and assess the impact of civic ecology approaches as an adaptive co-management strategy before and after disaster, we will further our understanding of how to build resilience in urban socio-ecological systems. Such action and research conducted as part of networks of diverse stakeholders both embodies multiple definitions of social learning and can also draw from social learning theory. Ultimately, these research, adaptive co-management, and social learning efforts should be directed to helping policy makers understand the role of civic ecology tools in building resilience in cities both before and after a disaster or conflict.

References

AFSC (American Friends Service Committee) (2006) Website: http://www.afsc.org/europe/bosnia/programoverview.htm.

Anderson, K. (2004) "Marginal nature: An inquiry into the meaning of nature in the margins of the urban landscape", University of Texas Department of Geography Urban Issues Program, http://www.utexas.edu/academic/uip/research/docstuds/coll/anderson.html.

American Community Gardening Association (2006) Website: http://www.communitygarden.org/

Armstrong, D. (2000) "A survey of community gardens in upstate New York: Implications for health promotion and community development", *Health & Place* 6(4): 319-327.

Berke, P. and Campanella, T. (2006) "Planning for post-disaster resiliency", *Annals of the American Academy of Political and Social Science* 604: 192-207.

Berkes, F., Colding, J. and Folke, C. (2000) "Rediscovery of traditional ecological knowledge as adaptive management", *Ecological Applications* 10: 1251-1262.

BPCG (Brook Park Community Garden) (2006) Website: http://www.friendsofbrookpark.org/

Columbia University/UNESCO Joint Program on Biosphere and Society (2006) Website: http://www.earthinstitute.columbia.edu/cubes/sites/southafrica.html.

Davis, G. (2005) "Biodiversity conservation as a social bridge in the urban context: Cape Town's sense of "The Urban Imperative" to protect its biodiversity and empower its people", in Trzyna, T., ed., *The Urban Imperative*, Sacramento, California: California Institute of Public Affairs, 168 pp.

Desfor, G. and Keil, R. (2004) Nature and the City: Making Environmental Policy in Toronto and Los Angeles. Tucson, AZ: The University of Arizona Press, 274 pp.

Dyball, R., Brown, R.A. and Keen, M. (2007) "Towards sustainability: Five strands of social learning", in Wals, A.E.J., ed., *Social learning towards a sustainable world – principles, perspectives, and praxis*, Wageningen: Wageningen Academic Publishers, pp. 181-194.

Folke, C. (2002) "Entering adaptive management and resilience into the catchment approach", in *Balancing Human Security and Ecological Interests in a Catchment – Towards Upstream/Downstream Hydrosolidarity*. Stockholm International Water Institute, Report 17. Stockholm, Sweden, pp. 39-43

Fulton, W. (2005) "After the unrest: Ten years of rebuilding Los Angeles following the trauma of 1992", in Vale, L.J. and T.J. Campanella, eds., *The Resilient City: How Modern Cities Recover from Disaster*, USA, Oxford University Press, pp. 299-312.

Gruenewald, D.A. (2003) "The best of both worlds: a critical pedagogy of place", *Educational Researcher* 32(4): 3-12.

Gutman, P. (1987) *Nutrition and Urban Agriculture Food and Nutrition Bulletin* 9(2).

Hartig, T., Mang, M. and Evans, G. (1991) "Restorative effects of natural environment experiences", *Environment and Behavior* 23: 3-26.

Helphand, K. (2006) *Defiant Gardens: Making Gardens in Wartime*. San Antonio, Texas: Trinity University Press, 303 pp.

Hough, M. (2004) *Cities and Natural Process*. Routledge, London and New York.

Hynes, P. (1996) *A Patch of Eden: Americas Inner City Gardeners*. Chelsea Green Publishing Co. White River Junction, VT. 185 pp.

IFRC (International Federation of Red Cross and Red Crescent Societies) (2004) *World Disasters Report: Focus on Community Resilience*. International Federation of Red Cross and Red Crescent Societies, Geneva, Switzerland, 231 pp.

Kaplan, R. and Kaplan, S. (1989) *The Experience of Nature: A Psychological Perspective*. New York: Cambridge University, 360 pp.

Krasny, Marianne E. and Keith G. Tidball (2006) "Civic ecology education: A systems approach to resilience and learning in cities", *Environmental Education Research* (under review) . Available at : http://www.dnr.cornell.edu/mek2/file/Krasny_Tidball_Civic_Ecology_Education.pdf .

Krasny, M.E., Tidball, K.G., and Najarian, N. (2006) Garden Mosaics website (www.gardenmosaics.org).

Kuo, F.E. and Sullivan, W.C. (2001) "Environment and crime in the inner city: Does vegetation reduce crime?", *Environment and Behavior* 33: 343-367.

Kuo, F., Bacaicoa, M. and Sullivan, W.C. (1998) "Transforming inner-city landscapes: Trees, sense of safety and preference", *Environment and Behavior* 30(1): 28-59.

Levin, S. (2005) "Self-organization and the emergence of complexity in ecological systems", *BioScience* 55(12): 1075-1079.

Lindow, M. (2004) "From rubble to revival: A South African man turns a dump into a cultural mecca", Christian Science Monitor, http://www.csmonitor.com/2004/0226/p14s01-lihc.html.

Norton, R.J. (2003) "Feral Cities", *Naval War College Review* LVI (4): 98.

Oliver-Smith, A. (2002) "Theorizing disasters: Nature, power, and culture", in Hoffman, S. and A. Oliver-Smith, eds., *Catastrophe and Culture: The Anthropology of Disaster*, Oxford: James Currey LTD, pp. 23-47.

Olsson, P., Folke, C. and Berkes, F. (2004) "Adaptive co-management for building resilience in social-ecological systems", *Environmental Management* 34: 75-90.

Page, M. (2005) "The City's end: Past and present narratives of New York's destruction", in Vale, L.J. and T.J. Campanella, eds., *The Resilient City: How Modern Cities Recover from Disaster*, USA: Oxford University Press, pp. 75- 93.

Patel, I. (1991) "Gardening's socioeconomic impacts", Journal of Extension 29(4): http://www.joe.org/joe/1991winter/a1.html.

Pinderhughes, R. (2001) "From the ground up: The role of urban gardens and farms in low-income communities", *Environmental Assets and the Poor*, Ford Foundation.

Rees, W. (1997) "Why urban agriculture?" Urban Agriculture Notes. Vancouver, BC: City Farmer, http://www.cityfarmer.org/rees.html.

Resilience Alliance (2006) Website: http://www.resalliance.org/1.php

Roling, N. and Wagemakers, A. (1998) *Facilitating Sustainable Agriculture*, Cambridge, United Kingdom: Cambridge University Press, 318 pp.

RUAF (Resource Centres on Urban Agricultural and Food Security) (2006) Website: www.ruaf.org.

Saldivar-Tanaka, L. and Krasny, M. (2004) "The role of NYC Latino community gardens in community development, open space, and civic agriculture", *Agriculture and Human Values* 21: 399-412.

SANBI (South African National Biodiversity Institute) (2006) Website: http://www.sanbi.org/frames/educatfram.htm.

Scheffer, M., Brock, W.A. and Westley, F. (2000) "Mechanisms preventing optimum use of ecosystem services: an interdisciplinary theoretical analysis", *Ecosystems* 3: 451-471.

Schmelzkopf, K. (1995) "Urban community gardens as contested spaces", *Geographical Review* 85: 364-381.

Serotta (no date) "Replanting trees, rebirthing hopes for peace", Union for Reform Judaism, http://www.seekpeace.org/Articles/serotta.shtml

Slater, R.J. (2001) "Urban agriculture, gender and empowerment: an alternative view", *Development Southern Africa* 18(5): 636-650.

Smit, J. and Bailkey, M. (2006) "Urban agriculture and the building of communities", in Van Veenhuizen, R., ed., *Cities Farming for the Future – Urban Agriculture for Green and Productive Cities*, Philippines: RUAF Foundation, IDRC and IIRR, 458 pp.

Smit, J. and Nasr, J. (1999) "Agriculture: Urban agriculture for sustainable cities: Using Wastes and idle land and water bodies as resources, in: Satterthwaite, D., ed., *Sustainable Cities*, London: Earthscan Publications Ltd, pp. 221-233.

Sullivan, W. and Kuo, F. (1996) "Do trees strengthen urban communities, reduce domestic violence?", USDA Forest Service Southern Region, Technology Bulletin No. 4, Forestry Report R8-FR 55, Athens.

Taylor, A., Wiley, A., Kuo, F.E. and Sullivan, W.C. (1998) "Growing up in the inner city: Green spaces as places to grow", *Environment and Behavior* 30(1): 3-27.

Ulrich, R. (1984) "View through a window may influence recovery from surgery", Science 224: 420-421.

UN Centre for Human Settlements (1999) "Cities as solutions in an urbanizing world", in Satterthwaite, D., ed., *The Earthscan Reader in Sustainable Cities*. London: Earthscan Publications Ltd.

Urban Security, Los Alamos National Laboratory (1999) Annual Report, LA-UR-99-5554, http://eesftp.lanl.gov/EES5/Urban_Security/FY99/

Vale, L.J. and Campanella, T.J., eds. (2005) *The Resilient City: How Modern Cities Recover from Disaster*, USA: Oxford University Press, 392 pp.

Walker, B., Carpenter, S., Anderies, J., Abel, N., Cumming, G., Janssen, M., Lebel, L. Norberg, J., Peterson, G.D. and Pritchard, R. (2002) "Resilience management in social-ecological systems: a working hypothesis for a participatory approach", *Conservation Ecology* 6(1): 14.

Weinstein, E. and Tidball, K. (2007) "Environment shaping: an alternative approach to development and aid", *Journal of Intervention and Statebuilding* 1: Spring.

Wells, N. (2000) "At home with nature – Effects of greenness on children's cognitive functioning", *Environment and Behavior* 32: 775-795.

Westphal, L. (2003) "Urban greening and social benefits: a study of empowerment outcomes", *Journal of Arboriculture* 29(3): 137-147.

Worden E.C., Frohne, T.M. and Sullivan, J. (2004) "Horticultural Therapy", University of Florida, http://edis.ifas.ufl.edu/pdffiles/EP/EP14500.pdf#search='horticulture%20therapy%20war.

Chapter 8

Reaching into the holomovement: a Bohmian perspective on social learning for sustainability

David Selby

Residual mechanism in the sustainability concept

The mechanistic worldview and the epistemological disposition it has engendered, reductionism, are informed by the idea that the world is constituted of entities which are outside of each other. They exist independently in space and time, and, while they interact, they do so in ways that do not effect changes in their essential nature (Bohm in Nichol 2003, p. 81). Similarly any composite body is ontologically reducible to its separate constituent parts while its lifetime trajectory (from generation to disintegration) is reducible to the sum total of the motions and interactions of those parts, "the motion, bump and grind of implacable particles"(Callicott 1986, p. 303).

Originating in the world of physical sciences, the mechanistic worldview has penetrated the assumptions and perceptions of mainstream Western thinking and, in so doing, has significantly influenced every area of human endeavour and enquiry. Western economic, cultural and political hegemony has, in turn, given mechanism global reach. Holistic interpretations of reality, in consequence, have been pushed to the countercultural and indigenous margins, receiving at best tokenistic and rhetorical recognition from those at the core of power and influence. The tacit and ubiquitous embrace of mechanism has been "made only stronger by the further assumption *that we have no metaphysical assumptions* – that we are more or less seeing reality 'as it is'" (Nichol 2003, p. 5). The mechanistic worldview has such a sway that it is pervasively assumed to capture the nature of things. Quantum physicist David Bohm (1917-1992) suggests that we have become rather like fish in a tank into which a glass barrier has been placed. The fish become conditioned to keep away from the barrier and once the barrier is taken away the fish never cross the line where the barrier was positioned (Horgan 1993, p. 42).

Compounding the obdurate hold of mechanism is a world of vested interests, including those within an academic world professedly committed to untrammeled enquiry, which would be threatened by any significant retreat from the mechanistic paradigm. Mechanism, writes Richardson (1990, p. 54), is "institutionalized in all sorts of structures and career patterns, and very many people have a material interest in fragmentation and false dichotomies, and depend enormously for their emotional security and identity, and indeed for what they believe is their very sanity, on fragmentation and false dichotomies".

Bohm and other critics of the mechanistic worldview have argued that its omnipresence is at the roots of the global ecological crisis and parallel and connected crises in the social, economic, cultural and personal spheres within human society. Influenced by Rene Descartes' fundamental distinction between *res extensa* (mechanical extended entities occupying space) and *res cogitans* (things of the mind neither limited by nor occupying space), mechanism arrogates mind and free will to human beings, and leaves mind to explore dispassionately what is conceived of as a machine-like world from which it is disconnected (Bohm 1980, p. 271). In that disconnection, mind, encased within but separate from its physical cladding, is held to be capable of arriving at objective truth (Callicott 1986, p. 304). By locating ourselves outside and above nature (viewed as having no mind), we have come to consider nature in primarily instrumental and objectified terms thereby denying ethical and moral status to the other-than-human life forms and to environments in general and, hence, allowing ourselves unfettered license to exploit (Bateson 1973, Bohm and Edwards 1991, Capra 1983, 1996, Evernden 1985, Merchant 1981). By understanding self, family, nation and other categories demanding of allegiance as absolutes rather than as abstract functional categories created entirely out of human thought, and on that account ephemeral, we have created the conditions for direct and indirect violence, rampant global competitiveness (with consequent destruction of the biosphere and ethnosphere), and ethnic, racial and religious strife (Bohm and Edwards 1991, p. 4, Capra 1983, p. 28). By uncritically embracing a worldview predicated upon dualisms – such as mind/body, masculine/feminine, production/reproduction, reason/emotion, ordinary/extraordinary, knowledge/experience, culture/nature, us/them – we have created a "store of conceptual weapons" that can be "mined, refined and redeployed" for negative and oppressive purposes (Plumwood 1993, p. 43). By creating a hierarchy within ourselves (for instance, intellect above emotions and body), we have straitjacketed our potential to respond to the crises spawned by mechanism, our 'solutions' oftentimes flawed from the outset in that they approach a problem rooted in mechanism through the further application of mechanism. The United Nations is identified by Bohm as a case in point in that its history has been to respond to global problems by creating more and more specialist organs, each of which concentrates upon the outward effects of a specific problem rather than taking the path of holistic and deep attentiveness to the whole (Bohm and Edwards 1991, p. 15). By separating mind from body and nature, the mechanistic worldview has also fuelled the hubris and dysfunction of human uniqueness, an underside of which is our modern sense of alienation and existential crisis. "We are distinct from everything around us and inexorably alone" (Zohar 1990, p. 34).

Suffering the illusion of experiencing ourselves as "isolated egos in this world" (Capra 1983, p. 29), we have cut ourselves off "from outer confirmation of our inner life" (Zohar 1990, p 217).

Sustainable Development and its educational manifestation, Education for Sustainable Development [ESD], both responses to the urgent global condition, can themselves be construed as adaptations that remain within the prevailing paradigm, the "crisis of perception" (Capra 1996, p. 4) that is mechanism. According to Bohm, there is the habituated tendency to perceive new experience and revealed phenomena from within the security long established by tacit processes and infrastructures of thought. The notion of sustainability seems for many to be in essence about maintaining security and a sense of normalcy and manageability in the face of potential ecological and social breakdown. Hence, there is an implicit acceptance of the growth principle in most renditions of sustainable development, there is a largely unchallenged anthropocentrism amongst sustainability propositions viewing nature as resource or 'natural capital', and there is an overarching managerial and technocratic thrust and tone to change proposals (Selby 2006). "The problem arises when the source of security is built upon sand. In such cases limited aspects of order are relevated (i.e. consciously lifted[21]) in importance to the point where they are taken as the touchstone of truth and the basis of belief" (Bohm and Peat 2000, p. 291). Faced with the challenge of new urgencies, there is a tendency of mind to "put new wine in old bottles" (*ibid.*, p. 22). "Unless the perceived rewards are very great, the mind will not willingly explore its unconscious infrastructure of ideas but will prefer to continue in more familiar ways" (*ibid.*, p. 23). In our reluctance to be attentive to, and hence begin to dismantle, habituated and essentially mechanistic ways of thinking and perceiving, we fail to engage in diverse and deep connection with the world around us as we go about enacting our unsustainable lives alongside our remedies for sustainability.

For Bohm, real creativity in response to the global condition calls for a reaching into the holomovement. Mechanistic thought stays at the level of what Bohm calls the *explicate order*, the world of separate, independent, stable and solid elements, between which there is linear and contiguous causality. Informing what is explicate is the deeper *implicate* or *generative order* of reality in which the totality of existence is enfolded (i.e. folded inwardly into) each region of time and space (Bohm 1980, p. 147-150). Bohm (*ibid.*, p. 48) uses the analogy of the stream

[21] Bohm (1980, p. 33) discusses the etymology of the word: "The word 'relevant' derives from the verb 'to relevate', which has dropped out of common usage, whose meaning is 'to lift' (as in 'elevate'). In essence, 'to relevate' means 'to lift into attention', so that the content thus lifted stands out 'in relief'."

to convey a sense of the fundamentally indefinable, non-measurable and deeply creative implicate order:

> On this stream, one may see an ever-changing pattern of vortices, ripples, waves, splashes, etc., which evidently have no independent existence as such. Rather, they are abstracted from the flowing movement, arising and vanishing in the total process of the flow.

Things at the explicate level of presence have no independent existence as such, but are temporary abstractions from the flowing movement. They unfold from an unbroken wholeness only to re-fold into the implicate order which at one level of presence they never leave (Selby 2001, p.9). The order and structure we discern is derived from the ground of the entire stream, although the influence of entities in their relative and transitory autonomy can and does feed back into the stream.

Another analogy employed by Bohm is that of the hologram. Unlike a normal photograph in which there is point-to-point correspondence between the object of the photograph and the image produced, a laser-produced hologram captures the code of the whole within its every part, a passably clear image of the whole being re-creatable from any part. A "*total order* is contained, in some *implicit* sense, in each region of time and space. ...in some sense each region contains a total structure 'enfolded' within it" (Bohm 1980, p. 145-147).

Thought as problem

If the global crisis and many of our 'solutions' are the outcome of reductive thinking and of actions consequent upon that thinking, and if reality is informed at a fundamental level by a generative implicate order or holomovement, can we change the way we perceive and address the former by reaching into the latter?

For Bohm thought is always an indissoluble mix of "the intellectual, emotional, sensuous, muscular and physical responses of memory" (*ibid.*, p. 50). Most thought, however, tends to be both mechanistic and mechanical in its operation. It has "developed in such a way that it has an intrinsic disposition to divide things up" (Bohm and Edwards 1991, p. 1). Compounding the problem, responses to data or stimuli draw repetitively from memory or, like new patterns in a kaleidoscope, involve new combinations and configurations possessing a surface novelty "resulting from the fortuitous interplay of elements of memory" but in which the "novelty is still essentially mechanical" (*ibid.*, p. 51). Within a self-referential process or closed loop, "there is no inherent reason why the thoughts that arise should be relevant or fitting to the actual situation that evokes them" (*ibid.*). The closed loop, moreover, leaves no space for attentiveness to the generative order.

Unless we address processes of thought, we cannot get to the root of the real problems we face "because the root is the overall process of thought itself" with its common underlying assumption "that thought is not limited" (*ibid.*, p. 17). In this regard Bohm was fond of recounting the Sufi story of the man who lost the key to his house:

> He was found to be looking for it under a light. He looked and looked and couldn't find it. Finally someone asked where he had lost the key. He answered, "Well, I did in fact lose it over there." And when asked why he didn't look for it over there, he said, "Well, it's dark over there, but there is light here for me to look." (*ibid.*)

For thought, an inherently limited and limiting medium, to extricate itself from a fixed, closed and often defensive loop, and see the territory rather than the map (Korzybski 1994), Bohm calls for the application of *intelligence*, the unconditioned and unprogrammed act of creative perception that, by definition, must draw energy and inspiration from the "undetermined and unknown flux, that is also the ground of all definable forms of matter" (Bohm 1980, p. 52). Social learning for sustainability thus involves reaching into the holomovement so that real creativity in response to crisis, liberated from habituation, adaptation and the limiting mental and psychological confines of disciplines, can be released. "If we don't do anything about thought, we won't get anywhere. We may momentarily relieve the population problem, the economic problem, and so on, but they will come back in another way" (Bohm and Edwards 1991, p. 25).

Dialogical social learning

From a Bohmian perspective, social learning for sustainability eschews debate and discussion. Interestingly, the etymology of the former derives from a lineage of Latin and Middle English words for doing battle while discussion is brother to 'percussion' and 'concussion,' each being concerned with shaking things and breaking things up (Bohm 1996, p. 6). Discussion as a learning process is thus rooted in the mechanistic fallacy of the motion, bump and grind of implacable particles. "It is like a ping-pong game, where people are battling ideas back and forth and the object of the game is to win or to get points for yourself". There is an essentially superficial quality to the engagement in that processes of thought, personal and collective assumptions, and perceptions and misperceptions of spoken interventions are not aired. It can be likened to addressing river pollution downstream while ignoring what is generating the pollution at source and, in so doing, introducing pollution of additional kind (Bohm 1998, p. 50). In the process, vulnerability to the radical proposition is avoided in an unspoken collusion between combatants that "all sorts of things are...held to be non-negotiable and not touchable" (Bohm 1996, p. 7).

The word 'dialogue' derives from the Greek *dialogos*, a composite of *logos*, 'the word,' and *dia* meaning 'through.' For Bohm, "the picture or image that this derivation suggests is of a *stream of meaning* flowing among and through us and between us" (*ibid.*, p. 6)

Dialogical social learning is, thus, about creating contexts, climates and personal and collective dispositions whereby a "flowing through" (Bohm 1998, p. 118) can occur, out of which radically new ways of seeing the world may emerge. Participants – ideally numbering about twenty so as to ensure a 'microculture' representing a range of sub-cultures but, for logistical reasons, not more than forty (Bohm 1996, p. 13) – engage in a free play of listening and sharing that draws upon intellectual, somatic and emotional sensibilities. There is empty space with no set agenda or purpose beyond communicating coherently in transparency and truth. Any perceived curriculum is emergent rather than planned, retrospective rather than anticipatory as anything but a framework. Finding resonance in Jiddu Krishnamurti's observation that "The cup has to be empty to hold something", Bohm writes: "As soon as we try to accomplish a useful purpose or goal, we will have an assumption behind it as to what is useful, and that assumption is going to limit us" (*ibid.*, p. 17).

At the heart of Bohmian dialogue is the idea of bringing attentiveness to bear on thought processes rather than on the thoughts themselves. In that way the abstraction, habituation and fragmentation of thought, spawning an ungrounded and mechanistic understanding of the world, can be challenged, so providing the opportunity and space for perceiving deeper implicate realities and, in turn, providing richer potential for personal and collective transformation. In this process thought as problem is exposed. "When we see a 'problem', whether pollution, carbon dioxide or whatever, we then say 'We have got to solve that problem.' But we are constantly *producing* that sort of problem – not just that particular problem, but that sort of problem – by the way we go on with our thought. ...The point is: thought produces results, but thought says it didn't do it. And that is a problem" (*ibid.*, p. 10). Thought, individual and collective, trapped within its self-referential processes, exonerates itself as a significant causal factor in unsustainability placing the blame beyond its responsibility, while falling easily into self-deception in that it defends accepted assumptions against evidence that they may be of suspect, diminishing or lost validity (*ibid.*, p. 11). For Bohm, proprioception of thought, the metacognitive ability to perceive one's own thought processes at work as well as their attendant repercussions for the planet, is a vital element in social learning for sustainability. "We could say", he asserts, "that all the problems of the human race are due to the fact that thought is not proprioceptive" (*ibid.*, p. 25). While physiological proprioception gives us immediate feedback on the activity of our bodies, and protects us from physical harm, we lack proprioception in terms of our thinking and, hence, fail to see that fragmented thought patterns and processes are fomenting unsustainability.

Applying attentiveness to thought and its processes of abstraction and fragmentation can, in Bohm's view, elevate thought to the level of 'participatory thought' in which discrete boundaries are sensed as porous, objects enjoy an underlying deep relationship or 'radical interconnectedness' (Selby 2001), and the flow of unbroken wholeness, inaccessible in literal terms, is sensed as informing the thought process. Here his thinking connects with a deep ecological perspective in which 'self', that is the individual human whose frontier is perceived as lying at the epidermis, co-exists alongside the extended or oceanic Self (*ibid.*, p. 10). Attentiveness involves a dynamic dialectic between the limited nature of the former and the unlimited extent of the latter (Bohm 1996, p. 84-95).

Following Bohm, participants in a dialogical social learning circle for sustainability – the circle expressive of symmetrical relationship, continuity and direct engagement with the whole community of learners – would individually and collectively commit to a range of things:

- Empathetic and alert listening in which each listener would make conscious efforts to be mindful of their refractive thought processes whereby others' ideas are selected, prioritized, aggrandized or belittled according to the degree of fit with the receiver's own *Weltanschauung*, participants being prepared to own to and discuss their listening difficulties in this regard.
- Attentiveness to their own emotional and somatic responses to the interventions of others and readiness to share and explore those reactions, inviting the reflections and insights of others.
- Pooling perceptions of what they construe to be the misperceptions on the part of others of their own – and other's – interventions.
- Suspending assumptions and opinions in the sense of suspending them in front of the group; that is, flagging them to participants, neither suppressing them nor allowing them to inhibit participation in an emergent pool of common meaning.
- Abandoning the "impulse of necessity", the assumption that something is so absolutely necessary that there cannot be any yielding on the issue, and, hence, being prepared for "new orders of necessity", however provisional, to emerge from the flow of dialogue (*ibid.*, p. 21-23).
- Engaging in open, transparent and mutual collaboration in applying proprioception to thought, bringing into conscious awareness, and thereby

seeking to dissolve, conditioned fragmentation in its intellectual, psychological, emotional and somatic manifestations.
- Bringing what is tacit (implicate) in individual responses, what is vaguely felt and normally not articulated, out into the open within the dialogical process and exploring whether and to what extent its articulation resonates across the group.

Group enactment of such commitments, argues Bohm, will have the effect of enabling *thought* to be responsive to *intelligence* initially through the perception of similar differences and different similarities (Khattar 2001, p. 30-32) within known orders of reality and then, as old understandings, beliefs and assumptions in consequence break up, through creative engagement with potentially new orders of understanding, almost certainly more complex, but that cohere in emotionally and intellectually satisfying and motivating ways:

> In a creative act of perception, one first becomes aware (generally non-verbally) of a new set of relevant differences, and one begins to feel out or otherwise note a set of similarities, which do not come *merely* from past knowledge, either in the same field or a different field. This leads to a new order, which then gives rise to a hierarchy of new orders, that constitutes a set of new kinds of structure. The whole process tends to form harmonious and unified totalities, felt to be beautiful, as well as capable of moving those who understand them in a profoundly stirring way (Bohm 1998, p. 16)

In Bohm's view, a dialogical group would (and should) have a limited life but one of regular meetings. It should last as long as it takes for a collectively shared meaning to emerge – a meaning that can be powerful in its intensity, with consequent personal and collective change, and change agency, implications. He cautions against premature abandonment for reasons of discomfort or because the process becomes becalmed:

> The frustration will arise, the sense of chaos, the sense that it's not worth it. The emotional charge will come. ...It's going to happen that the deep assumptions will come to the surface, if we stick with it. ...Now, dialogue is not going to be always entertaining, nor is it doing anything visibly useful. So you may tend to drop it as soon as it gets difficult. But I suggest that it is very important to go on with it – to stay with it through the frustration (Bohm 1996, p. 19).

Going through frustration and lack of obvious direction towards shared meaning, he compares to moving from a state of ordinary, incoherent, light to the intensity and wave coherence of the laser beam (*ibid.*, p. 14), an analogy that will be returned to shortly. During the period of incoherence, of whatever duration, a group facilitator will be necessary to school participants in dialogical practice, bringing attentiveness to their conditioned inattentiveness (Katthar 2001, p. 149), help hone listening skills, help the group negotiate conflict, build intra-group empathy and bonding through shared storying and experience, and foster self-knowing through embodied learning modalities such as artful self enquiry, body/mind relational work, contemplative and therapeutic art (Selby 2001, p. 12). But as the shift to coherence is increasingly manifest s/he will take a commensurately lower profile and ultimately become more or less redundant. In a successful dialogical group, the facilitator morphs into participant.

There is correspondence between Bohm's conception of dialogue as a means of fomenting deep learning and personal and collective transformation by opening learners to the flow of the implicate order, and the notion of third-order learning. While first-order learning is adaptive, leaving basic values and assumptions unchallenged and unchanged, and second order learning takes us to the level of active reflection on thinking and learning processes, third order learning is profoundly concerned with embracing epistemic and paradigmatic challenges to the way we see the world with the conscious goal of transformation (Sterling 2001, p. 15). A dialogical learning circle for sustainability is, par excellence, and not least in its resistance to habituation on account of its intentional transience, a learning community or organization (Senge 1990). In this regard, it may well be useful to revive Alvin Toffler's proposal (1970, p. 425-440) for 'anticipatory democracy', the spawning of transient, grass roots 'social future assemblies' within communities which, under the broad heading of considering future directions and anticipating alternative futures, and open to anyone to join, might become dynamical nodal points of learning and change for sustainability. It is not inconceivable to bring Bohmian dialogical processes to such assemblies.

Social learning and the wave/particle duality

Bohm, like other quantum physicists, was much engaged with the so-called wave/particle duality. Sub-atomic entities, he and others found, manifest themselves simultaneously as both particles and waves, so that a full description of any entity calls for a description of both particle and wave functions and their complementarity. That full description is ever elusive, however, in that scientists cannot design an experiment enabling them to see both aspects of the duality at one and the same time. We identify and measure waves or particles while the whole picture, writes Danah Zohar (1990, p. 11), "remains indeterminate, somewhat ghostly, and just beyond our grasp".

Dialogical learning, as understood by Bohm, seems to bring insights from the micro world of sub-atomic relations to the macro world of learning for transformation. His view of the social learner portrays a dynamic and essentially fuzzy complementarity between their particle (individual) and wave (relational) aspects, between endogenous and ecological learning.

Engagement in learning with transformative intent necessarily involves an introspective process in which an individual confronts their ideas, understandings, assumptions and windows on the world as well as their emotional and somatic responses to external stimuli. For transformation to occur, however, the inner learning journey has to happen in dialectical relationship with an outer journey that happens alongside other learners and is set within the context of an immediate and wider learning environment. Bohm's proposals for dialogical social learning are articulated in response to this insight, and within an understanding that attentiveness to the implicate or generative order, breaking loose from the shackles of habituation while releasing the creative impulse, happens both within the individual and within the learning group and at its transformative best when there is a confluence of inner and relational (particle and wave) learning dimensions.

Bohm spent prolonged periods in empathetic dialogue with Jiddu Krishnamurti (Krishnamurti and Bohm 1986, 1999) and was much influenced by his friend. For Krishnamurti, individuals are conditioned into dependencies on various forms of authority – parents, wider family, school, society – to the degree that self becomes a closed and internally referential loop:

> Most of us are not creative; we are repetitive machines, mere gramaphone records playing over and over again certain songs of experience, certain conclusions and memories, whether our own or those of another (1954, p. 48).

The conditioned self, comfortable and secure as it feels, is denied full self-awareness, hence freedom and creativity, by the sway authority has had in its formative development (*ibid.*). For Krishnamurti, liberation from authority and liberation of self is one and the same thing and liberation *of* self initially calls for liberation *from* self – that is from the accumulation of habituated modes of encountering new experience. Such liberation is achieved through a periodic but momentary emptying or suspension of thought, a distancing from the unending pursuit of purposefulness, combined with a passive attentiveness to the empty self. "We *must* be lost before we can discover anything", he observes (1953, p. 124). "Discovery is the beginning of creativeness". Or, as he puts it elsewhere (1997, p. 131): "Whatever the mind creates is in a circle, within the field of self. When the mind is noncreating, there is creation."

The process of attentiveness to self as proposed by Krishnamurti coincides with Bohm's urging that intelligence should illumine thought. Both proposals speak to an ending of fragmentation of thought and fragmentation of self and world. Both connote endogenous as well as ecological learning. Both embrace a recognition that inner learning and learning in relation to others may be usefully distinguished at particular moments in the learning process, but also that they are ultimately inseparable. Bohm particularly emphasizes the transformative learning potential arising from a group whose individual wave functions increasingly coincide as the result of an ongoing dialogical process. "The light waves build up strength because they are all going in the same direction. This beam can do all sorts of things that ordinary light cannot" (Bohm 1996, p. 14). In this regard it is interesting that Bohm identifies communication skills as essential to dialogue – interpreting 'communication' as "to make something common" (*ibid.*, p. 2) – even though much that he says about wave synchronicity and synergy is rather about skills and processes of communion. Within the unfolding of a dialogical social learning process, as conceived by Bohm, a turning point or 'sea change' will happen within the group sooner or later during which communication transmutes into communion.

Far-from-equilibrium social learning

Alongside Bohm and Krishnamurti, Ilya Prigogine holds that the impulse to creativity is a constant amongst entities that have an unblocked and uncluttered connection to the flow of life. For him, a significant block is the hegemonic determinism that informs mainstream understandings of nature. He asks us (1989, p. 396) to think of a pendulum. If we agitate the pendulum, we can predict that it will move inexorably towards minimal then no swing with its centre of gravity as low as possible. We can be certain what will happen. But what, he asks, if we turn the pendulum on its head? It is difficult to predict what will follow. Fluctuating forces may make it fall to left or right, become entangled or even break. The notion of the upturned pendulum, Prigogine avers, has been "ideologically suppressed" (ibid) in that its message of instability is inconvenient for a culture that seeks to dominate and exploit nature.

> In a deterministic world nature is controllable, it is an inert object susceptible to our will. If nature contains instability as an essential element, we must respect it, for we cannot predict what may happen (*ibid.*, p. 397).

There are serious implications for 'stasis' interpretations of sustainability here, just as there are serious implications for sustainability-related social learning in Prigogine's concept of 'dissipative structures' within self-organizing systems.

Prigogine distinguishes between systems at equilibrium or near-equilibrium where huge disturbances would be required to effect radical change, and hence where creativity is low, and far-from-equilibrium systems. In the case of the latter, a fluctuation can induce movement into disequilibrium – dissipation – at which point the system responds by bringing to bear on the situation as wide and coherent a range of forces as is necessary to effect a new level of (complexified) equilibrium. It is at the far-from-equilibrium, then, that deep creativity, enhanced by strange non-linear, non-contiguous and unpredictable causal relationships, flourishes (Capra 1996, p. 180-183).

In terms of social learning for sustainability, questions follow concerning the ethics, nature, viability and potential of learning communities as dissipative structures. If our intent is transformation, should facilitators of learning purposefully seek to tilt learning experiences towards the disequilibrium that will effect radical changes in relationships and perceptions leading to new, more complex, configurations and equilibrium between the learners? Should they do so without or only with the permission of the learners? How would they create dissipative structures in the learning community? Would a cocktail of modalities of paradigmatic intellectual challenge, experiential learning, systemic learning, embodied, emotional and somatic learning (Selby 2001, p. 12) suffice? Do Bohmian dialogical processes inherently carry disequilibrium potential within them? Would formal learning institutions countenance their classrooms, and their institution taken as a whole, as dissipative structures? For Prigogine, disequilibrium within and across the radically interconnected levels of the individual, the collective and the systemic is a *sine qua non* of holistic, systemic and transformative perception:

> Coherence far from the state of equilibrium acquires huge dimensions in comparison with what happens in a state of equilibrium. In equilibrium each molecule can only see its immediate neighbour. Out of equilibrium the system can see the totality of the system. One could almost say that matter in equilibrium is blind, and out of equilibrium starts to see (Prigogine 1989, p. 399).

The powerfully emergent, multi-dimensional and globally ubiquitous phenomenon that is the sustainability movement is interpretable as the harbinger of a dissipative structure in response to a global condition tilting over the edge.

Virtual transitions

Within the sub-atomic world 'quantum leaps' occur when an electron suddenly and without apparent reason moves into higher or lower energy orbits around the nucleus. Before a 'leap' happens (a 'real' transition), an electron, as it were, smears itself everywhere simultaneously 'feeling out' all possible states, directions and journeys ('virtual' transitions). The wave aspect, thus, enables a "spreading out across the boundaries of space, time, choice and identity" (Zohar 1994, p. 111).

Social learning for sustainability, transformative in its intent, is suggestive of learning contexts and learning processes that offer fertile potential for virtual transitions on the part of the learner, virtual transitions which, rather like grit in the oyster, can be a prelude to real transition by eliciting a sense of discomfort and unease combined with an intimation of what is thinkable and what is realizable. The Bohmian dialogical process itself presents manifold opportunities for participants to 'feel out' and empathise with different selves, different worldviews and, not least, different ways of relating to others and the planet.

Opportunities for virtual transitions that can be incorporated into dialogical processes as opportunities present themselves are many and varied. Alistair Martin-Smith has written extensively (1993, 1995) on the use of storying and role play within 'quantum drama' contexts as means of enabling learners to feel inside different perspectives and different realities. Techniques such as role swapping and the use of an alter ego shadow during a role play can be very helpful in this respect. Imaginal exercises such as guided fantasies and visualisations also offer useful modalities for transitioning learners into different worlds (Pike and Selby 1988, p. 184-193), as do real or simulated cross-cultural learning experiences with their potential for culture shock (Batchhelder and Warner 1977). Futures-oriented activities through which learners envision alternative future scenarios on a spectrum from the utopian to the dystopian can impact deeply on learners' perceptions (Pike and Selby 1999, p. 217-247).

Quantum learning for quantum leaps

The leitmotiv of this chapter is that social learning for sustainability calls for learning processes divested of mechanistic influences. The mechanistic worldview lies behind the global mega-crisis while efforts to realize a sustainable world are themselves hampered by our inability to remove residues of mechanism from our sustainability proposals in which project we are straitjacketed by our failure to see, let alone address, mechanism within our processes of thought.

What is proposed is dialogical social learning based upon, but extending, David Bohm's conception of dialogue, a process intended to enable communities of learners to see beyond the world as it immediately and outwardly seems, and to experience and draw upon the dynamic flow of the whole, what Bohm refers to as the implicate or generative order of reality. As they experience dialogue, learners will experience endogenous and ecological learning as they successively and oftentimes seemingly simultaneously focus upon the particle and wave aspects of their identity. The place of far-from-equilibrium learning in which the learning community, and even the individual, becomes a dissipative structure heralding transformation is considered as is the potentially dissipative and transformative effects of virtual transitions on the part of individual learners.

All in all, what is being proposed is quantum learning for sustainability. Amongst sustainability proponents there is broad agreement that the world as we know it as well as our place in it are at risk, the argument being about where along a continuum between redeemability and irredeemability we presently stand. There is common agreement amongst proponents that transformation, and hence transformative learning, are vital if we are to achieve sustainability. The problem is that, while we have a goal of transformation through learning, the processes and modalities of learning we employ carry more than a residue of what has fanned the flames of unsustainability. If we are to effect a quantum leap towards sustainability, we need quantum learning.

Dialogue is "a quantum process, a means of doing and using quantum thinking" (Zohar 1997, p. 136). It "can be a crucial infrastructure for any thinking or learning organization" (*ibid.*). It can also be the key to bringing participatory democracy to formal, non-formal and informal learning communities and to society at large. "Imagine", writes Jane Vella (2002, p. 82), "a society where teaching as dialogue is the norm. Consider the possibilities for inclusion in decision making, program design, and collaborative work. This applies to family, university, and corporate organizations. This is a move toward a more honest and comprehensive democracy. The educational practices of a time are a clear and efficient mirror of the time. Quantum learning moves us toward a quantum society in which no one is excluded."

References

Batchelder, D. and Warner, E.G.(1977), eds., *Beyond Experience. The Experiential Approach to Cross-cultural Education,* Brattleboro VT.: Experiment Press.

Bateson, G. (1973) *Steps Towards an Ecology of Mind,* London: Granada.

Bohm, D. (1980) *Wholeness and the Implicate Order,* London: Ark.

Bohm, D. (1996) [Ed: L. Nichol] *On Dialogue,* London: Routledge.

Bohm, D. (1998) [Ed: L. Nichol] *On Creativity,* London: Routledge.

Bohm, D. and Edwards, M. (1991) *Changing Consciousness. Exploring the Hidden Source of the Social, Political and Environmental Crises Facing Our World,* San Francisco CA.: Harper.

Bohm, D. and Peat, F.D. (2000) *Science, Order and Creativity,* Toronto: Bantam.

Callicott, J.B. (1986) "The Metaphysical Implications of Ecology", *Environmental Ethics,* 8 (Winter): 301-315.

Capra, F. (1983) *The Turning Point. Science, Society and the Rising Culture,* London: Flamingo.

Capra, F. (1996) *The Web of Life: A New Scientific Understanding of Living Systems,* New York: Anchor/Doubleday.

Evernden, N. (1985) *The Natural Alien. Humankind and Environment,* Toronto:University of Toronto Press.

Horgan, J. (1993) "Last Words of a Quantum Heretic", *New Scientist,* 27 February: 38-42.

Khattar, R. (2001) "Creativity as the Impulse of Life: Scholarly Philosophies and Thoughts for Education", unpublished Master of Education thesis, Graduate Programme in Education, York University, Toronto, Canada.

Korzybski, A.O. (1994) [1933] *Science and Sanity. An Introduction to Non-Aristotelian Systems and General Semantics,* Englewood NJ.: Institute of General Semantics.

Krishnamurti, J. (1953) *Education and the Significance of Life,* New York: Harper.

Krishnamurti, J. (1954) *The First and Last Freedom,* New York: Harper & Row.

Krishnamurti, J. (1997) *Reflections on Self,* Chicago: Open Court.

Krishnamurti, J. and Bohm, D. (1999) *The Limits of Thought. Discussions,* London: Routledge.

Krishnamurti, J. and Bohm, D. (1986) *The Future of Humanity. A Conversation,* San Francisco CA.: Harper Row.

Martin-Smith, A. (1993) "Drama as Transformation: A Conceptual Change in the Teaching of Drama in Education", unpublished Ph.D. dissertation, University of Toronto, Toronto, Canada.

Martin-Smith, A. (1995) "Quantum Drama: Transforming Consciousness Through Narrative and Roleplay", *The Journal of Educational Thought,* 29 (1): 34-44.

Merchant, C. (1981) *The Death of Nature. Women, Ecology and the Scientific Revolution,* San Francisco, CA.: Harper & Row.

Nichol, L., ed, (2003) *The Essential David Bohm,* London: Routledge.

Pike, G. and Selby, D.E. (1988) *Global Teacher, Global Learner,* London: Hodder

Pike, G. and Selby, D.E. (1999) *In the Global Classroom 1,* Toronto: Pippin.

Plumwood, V. (1993) *Feminism and the Mastery of Nature,* New York: Routledge.

Prigogine, I. (1989) "The Philosophy of Instability", *Futures,* 21 (4): 396-400.

Richardson, R. (1990) *Daring to be a Teacher,* Stoke-on-Trent: Trentham.

Selby, D.E. (2001) "The Signature of the Whole. Radical Interconnectedness and its Implications for Global and Environmental Education", *Encounter. Education for Meaning and Social Justice,* 14 (4): 5-16.

Selby, D.E. (2006) "The Firm and Shaky Ground of Education for Sustainable Development", in J. Sattherwaite, L. Roberts and W. Martin, eds, *Discourse, Resistance and Identity Formation,* Stoke-on-Trent: Trentham.

Senge, P.M. (1990) *The Fifth Discipline. The Art and Practice of the Learning Organization,* London: Century Business.

Sterling, S. (2001) *Sustainable Education. Re-visioning Learning and Change,* Totnes: Green Books.

Toffler, A. (1971) *Future Shock,* London: Pan.

Vella, J. (2002) "Quantum Learning: Teaching as Dialogue," *New Directions for Adult and Continuing Education,* 93 (Spring): 73-83.

Zohar, D. (1990) *The Quantum Self. A Revolutionary View of Human Nature and Consciousness Rooted in the New Physics,* London: Bloomsbury.
Zohar, D. (1994) *The Quantum Society. Mind, Physics and a New Social Vision,* New York: Quill/ William Morrow.
Zohar, D. (1997). *Re-wiring the Corporate Brain. Using the New Science to Rethink How We Structure and Lead Organizations,* San Francisco CA.: Berrett-Koehler.

Chapter 9

Towards sustainability: five strands of social learning

Robert Dyball, Valerie A. Brown and Meg Keen

Introduction

Enhancing the sustainability of any human managed environment inevitably involves a process of social learning, yet the nature of environmental problems presents particular obstacles to achieving that learning. Environmental problems invariably demand co-operation between a number of different groups operating at a number of different levels, including individual, community, specialist, and government. Although each group is in a position to make important contributions to the resolution of the problem, their slightly different backgrounds and relation to the problem results in them constructing the problem in different ways. Environmental problems are multi-dimensional and cannot be fully grasped using current analytical frameworks. Typically, environmental problems are what Rittel and Webber call 'wicked problems'. Our processes of conceptualizing wicked problems determines what is seen as a problem, and thus what kinds of actions might lead to solutions (Rittel and Webber 1973, p. 161).

To deal with this level of complexity, environmental management strategies that are intended to improve the situation must be able to collectively learn from their successes and failures, so that over time they can be redesigned, improved and refined. If we are to do this successfully we need to acknowledge and negotiate issues that arise from the innate complexities of environmental problems. Not only must we acknowledge that different actors and groups see common problems from different perspectives, but we must recognize that these actors have different social power bases. These power relations change with contexts. A particular farmer, for example, may exercise a great deal of power and authority in a local community context, but be disempowered and marginalized in the context of a board-room meeting in the offices of a government agricultural or conservation agency.

Even where broad agreement can be reached about the kinds of environmental changes that are occurring in a particular situation, the value judgments that different actors place on that change often differs widely. Different parties to a proposal may broadly agree on what changes will likely occur if a wetland is allowed to be developed into a marina, whilst disagreeing vehemently on whether that is a good thing or not. When coupled with dimensions of social power, the values and ethics of one group can regularly dominate the outcome of decision making processes.

Retreating from the challenges that environmental problems pose for social learning by adopting 'individualist' approaches is no solution. The history of government and other institutional intervention into environmental problems is littered with fragmented and often contrary agendas, including within the same organization. Attempts to 'import' solutions to a local problem from some remote situation are common, despite the frequent failure of those solutions to work in the environment upon which they are imposed. Failure to develop coherent learning approaches to incremental change within institutions results in what Dovers calls 'policy amnesia' (Dovers 2001).

These kinds of challenges present some of the key issues affecting social learning for sustainability. In November 2003 the authors of this chapter coordinated a workshop to explore the different ways that different environmental practitioners learnt from their different experiences of these challenges. The workshop drew together practitioners from various environmental management domains, including those with backgrounds in government and policy making, higher education, science and the community. We were particularly interested in drawing lessons from adaptive approaches to environmental management, aimed at creating social learning partnerships, by building platforms that support sustainability across multiple scales. Using an open social learning structure that applied these principles in practice, the workshop sought to bring together the workshop participants' diverse experiences in incorporating sustainability into environmental management. The workshop dialogues identified five 'strands' to successful social learning. These strands, with examples of their application in practice, are discussed in full in *Social Learning in Environmental Management: Towards a Sustainable Future* (Keen *et al.* 2005), and are summarized in the balance of this chapter.

Five braided strands of social learning

The five braided strands of social learning distilled from workshop participants' collective experiences were seen as crucial to good environmental management in times of uncertainty and change (see Figure 9.1). They are braided in the sense that they interact and overlap, yet each has an important role on its own. We discuss each of the five strands (reflection, systems orientation, integration, negotiation and participation) in the subsections to follow, and then combine the strands to provide insights into the challenges ahead for environmental managers everywhere.

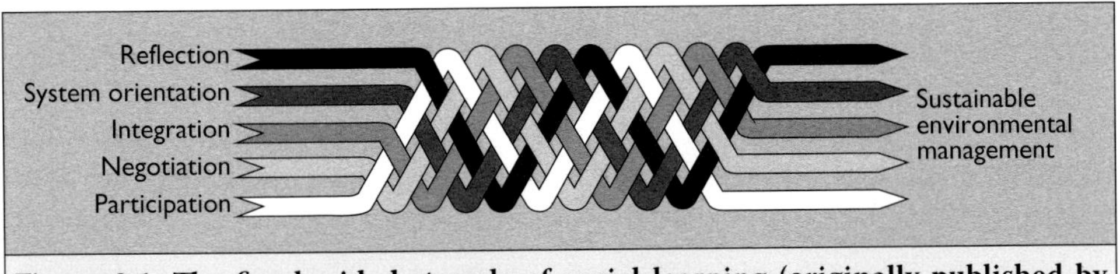

Figure 9.1. The five braided strands of social learning (originally published by Earthscan in Keen *et al.* 2005).

Strand one: reflection and reflexivity

Social learning is a process of iterative reflection that occurs when we share our experiences, ideas and environments with others. The importance of reflexivity – reflecting on the value of what we know and how we know it, leads to new understandings and is a crucial component of successful social learning. Drawing on organizational psychology and adult learning theory (Knowles *et al.* 1998, Kolb *et al.* 1995), the reflective learning process can be depicted as a learning cycle (see Figure 9.2). The cycle provides a framework for continuous reflection on our actions and ideas, and the relationships between our knowledge, behaviour and values.

The simple sequence follows the steps of diagnosing what matters, designing what could be, doing what we can and then developing a deeper understanding from reflecting on and evaluating that practical experience. Where you start in the cycle and the direction the learning takes depends on you as an individual, or your group's needs and goals. For the environmental manager, the cycle can be used as a planning process for bringing about change and stimulating transformative learning among all the participants.

Critical awareness and reflective processes, such as the one depicted in Figure 9.2, are a part of daily activities. Schön (1983, 1987) proposes that the 'reflective practitioner' engages in a learning process that continually reviews models, theories and ideas applied to the context. In practice, these reflective processes are at the:

- personal level, through setting goals and critically monitoring processes and outcomes;
- interpersonal level, through briefing and debriefing within groups;
- community level, through creating a common vision, identifying priorities and setting performance indicators to be assessed;

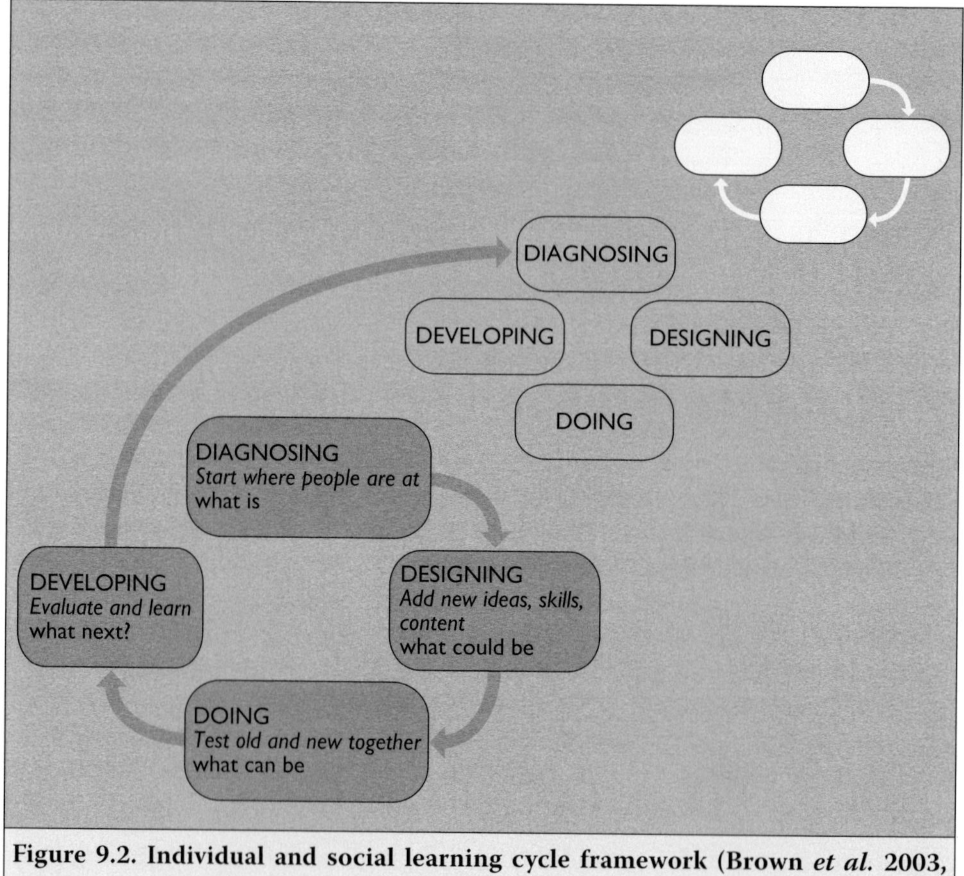

Figure 9.2. Individual and social learning cycle framework (Brown *et al.* 2003, Kolb *et al.* 1995), (this illustration originally published by Earthscan in Keen *et al.* 2005).

- social level, through evaluating and auditing the impacts of laws, regulations and markets.

These types of reflective learning processes form the foundation of a number of social learning approaches in use in environmental management. Examples include participatory rural appraisal (Chambers 1992, 1994, 1997), participatory learning and action (Bass *et al.* 1995, Pretty and Chambers 1994), participatory monitoring and evaluation (Estrella and Gaventa 1997) and adaptive management (Gunderson and Holling 2002, Holling 1978, Lee 1993).

Reflexivity in environmental management is an important lever for social change because it can reveal the ways in which theoretical, cultural, institutional and political contexts affect our learning processes, actions and values (Alvesson and Skoldberg 2000, Harris and Deane 2005). To reflect on ourselves and our practices, we need catalysts that can help us see what would otherwise be invisible to us. In some cases, this is achieved through monitoring and evaluation, for example adaptive management approaches (Dyball *et al.* 2005). In other cases, collaboration can provide a catalyst for recognizing difference, challenging us to consider new knowledge and insights, or to rethink our assumptions (Measham and Baker 2005, Keen and Mahanty 2005).

Strand two: systems orientation and systems thinking

Systems thinking offers a powerful way of understanding the dynamics of change in complex situations, typical of human interactions with their environments. Systems thinking is powerful because it uses a few simple descriptions that capture important generic change processes that characterize the system's behaviour.

A system is a mental abstraction made by an observer from a complex real situation. Systems are really 'systems of interest' in that they are "the product of distinguishing a system in a situation in which an individual or group has an interest or stake" (Open University 2005). What different people find 'interesting' and thus what systems they identify, differ from one group to another, and stems in part from how they find themselves in relation to the problem situation. Different individuals may also regard the changes to the value of the same variable as either desirable or undesirable or of no concern at all. Recognizing this subjectivity is an important step towards unearthing conflict. By clearly identifying what relevant 'parts' or variables, bounded across which dimensions of time and scale, each individual or group has selected, it becomes possible to identify points of disagreement.

Systems thinking is concerned with the state of the variables that comprise the system, and with the processes that account for the change in the value of the variables across a given period of time. An example of a variable might be 'toxin in a river'. If the value of this variable rises to a level at which it is of concern to someone, a systems approach would look to what other variables in the system were influencing that rise. This would then tend to favour solutions that addressed these relationships, such that the value 'level of toxin' returned to an acceptable state. Systems thinking favours solutions that are self-sustaining, in that they arise from the structure and properties of the system as a whole.

When striving to understand systems and our place in them, we are compelled to look for patterns rather than events and for processes rather than end points. Our understanding of system behaviour must be contingent on incremental, experiential learning and decision making, supported by active monitoring of, and feedback from, the effects and outcomes of decisions (Jiggins and Röling 2002). Where we assign a goal or a purpose to systems, we must again recognize that this purposefulness is the product of subjective human values and thus always open to ongoing re-validation and negotiation.

We also have to accept that surprise and change are endemic to the dynamics of many of the systems that concern us, and a system may change its fundamental behaviour quite suddenly (Holling and Gunderson 2002, see also Ison 2005, Dyball *et al.* 2005). A belief that complex systems can be manipulated with a high degree of certainty is simply a delusion. Often systemic change may be inevitable and the only appropriate response is adaptive change in the practice and expectation of environmental managers, decision makers and the public alike. In other words, the inherent behaviour of the systems that environmental managers seek to manage necessitates a commitment to ongoing social learning across diverse groups.

Strand three: integration and synthesis

The pursuit of sustainability in environmental management requires holistic and integrative frameworks from which to investigate the world, rather than ones that divide observations into a selected set of elements. Frameworks that represent the patterns linking people, roles and relationships, such as population flow charts, social mapping, professional relationships and informal networks, deal with forms of horizontal integration. Frameworks representing scales of governance and levels of management systems describe the avenues for vertical integration. Vertical, horizontal, place and issue-based integration are equally necessary in creating social learning processes. Integration is so central a concept in environmental management that it has become a *portmanteau* word, covering a range of very different processes. Under some circumstances, integration has become synonymous with processes and concepts as different as coordination, collaboration, cooperation, systems, synthesis, holism, unity and consensus. The goal is not a single consensus, nor the lowest common denominator, but a search for a rich tapestry that weaves together diverse ideas to reveal the nature of the complexity. Ison calls for transparency and coordination of traditions, noting that "traditions in a culture embed what has been judged to be useful practice. The risk for any culture is that a tradition can become a blind spot when it evolves into practice that lacks any avenue for critical reflection" (Ison 2005).

For integrated decision-making, the participants have to consider the traditions of understanding already established among the sets of contributors. Age cohort, gender, and expert groups have their own internally agreed interpretations of the way the world is. For instance, Australia is characterized by the advanced age of its farmers and the disappearance of the family farm. The generation taking over is more likely to be corporate managers of large holdings, with quite a different interpretation of their roles. The key decision-making groups in environmental management of any site, namely, the local community, the specialist advisors, and the influential organisations (including governments) each work from their own traditions. Communities have their own shared memories and first-hand experience. Specialists work from the particular ethical positions and skills instilled during their training. Organizations have internal loyalties and types of expertise determined by management (Brown and Pitcher 2005).

In a shared integrated understanding, remedial action will be necessary to reconcile these very different interpretations of the same reality. The value of reflectivity and systems thinking for this purpose has been discussed, but a generic approach is to apply Bohm's principles of dialogue (Bohm 1996). Designed to maximize learning through difference, Bohm's principles are to: commit yourself to the integrative process; listen and speak without making a judgment; identify your own and others assumptions; acknowledge and respect the other contributors and their ideas; recognize the difference between inquiry and advocacy; relax your need for any particular outcome; listen to yourself and speak for yourself when you need to; and to go with the flow.

Strand four: negotiation and collaboration

So far we have discussed the benefits of reflexive, systemic and integrative approaches to the social learning process. This could bring with it a mistaken idea that, under the right conditions, different communities, professions and agencies, with their associated values, knowledge and sets of skills, can come together easily and work seamlessly in environmental management. Nothing could be further from the truth. Negotiation is still needed at every interface within and between these elements of social learning. Each group has its own identity, created by defining a core area of interest and establishing boundaries that distinguish it from the others.

A constructive approach to negotiation assumes that conflict generates opportunities for learning. Competing opinions and evidence are to be welcomed as creating the conditions for generating new knowledge. Every stage of the social learning cycle requires participants to embrace dialogue that addresses conflicts over ideas, potential solutions and proper practice. Brown *et al.* (1995, p. 36) take a positive perspective of conflict management as follows:

- Conflict is an inevitable part of change – it is not a sign of failure of people or the system.
- Conflict is a step towards a solution – it is not the signal to give up.

- Conflict is shared – it is not the sole responsibility of any one person or group.
- Conflict is part of a process – it is not an outcome, a barrier or an excuse.
- Conflict is a matter for negotiation – it is not the end of the line.

Negotiation processes are actually built into the very fibre of society, with set terms for who consults with whom, under what conditions and according to what ground rules. Avenues that are taken for granted include voting, arbitration, commissions of inquiry, lobbying and regional development associations. At present the ground rules are shifting, since it is recognized that achieving sustainability will require the collaboration of all decision making sectors. Community consultation by researchers and government and community conferencing in the law, have become standard practice, although the objective of full collaboration is rarely met.

Social learning directed towards wicked problems involves collaboration between knowledge traditions – for example such as held by a scientist with expert knowledge of the problem, the local community that lives with direct experience of the problem, and a government organisation charged with funding and administration of the problem's solution. Each contributing group will need to recognize and respect the forms of evidence held by the other knowledge traditions. Science has long considered that peer review within scientific journals is all the validation its contribution requires. However, this excludes the contribution of the community, whose standard of local consensus might be born of year's of lived experience and which may supply more accurate and longstanding evidence than is available to external observation (Wynne 1996). The knowledge tradition of the government organisation is typically the result of their habit of consulting internally, and presenting the outcome to their collaborators as a *fait accompli.* The fifth strand of social learning is required in order to provide pathways that permit an overarching synthesis between these traditions.

Strand five: participation and engagement

Collaboration processes require that communities engage in learning partnerships, as much as learning requires collaboration. Typologies of participation indicate that when diverse social actors engage in environmental management activities, the outcome can range from coercion to co-learning (Arnstein 1969, Cornwall 1995, Parkes and Panelli 2001, Pretty 1995, Pretty and Chambers 1994) (see Table 9.1).

Participation typologies used in environmental management tend to break participation into discrete categories, rather than acknowledging that different forms of participation can contribute to social learning and a mix of approaches

Table 9.1. Types of participation. (Arnstein 1969, Cornwall 1995, Parkes and Panelli 2001, Pretty 1995).	
Type of participation	**Power relationships**
Coercing	The will of one group is effectively imposed upon the other
Informing	Information is transferred in a one-way flow
Consulting	Information is sought from different groups, but one group (often the government) decides on the best course of action
Enticing	Groups jointly consider priority issues, but one group maintains power by enticing other groups to act through incentives (such as grants)
Co-Creating	Participants share their knowledge to create new understandings and define roles and responsibilities, within existing institutional and social constraints
Co-acting	All participants set their own agendas and negotiate ways to carry it out collaboratively. Power tends to shift between participants depending on the actions negotiated

may be needed over the life of a project or program. In her review of Australian rangeland management programs, Kelly (2001, see also Kelly in Andrew and Robottom, 2005) found that landholders actually preferred different types of participation at different stages of the programs, depending on their learning and management objectives.

From a social learning perspective, the process of participation and engagement can be referred to as single-, double- and triple-loop learning (see Figure 9.3). Single-loop learning refers to developing skills, practices and actions. This is typically within a project team. Double-loop learning facilitates the examination of underlying assumptions and models driving the different actions and behaviour patterns. This is necessary where different knowledge traditions need to come together, as they do in nearly all environmental management issues. Triple-loop learning allows us to reflect on and change values and norms that are the foundation for our operating assumptions and actions. Participatory approaches consistent with multiple-loop learning thus provide a deeper understanding of the contexts, power dynamics and values affecting environmental management.

Social learning is, by definition, based on existing ethics and values about how the world should be. The five braided strands of social learning provide an integrated

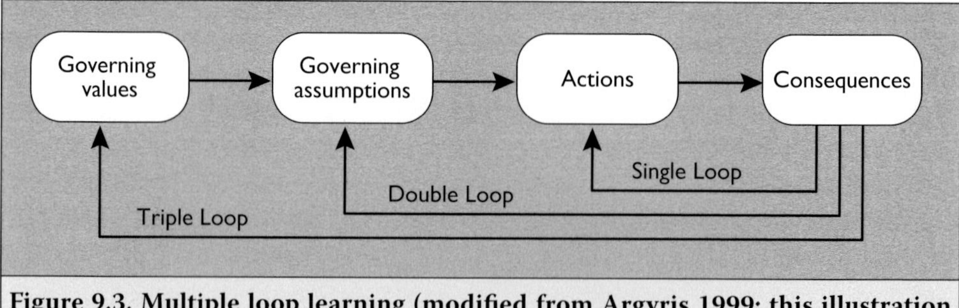

Figure 9.3. Multiple loop learning (modified from Argyris 1999; this illustration originally published by Earthscan in Keen *et al.* 2005).

set of processes that can guide social learning practice in areas of social change, such as sustainability goals. To encourage reflexivity, we suggest you ask yourself the following questions, in order to help to ground the theory in your personal and professional practice:

What are the social learning processes embedded in current environmental management policies and programs, and how do they relate to different ways of knowing and engaging?

- How can environmental management approaches facilitate the creation of learning opportunities that bridge different disciplines, subgroups within society and levels of governance?
- Do our present dialogues, negotiations and participation processes enable a wide variety of social learning opportunities in environmental management?
- How is our ability to act and adapt environmental management approaches affected by social structures and relationships?
- Are our processes of reflection and learning in environmental management fragmented and unable to discern the more subtle patterns of change over time and space?

Questions such as these are intended to help stimulate dialogues and critical reflections on social learning by unearthing some of the hidden assumptions, values and social structures that have long affected social learning in environmental management, but are not often discussed openly given the challenges of the 21st century.

Conclusions

Social learning in environmental management is essentially about managing change. Every environmental management practitioner is involved in larger questions such as: What is the purpose and direction of the change? How do we as a society create more equitable processes to share knowledge and engage in decision making that leads to a more sustainable environment, locally and globally? In this sense, everyone is an environmental manager, since we are all influencing and being affected by the answers. Changing social and organizational structures lead to the need to reflect critically on the cultures and values on which our decision making processes are based. Part of this critical reflection is accepting that there is not only *one* sustainability solution based on a single knowledge set.

Each environmental context will encompass different relationships between people and place. There is an array of possible sustainability pathways. These pathways will be affected by knowledge 'matrices', that is, the mix of understandings that are a product of the diverse experiences, values and principles of those in a particular place. Social learning processes allow us to better share our understandings and to negotiate social change in a way that takes account of a diverse range of worldviews. The more we build up our knowledge matrix through shared understandings, the greater the insights we can gain.

The management of issues for sustainability requires the integration of our thinking across disciplines, sectors and knowledge groups. It is not about one way of knowing or one way of doing. Sustainability is about relationships, dependencies and networks that can facilitate such integration in environmental management. Ultimately this systems orientation is intended to lead to greater equality between social groups, as well as a holistic approach to decision making that affects social and ecological systems. Core principles of a social learning approach that have emerged from our work are described in Box 9.1.

Social learning is about development, but it must also allow a collective 'letting go' of ideas, practices and values that no longer contribute to a sustainable future. The learning process is essentially social, because sound environmental management requires us to link our personal and local behaviours to outcomes at broader scales. This vertical and horizontal integration of ideas and practices helps us to gain a deeper understanding of different traditions of knowing. This, in turn, can help shift our focus from constraints and artificial jurisdictional and disciplinary boundaries to the opportunities for creative new approaches to action and learning that support sustainability. Environmental managers are leaders, not followers, of change.

Box 9.1. Principles of social learning for environmental management.

1. Reflexive processes that critically consider actions, assumptions and values are integral to all social learning processes in environmental management.
2. A systemic learning approach takes account of the interrelationships and interdependencies between social and ecological systems and is essential to achieving progress towards sustainability.
3. Social learning in environmental management is a commitment to integrating ideas and actions across social boundaries, including those that divide professions, communities, cultures and ecosystems.
4. The negotiation of learning agendas and indicators of success across the whole community is essential.
5. Conflict and tensions arising from synthesizing different types of knowledge should not be avoided, but do require facilitated negotiations.
6. Social learning is participatory and adaptive, and fundamentally about a commitment to equitable decision making on social and environmental issues.
7. Social learning in environmental management takes into account social and environmental relationships and structures, particularly those pertaining to power relations.
8. Social learning is about supporting social change processes by transforming organizations, institutions, and individual and group identities in a way that increases sustainable environmental management.
9. Social learning promotes a culture that respects and values diversity, transparency and accountability in working towards a sustainable future.

Social learning is rather like a spider web, with many different components all interacting and affecting movements towards social action and change. While it is impossible to untangle and dissect a web and still maintain its essential character, we can embark on an experiential and adaptive process of learning that strengthens rather than weakens the web. Each time we find a new web of social learning, we need to work with it gently, probing to see how the parts are connected and the strands are related. We encourage you to help weave this web and learn from it. Most importantly, we hope you'll join us in our efforts to establish social learning processes that support sustainability in environmental management.

References

Alvesson, M. and Skoldberg, K. (2000) *Reflective Methodology: New Vistas for Qualitative Research*, London: Sage.

Andrew, J. and Robottom, I. (2005) "Communities Self-Determination: Whose Interests Count?" in M. Keen, V.A. Brown and R. Dyball, eds., *Social Learning in Environmental Management: Towards a Sustainable Future*, London: Earthscan.

Argyris, C. (1999) *On Organizational Learning*, 2nd ed, Malden: Blackwell Business.

Arnstein, S. R. (1969) "A ladder of citizen participation", *Journal of American Institute of Planners*, 35: 216–224.

Bass, S., Dalal-Clayton, B. and Pretty, J. (1995) "Participation in Strategies for Sustainable Development, International Institute for Environment and Development", International Institute for Environment and Development, Environment Planning Issues No. 7, London: IIED.

Bohm, D. (1996) *On Dialogue*, London: Routledge.

Brown, V. and Pitcher, J. (2005) "Linking Community and Government: Islands and Beaches" in M. Keen, V.A. Brown and R. Dyball, eds., *Social Learning in Environmental Management: Towards a Sustainable Future*, London: Earthscan.

Brown, V.A., Smith, D.I., Wiseman, R. and Handmer, J. (1995) *Risks and Opportunities: Managing Environmental Conflict and Change*, London: Earthscan.

Brown, V.A., Grootjans, J., Ritchie, J., Townsend, M. and Verrinder, G. (2003) *Sustainability and Health: Working Together to Support Global Integrity*, Griffith University, Nathan, Queensland.

Chambers, R. (1992) *Rural Appraisal: Rapid, Relaxed and Participatory*, University of Sussex, Brighton, UK.

Chambers, R. (1994) "Participatory rural appraisal (PRA): Challenges, potentials and paradigm", *World Development*, 22 (10):1437–1454.

Chambers, R. (1997) *Whose Reality Counts: Putting the First Last*, London: Intermediate Technology Publications.

Cornwall, A. (1995) 'Towards participatory practice: PRA and the participatory process', in K. de Koning and Martin, M., eds., *Participatory Research in Health*, London: Zed Books, pp. 94–107.

Dovers, S. (2001) "Institutions for Sustainability", Australian Conservation Foundation, Fitzroy, Victoria.

Dyball, R., Beavis, S. and Kaufmann, S. (2005) "Complex Adaptive Systems: Constructing Mental Models" in M. Keen, V.A. Brown and R. Dyball, eds., *Social Learning in Environmental Management: Towards a Sustainable Future*, London: Earthscan.

Estrella, M. and Gaventa, J. (1997) *Who Counts Reality? Participatory Monitoring and Evaluation – A Literature Review*, London: IDS.

Gunderson, L. and Holling, C. (Eds) (2002) *Panarchy: Understanding Transformations in Human and Natural Systems*, Washington, DC: Island Press.

Harris, J. and Deane, P. (2005) "The Ethics of Social Engagement: Learning to Live and Living to Learn" in M. Keen, V.A. Brown and R. Dyball, eds., *Social Learning in Environmental Management: Towards a Sustainable Future*, London: Earthscan.

Holling, C.S. (1978) *Adaptive Environmental Assessment and Management*, Chichester, UK: John Wiley & Sons.

Holling, C. and Gunderson, L. (2002) "Resilience and adaptive cycles", in L. Gunderson and C. Holling, eds., *Panarchy: Understanding Transformations in Human and Natural Systems,* Washington, DC: Island Press, pp. 25–62.

Ison, R.L. (2005) "Traditions of Understanding: Language, Dialogue and Experience" in M. Keen, V.A. Brown and R. Dyball, eds., *Social Learning in Environmental Management: Towards a Sustainable Future,* London: Earthscan.

Jiggins, J. and Röling, N. (2002) 'Adaptive management: Potential and limitations for ecological governance of forests in a context of normative pluriformity', in J.A.E. Oglethorpe, ed., *Adaptive Management: From Theory to Practice,* Gland, Switzerland: IUCN, pp. 93–104.

Keen, M. and Mahanty, S. (2005) "Collaborative Learning: Bridging Scales and Interests" in M. Keen, V.A. Brown and R. Dyball, eds., *Social Learning in Environmental Management: Towards a Sustainable Future,* London: Earthscan.

Keen, M., Brown, V.A. and Dyball, R., eds. (2005) *Social Learning in Environmental Management: Towards a Sustainable Future,* London: Earthscan.

Kelly, D. (2001) *Community Participation in Rangeland Management,* Rural Industry Research and Development Corporation, Canberra.

Knowles, M., Holton III, E. and Swanson, R. (1998) *The Adult Learner: The Definitive Classic in Adult Education and Human Resource Management,* Woburn, MA: Butterworth-Heinemann.

Kolb, D., Osland, J. and Rubin, I. (1995) *Organizational Behaviour: An Experiential Approach,* Englewood Cliffs, NJ: Prentice Hall.

Lee, K. (1993) *Compass and Gyroscope: Integrating Science and Politics for the Environment,* Washington, DC: Island Press.

Measham, T. and Baker, R. (2005) "Combining People, Place and Learning" in M. Keen, V.A. Brown and R. Dyball, eds., *Social Learning in Environmental Management: Towards a Sustainable Future,* London: Earthscan.

Open University (2005) *Systems Practice Glossary,* available from http://www.open2.net/systems/glossary/index.html

Parkes, M. and Panelli, R. (2001) "Integrating catchment ecosystems and community health: The value of participatory action research", *Ecosystem Health,* 7 (2): 85–106.

Pretty, J. (1995) "Participatory learning for sustainable agriculture", *World Development,* 23 (8): 1247–1263.

Pretty, J. and Chambers, R. (1994) 'Toward a learning paradigm: New professionalism and institutions for agriculture', in I. Scoones and J. Thompson, ed., *Beyond Farmer First,* London: Intermediate Technology Publications.

Rittel, H. W. J. and Webber, M. M. (1973) "Dilemmas in a General Theory of Planning" *Policy Sciences,* 4: 155-169.

Schön, D. (1983) *The Reflective Practitioner,* New York: Basic Books.

Schön, D. (1987) *Educating the Reflective Practitioner,* San Francisco: Jossey-Bass.

Wynne, B. (1996) "May the Sheep Safely Graze? A Reflexive View of the Expert-lay Knowledge Divide", in S. Lash, B. Szerszynski and B. Wynne, B., eds., *Risk Environment and Modernity. Towards a New Ecology,* London: Sage Publications.

Part II

Perspectives

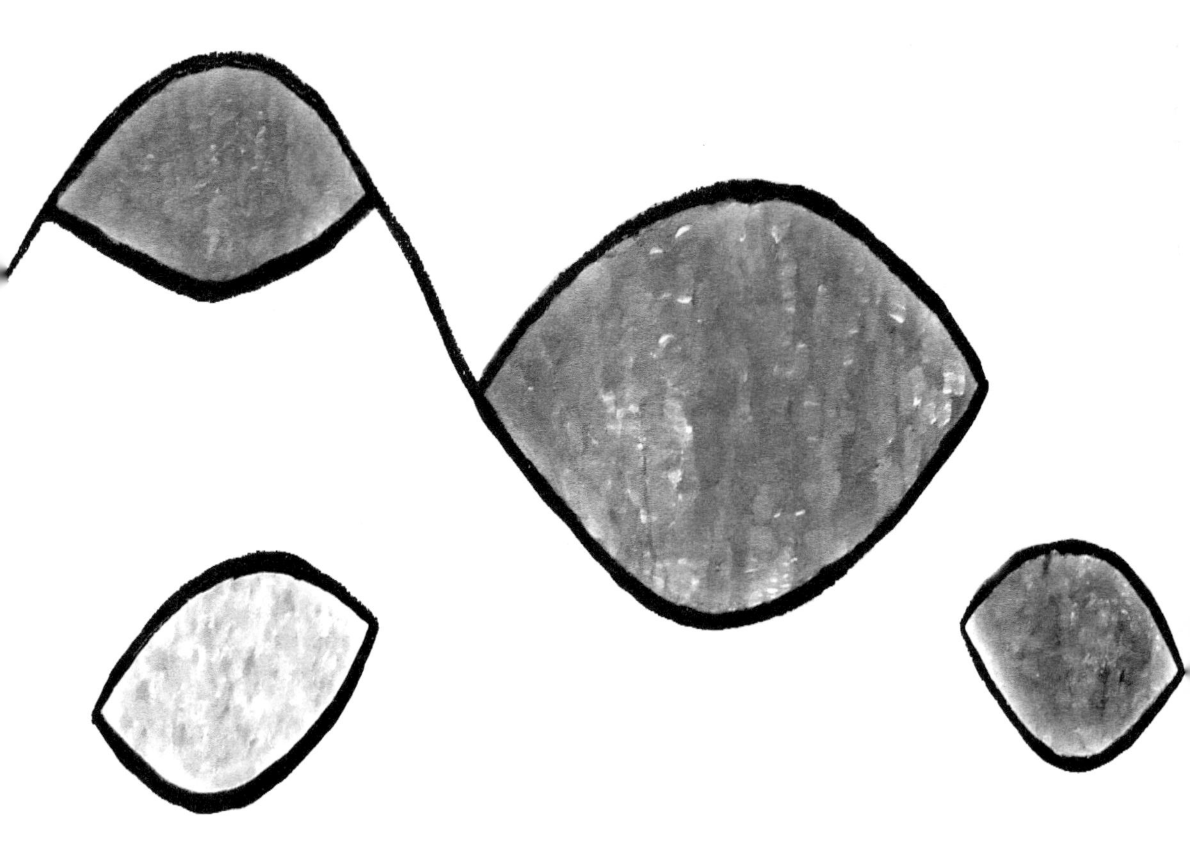

Chapter 10

Participatory planning in protected areas: exploring the social-science contribution

Joke Vandenabeele and Lieve Goorden

Nowadays 'participation' is an inevitable theme in planning and policy-making processes. One can make a distinction between three different arguments in favour of participation. An often-heard argument has a pragmatic and instrumental character: participation results in a greater legitimacy of policy and thus enhances the effectiveness of governance. The pragmatic element lies in the creation of a broad public base. If people are involved in policy making, they have a joint responsibility for the results achieved. In this manner, one can avoid policies being subsequently contested. In the case of natural resources different interests are at stake, and a diversity of local actors should therefore participate in the policy-making process with regard to biodiversity. Fishermen, for example, want to improve their access to the riverbank, but the riverbank vegetation, planted especially to enhance ecological diversity, can constitute a major obstacle for them. Hunters want to operate during the prescribed hunting season, but in doing so they also want to enter wetlands with breeding places for birds. Farmers want to make a living, and thus develop their farm according to the needs of an economic logic, but in doing so they are often at odds with ecologists. More generally, people who live and own property stick to their own opinion on how to use their land.

There are two reasons why a pragmatic argument is an inappropriate basis for encouraging the involvement of this variety of local actors in decision-making. First, if one creates the impression that people are merely involved in policy making with a view to legitimising policy outcomes, the public will quickly lose its faith in the participatory process. Furthermore, the participatory process cannot possibly guarantee that conflicts will not emerge in the long term. After all, taking into account the views of local actors on natural resources, and gearing policy optimally to public opinion, does not alter the fact that the perception of risks and benefits of management can change very quickly (e.g. as a result of changing economic constraints for farmers).

The second argument has a moral character: participation is conducive to more democratic policymaking. This argument relates not so much to the effectiveness of governance as to the substance of democracy. It is based on the assumption that decisions on natural resources are usually taken in a technocratic fashion, within a closed circle of experts, politicians and managers. According to this argument, the management of natural resources is everybody's business because its development is essential for our future. An often-heard response to this argument is that people cannot be forced to contribute to debates that are exceedingly technical and complex. If a representative democracy works well in this respect, then it may well be the case that people will delegate their vote to experts whom they trust. However, the question arises as to whether this reasoning perhaps focuses too strongly on a particular perspective on democracy, i.e. democracy as 'representation', whereby citizens choose a group of representatives to whom they entrust decision-making on social issues. Another, equally important aspect of democracy is deliberation or consultation and debate between local actors as individual citizens holding divergent views and opinions (Hajer 2000).

In this light, one sees a third important argument for stimulating the participation of local actors in decision-making on natural resources, namely a content-related argument. This third argument is in favour of processes of social learning. By mobilising and confronting a greater diversity of experiences, forms of knowledge, insights and perspectives on issues, one increases the likelihood that new, original ideas will emerge; ideas that might not surface in a technocratic decision-making process. In this respect, the participation of local actors can provide a more solid foundation for decision-making. A public debate is successful if it results in a good idea, 'just that, a good idea' (Nussbaum 1999); if it prevents ideas from being defended with weak arguments, and becoming compelling through lack of anything better (Putnam 2001). As social scientists we argue for this third, content-related argument. But it is especially in this respect that officials encounter many difficulties in stimulating the participation of local actors who live and work in the area of ecological networks. We can make several observations here regarding the regular political procedures for handling the issue of biodiversity.

Protected areas and ecological networks

Protected areas and ecological networks are priorities of Europe's Biodiversity Policies. They provide a legacy to ensure that the generations after us have access to nature and all the material and spiritual wealth it represents. The adaptation of the Convention on Biological Diversity (the CBD) in the nineties was a milestone. It was the first global comprehensive agreement to address biodiversity at different levels of biological organization (genes, species and ecosystems). It is an international treaty that identifies a common problem, sets overall goals and general obligations, and organises technical and financial cooperation.

Two main characteristics are:

- the definition of 'biodiversity' takes on a very fixed and scientific meaning;
- execution of the CBD is largely in the hands of the politicians and administrators of the European countries. They have to make the decisions regarding biodiversity and natural areas in such a way that CBD's objectives are served.

The very fixed and scientific meaning of biodiversity leads to the (implied) claim that this kind of knowledge is more legitimate, because it refers to 'actually existing' natural laws. This scientific knowledge then becomes a powerful source of normative justification for particular actions (Tazim *et al.* 2002). In nature restoration there is also a dominant view on multi-level government. Abstract principles are defined at the top and then deduced to more specific principles and concrete action for the local level of environmental policy. But having a privileged or absolute position can be detrimental to dialogic processes, especially in areas where there is a wide range of interests, values and goals and a set of different actors who live and work in the area.

'Directive Nature Plans' in Flanders

The proper management of a series of protected areas in Flanders, the so-called 'Directive Nature Plans' (DNP) is an example of the execution of this European policy on biodiversity. The aim of a DNP is tailor-made nature conservation in special sites of Flanders (e.g. ecological network, international habitats, parks and woodlands). At the end of the process, a Directive Nature Plan contains both a vision and measures and devices. The Flemish government took up the challenge to develop and execute this plan with the support of owners and land users of the area. The planning process is based on three kinds of consultation:

- The consultation with a planning group during the planning process. This is an interdisciplinary cooperation of government officials from the Flemish Region.
- The consultation with a steering committee during the planning process. This is an advisory board of local authorities and stakeholders.
- The consultation with the public at the end of the planning process by means of a 'public hearing'.

A useful image to describe such a planning process has been introduced by Schön (1990): a highland with a view of the swamp. The official who has to facilitate this process is confronted with a choice. Will he/she stay on the safe floor of the cooperation between civil servants? According to Schön this is a rather simple enterprise. Or will he/she also choose to come down to the swamp of local actors and their particular uses and definitions of the area? According to Schön, debates within the 'swamp' of an area are the only way to enhance deeper learning. Only in this way can one really learn to handle the basic and most important issues in nature conservation. It is within this swamp that learning becomes a social or interactive enterprise, developing the competence to discuss different meanings and interpretations in the area.

The planning process of DNP follows a very strict legal procedure. It all starts with a signature from the minister responsible for nature policy – a signature by which authorisation is given for the composition of the steering committee and planning group. The rules for the composition of these groups are very strict, and deal with the question of who can or cannot be included. Furthermore, there is the intention to collaborate with 'working' groups in sub-areas of the Nature Plan Site, giving owners and land users the opportunity to formulate their remarks as early as possible in the planning process. With regard to these working groups, we, as social researchers, also play an important role in observing and giving suggestions. From the moment a provisional draft is written, time schedules start to run, as indicated in the flowchart shown in Figure 10.1. The Steering Committee has to deposit its advice within thirty days. The Minister has to approve of the final draft within another thirty days. Then the announcement of the public hearings can start. The legal procedure defines a time period of thirty days for the communication to the public and a period of sixty days for the public hearing itself. The Planning Group gets sixty days to incorporate the remarks of the public hearing in the final text of the plan. Within sixty days the minister has to have the plan approved.

Our own involvement as social science researchers started at the moment civil servants were composing the membership list of the planning group and the steering committee. The composition of these lists took a long time. Other Ministries were not very eager to contribute to a DNP in different areas of Flanders. And there exists an inflationary spiral of too many deliberation processes between the Ministries in the Flemish administration. The Ministry in charge of nature policy also has a doubtful reputation in relation to processes of deliberation and consultation. Leroy and Bogaert (2004) analyse how in the early 1990s and again in 2002 Flemish nature conservation policy launched the initiative for a Flemish Ecological Network (FEN), aimed at the implementation of European policies, and particularly at the realisation of the European Natura 2000 network. That is a network created for the conservation of wild plants, animals and habitats of community interest across national boundaries. Twice the initiative failed, as it faced firm societal and political opposition, mainly from the agricultural sector. Leroy and Bogaert (2004) assess that officials developed messy processes of communication and consultation and because of this provoked societal and political obstruction of the FEN, rather than support. (Leroy and Bogaert 2004)

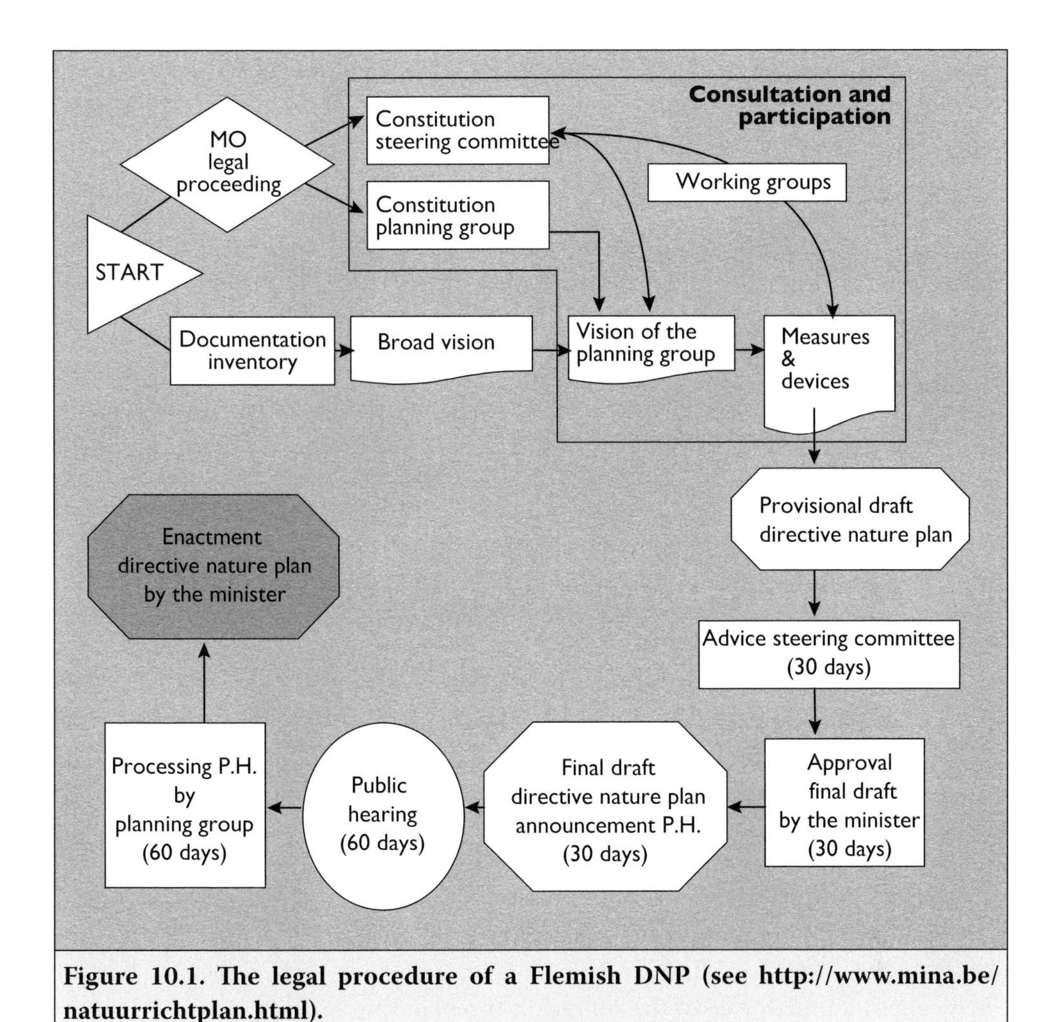

Figure 10.1. The legal procedure of a Flemish DNP (see http://www.mina.be/natuurrichtplan.html).

The composition of the membership list of the steering committee, an advisory board of local authorities and stakeholders, caused additional problems. The legal procedure prescribes that the representatives of the different stakeholders, e.g. farmers, hunters, nature conservationists, have to live in the area of a DNP. The advantage of this directive is that representatives living in the area of a DNP, are better acquainted with the particular features of the regional area. However, the official organisations of the stakeholders were dissatisfied with this and tried to put forward a candidate who was used to defending the interests of the group, but who was not living in the area. Furthermore, the minister responsible for nature conservation took his time to check every membership, and waited for a considerable time before agreeing with the lists. Critical servants drew the conclusion that the minister was not supporting this DNP process and in doing so, he was favouring the farmers, who constitute an important share of his voters. Some civil servants were discouraged from making further arrangements for the composition of the working groups in the sub areas of a DNP, but others used the available time to get an overview of the local actors in the area and to develop a participation plan, listing those who could participate when and how. Some civil servants even started consultation with owners and land users in informal meetings.

Exploring the social-science contribution

If we look at the planning process from the point of view of the officials, the DNP involves four different tasks:

- to protect and develop nature in protected areas of Flanders;
- to put forward a project and policy line for civil servants;
- to stimulate the commitment of owners and land users in the area;
- to edit the policy plan and incorporate the suggestions and remarks of all kinds of actors.

We can relate these tasks to four different roles the official has to assume in relation to the DNP. First, the official will implement the plan and take care of its proper management. Second, he/she can bring in expert knowledge during the development of the plan and/or be a manager who focuses on the efficiency and efficacy of the plan. Third, the official is a communicator who increases the commitment of owners and land users. These three roles are rather obvious and well known by officials. However, as social researchers, we observed a lack of clarity about a fourth role of the official in the planning process, namely the role of editor of the plan, who has to take into account the many remarks from the very different types of actors. This seems to be a very difficult task, and civil servants are searching for directive guidelines. They have a lot of questions about this role. Is it about rendering a service to a particular fisherman for example? Or is it only a matter of power and the strategic capacity of, for example, the official organisations of farmers who are able to impose their own view?

What is the contribution of social researchers, in trying to handle the tension between the activities as an expert, a manager and a communicator on the one hand, and the facilitation of a community-based planning process on the other hand? At this point, we can return to our content-based or social learning argument. By mobilising and confronting a diversity of experiences, forms of knowledge, insights and perspectives on issues, one increases the likelihood that new, original ideas will take shape; ideas that might not surface in a technocratic decision-making process. Social learning also enhances the development of responsible adults able to contribute to biodiversity in their living and working environment. The learning is about the question 'what's the issue here'.

During our research, we were able to make three observations about the way the 'issue of biodiversity' is usually defined. First, biodiversity is a 'data demanding' issue. The variation in and among genes, species and ecosystems, makes biodiversity a rather complex phenomenon. By defining biodiversity as a complex issue, the type of deliberation the official is seeking is the debate with and among experts. They can assess the status of species and habitats and also predict the effects of a project. A clear distinction is made between experts who can legitimately speak for biodiversity and lay people who don't have the authority to do this. Lay people would even destroy biodiversity without a second thought. Second, it is increasingly recognized that biodiversity, much like sustainability, is driven and caused by a dynamic interplay between various factors and processes. Many of them are far beyond any certainty, control and predictability. An additional focus is, therefore, on the bargaining with stakeholders with their own particular sensitivities. Third, in practice we can observe that a working category has emerged, called 'likely presence' (Hinchliffe 2004). Recording the presence of a rare species is far from straightforward. A likely presence claim is enough to start a survey on the presence or the absence of, for example, a bird called black redstarts. But here we can make another observation: it turns out to be more effective to act as if there was a presence and generate suitable habitats. A policy for biodiversity is then founded on the recognition of 'matters of concern' in combination with the search for precise, factual evidence.

Process-content management

It is in the search for the elaboration of biodiversity as a matter of concern that social scientists have a role to play. They should look for concepts to understand better the series of social and political conflicts one witnesses throughout Europe with regard to nature policy. Loots and Leroy (2004) assess that in many respects 'nature conflicts' remind us of classical sighting issues regarding the location of hazardous activities, waste disposal sites, nuclear power plants and others. The latter have, albeit in different ways, been conceived by many scholars as a particular form of the NIMBY-syndrome ('Not in my back yard'), namely the LULUs: the locally unwanted land use. (Loots and Leroy 2004). As we mentioned before fishermen, hunters, farmers, etc, people who live and own property, stick to their own opinion on how to use their land. The implementation of nature policies is inevitably faced with these different views on land-use implication. In Figure 10.2 we distinguish three questions or three issues in nature sighting: the issue of allocation, the issue

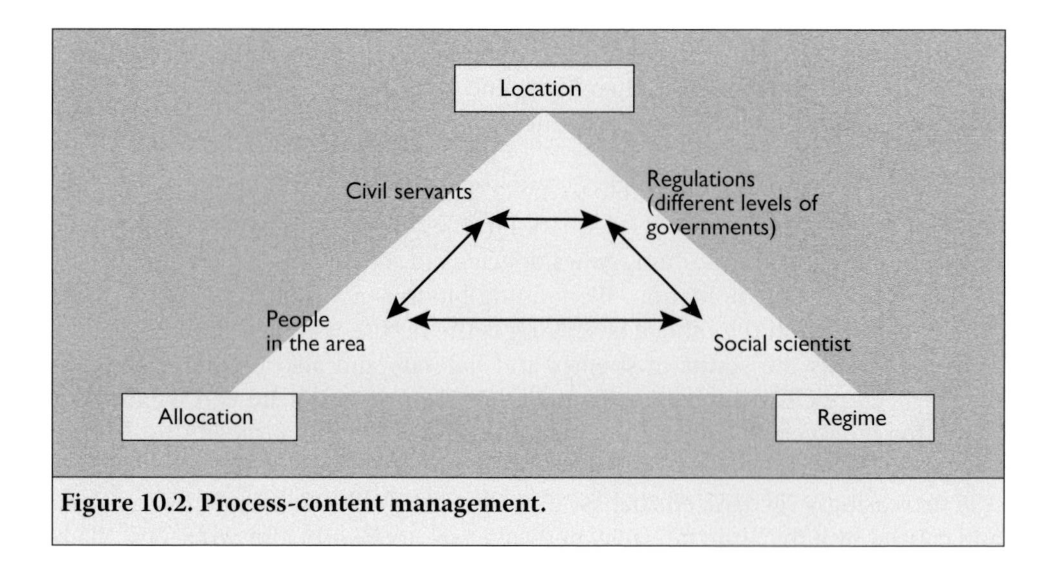

Figure 10.2. Process-content management.

of location, and the issue of regime of exclusive or joint use of the nature area involved. (Loots and Leroy 2004).

The question of allocation is a main issue in nature conservation; it deals with the question of how much of the area is intended for nature conservation. It is a matter that is raised on different levels of government, from the European to the local level, and where power relations between, for example, farmers, industry and nature conservation, play a decisive role. Next, the issue of location is also a main question in nature conservation. It deals with the issue of tracing boundaries around and also within protected areas. The issue of regime, finally, has to do with the kind of nature one wants to obtain: is the area exclusively meant for ecological purposes, or is it a joint area that has to be shared with, for example, farmers, hunters, etc.

Research into the debates, for example, in Flemish newspapers during the period when the Flemish Ecological Network (FEN)) was launched (2001-2003), shows that local actors put forward the issue of location and the issue of regime of joint land use to act against this initiative (De Zitter *et al.* 2003). Arguments for the Flemish Ecological Network are in line with the allocation issue and came from the minister, the nature movement and officials, saying that nature policy need to have its areas in Flanders. In the end the 'not in my backyard' reactions of local actors in particular regions became a 'not in any backyard' reaction that was shouted loud at a big manifestation of hunters, fishermen, farmers, etc, against the Flemish Ecological Network.

In this respect it is important to see the interconnection between the three questions. Conflicts about the allocation issue (how much area is for nature) can be made easier when one can rearrange the location issue (where is the area for nature). Joint land use (the issue of regime) can sometimes help to solve the other two issues. By trying to move from one issue to another, social learning becomes possible. A content-rich deliberation between the different actors is then an important condition for a tailer-made debate.

Process-criteria of participative planning

The issues of allocation, location and regime bring to the fore the social and political conflicts of a nature policy in densely populated areas like the region of Flanders. Next to the analysis of content social scientists could engage in so-called, 'onto-political proceedings' (Hinchliffe 2004). These are ways of dealing with the issues in more open discussions where the inherently political nature of nature conservation is acknowledged and the debate is conducted on the basis of both practical and theoretical knowledge. Our focus as social researchers on proceedings is inspired by the research of Bruno Latour (2004) who tries to elucidate the shifts in the set of rules for nature policy. For Bruno Latour (2004) the public question of nature policy is about how different actors can live together in relation to non-humans. He makes a distinction between four process criteria: 'perplexity' (holding on to a broad horizon); 'consultation' (excluding nobody arbitrarily), 'hierarchy' (trying to understand the relationship between new values and what is prevailing now); 'institutionalisation' (closing the debate for the time being).

With these four process criteria Latour suggests that we should not separate the discussions about facts from the discussions about values. He refers to Plato's allegory of the cave, by which he tries to describe the common way of dealing with the differences between facts and values. At the bottom of the cave, people have been chained to the cave floor – a lifelong captivity. There is a fire in the cave, but the prisoners cannot see the fire, only its shadows dancing on the walls. Between the fire and the prisoners there is a parapet, along which puppeteers can walk. The puppeteers, who are behind the prisoners, hold up puppets that cast shadows on the wall of the cave. The prisoners are unable to see these puppets, the real objects that pass behind them. What the prisoners see and hear are shadows and echoes cast by objects that they do not see. At a certain moment, only the scientist is able to free himself and to leave the cave. He is blinded by the sunlight. It takes a while for his eyes to adjust. After having experienced this wonderful light, he re-enters the cave. The man tries to tell the other people about the world outside the cave. He is driven to do this. As he is telling the others about the reality outside the cave, there are two possible reactions. Intimidated by science, the others confess in chorus, "the more we talk about social construction, the further away we actually move from the real unified things in themselves" (Latour 2004, p. 40). The other possible reaction is that the scientist and the facts he is relating, are simply ignored. It can get even worse: the scientist is considered as being insane and/or is killed.

To escape from this allegory of the cave – the disadvantages of the separation between facts and values, between the world of nature and the world of politics – Latour makes a distinction between 'the power to take into account' and 'the power to arrange in rank order'. Each power has its own question and the search for an answer for each of them requires both facts and values.

- Question 1: Which options do the involved actors take into account? In other words: which opportunities do they themselves put forward to acquire their right to existence in a future society? Latour speaks here of the power to take into account.
- Question 2: Which options do the actors experience as being useful? In other words: which of these innovations or options allow us to live meaningfully together? Latour speaks here of the power to arrange in rank order.

The four process criteria we mentioned above can foster the two powers. The research into the first question – which options do the actors take into account or in the words of Latour 'how many are we'- will have to meet with the requirements of 'perplexity' and 'consultation'. The research into the second question – how are these options to be arranged in rank order or in the words of Latour 'can we live together' – will have to meet with the requirements of 'hierarchy' and 'institutionalisation'.

Onto-political proceedings are, according to Latour, based then on a process of answering the two questions successively, dealing with the focus of one particular process criteria one after the other. The process usually starts with the question of perplexity and then moves forward to the question of consultation, the question of hierarchy and finally to the question of institution. As a facilitator the official can stimulate the commitment of experts, owners and land users in the area from the perspective of 'holding on to a broad horizon' and 'excluding nobody arbitrarily'. As an editor of the plan the official can look for the compatibility between new values and what is prevalent now, and finally look for projects and agreements that can close the debate for the time being. Stimulating an answer on the two questions means that decision-makers try to understand 'the matters of concern' and search for precise, factual evidence for the different options nature policy has in a particular area. With this kind of procedure, two dichotomies can be resolved in relation to expert knowledge (Buttel and Taylor 1992, Douglas 1999). The first one is the dichotomy between on the one hand, the exclusive reference to the scientist's authority and, on the other hand, the demystification of the scientific

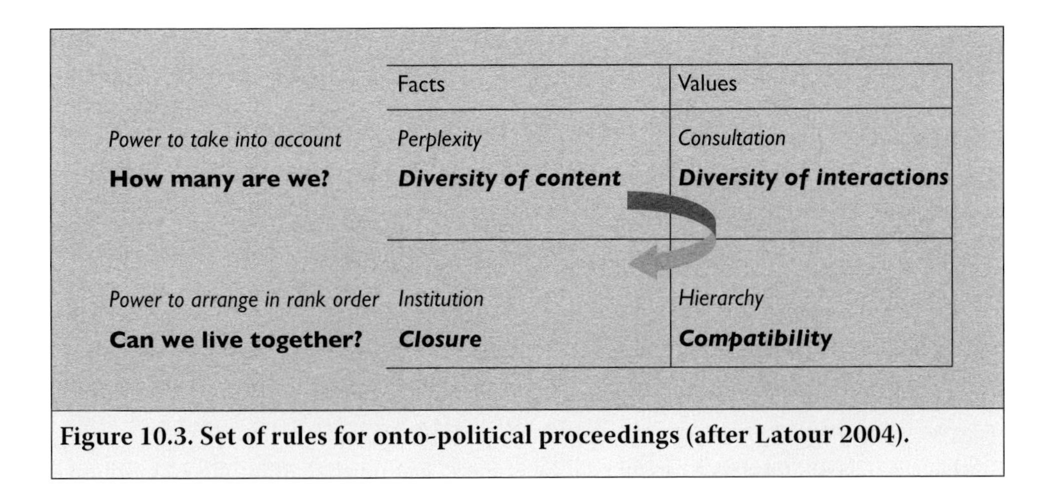

Figure 10.3. Set of rules for onto-political proceedings (after Latour 2004).

enterprise. Dichotomy two is that of seeing science either as the development of abstract ideas, or seeing science as serving particular forces of power in society.

A plea for public deliberation

Policymakers can stimulate a public debate on biodiversity if a number of conditions are fulfilled. First and foremost, natural science cannot be attributed the role of a referee in such disputes. Natural scientists should be explicit about their position in the societal debate and they should negotiate about 'robust knowledge in the making'. Good scientific practice also takes into account the particular context of the actors involved. Second, citizens, be it organised or as individuals, may contribute their own knowledge and expertise and participate in the debate. In this model, the government is not a 'customer' who commissions a study and cuts the Gordian knot, but rather an audience that wishes to be convinced by a diversity of arguments. Direct interaction and debate will be conducted on the basis of practical knowledge about a good life and on the basis of expert knowledge. The starting point for the debate is an interest in the concrete manner in which nature is shaped in our society.

References

Buttel, F.H. and Taylor, P.J. (1992) "Environmental Sociology and Global Environmental Change: A Critical Assessment", *Society and Natural Resources*, 5:221-230.

De Zitter, M., Wymeersch, T., Bogaert, D. and Cliquet, A. (2003). *Eindrapport van de deelopdracht 'Opstellen van een korf van indicatoren voor het meten van het maatschappelijk draagvlak in Vlaanderen,* Gent: Artveldehogeschool-Universiteit Gent. 293 pp.

Douglas, C.W. (1999) "Fisheries Science Collaborations: The Critical Role of the Community", keynote presentation at the Conference on Holistic Management and the role of Fisheries and Mariculture in the Coastal Community, Tjärnö Marine Biological Laboratory, Sweden, 11-12 November.

Hajer, M. (2000) *Politiek als vormgeving,* oratie uitgesproken, 16 juni.

Hinchliffe, S. (2004) "Towards a careful political ecology", first rough draft for the Reconstituting natures workshop, OU December.

Latour, B. (2004) *Politics of Nature. How to Bring the Sciences into Democracy*, Cambridge: Harvard University Press.

Leroy, P. and Bogaert, D. (2004) "Het Vlaams Ecologisch Netwerk als natuurconflict", *Landschap*, 4: 211-224.

Loots, I. and Leroy, P. (2004) "Conflicten bij de uitvoering van gebiedsgericht natuurbeleid", *Landschap*, 4: 199-209.

Nussbaum, M. (1999) "In defense of universal values", occasional Paper Series, Women and Human development, The fifth annual Hasburgh Lectures on Ethics and Public Policy, University of Notre Dame.

Putnam, H. (2001) *Pragmatisme, Een Open Vraag*, Amsterdam: Meppel

Schön, D.A. (1990) "Handelend leren. Zoeken naar een andere epistemologie van de professionele praktijk", *Filosofie in bedrijf*, 2 (1): 12-17.

Tazim, B. J., Stanley, M.S. and Harper, T.L. (2002) "Beyond Labels. Pragmatic Planning in Multistakeholder Tourism-Environmental Conflicts", *Journal of Planning Education and Research*, 22: 164-177.

Chapter 11

Social learning amongst social and environmental standard-setting organizations: the case of smallholder certification in the SASA project

Rhiannon Pyburn

Introduction

This chapter explores the (social) learning process amongst organizations already working for a more ecologically and socially sustainable world – *Fair-trade Labelling Organizations International* (FLO), *Social Accountability International* (SAI), the *Sustainable Agriculture Network of the Rainforest Alliance* (RA) and the *International Federation of Organic Agriculture Movements (*IFOAM). The *Social Accountability in Sustainable Agriculture* (SASA) Project was a two-year initiative (February 2002-April 2004) undertaken by these four social and environmental international standard-setting organizations and was an initiative of the *International Social and Environmental Accreditation and Labelling* (ISEAL) *Alliance,* of which the organizations mentioned above are members. The case involves multiple levels of learning – platform (ISEAL), project (SASA), organization (FLO, IFOAM, RA and SAI), and individual. Through pilot audits in nine different countries, steering committee meetings and workshops, the organizations and their representatives broadened their understanding of their own and each other's systems. Each organization and the individuals engaged in the project had different values, actions, theories and perceptions of the context (sustainable agriculture) as starting points and a range of vantage points from which they addressed critical issues related to social certification in agriculture. Over the course of the research, the understandings of the organizations evolved, merging in some cases and diverging in others. The chapter examines current theoretical literature on social learning in relation to the experience of the SASA project, by analyzing the progression of one of the project's sub-objectives: to address the particular needs of smallholder producers in the development of social guidelines for sustainable agriculture including the challenge of smallholder access to certification in developing countries. My role was to research and document the SASA Project's learning on this issue over the two year project period. Some questions related to the espoused theoretical framework are identified in this chapter and factors supporting and challenging cognitive convergence amongst the organizations are identified in the conclusions.

Background

Social and environmental standards, certification and labelling are an example of private sector and non-governmental organization (NGO) initiatives to promote sustainable development. Eritja *et al.* (2004, p. 32) consider sustainability labelling and certification schemes to be "market-based tools *[that]* may contribute to sustainable development insofar as they impose, encourage, and/or promote actions to introduce environmental and social aspects and concerns within the decision-making processes of the actors involved." Courville concludes that, "at the global level, social certification systems and the constituencies that they represent are voicing a powerful message that workers' rights and fair international trade are important values and can be incorporated into business practice" (Courville 2003, p. 294). In recent years there has been a proliferation of such efforts including corporate social responsibility codes of conduct, human rights, environmental, trade and labor standards. As Tiffen notes in her study of a fair trade chocolate business model, "codes of conduct and social auditing, while not 'solving' all problems, do provide a point of entry for continuous improvement and dialogue" (2003, p. 169). Rigor and levels of accountability vary from one program to the next. The organizations that are members of the ISEAL Alliance are distinct from governmental standards/regulations and other private but less rigorous certification systems in that their standards are private, third-party verified, voluntary, and the standard-setting processes adhere to the ISEAL *Code of Good Practice for Setting Social and Environmental Standards* (ISEAL 2006) involving a broad range of stakeholders. ISEAL members' work covers the ornamental fish, ocean fisheries, forestry, sustainable agriculture and organic agriculture sectors.

The SASA Project was an innovative initiative of the ISEAL Alliance that sought to address some key auditing issues drawing on the experience and expertise of leaders in the field. Four ISEAL Alliance members initiated the project and were the participating organizations: *Fair-trade Labelling Organizations International* (FLO), *Social Accountability International* (SAI), the *Sustainable Agriculture Network of the Rainforest Alliance* (RA) and the *International Federation of Organic Agriculture Movements (*IFOAM). The SASA project's main objectives were to "seek to improve social standards setting and auditing methodologies in the agricultural sector worldwide and to foster cooperation between the four participating social and environmental verification initiatives" (SASA 2004). One of the sub-objectives was to address the particular needs of smallholder producers in the development of social guidelines for sustainable agriculture. Exploring smallholder group certification and internal control systems (ICS) was the main way in which this sub-objective was addressed. I documented the learning process within the SASA project on this topic and it will be the basis of discussion in the chapter. The next sub-section provides background as to organic group certification model, which was the foundation for initial project discussions relating to smallholder certification.

Smallholders and group certification

Group certification via internal control systems was initially developed in Latin America in the 1980s to allow small farmers to be able to build up their systems in order to access organic markets in the USA, Europe and Japan. It is a mechanism by which small farmers in developing countries come together in a cooperative or an association in order to facilitate market access to the 'just market' – in this case, organic markets. In 1994 IFOAM published specific criteria for grower group certification and it was first regulated in 1996 (Elzaaker and Rieks 2003, p. 6). In 2001, IFOAM estimated that there were more than 350 smallholder groups representing over 150,000 organic farmers (IFOAM 2003, p. 1). Small farmers produce an estimated 60-70% of organic exports to Europe (*ibid.*).

IFOAM began a process in 2000 that included smallholder farmers, auditors, standard-setting and certification bodies as well as competent authorities, to set more exact criteria and definitions for ICS inspection and certification. In 2003 after three years of stakeholder meetings the IFOAM-led consortium re-defined the main mechanism used to enable group certification – Internal Control Systems (ICS): "An Internal Control System is a documented quality assurance system that allows an external certification body to delegate the annual inspection of individual group members to an identified body/unit within the certified operator. (As a consequence the main task of the certification body is to evaluate the proper working of the ICS)" (Elzaaker and Rieks 2003, p. 11).

A key component of group certification schemes are the internal monitoring and internal standards in place. The IFOAM-led meetings addressed many other contentious issues including sample sizes for re-inspection, definitions of smallholder, the agreed upon minimal elements of an ICS, non-compliance related sanctions, risk assessment, evaluation protocol and conversion (*ibid.*). The EU has since taken up the stakeholder consensus on re-inspection (sampling) rates (European Commission Agricultural Directorate General 2003) and all IFOAM members working with smallholder groups have agreed to use the outcomes of the workshops as a basis for their certification programs.

All of the organizations involved in the SASA Project are mission driven, and provide a different vantage point for reflection and action related to smallholders and agricultural certification. IFOAM's integral role in the development of ICSs is outlined above. It is the most experienced organization of the four in terms of group certification and ICSs. FLO focuses explicitly on improving the position of poor and disadvantaged farmers in the developing world. At the outset of the Project, FLO was not implementing group certification or using an internal control system, though many FLO-certified farmer cooperatives were also organically certified via an ICS. The Rainforest Alliance (RA) standard covers social, environmental and health and safety issues. From the beginning of the SASA Project, it was interested in developing group certification both for small farmers groups and other situations that might be suitable. SA8000 is human rights based facilities certification based on International Labor Organization (ILO) standards. SAI, the organization that develops the SA8000 standards, was new to the agricultural sector at the outset of the Project and was keen to learn how other social certifiers were addressing issues specific to the agricultural context. SA8000 standards do however include a Management System component, which proved to be comparable in function to an ICS.

Social learning: theoretical touchstones

In developing a conceptual framework to study this case, several bodies of literature were explored. Organizational learning and learning organization theory offer some interesting elements that may demand further consideration. However, this body of literature generally applies more to individual organizations rather than to platforms or multi-organization projects like the one presented here. Collaborative learning, as defined by Keen and Mahanty is another source of valuable input: "collaborative learning involves a range of social actors negotiating and agreeing on the nature of the required learning and action, their respective roles and responsibilities and the process of reflection that will occur over space and time. It is best viewed as an iterative process of collaboration and negotiation between actors that is strongly affected by dynamic social networks, relationships and structures" (2005, p. 105). The '*Communities of Practice*' (Wenger 1998) literature was also considered. The notion that collective wisdom is superior to individual knowledge – as expert as it may be – (Surowiecki 2004) is critical to valuing social or shared learning. The ideas found in these various approaches provide context to the espoused framework – that of social learning.

Social learning has been defined in many ways by different authors and has evolved over time as a notion referring specifically to development and to ecological sustainability drawing on soft systems thinking, which suggests that joint learning amongst interdependent stakeholders is a key mechanism for sustainable development. Facilitation and participation are key attributes in a social learning process (Leeuwis and Pyburn 2002, p. 11), as are reflection and reflexivity. Keen *et al.* define social learning as: "the collective action and reflection that occurs among different individuals and groups as they work to improve the management of human and environmental interrelations" (Keen *et al.* 2005, p.

4). Shared learning that leads to joint or concerted action encapsulates the heart of what social learning has come to mean. Joint or collective action occurs when everybody involved is working on the same things whereas concerted action suggests that the actors are working for a common purpose but not necessarily contributing in identical ways.

Röling brings in the concept of cognition stating that: "social learning can best be described as a move from multiple to collective and/or distributed cognition" (Röling 2002, p. 35). In Röling's model, (adapted from Kolb, Maturana, Varela and Bawden) cognition is made up of the following elements: theory; action; perception of the context, and; values, emotions, goals (2002, p. 33). These elements tend towards coherence. Multiple cognition refers to the various perspectives made up of different sets of the coherent elements (see above) of different stakeholders in a given situation. Collective cognition refers to attributes shared by different actors and distributed cognition emphasizes different but complementary contributions that allow for concerted action (*ibid.*). Distributed cognition is described by Leeuwis as a situation wherein ideas, values and aspirations need not be shared but are overlapping or mutually supportive (2004, p. 145). I would add to this description that distributed cognition also entails entangled and concerted efforts that, while separate, contribute to overall movement in a particular direction. The categories of multiple and distributed cognition are used to frame analysis of the SASA learning process in this chapter. Two drivers of cognitive processes are *correspondence* (the applicability to context or the environment) and *coherence* (internal cognitive consistency) (Gigerenzer 1999 in Röling 2002, p. 33). Through these 'drivers', the evolution from multiple to distributed/collective cognition unfolds. These concepts are also applied to the SASA case.

(Social) learning in the SASA project

The SASA Project was a 26-month project that consisted of nine pilot audits in eight different countries with nine different types of production systems. Teams comprised of 6-10 participants including auditors and/or staff representing the participating organizations, a facilitator and in some cases a researcher, undertook intensive learning audits to explore pre-defined and emerging social auditing issues. The organizations participating in the SASA project recognized the unique needs of small producers and in the case of FLO, small farmers were a target group for the certification scheme. Market access (via certification), technical capacity building and access to resources are amongst the challenges facing small producers. The SASA Project sought to examine a constellation of concerns related to small producers in developing countries primarily through examining group certification and internal control systems. Four of the nine audits directly addressed smallholders and group certification or internal control systems (ICS):

Thailand (Fairtrade/organic rice), Burkina Faso (Fairtrade/organic mangoes), Costa Rica (certified coffee), and Uganda (Fairtrade/organic cotton). In addition a stakeholder workshop (Nuremburg, Germany in February 2003) was held in order to broaden stakeholder input. SASA Project steering committee meetings and the final retreat provided further opportunities for discussion and debate. Table 11.1 provides an overview of the learning events and key aspects related to group certification that were addressed in each.

The initial learning events of the SASA Project (Thai pilot audit and the stakeholder workshop in Germany, in particular) in terms of smallholder access and group certification, focused on learning from the organic Internal Control System and on considering its viability *vis à vis* social certification. The Burkina Faso audit

Table 11.1. SASA project learning events and the issues addressed.

	Learning event and date					
Issue addressed	**Thai Rice Audit** 11.2002	**Biofach Stake-holder Meeting** 02.2003	**Burkina Faso Mango Audit** 06.2003	**Costa Rica Coffee Audit** 11.2004	**Final Project Retreat** 02.2004	**Uganda Cotton Audit** 03.2004
1. Smallholder access						
Organic approaches	X		X			
Strengths, weaknesses, nuances of the organic ICS	X	X	X			
West African context			X			
Fairtrade and RA Models	X			X		
2. Social certification						
Organic model		X				
Organic stakeholder Perspectives		X				
RA, FLO SA8000 perspectives	X	X				
SA8000 management systems input	X			X		
3. Internal control elements within SASA systems				X		X
4. Organisational coordination			X	X	X	X

continued on this trajectory, considering in more depth the potential coordination opportunities between Fairtrade and organic certifiers. The Costa Rica and Uganda audits reflected a turning point in the project's conceptualization of group certification and internal control systems wherein the participating organizational representatives began to recognize and distinguish common ground in terms of the internal control or management system requirements of each system. The final audits and project meetings were directed towards developing a tool that would provide a basis for smallholder groups who aspired to any of the certifications. This tool was intended to be foundational; specific requirements of each system would be additional modules.

Over the course of the project, the discussions moved from auditing 'how-to' questions to standard alignment and organizational cooperation. Four main areas of learning emerged related to smallholder farmers and group certification:

- Internal Control Systems and smallholder access to certification.
- Internal Control Systems and social certification.
- Internal control elements in SASA organizations' requirements.
- Organizational coordination opportunities related to internal control (Pyburn 2004, p. 14).

Each pilot audit and steering committee meeting or workshop addressed different and overlapping aspects of these learning areas. Pilot audit and workshop learning (points 1-3 above) drew from earlier SASA audit experiences and reports, the practical knowledge and experience of auditors on each team and the local conditions and context. In some cases international and local stakeholders also provided input both prior to and during the audits themselves (*ibid.*). The learning process amongst the four organizations can be described as semi-linear as later learning events built on the foundation of initial audits but also addressed local, context-specific issues (e.g. on the Burkina Faso audit, the issue of lack of models and capacity building resources regionally emerged as a concern). The fourth area of learning culminated at the project steering committee level where the organizational representatives grappled with how to improve dual or multiple certification situations for smallholder groups (i.e. where a smallholder group seeks to have both Fairtrade and organic certification and therefore access to both markets) and developed a template for group certification manuals for smallholder groups that identified elements of internal control common to all four systems.

The case of group certification is a compelling one as it is widely recognized as a challenge demanding immediate attention. This urgency was a stimulating factor catalyzing the project participants towards collective agreements and action. The purpose of this chapter is not to delve into the details of group certification, smallholders, internal control systems and management systems. Instead it is the learning process amongst the four participating organizations that is of interest. That said, the learning process is embedded in the subject matter. In order to explicate the learning process, learning content specific to the topic is necessary and hopefully will prove to be illuminating.

Multiple cognition vis à vis group certification

Interviews with key informants from the organic and Fairtrade sectors in August and September 2002 at an IFOAM conference in Victoria, Canada, revealed a wide array of perspectives on ICS and key issues related to group certification. Excerpts from several of the interviews with key informants are found in Box 11.1. The issues raised include the credibility of group certification as compared to individual certification, the burden of additional documentation required for certifying a 'system' as well as a farm, and the applicability of the organic internal control system to social (Fairtrade) standards. These excerpts do not cover the whole range of issues related to internal control systems, but introduce some of the divergent views that characterized the context at the outset of the SASA Project. The varying perspectives and concerns about ICS expressed in interviews at the IFOAM conference reflect some of the values, theories, emotions, perceptions of the context in the sector that were present as the SASA project began its research and learning endeavour.

With the multiple cognitions on ICS and smallholder certification in the broader sector, it is unsurprising that the organisations involved in the SASA Project had different starting points for initiating the learning process on this topic. A preliminary step towards developing distributed cognition amongst the FLO, IFOAM, SAI and RA was to document and compare the (then) current standards of each organisation on each project issue. This was done in a document entitled *Joint Audit Template* (Courville 2002) that was compiled by the project coordinator. By comparing the standards, policies activities, and requirements of each organisation on each issue, a foundation for discussion was laid. On the smallholder certification topic, six elements were compared: responsibility for ensuring the management system/ICS is functioning; training/capacity of responsible persons; training and/or capacity building of workers and/or producers; internal audit and monitoring; social control, and; outcomes related to continual improvement of the system in place. Table 11.2 is an excerpt drawn from the document with two examples of issues addressed in the requirements of the organisations: training and/or capacity building of workers and/or producers; internal audit and monitoring. Table 10.2 notes the requirements, activities, indicators or documents of each organisation in reference to each particular issue addressed. It indicates where each component

Box 11.1. Excerpts from Interviews with Key Informants at IFOAM Conference 2002.

Nabs Suma, Twin Trading, UK:
"I wasn't just saying that ICS was as credible as the current system that's in place, I think for the context of smallholders it is perhaps more credible than any other system.....I don't think organic certification should be about policing – its one of the big differences I have from my Fairtrade perspective that I have with the organic movement. I think that certification shouldn't just be about guaranteeing something for one end of the chain. It should be just about adding credibility to the chain and the benefits from organic production are not just at one end of the chain. There is benefit at the Southern end – the producers end as well."

Jochen Kreubel, Certification Coordinator, FLO International, Germany:
"You are saving inspection costs of the certification body because they *[the inspectors]* have to travel less, but you are not saving on communication, you are not saving on documentation because you are adding to the system – you are adding another layer. The work has to be done. I mean the advantage of an ICS is that they can do it in a harmonized way for every producer. If this is working, then they create their formats ...in the same way for every producer – this is a big advantage. And they might be able to reduce documentation loads on the side of the producers. But in the end, the big danger is that you are creating a double system – the ICS documents and the producer documents."

Thomas Cierpka, Managing Director IFOAM Head Office:
"....we have to work with trust. If you have 100% third party inspections with ICS – what are the chances of [*people*] taking advantage? If you are a good administrator – you can get ready, make selections and be very quickly certified - you know how to guide your inspectors through the inspection so they will not see the bad points. The 'third party' person coming from overseas or wherever, he doesn't understand. It is not a guarantee. More guarantee could be done with the neighboring village farmer – who'd have a very good background as to what's going on in the system. He knows what's going on in the system – he keeps aware of it."

Olaf Paulsen, FLO International:
"Are ICS an appropriate means at all or do we need other mechanisms to ensure that the core standards [*Fairtrade*], core requirements are being complied with? For instance, if you talk about democratic decision-making processes, its something you can't verify through an ICS – because the structure as such is a controlling system... If you talk about the requirements that a cooperative needs to have – a clear, transparent way of decision-taking. Then, there's no need for an ICS."

Table 11.2. SASA *Joint Audit Template* excerpt from ICS and Management Systems section (*ibid.*, p. 51) modified by Pyburn.

Issue covered	Requirement/Activity/Indicator/Document	
Training and/or capacity building of workers or producers	SAI:	Are new employees trained on SA 8000 upon being hired? Are periodic training course given to existing employees to promote awareness? (SAI 9.5).
	RA:	Refer to social policy and training/awareness requirements.
	FLO:	Participation of members in co-op admin and internal control is promoted through training and education and improves as a result – inspectors check capacity building/training within last year (PROCESS) (FLO 1.3.2.2).
	IFOAM:	Do farmers have sufficient knowledge of the internal regulations? Are the farmers aware of the contents of the organic farmers' contract they signed? Are the farmers prepared to apply the internal regulations and are they doing so? (Proposal IFOAM smallholder protocol).
Internal audit and monitoring	SAI:	Are management review meetings held periodically in accordance with procedures? Is there evidence of senior management involvement in these meetings? Are the meetings documented? (SAI 9.2). Is there evidence that matters have been referred for consideration by the non-management representative? (SAI 9.3). Are the results of internal audits and corrective actions reviewed for effectiveness and follow up? (SAI 9.2). Does the company have a means of ensuring that the requirements of this standard are met? (SAI 9.5). Is there a system for continuous monitoring of activities and results to demonstrate the effectiveness of the company's SA 8000 policy? (SAI 9.5).
	RA:	RA check that an internal audit system has been set up. Examine frequency of internal audits, checklist used, auditor's report, observations of non-compliance, corrective action plans as a response to observations, implementation of corrective actions. (SAN 9 #7).
	IFOAM:	Who is in charge of internal inspections and who realises them? (Proposal IFOAM smallholder protocol).
	FLO:	No reference to internal audits or monitoring.

is found within the organisations, which each have their own logic, structures and standards in relation to similar concerns.

While the Joint Audit Template was an exhaustive piece of work that provided a wealth of information, it must be noted that in fact, it was not explicitly used by participants as a reference during the four audits discussed in this chapter. Instead it was circulated as background documentation within the project and provided to participants prior to the pilot audits as reference material. The Joint Audit Template essentially mapped the various starting points for each organisation on each aspect of internal control or management systems and is a visual representation of the multiple cognitions at the organisational level present at the outset of the SASA project with references to specific standards and policies. It should be noted that by the end of the Project, the indicators, requirements, documents and activities related to management systems and internal control were significantly more developed than at the outset as presented in Table 11.2.

The first step within the project was to examine the organic ICS, especially in terms of smallholder access to certification and then consider social certification potentials using the ICS mechanism. After the first round of pilot audits in the SASA Project (including the Thai rice audit with its focus on group certification), a stakeholder meeting on smallholders and group certification was organised to take place at the annual Biofach Organic Trade Fair in Nuremberg, Germany. This event provided an opportunity for issues surfacing within the SASA Project to be further elaborated and discussed by key stakeholders, primarily though not uniquely, from the organic sector. Many of the concerns in the broader group certification discussions (i.e. in Box 11.1) were raised. Through the Biofach meeting and subsequent report, these concerns became a part of the project history and shared understanding.

Whereas Boxes 11.1 and 11.2 above are examples of some of the varied perspectives on the organic Internal Control System that provide insight into the context in which the SASA Project took place, Table 11.2 illustrates the organisation-specific starting points. In theoretical terminology, they are examples of the multiple cognitions feeding the starting points for organisations participating in the SASA Project. As part of a project document that was read by the steering committee and participants in subsequent audits, Box 11.2 (and the full report the Box is drawn from) directly fed into the shared learning of the Project.

Box 11.2. Biofach Stakeholder Meeting Perspectives on the organic ICS (excerpted and adapted from Pyburn 2003).

"ICSs are not monolithic"
–David Gibson, Practice Director, Natural Resource Management, Chemonics USA.
Despite the multiplicity of ICS possibilities the term 'ICS' is becoming a reference point as though it defined only one specific way of organizing for group certification. External regulations and structural requirements are imposed. A fear is that the local, unique qualities and character of an ICS, and the internalized ideals widely upheld as an intention, may diminish under such imposition. It is crucial to accept different entry points and levels for internal control systems. Two different models or kinds of ICS emerged, which nuance the dimensions of ICSs: a. endogenous ICS - farmer associations with well-developed and active internal systems, and; b. out-grower schemes developed for economic objectives as opposed to internal support and development. Each scenario has different needs. The first is well-organized internally, not requiring elaborate external criteria, the other is without internal organization and requires external guidelines. An ICS of the first type has its own standards, its own system – it reflects ownership. In the second scenario farmers are suppliers to a buyer. The buyer controls the ICS to regulate his supply chain and the outsourced farmers.

"An ICS is a living organism that will externally reflect the nature of the group"
– Pedro Landa, Organización Internacional Agropecuaria (OIA), Argentina.
In the first system [an endogenous ICS] farmers need to think about how the ICS works for them and how documentation is beneficial. There is an acceptance of the ICS internal culture and rules, and criteria are internally developed and understood through extension. An ICS in this category will evolve under the direction of the producer group without the imposition of external criteria. If we create extensive rulebooks then the dynamism of the ICS will not have the space to flourish. Documents or rulebooks should be seen as a target of ideals for an ICS as opposed to a set of regulations. An ICS reflects the internal aspects of a group that is evolving but it is also demands cultural change within the group – this is an evolutionary process.

"The Big Insult"
–Vitoon Panyakul, Thailand Green Net.
When external criteria are imposed on an endogenous ICS, it is insulting and does not work. The insult gets even worse when you have multiple certification schemes from different certifiers with varying ICS criteria. When the EU and US regulations change, it is very difficult for producers. Harmonization is important. A bigger question is: when there is a situation of 30,000 producers, why can a local certification body not suffice? Yet local certification bodies are not generally recognized internationally.

»

"A Science for the Art"
– Ong Kung Wai, Humus Consultancy, Malaysia.
Disagreement arises when trying to define which part of the ICS is the most important – extension versus inspection. As the market place has specific concerns like agro-chemical use, one issue is accountability and credibility of the system. Concerns about conflict of interest are dividing inspection and technical assistance functions. Certifiers are no longer allowed to help producers through providing advice and technical support meaning that a learning or pedagogical approach in their field inspections cannot be employed. Instead inspectors must describe an ICS in concrete and quantitative terms. In this way, actors engaging with the ICS are fragmented. This approach is imbalanced and incorrect – the monitoring part is overstressed to the detriment of extension, while the latter is vital to the creation of a credible system.

"Romanticism"
– Alan Tulip, Agro-Eco EPOPA Country Manager, Uganda.
The idea of a group of producers developing an ICS for export, while perhaps ideal, is in many ways quite romantic and impractical. A Ugandan example was described wherein an average family cultivates three to four acres with family labor, one to two acres of those being for subsistence. The harvest of a 40-acre producer group will not fill a container for export. Top-down control is then needed as small producer groups lack the organizational and production capacities to survive. Much of the ICS discussion romanticizes farmer associations. Permitting grower groups to develop their own standards is simply not possible. Standards are imposed on producers, which allow quality assurance to the consumer. They are also necessary for ICS smallholder group certification. While the desire for a more participatory approach is commendable, standards are a threshold and a group is either on one side or the other.

Developing distributed (or collective) cognition

The development of shared perspectives and understanding on group certification began with the recognition that smallholders had specific needs that warranted attention within the project. Actual exploration of the issue began with the audits themselves. The audits were generally about one week of intensive shared learning experiences in a particular country and production context. Audit teams were made up of 6-10 participants representing the SASA organisations – auditors, steering committee members or Board members of the organisations involved and a researcher (the author of this chapter). In the spirit of the learning discussed by Hailey and James (2003, p. 195), the teams, "relied on village-based processes of dialogue to spearhead internal learning about the authentic needs of the communities". In the SASA case, the dialogue included representatives of farmer cooperatives, farmers, processing plant management and workers, internal inspectors and certification body representatives both at and beyond the village level. "The primary source of learning for most successful NGOs is the conscious reflection and analysis of their own implementation experiences (particularly where things have gone wrong) in order to learn and improve" (*ibid.*). The SASA audits provided a venue for this reflection and shared learning amongst organisations facing similar challenges and with complimentary missions.

The first round of SASA pilot audits was intended to introduce each organisation's auditing methodologies and examine difficult-to-audit issues in a particular crop and/or country context amongst other audit-specific goals. The Thai audit was the first that specifically targeted the organic internal control system. By meeting with farmers, internal inspectors and certification body auditors, the team was provided with a picture of how the ICS was operating in the cooperative being studied. Team discussion had two core themes: (1) the organic ICS; (2) social certification potential using the ICS structure. The initial discussion examined the organic ICS the team had audited and then representatives from the other (more social certification focussed) organisations presented their own systems and how they operated, which led to more in-depth consideration of how the ICS might be applicable to social certification.

Several months later, the Biofach stakeholder meeting provided both an opportunity to understand the many perspectives on and experiences with ICS within the organic sector (Box 11.2 above), and a venue for probing the potential for ICSs to be used for social certification. The learning of the SASA Project from the Thai audit was verified and used as a basis for broader stakeholder discussion. Related to social certification, the Biofach meeting surfaced key issues including some that were verified from the Thai audit. Box 11.3 reviews some of the issues raised in that meeting related to using ICSs for social certification.

In the second round of SASA audits the focus shifted from examining the organic ICS and considering the structure for social certification, to questions of organisational coordination for improved efficiency in the case of dual or multiple certifications and an examination of internal control mechanisms at work in each of the four systems involved in the project. This went beyond the initial objectives and expectations of the project.

Correspondence and coherence

As one of the two fundamental drivers of the cognitive process, *correspondence* – the match between an agent and its environment or context (Röling 2002, p. 33) – is one of the steps distinguished in the SASA learning trajectory. The pilot audits

Box 11.3. Biofach Workshop ICS and Social Certification Issues (excerpted and adapted from Pyburn 2003).

- Definition of social issues relevant to smallholders using ICS.
- Development or process dimensions to meeting standards within the ICS.
- ICS Involvement of smallholders in social standard setting.
- Methodologies for social certification using ICS.
- Challenges to using the ICS model.
- Redirecting the focus on social issues to social control and capacity building as opposed to standards, certification and inspection.
- Training the trainer as an approach to building capacity on meeting social standards.
- Smallholders and their role in the creation of more innovative systems (stimulating and creating the space for more input by smallholders).

provided this opportunity – particularly the second round of audits (Burkina Faso – organic Fairtrade mangoes, Costa Rica – certified coffee and Uganda organic Fairtrade organic cotton (see Table 11.1 for an overview of the pilot audit site, production systems and key learning), which were more directed towards specific issues as opposed to the first round (Thai rice audit), which were introductory in terms of the auditing methods, standards and structures of the four systems.

When the need for coordination to streamline smallholder group certification became apparent – especially between Fairtrade and organic certification bodies – the logistics and feasibility of joint auditing was tested on the Burkina Faso audit. The auditors for FLO and the organic certifier worked to develop an integrated audit template and to address the overlaps and unique aspects of the two sets of standards.

At certain points it was clear that despite the need for a streamlined audit process in order to better meet the needs of smallholder farmer groups, the reality was that the two sets of standards being verified (organic and Fairtrade) were very different and not very compatible in terms of a joint effort. That said, FLO has (post-SASA) addressed this issue by beginning to train some local organic auditors in Latin America, Africa and India in the FLO system so that they can coordinate audits not by having joint check-lists or templates, but by individual auditors who know the group via the other audit (Ruediger Meyer, FLO-Certification Ltd., pers. comm. May 2005). In this way FLO is capitalizing on the learning from the SASA Project, not for collective action or further work on standards coordination with the other organizations, but to directly address the 'on the ground' problem of dual certifications for producers.

The second driver of the cognitive process – *coherence* – is a concept used to understand the struggle and success of the SASA Project partners in their effort to move towards collective visions and project conclusions, including agreement on future steps. The final two audits (Costa Rica and Uganda) saw a very directed focus on understanding the elements of internal control in each SASA organisation and collective consideration on what the basic, common elements were amongst the four systems. From that, a baseline management system manual template was developed that would be applicable to smallholders aspiring to any one of the four certification systems. This demanded agreement amongst the participants on the essentials of each of their systems, and then with the shared concern for meeting smallholder challenges in mind, negotiation of a common set of criteria. These common criteria were captured in the *Generic Management System for Small Producer Groups*, one of the key SASA Project outcomes (Pyburn 2004).

In addition to this formal project outcome, bilateral learning and organisational learning were also observed. For example, the *Rainforest Alliance* as a result of learning about the organic ICS, through the SASA Project, transferred the system requirements to its own standards relating to producer groups. The *Rainforest Alliance* adopted the organic ICS almost verbatim. In addition the *Rainforest Alliance* then distinguished several kinds of producer groups and the associated risks with each kind of group for their own certification program. The ISEAL Alliance platform followed up the SASA Project learning on the ICS topic by commissioning a broader study on group certification that included in addition to the SASA organizations, also the *Forest Stewardship Council, the Marine Stewardship Council and* the *Marine Aquarium Council* (Pyburn 2005).

Lessons about social learning from the SASA case: theory and practice

Cognition has been used in this chapter as a basic theoretical foundation to try to capture the complexity of the different levels of learning involved in the project. The organizations involved in the SASA Project began with very different interests in terms of their involvement in the collaborative action-research on social certification in sustainable agriculture. In addition they each entered the Project embedded in the missions, politics, commitments and values of their own individual organizations. Over the two year period, considerable convergence amongst participating organizations can be seen as was described in this chapter using the example of Project learning on the topic of group certification and ICSs. Collective directions were agreed and the final outcome (the generic management system for small producer groups) represents a shared effort to streamline requirements to ease the burden of dual or multiple certifications for smallholder groups.

While cognition does provide an analytical framework to examine the process of inter-organizational change through learning, this case reveals some important questions. For example, the lines between multiple and distributed cognition are not absolute: at what point does multiple cognition begin to develop into distributed or collective cognition? Box 11.2 both illustrates some of the breadth of multiple cognitions expressed on the topic of organic group certification, and is representative of the shared learning of the project participants who organized and took part in that workshop. Despite some potential shortcomings of the theory-in-use, the four participating organizations did converge considerably through the shared learning experience of the SASA Project. Factors supporting convergence include: mutual concern for the plight of small farmers; a sense of urgency with regards to easing the burden of (dual/multiple) certification on small farmers; commitment to the missions and underlying values related to the organizations' standards; and, a sense of mutual interdependence – the perceived potential benefits of cooperation. Challenges to organizational convergence include: the embeddedness of each organization in its own mandate, mission and values; concern about losing a particular niche in the market; in-depth understanding of one's own system and its development and less comprehensive understanding of the other systems (though mutual understanding improved markedly throughout the project cycle); and the reality that people who learned most were actually on the pilot audits and not necessarily members of the Project steering committee – it is difficult to fully communicate the audit level learning to the project as a whole, particularly given the rotating participation of organizational representatives within the project.

This chapter presented the case of social and environmental standard-setting organizations learning their way towards more accessible yet credible certification systems for smallholder farmer groups in developing countries. Project learning was encapsulated in and expressed through:

- Audit and workshop reports.
- Action within each organization which may or may not be credited to participation in the SASA project itself.
- Collective action amongst project participants (bilateral or multilateral) as a direct result of the SASA project.
- Collective action at the ISEAL platform level.

Learning took place at multiple levels: individual, audit, project, organization and platform. What did not necessarily happen in the SASA project was the institutionalization of channels for absorbing individual learning into the project and organizations. The "challenge of sharing learning internally so that individual learning becomes organizational learning" (Hailey and James 2003, p. 196) – in this case, multi-organizational learning – was not universally met. Audit reports were the main artifacts that communicated audit learning. However, not everyone in each organization read these sometimes very detailed documents. Steering committee meetings and workshops allowed an integration of some audit learning. During those meetings, future actions of the project and for participating organizations post-project were decided upon. The transition from individual to organizational learning is an area for further exploration and research. That different individuals from the four organizations or their members/colleagues participated in the audits was both a strength in terms of broader participation, and a weakness in terms of continuity from one audit to the next and communication of the learning within each organization. Where one individual participated in two or more audits, a continuity and better understanding of the project process and objectives often were the outcome. The tradeoff between broad participation and project continuity is an interesting point of attention for future efforts in multi-organizational (and multi-level) social learning.

References

Courville, S. (2002) "SASA Project Pilot Audit Template" Social Accountability in Sustainable Agriculture Project, third draft 28/08/2002. available from http://www.isealalliance.org/sasa/documents/SASA_pilotaudittemplate_draft3.pdf

Courville, S. (2003) "Social Accountability Audits: Challenging or Defending Democratic Governance?" *Law and Policy,* 25 (3 July): 269-297.

Elzaaker, B. and Rieks, G. (2003) "Smallholder Group Certification – Compilation of Results – Proceedings from three Workshops February 2001, 2002, 2003", Tholey-Theley, March 2003.

Eritja, M.C., van der Grijp, N. and Gupta, J. (2004) "Sustainability Labelling and Certification in the Context of Sustainable Development" in Ertija, M.C., ed., *Sustainability Labelling and Certification,* Madrid: Marcial Pons, Ediciones Jurídicas Y Sociales S.A pp. 27-37.

European Commission Agricultural Directorate General 2003, *Guidance document for the evaluation of equivalence of organic producer group certification schemes applied in developing countries AGRI/03-64290-00-00-EN,* 6 November 2003.

Hailey, J. and James, R. (2003) "Learning leaders: the key to learning organizations" in Roper, L., Petit, J. and Eade, D., eds. *Development and the Learning Organization,* Oxfam, UK. pp. 190-204.

IFOAM (2003) *IFOAM's Position on Smallholder Group Certification for Organic Production and Processing. Submission to the European Union and member states 2003-02-03.* available from http://www.ifoam.org/press/positions/Small_holder_group_certification.html

ISEAL Alliance (2006) *ISEAL Code of Good Practice for Setting Social and Environmental Standards P005 – Public Version 4 – January 2006,* Oxford UK p. 6. available from http://www.isealalliance.org/documents/pdf/P005_PD4_Jan06.pdf

Keen, M., Brown, V.A. and Dyball, R. (2005) "Social Learning: A New Approach to Environmental Management" in Keen, M., Brown, V.A. and Dyball, R., eds., *Social Learning in Environmental Management – towards a sustainable future*, London, UK: Earthscan, pp. 3-21.

Keen, M. and Mahanty, S. (2005) "Collaborative Learning: Bridging Scales and Interests" in Keen, M., Brown, V.A. and Dyball, R., eds., *Social Learning in Environmental Management – towards a sustainable future*, London, UK: Earthscan, pp. 104-120.

Leeuwis, C. and Pyburn, R. (2002) *Wheelbarrows full of frogs – social learning in rural resource management: international research and reflections,* Assen: Koninklijke Van Gorcum.

Leeuwis, C. (2004) with contributions from A. van den Ban, *Communication for Rural Innovation – rethinking agricultural extension*, Third Edition. Oxford, UK: Blackwell Science.

Pyburn, R. (2003) "SASA Smallholder ICS Workshop Report", 15 February, unpublished internal report.

Pyburn, Rhiannon (2004) "SASA Final Report on Internal Control Systems and Management Systems", ISEAL Alliance, Bonn, Germany, 3 August, p. 94, available from http://www.isealalliance.org/sasa/documents/SASA_Final_ICS.pdf

Pyburn, Rhiannon (2005) "Towards Best Practices in Group Certification – a report for the ISEAL Alliance", ISEAL Alliance, Bonn Germany, January, unpublished internal report.

Röling, N. (2002) "Beyond the Aggregation of Individual Preferences. Moving from multiple to distributed cognition in resource dilemmas", in C. Leeuwis and R. Pyburn, eds., *Wheelbarrows full of frogs – social learning in rural resource management: international research and reflections,* Assen: Koninklijke Van Gorcum, pp. 25-47.

SASA Project (2004) *Social Accountability in Sustainable Agriculture (SASA) Project Main Report for Funders*, Bonn, Germany, 15 September, unpublished.

Surowiercki, J. (2004) *The Wisdom of Crowds – why the many are smarter than the few and how collective wisdom shapes business, economies, societies and nation, USA:* Doubleday.

Tiffen, P. (2003) "A Chocolate-coated case for alternative international business models" in Roper, L., Petit, J. and Eade, D., eds., *Development and the Learning Organization*, Oxfam, UK. pp. 169-189.

Wenger, E. (1998) *Communities of Practice – learning, meaning and identity*, Cambridge, UK: Cambridge University Press.

Chapter 12

Social learning processes and sustainable development: the emergence and transformation of an indigenous land use system in the Andes of Bolivia

Stephan Rist, Freddy Delgado and Urs Wiesmann

Introduction

The first part of the present paper gives a brief overview of the context in which the 'social learning approach' emerged as a steadily growing stream of theory and practice aiming to integrate societal development with the normative principles of sustainability. The second part summarises the key findings of a research partnership between the Agroecology Program of the University of Cochabamba (AGRUCO) in Bolivia, the Centre for Development and Environment (CDE) at the University of Bern, Switzerland, and the Swiss National Centre of Competence in Research (NCCR) North-South, concerned with the elaboration of concepts and instruments designed to co-produce knowledge for sustainable development through the enhancement of social learning processes. Analysis of the emergence and transformation of an indigenous land-use system in the Bolivian Andes will serve to illustrate how the idea of social learning processes is used by local people when interpreting their own historic developments. Against this background it will be shown what kind of potential this has for the development of a more sustainable land-use system which, while based on basic principles of the Andean cosmovision, incorporates new insights gained through the interaction with external actors.

Sustainable development, social learning and communicative action

A major challenge of sustainable development is its essentially normative character. It defines *what* to aim for, without saying *how* to achieve this aim in specific social, ecological, economic, cultural or historical situations. Making the concept operative means translating it into a set of concrete action-guiding ethical values and norms for social actors (Wiesmann 1998). As a consequence, sustainable development cannot be limited to the development of new technologies. Instead, it depends directly on substantially changing the norms, rules and regulations according to which social actors interrelate. Consequently, 'governance for sustainability', which focuses on collective learning about how to reform policy-making to promote sustainability, is becoming increasingly important (Meadowcroft 2004). In this view, instead of participation being a blueprint for the implementation of preconceived policies, it becomes a linchpin of societal processes that attempt to reshape present models of governance of human-nature relationships.

Conceptualisation and problem-solving in the case of complex and highly uncertain socio-environmental dynamics and problems can be understood as a 'collective experiment' (Latour 1998). The principle of certainty in pure science is replaced by a 'culture of research', meaning that science and society engage in a collaborative process through which questions are asked and solutions are sought collectively. Such a 'social learning approach' has become increasingly important in environmental policy-making, both in developed and developing countries (Parson and Clark 1995, Wollenberg *et al.* 2001). The 'social learning approach' represents a philosophy focusing on participatory processes of social change; it is based on an actor-oriented approach that forms part of a theoretical framework in which social processes are defined as non-linear and non-deterministic (Woodhill and Röling 2000).

The shift from multiple to collective cognition is defined as a key feature of the social learning approach (Röling 2002, p. 35). Multiple cognitive agents are understood as actors who tend to maintain mutual isolation. But the more they become interdependent, e.g. in the context of resource management, the more likely they are to be caught in conflicting or competitive, power-driven relationships that emphasise the differences between them. In such situations, action research-based initiatives make it possible to broaden the space for communication, allowing for more reflexive communication, e.g. through platforms or forums for deliberation about the use of natural resources and negotiation and coordination of relevant aspects and activities. Such reflexive communication makes it possible for multiple, but separate, cognition to develop into collective and consensual cognition.

In order to become operational, this socio-psychological aspect of social learning must become part of a theory and practice of action. Röling and Maarleveld (1999) point out that Habermas's 'theory of communicative action' (1984) has done much to place social learning and collective action on the agenda. By focussing on the interrelationships between the 'system' (structures), and the specifically possible type of action of 'strategic action' (defined as oriented towards egocentric goals) to 'communicative action' (defined as oriented towards mutual understanding of situations involving interaction) among social actors, Habermas' theory offers a way to link the idea of 'collective cognition' with action theory and practice: social learning processes can be understood as 'collective experiments' through which social actor categories aim to broaden spaces, allowing for a shift from 'strategic' to 'communicative action' (Röling 2002).

Research on the nature of social learning processes is still incipient. Several studies (Dewulf *et al.* 2005, Woodhill and Röling 2000) show that investigating the patterns of communication that emerge during collective learning processes might make it possible to shed light on the arenas, conditions and dynamics that either enable or hinder a shift from strategic to communicative action. Diving more deeply into the dynamics and characteristics of social learning processes, Rist *et al.* (2006) demonstrate that they cannot be understood exclusively as a more reflexive or science-based treatment of the unintended outcomes of current forms of land-use or development. Besides these important factors, the social learning process can only emerge if they allow for restoration of trust, cooperation, empathy, intuition and inspiration between different groups characterised by multiple cognition. In order to support a shift to collective cognition, it became evident that jointly developed social capital, cognitive, social and emotional competences represent four fundamental key dimensions of social learning processes.

The study area

The *Ayllu Majasaya Mujlli* is a traditional mini-federation of 16 indigenous Aymara communities, living in a territory of about 18,500 hectares, situated in the eastern cordillera of the Bolivian Andes at an altitude of 3,800–4,500 m. There are some 3,500 people living in this area whose livelihoods depend on a complex system of collectively synchronized strategies related to agriculture, livestock, food processing, handicrafts, off-farm activities, and education.

The territory of the *Ayllu* is managed as a common property resource, comprising four sectors of arable land known as *Ayta.* They consist of a total of 12 smaller adjacent cultivated areas called *Aynokas,* and are cultivated according to a three-year crop rotation cycle of potatoes, quinoa/canahua/oats, and barley. After three years of cultivation the fields enter a fallow period, normally for 9 years, allowing soil fertility and natural vegetation to be restored. At the same time, the fields are used as a grazing area. Although coordination of crop rotation, fallow periods, and redistribution of land are organized and controlled collectively, crop management and external inputs, as well as use of production, remain the responsibility of the families. Every year the community names two young men and their families as *Jilakata* (protectors of *Aynokas*). As 'major brothers' they are in charge of 'shielding' the *Aynokas* by monitoring correct adherence to the rules of the land-use system, as well as by motivating people to perform the rituals necessary for assuring the important contributions of spiritual beings- '*Pachamama*' (earth's mother), and the '*Achachilas*' (ancestors) – to good crop production.

The *Ayllu Majasaya Mujlli* is particularly interesting for the study of social learning processes in light of the following findings:

- Tapia (2000) and Pestalozzi (2000) show that the *Aynoka* land-use system displays high levels of ecological and social sustainability, despite a more than 4-fold increase in population since the beginning of systematic settlement. Its main advantages are: compulsory crop and fallow cycle management throughout the entire territory; a lower incidence of soil-borne pests (nematodes); more efficient use of family labour through collective maintenance of basic infrastructure (access roads, barrages, stonewall-fencings; and clear separation of grazing and cultivated areas, allowing for significant reduction of herding labour as well as facilitation of community-based redistribution of access to agricultural land from 'old' (less numerous) to young, newly established families.
- People have always been, and still are, a part of social networks linking them with historically changing poles of political and economic development (Delgado 2002). Consequently, local history reflects transformations since pre-colonial times, whether in terms of the co-existence of historically different forms of social organization, native crops and husbandry and those introduced from Europe, or indigenous and Christian forms of religiosity (Rist 2002).

Emergence and transformation of the *Aynokas* as a result of social learning processes

Analysis of the evolution of social organization clearly showed that people perceive the present land-use system as the result of co-evolution between society and nature. In collective memory, co-evolution is an essentially open, long-term social learning process that ranges from prehistoric periods to the present, as evidenced by a systematisation of the main historical periods shaping the chronological grid of local history (see Table 12.1).

People interpret the emergence and transformation of social organization and related forms of land use as a result of the gradual development of certain social and cognitive competencies in different historical periods (see Table 12.1). For the *Pre-Inca period,* they highlight the development of social competencies directly related to respectful obedience and duty towards spiritual entities e.g., *Pachamama* or 'Earth's mother', *Achachilas*, or 'ancestors', and *Mustrama*, or 'ancient Sun'. Cognitive competencies are described as impersonated by relatively few *Yatiris* (shamans), who represent 'ordinary people' in direct dialogue with ancestors.

The *Inca Period* is characterized by a weakening of unconditional acceptance of the ethical principles of respect and obedience toward ethnic authorities. These principles were recognized only to the extent that local representatives succeeded

Table 12.1. Presently existing local institutions and their correlation with different historical periods. For details of each period, see Rist *et al.* 2003 (Table first published in Journal of Mountain Research and Development).

Institutions of social organisation	Historic period	Approximate dates
Ritos para Pachamama, Achachilas	Pre-Inca	± 30,000 BC
Ayllu	Pre-Inca	1400 AD
2 Tambos and 3 sacred places	Inca	1400 – 1534
3 Chapels; pasantes; ayta & aynoka	Early Colony	1534 – 1781
Corregidores	Late Colony	1782 – 1826
School	Independence	1934
Sindicato	Land reform	1954
Municipal councillor	Decentralisation	1994

in transmitting the wisdom of the Incas to local authorities and their 'vassals'. The first element of ethical self-determination arises at this time, expressed through interest in learning, understanding, and disseminating the wisdom of the Incas. In terms of cognitive competencies, a reflexive and more personalized element was emerging as a part of social and spiritual organization. The skill of reflexivity is perceived as emerging from the 'heart', representing reflection as a latent rather than an intellectual quality. In terms of social competencies, cooperation and interaction between authorities and 'ordinary people' are perceived as becoming gradually less vertical.

During the *Colonial Period*, the region became more densely populated, leading to the creation of *Aynokas*, implying more explicit definitions of norms and rules for regulating access to and distribution of land. This was possible because the people were recognised by the colonial administration as an 'original community', provided with relatively high degrees of internal autonomy in exchange for forced labour in the mines of Potosi and taxes paid to the church and the colonial administrators. The development of social competence required in the struggle for autonomy and internal self-determination is seen as being closely related to the main structural innovation in this phase of the collective learning process: the creation of a system of communal duties was based on a yearly renovation of all duties. This period is perceived as giving rise to the origins of community organisation, which, by relating to pre-colonial principles, aimed to achieve self-governance based on an ongoing personalized effort to overcome selfish behaviour. Social competencies, such as the ability to communicate, empathy, social mobility, personal autonomy, and self-esteem, were becoming important.

With regard to cognitive competencies, the clash of indigenous and colonial norms and rationalities led to higher levels of self-reflection, resulting in a substantial transformation of knowledge that had previously been tacit to knowledge that was explicit. As a consequence, people started to define their identities in terms of active participation in social organization, rather than on the basis of ethnic aspects. This gave them insight into and experience with a collective processes of co-production of explicit indigenous knowledge, mainly concerned with the internalisation of procedural and normative aspects ('knowing how to do things' and 'what is good and bad').

In this period people focused on the emerging differentiation of social action with regard to internal and external spheres: internal social organization was characterised by a space motivating people to overcome egocentric behaviour or 'strategic action', in favour of 'communicative action'. The latter is perceived as an ongoing process based on an inter-subjectively validated understanding of one's own situation. A key mechanism for this can be seen in the compulsory and rotating integration of all members of the communities in self-governance of the *Ayllu*. With regard to external relationships, communicatively (internally) defined norms and values were defended by adapting to the rationality of the colonisers, which emphasised the egocentrically defined rights and duties of the subordinated, thus assuming a predominance of 'strategic action'.

During the *Republican Period* internalization of ethical values related to reciprocity, autonomy, equity, and personal responsibility was important. Community members sharply opposed the established political structures. A clear sign was the rejection of privatization of community land in 1882, promoted by the first liberal government, which also points to a substantial increase in the ability to negotiate with external actors. Internally, it points to a broader capacity for deliberative reasoning about the advantages and disadvantages of the community-based use of natural resources. In this context, the first school in the area was founded in 1934. It was a private school supported by the community itself, because public education was not available to indigenous people. As a consequence, reflexivity was further developed by increasing the number of people involved in this process, as well as in its scope: besides covering procedural and normative aspects, the process of knowledge production in the communities was further deepened by giving attention to the ontological and epistemological differences underlying the rationalities of social organization of the *Ayllu* and the liberal-capitalist state.

The *Modern Nation-State Period* is marked by the Land Reform of 1952, leading to the creation of 5 agrarian *sindicatos* (peasant syndicates) run by a democratically elected council; the *Ayllu* then witnessed a new form of socio-political organization. However, this did not mean that traditional duties were abandoned. In 1962, public administrators granted full recognition to the *Ayllu*, consisting of 5 *sindicatos*. The number of *sindicatos* has since increased from 5 to 16, mainly due to population growth and conflicts over land use in the *Aynokas*.

Further analysis of oral history revealed that distinctions between chronologically separate periods are not identical with a diachronic perception of time. They merely express the temporal distance between different periods and the present. According to the worldview of the people in the areas, the past, the present and the future are interconnected through the spiritual entities of *Pachamama* and the *Achachilas*.

This synchronistic view of time was identified as a major explanation for the present co-existence of different forms of organisation, agricultural resources, and religious practices. People conceive of the co-existence of endogenous and exogenous elements as the basis for a dynamic that, rather than being dialectical, is based on reciprocity and complementarity. This synchronistic notion of time, rather than just representing an important underlying aspect of the local cosmovision, expresses a key feature of the currently dominant social and political organization, which is based on interaction between institutions originating in different historical periods (Figure 12.1).

The current form of social organization is based on four interrelated spheres of self-governance. Formally and informally, all four spheres depend on the general assembly of the *Ayllu*, normally held annually. The most remote historical reference is the 'native organization'. Through a complex network of annually rotating authorities, it focuses on the coordination of issues related to the whole *Ayllu*, paying special attention to the maintenance and revitalization of the cultural and spiritual relationships of living members of the *Ayllu*, while ancestors and other spiritual entities live beyond a diachronic notion of time.

Within the context of the native organization, there are more recently created *sindicatos*. These are in charge of issues relating to the communities and also have annually rotating authorities. Since they were formed as part of a formal requirement to obtain land rights recognised under the land reform law, they were initially closely related to the outside world of politics and political parties. Later they became partners of external NGOs and local bodies of public administration. Since 1979, all *sindicatos* have been affiliated with the powerful National Confederation of Sindicatos of Bolivia (CSUTCB). This has permitted integration of the communities into powerful social movements. These movements were deciding factors in putting an end to more than 500 years of colonization and neo-colonization. The creation of this historically unique situation meant that in the elections of December 2006, Evo Morales Ayma, a representative of the vast

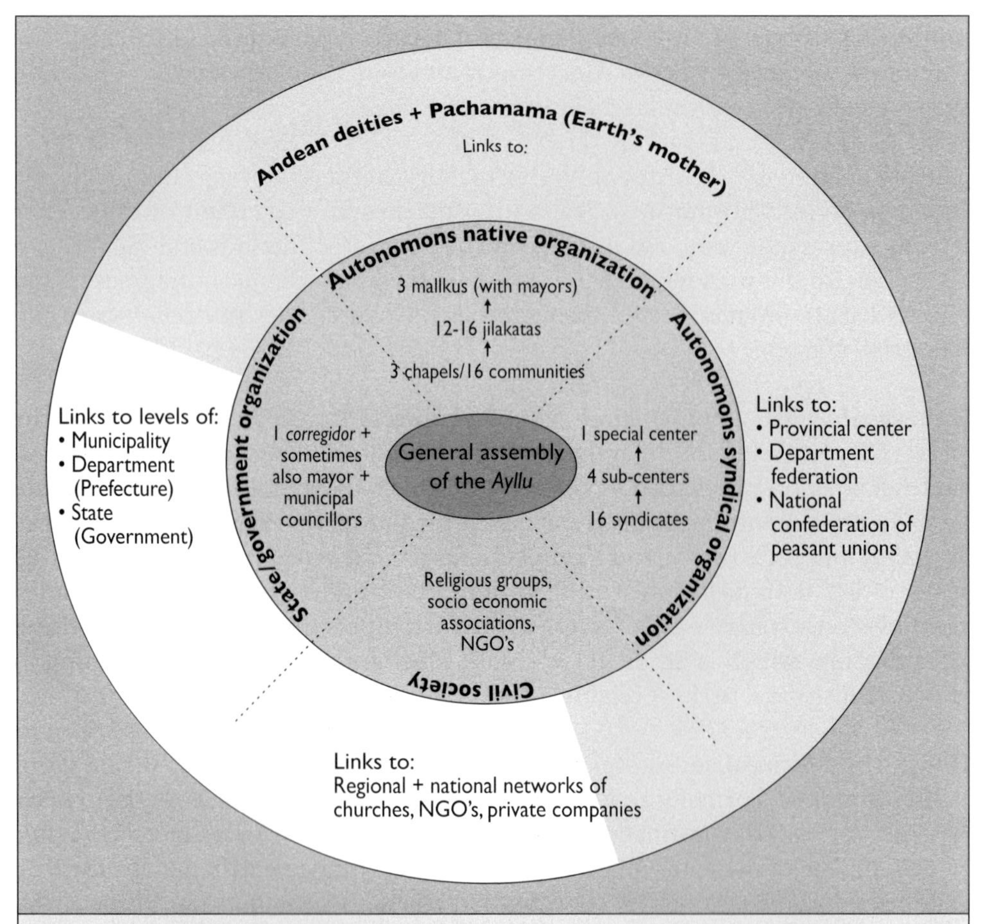

Figure 12.1. Main institutions related to local governance in the Ayllu Majasaya de Mujlli. (Figure first published in Journal of Mountain Research and Development, Rist *et al.* 2003).

majority of indigenous people, became president of Bolivia. The involvement in this social movement led progressively to an emancipation of the communities from the state and over-dominant NGOs. The authorities of the *sindicatos* are also strongly committed to representing their needs in the municipalities, which, due to substantial progress in decentralisation of central government responsibilities and financial resources, are becoming important partners in the development of the *Ayllu*.

Relationships to religious groups or NGOs involved in social work are considered to be 'free', meaning that the definition of this dimension of social organization takes place mainly at the family level. It is important to note that the annual rotation of the authorities is compulsory. People are generally named through assemblies of the *sindicatos* or the *Ayllu.*

This system of self-governance means that every family forms part of the local 'government' for several years by executing one of the 13 duties within the *sindicatos* or one of the 21 duties in the native organization. The election of the authorities is decided by considering the experience of people in the management of community affairs. Younger people move along a path of community duties, starting with less important responsibilities such as being secretary of sports, roads or education in the *sindicato,* or fulfilling the duty of *Jilakata* within the native organization. The more experienced people become in the management of community affairs, the more frequently they become candidates for higher positions in the social organization, such as the secretary for relations and the secretary general of a *sindicato.* If already older, an individual can be appointed to the very prestigious position of *Mallku,* which means 'father and ancestor' of the *Ayllu.*

By moving along this 'path of duties,' individuals follow a pattern of social learning based on competency development that is also characteristic of individual socialisation: starting with learning about procedures (i.e. learning *how* things are to be done), the process moves further to acknowledge its normative basis (i.e. learning to know the values and principles of what is *good and bad*) ending up with an understanding of the underlying patterns of interpretation (or *why* things have to be done in certain ways) (see Figure 12.2).

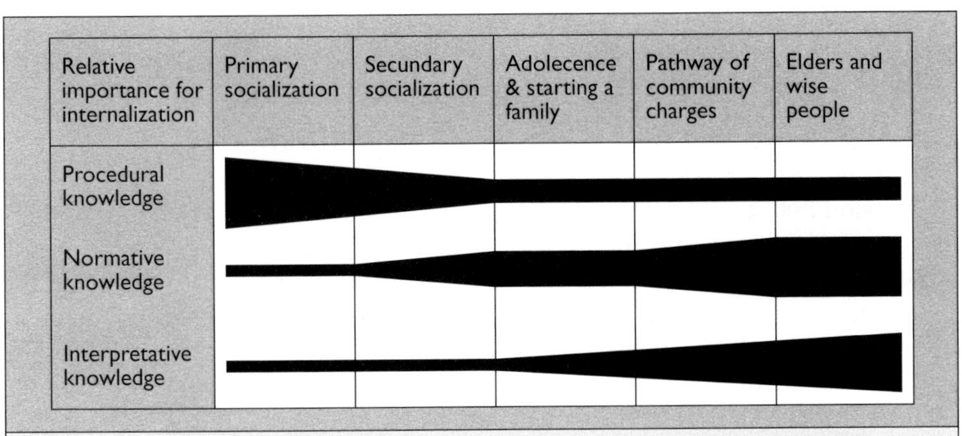

Figure 12.2. Relative importance of procedural, normative and interpretative knowledge related to individual socialisation in the Ayllu Majasaya Mujlli. Modified version from Rist 2002.

Privatisation of the *Aynoka*?

The following section will examine the currently ongoing social learning process related to the question of whether privatisation of community-owned land is likely to lead to more sustainable use of the territory. The aim is to visualise some additional features of the social learning process related to the search for more sustainable development in the context of Andean communities. The promoters of privatisation were mainly young peasants who so far have learnt to serve their communities through duties in the *sindicatos*. Inspired by other experiences and the increasingly dominant discourse on 'modernisation' of political parties, NGOs and public administration, they argue that collectively-owned land prevents them from getting access to credit from banks (due to the lack of individualised land titles), and that it requires too much investment of time in the complex network of duties that have a constitutive character for maintaining the *Aynoka*.

Due to interaction with AGRUCO, which began to work in the *Ayllu* in 1986, the increasingly visible signs of overexploitation of natural resources (soil erosion, overgrazing, loss of cultivated and spontaneous biodiversity) also became important. The promoters of the 'privatisation' strategy took up these aspects, arguing that the required improvements in land use through soil conservation, and improved management of pastures and animals combined with a shortening of the fallow period, would be more easily achieved. Accordingly, the *Aynokas* should be distributed to each family, making it possible to have plots and pastures near the houses. However, as the land titles are officially held by the native organization of the *Ayllu*, privatisation could only progress if the authorities who represent the whole *Ayllu* could be convinced. But, as they were against privatisation, a sharp conflict was generated between 'traditionalists' and 'privatisers' concerning the future of the *Aynoka* and that of the *Ayllu*.

Religion, land use and the rage of Pachamama

The period 1982 to 1992, when the area was repeatedly struck by drought, hail, storms, and heavy frosts related to the 'El Niño' phenomenon, was the time during which the conflicts began. Conflict was gradually transformed into debate, to the extent that people began to recognise that the breakdown of consensus about the appropriateness of community-based management of the Aynoka system had its roots in changes in religious life. This started in the early 1970s, when the Catholic Church launched a new form of Christian instruction aimed at 'soft' reformation of native beliefs. This simultaneously led people to a more individualised way of relating to religion and selective disappearance of native feasts and rites. Personal choices in religious and spiritual matters became possible. Several evangelical sects that arrived a few years later increased the religious options available to local people. By radically rejecting all types of symbols, whether Catholic or Andean, Protestants nevertheless contributed to a substantial increase in reflection on religiosity. In a short time the majority of the communities belonged to one of the many evangelical confessions, leading to a massive abandonment of rituals, feasts, and other Andean customs. The confrontations between different religious confessions refusing to fulfil communal duties significantly weakened social organization.

Only with this insight were people able to engage in a search for alternatives. Doubts about the relevance of current religious life were the starting point for emerging learning processes. People began to ask themselves whether their difficult situation was a sign from *Pachamama*, who was 'angry' because she had been forgotten. According to their belief, *Pachamama* makes 'Nature' act in response to the degree of understanding (cognition) and fulfilment of Andean ethical values in communities. This led to a dialogue between the different and contradictory Christian confessions, resulting in an agreement to return to the practices of some of the lost collective rites that express gratitude and respect for *Pachamama*. Consequently, the reflective process on religion, which had been significantly enhanced in the past 25 years, now began to focus on existing relationships between spiritual life and current social, economic, and ecological conditions. This has restored a pattern of interpretation that is eminently Andean. In terms of discourse, the main difference with the past is that "today we want to *understand* about *Pachamama* and religion. Just *believing* is not enough".

Once Christian religiosity was re-conceptualised as being tolerated by *Pachamama*, the fulfilment of community duties and the performance of rituals and feasts turned out to complement rather than to contradict an indeed autonomously redefined understanding of Christianity. As a consequence, social connectivity, trust and cooperation became fundamental in a collective experiment through which people explore new alternatives for land use. While some communities search for alternatives within the existing *Aynoka* system, others have been dividing the territory family-wise, which would have been the case had they been allowed to privatise the land. The common denominator is a basic agreement that every such attempt must respect the rules of collective land tenure in the entire area, guaranteeing that evaluation of the results and experiences remains the task of the entire *Ayllu*.

Another aspect credited with great value in finding a way out of the problems was related to restoration of trust, cooperation, empathy, intuition and inspiration between the different divided groups. Accordingly, it became evident that besides social and cognitive competence, emotional competence and social capital also play a fundamental role in the development of social learning processes. This aspect, which has also been confirmed in case studies from Peru, Mali, India and Madagascar (Rist *et al.* 2006), made it possible to perceive social learning processes in terms of four fundamental dimensions: social capital, and cognitive, social and emotional competencies (see Figure 12.3).

Concluding discussion

The main stages through which the members of the *Ayllu Majasaya Mujlli* interpret their history were analysed from the perspective of a social learning process. It was shown that social learning processes are principally related to the development of four dimensions. Within a steadily growing social network of local and external actors the joint development of social capital, and cognitive, social and emotional competencies are deciding factors for the emergence of social learning processes. The more fully these four dimensions were developed, the more important it became to integrate the interrelationships and consistency of procedural, normative and interpretative forms of knowledge as fundamental resources in the social learning processes.

Cognitive competencies evolved to ever-deeper levels of reflexivity, leading people to recognize them as part of the unity of humans and Nature, which in turn allowed a reinterpretation of the ecological situation in relation to social and religious–spiritual life. The connecting element between Nature and society is morally correct action. However, in this particular case, acting 'correctly' not only means fulfilling a set of defined moral norms; it also implies that the community

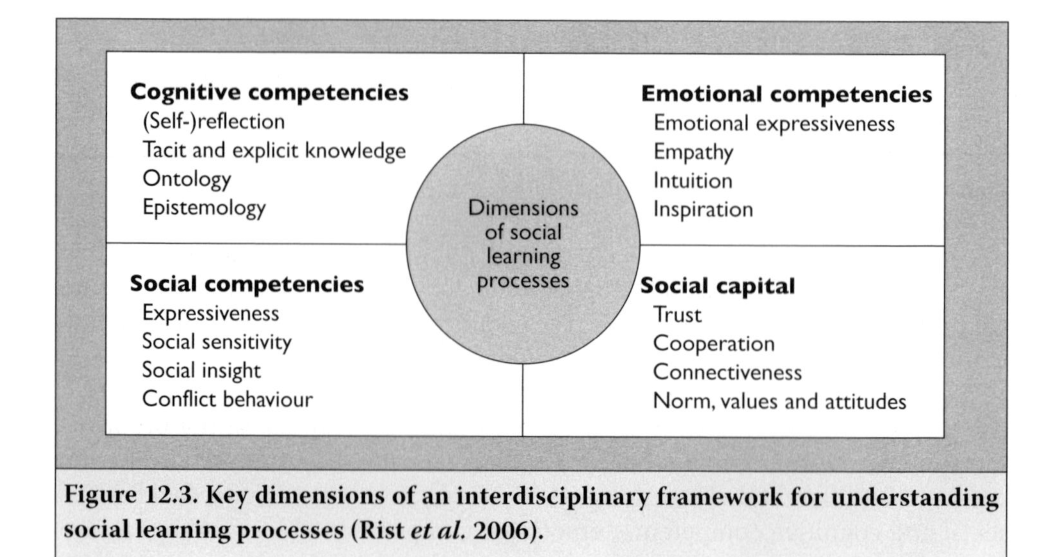

Figure 12.3. Key dimensions of an interdisciplinary framework for understanding social learning processes (Rist *et al.* 2006).

must *learn* to understand the world according to Andean concepts of life, nature, and society.

Thus, solutions to current social, ecological or economic problems are essentially perceived as a social and cognitive issue that requires a combination of deep reflection with an ongoing effort to decode, differentiate, and internalize ethical principles in everyday actions. This results in an increase in social and emotional competencies that permits internally driven transformations of social organization. This leads in turn to a dialogue, where nature mirrors the state – or 'health', as expressed by local people – of the society. Meanwhile, the society mirrors the health of the natural environment. This points to an important dimension of sustainable development that has not been adequately considered: a non-dualistic view of the relationship between human beings and nature seems to be closely related to the existence of a land-use system that displays a high potential for sustainability. This means that sustainable development cannot be separated from consideration of specific ontological and epistemological dimensions related to basic questions about the nature of human beings, the time-space continuum, mind, and matter.

As demonstrated by the conflict over privatisation of community land, there was a clear tendency to strengthen the role of multiple cognition within the social organization of the *Ayllu*. This tendency initially manifested itself in the emphasis on procedural knowledge related to the need for acceding to credits. By interrelating this procedural requirement to its consequences for the normative and interpretative forms of Andean knowledge, multiple cognition was gradually re-integrated into more collective cognition. This, in turn, made it possible to overcome periods of crisis allowing time to experiment and introduce innovations in the land-use system as part of a social learning process.

The observed increase in autonomy and self-determination in social and religious-spiritual life was a decisive factor in successful collective learning. However, the close relationships maintained between the development of social competencies and social capital, through the system of self-governance and the expansion of cognitive competencies that is only possible at an individual level, did not lead to individualisation. Rather, it led to a process of individuation, in the sense of the incorporation of basic features of a collectively, created cultural identity into persons belonging to a certain group, with the aim of transforming egocentric behaviour into communicative action, leading to support community building 'from within' the persons involved.

The focus on a simultaneous development of social and cognitive competencies as part of social learning processes represents another important element. This prevents people from conceiving of 'development' as a mere issue of substitutable transfer and application of new practices or norms, making it possible to re-connect any innovation as part of a social learning process that aims at the re-creation of coherence between procedural, normative and interpretative forms of knowledge.

Against the background of the still strongly growing influences of a one-sided and economically driven 'globalization', this case study reveals some important lessons regarding social learning as a way towards a more sustainable world. First, the case demonstrates that more sustainable practice is only possible when people work to create social spaces in which it is possible to collectively learn how to organize life on the basis of communicative, rather than on strategic action allowing for a gradual shift from multiple to collective cognition. A second lesson is that social learning processes are more likely to overcome multiple cognition and strategic action if they are able to interconnect the levels of procedural (practical), normative (ethical-moral) and interpretative (philosophical) dimensions of the knowledge of the actors involved in the search for more sustainable development. It is suggested that this also represents an important resource for uncovering the normative and philosophical assumptions, which are behind what appears as 'modern development' based on 'globalization', e.g. related to the supposed superiority of privatization and 'free-market' societal organization. Through this, the promoters of 'globalization' are forced to engage in a debate about their normative and interpretative assumptions; instead of presenting them as quasi 'natural laws' they turn out to be just one – and indeed a highly reductionist and therefore questionable – position in the societal debates on the orientation of future human development. A third lesson refers to the importance of perceiving social learning as based on individually and collectively developing social capital, cognitive, social and emotional competences. Social learning, rather than being a passive expression of new 'sustainable' policies, becomes the result of increasing spaces for, and vigour of agency that aims to influence the 'process of structuration' of the future societies.

Acknowledgements

The authors acknowledge the support of the Swiss National Centre of Competence in Research (NCCR) North–South: Research Partnerships for Mitigating Syndromes of Global Change, co-funded by the Swiss National Science Foundation (SNSF) and the Swiss Agency for Development and Cooperation (SDC), making possible the joint elaboration of the present paper.

References

Delgado, F. (2002) *Estrategias de autodesarrollo y gestión sostenible del territorio en ecosistemas de montaña – Complementariedad ecosimbiótica en el ayllu Majasaya Mujlli, departamento de Cochabamba, Bolivia*, La Paz: Ediciones PLURAL – AGRUCO.

Dewulf, A., Craps, M., Bouwen, R., Abril, F. and Zhingri, M. (2005) "How indigenous farmers and university engineers create actionable knowledge for sustainable irrigation", *Action Research,* 3(2): 175-192.

Habermas, J. (1984) *The theory of communicative action – Volume 1* (Translated by T. McCarthy), Boston: Beacon Press.

Latour, B. (1998) "Essays on Science and Society: From the World of Science to the World of Research?", *Science,* 280: 208-209.

Meadowcroft, J. (2004) "Participation and sustainable development: modes of citizen, community and organisational involvement" in W. Lafferty, ed., *Governance for sustainable development: the challenge of adapting form to function*, Cheltenham, UK; Northampton, MA, USA: Edward Elgar, pp. 162-190.

Parson, E. and Clark, W. (1995) "Sustainable Development as Social Learning: Theoretical Perspectives and Practical Challenges for the Design of a Research Programme" in Gunderson, L., Holling, C. and S. Light, eds., *Barriers and bridges to the renewal of ecosystems and institutions*, New York: Columbia University Press, pp. 428-460.

Pestalozzi, H. (2000) "Sectoral Fallow Systems and the Management of Soil Fertility: The Rationality of Indigenous Knowledge in the High Andes of Bolivia", *Mountain Research and Development,* 20(1): 64-71.

Rist, S. (2002) *Si estamos de buen corazón, siempre hay producción – Caminos en la revalorización de formas de producción y de vida tradicional y su importancia para el desarrollo sostenible*, La Paz: Ediciones PLURAL – AGRUCO – CDE.

Rist, S., Delgado, F. and Wiesmann, U. (2003) "The role of Social Learning Processes in the Emergence and Development of Aymara Land Use Systems", *Mountain Research and Development,* 23(3): 263-270.

Rist, S., Chiddambaranathan, M., Escobar, C. and Wiesmann, U. (2006) ""It was hard to come to mutual understanding..." Multidimensionality of social learning processes in natural resource use in India, Africa and Latin America", *Journal of Systemic Practice and Action Research,* 19(3): 219-237.

Röling, N. (2002) "Beyond the aggregation of individual preferences. Moving from multiple to distributed cognition in resource dilemmas", in Leeuwis, C. and R. Pyburn, eds., *Wheelbarrows full of frogs – Social learning in rural resource management*, Assen: Van Gorcum, pp. 25-47.

Röling, N. and Maarleveld, M. (1999) "Facing strategic narratives: In which we argue interactive effectiveness", *Agriculture and Human Values,* 16: 295-308.

Tapia, N. (2000) *Agroecología y agricultura campesina sostenible en los Andes bolivinos – El caso del Ayllu Majasaya Mujlli, Cochabamba-Bolivia*, La Paz: Ediciones PLURAL – AGRUCO.

Wiesmann, U. (1998) *Sustainable Regional Development in Rural Africa: Conceptual Framework and Case Studies from Kenya*, Bern: Geographica Bernensia.

Wollenberg, E., Edmunds, D., Buck, L., Fox, J. and Brodt, S., eds. (2001) *Social Learning in Community Forests*, CIFOR and East-West Centre.

Woodhill, J. and Röling, N. (2000) "The Second Wing of the Eagle: The Human Dimension in Learning our Way to more Sustainable Futures", in Röling, N. and A. Wagemakers, eds., *Facilitating Sustainable Agriculture. Participatory Learning and Adaptive Management in Times of Environmental Uncertainty*, New York: Cambridge University Press, pp. 46-71.

Chapter 13

From centre of excellence to centre of expertise: regional centres of expertise on education for sustainable development

Zinaida Fadeeva

Introduction

Accelerated changes in the growth and economic performance of nations and regions, trends in trade and investment, media and information flow, cause shifts rapid enough to prevent accumulated knowledge and expertise from more accurately forecasting development. This calls for the knowledge systems to prepare people not only for the adequate acquisition of skills for coming to grips with upcoming changes but to generate solutions that would anticipate and shape future development trajectories. Proactive strategies require more, and different, skills than reactive ones. While it is proven to be true as a response to the conditions of uncertainties caused by rapid development (Wolfe and Gertler 2002), it is essential in the circumstances of slow or no progress. The need to initiate changes could come from economically underdeveloped nations suffering from the lack of investment in economic and social infrastructure. In these regions, the call for social mobilization comes not from the need to respond quickly to the forces of globalization but from the necessity to initiate positive changes. Both societies with economic disparities and countries that are perceived as being economically developed and generally well-off could benefit from initiating changes in their learning systems. Challenges of unsustainable production and consumption, competition from other regions, and a variety of other problems, place questions on different forms of learning, unlearning, innovation and the role of various actors in the new learning order high on the agenda.

The capacity to learn is an advantage for institutions wishing to maintain their development. This capacity becomes even more essential if this development is to become sustainable. More learning will happen while inventing and testing systems that are expected to prove adequate for balanced development for present and future generations. The complexity of questions and uncertainties associated with long-term development perspectives explains the need for repeated (re)design and judicious testing of sophisticated solutions for sustainable development.

The challenge of creating a new learning order is recognised by multiple disciplines and political processes. In this chapter, I would like to present one of the opportunities for developing regional learning systems provided by Regional Centres of Expertise (RCEs) for Education for Sustainable Development (ESD) – a strategy developed in response to the United Nations Decade of Education for Sustainable Development (DESD) by the United Nations University (UNU).

Political processes are initiated to contribute to socio-political changes. DESD is one such process. With its attention to the importance of regional and local characteristics and emphasis on the socio-cultural aspects of required change for sustainable development, including changes of attitudes and values, DESD provides a unique opportunity to initiate regional social learning processes and to make ongoing processes legitimate.

The interpretation of DESD goals and ambitions takes place at and is affected by regional and national institutions. Regional Centres of Expertise is a strategy for translating global DESD agenda into local realities.

United Nations Decade of Education for Sustainable Development (DESD)

The year 2002 was a significant milestone for those who believe in the importance of education for sustainable development. The United Nations General Assembly, at its 58th Session, adopted a resolution to start the Decade of Education for Sustainable Development (DESD) from January 2005.

Education and its significance for development have been emphasised throughout the history of development and sustainable development. The importance of education for development is reflected in numerous international declarations, national and regional policies and development programmes. Several international political processes, such as the Education for All movement, United Nations Literacy Decade, Millennium Development Goals highlight education as a means of addressing specific and general challenges for development.

The vision of the Education for Sustainable Development (ESD) conveyed by UNESCO is that of a world where everyone has the opportunity to benefit from quality education and develop the values, behaviour and lifestyles that appear more promising in creating positive societal transformation. It is this reference to *change in the values, behaviour, and lifestyle* that make DESD remarkably different. It clearly recognizes that in order to overcome lack of motivation and political will for required transformation, the very way of viewing the world, and the conventional ways of thinking and acting should be changed.

The DESD agenda seems to be much broader than that of other (educational) movements. Dependent on the specific regional challenges of sustainable development, it could embrace questions of human rights, environment, gender equality, democracy, poverty, health and diversity of other topics (see Table 13.1).

The breadth of the agenda and attention to the foundations of human actions are not the only distinctive characteristics of the DESD. Another important feature is the emphasis on the importance of context in defining the style and subjects of learning. In all its diversity the content of ESD will vary depending on the regional, national or local priorities. The vision of the DESD "will find expression in varied socio-cultural contexts – where 'positive societal transformation' will be articulated in different ways." (UNESCO 2005).

The DESD is concerned not only with granting *access to the educational opportunities* for everybody, including the illiterate, which is a concern of the UN Literacy Decade and Education for All (EfA) movement, but with what the *content* of such education should be. The major question of DESD is: *What should be learned* by each and every member of society in order to pursue a sustainable future?

ESD advocates a need for education that is broader than any topical education. It is concerned with learning for, and not *about*, sustainable development. The four major thrusts of ESD – improving access to quality education, reorienting existing education to address sustainable development, developing public understanding and awareness, and providing training programmes for all sectors of society (UNESCO 2005) – make obvious the relevance of the ESD for all regions irrespective of the dynamics of their development.

Table 13.1. Strategic perspectives of ESD (UNESCO 2004).

Socio-cultural	Environmental	Economic
1. Human rights	9. Natural resources	13. Poverty reduction
2. Peace and human security	10. Rural transformation	14. Corporate responsibility and accountability
3. Gender equality	11. Sustainable urbanization	15. Market economy
4. Cultural diversity	12. Disaster prevention and mitigation	
5. Intercultural understanding		
6. Health		
7. HIV/AIDS		
8. Governance		

Education for sustainable development and social learning

While from the outset the ideology of DESD, as any global movement, might seem to contradict the idea of local bottom-up mobilization warranted by social learning, closer examination of some DESD characteristics presents a different picture.

Many international initiatives, the DESD among them, are large in scale and general, because of the global nature of their activities. This general quality often allows for identification of the problematic areas and statement of the main principles, but does not necessarily call for specific actions and outcomes. The spirit of the DESD movement suggests an idea and a direction. The openness allows a translation space. One may argue, however, that it is this breadth of the DESD agenda and principles that makes it non-restrictive for local practices. Like many global discourses DESD is an 'empty vessel' that has its universal shape, but which is to be filled with a meaning generated through the interaction of local and global processes.

While DESD appears inspirational for many around the globe, the ways and principles of such mobilization should be further refined. In other words, the translation of the DESD agenda into regional realities needs to be informed by appropriate considerations and principles. The emphasis of social learning on *bottom-up formulation of the sustainability agenda* and the importance of *multiple ways of knowing and learning*, as opposed to the necessarily consensus-driven agenda, re-emphasises and refines DESD's call for collaboration and the importance of different perspectives. I believe that it is only possible to implement the idea of socially-relevant and culturally appropriate learning systems highlighted by DESD, when ideas of social learning are fully considered. Choices that are to be made while dealing with ESD – from a selection of the most pressing sustainability topics to the ways of teaching and learning – are considerable. Recognition of styles of knowing and learning inspired by the social learning ideology might help to secure the necessary sophistication of approaches for solving sustainability challenges.

Regional centres of expertise

The goals of DESD are broad, which makes it easy for nearly everyone to relate to. The four thrusts of the Decade are easily recognisable by people and organisations in all regions and sectors of society. In spite of the appeal of ESD goals, the task of bringing the global DESD agenda closer to the everyday life of people is enormous. In order to address the challenge, the United Nations University introduced a strategy of creating Regional Centres of Expertise (RCE) for Education for Sustainable Development.

The RCE strategy was also inspired by the realisation that there are several institutional problems preventing learning for sustainable development. There are frequently identified failures to introduce the latest achievements of science and technology into education and learning, and to bridge the gaps between different levels of education and between formal, informal and non-formal education.

With the popularity of the RCE strategy – in December 2006, we at UNU were aware about more than fifty initiatives to develop RCEs in Africa, Asia, Northern and Latin America, Middle East and Europe – clarifications of the RCE concept and discussion about its development implications becomes a matter of priority. That priority becomes even more important if RCEs are to be developed to their full potential in the regional social learning systems.

The scope of this chapter does not allow us to address all questions of importance for an RCE as a networking and learning approach. It is restricted to the considerations that are most urgent for the initial stages of RCE mobilisation and strategising. In the search for solutions, these initial issues were brought forward between RCE actors and the RCE Service Centre of UNU-IAS.

Translating DESD goals into regional realities

An RCE is seen as a network of existing organisations representing formal, informal and non-formal education interested in joint implementation of DESD goals and aspirations at the regional level. RCEs call for as many organisations as possible, including schools, universities, organisations representing civil society, local government and the private sector, to be involved in designing and implementing ESD in the region (Figure 13.1 presents, schematically, members of RCEs and the links that RCEs aspire to facilitate among them). The geographical range covered by RCEs should be large enough to contain a sufficient diversity of members from each category of actors but small enough to allow relatively easy interactions among them.

In addition to translating the goals of the DESD to the regional level through networks of regional stakeholders, the ambition of RCEs is to address often observed gaps in the regional formal, informal and non-formal education and learning sub-systems. It intends to re-orient curricula of the formal educational system while aligning different levels of the educational column, from pre-school to tertiary education, with each other. It also aspires to bring together actors whose role is traditionally associated with knowledge-generating activities and educators,

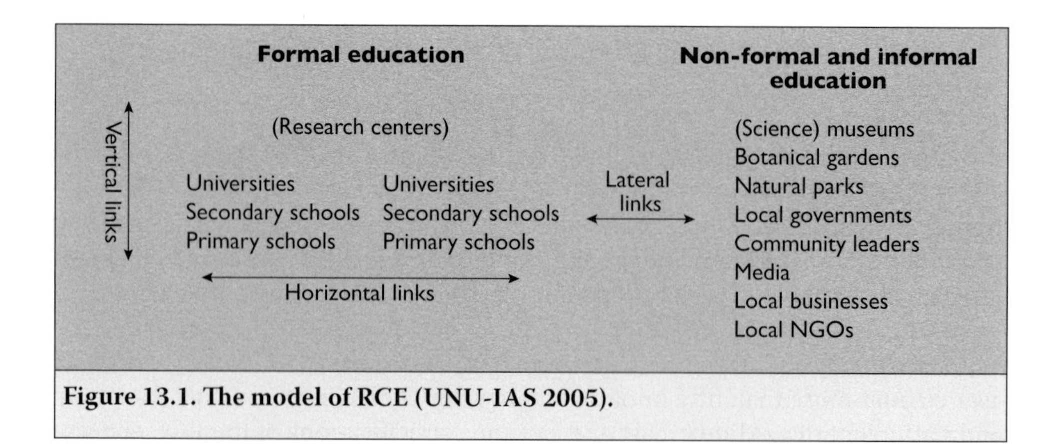

Figure 13.1. The model of RCE (UNU-IAS 2005).

including those working with and within non-formal and informal educational sectors. The aim of this undertaking is not only, or not simply, to initiate the flow of information between partners but to stimulate *knowledge co-creation* in the area of ESD. Finally, those actors who play a crucial role in supporting institutions of education and learning are encouraged to join the RCE as, once again, co-creators of the new learning order.

In its vision, the RCE concept aims at the establishment of a new governance structure that makes it possible for experts and laymen to contribute equally to the creation of regional learning systems for sustainable development. The RCE strategy calls upon actors in the formal, non-formal and informal educational sectors, providers of educational content, supporters of ESD delivery, students and learners to generate expertise for ESD. An RCE, by creating a network of organisations, assigns new roles to partners and encourages them to employ a reflective approach in their actions, and contributes to the social learning in the region.

RCEs are expected to carry out several major functions – *providing a governance mechanism for ESD* in the region, including *coordination of collaborative RCE stakeholders network, coordinating research and development beneficial for ESD* and supporting and *implementing transformative educational practices* at all levels of society.

Research and development activities of RCEs address the need for reflexivity in the coordination of RCEs. RCEs are networks of actors that aspire to ambitious goals beyond their traditional organisational agenda. The strategies of such undertakings could be highly complex and might need continuous readjustment.

Research and development activities are also called upon for specifying the goals of an RCE initiative in the context of the region. This reflective activity assists the function of social learning by providing the RCE network with an analysis of the complex problems and detailed solutions for sustainability. The process of reflection is essential when building an RCE because there is no prior knowledge on how best to proceed. With little direction and limited rules of organisation from the start extra analysis might be beneficial. There is also a need for strategic alterations due to the continuously shifting conditions within and outside the system.

The transformative education activities of RCEs are expected to affect the reorientation of educational programmes including formal educational curricula, as well as learning programmes of non- and informal educational sectors, in responding to the requirements of sustainable living.

RCE strategy: status of implementation and vision ahead

The UNU started working on the RCE strategy from early 2003 with the first few regions willing to implement the initiative. To some extent, the development of the concept was guided by the experiences from this engagement. In June 2005, the first seven RCEs were launched during the International Conference on Globalization and Education for Sustainable Development – Sustaining the Future, held in Nagoya, Japan. They were followed by two RCEs launched by the end of 2005. Another twenty three RCEs were launched in December 2006. RCEs, located in Asia, North America, Middle East and Europe, are at different stages of development. They differ in the scope, aims and focus of their activities. Table 13.2 lists the launched RCEs in various regions of the world.

The interest in the RCE strategy and in developing international RCE network entices other regions to join the initiative. At the moment, nearly thirty RCE initiatives are in various stages of development all over the globe. With their newly acquired experiences of multi-stakeholder collaboration, potentially innovative designs of learning systems and ways of addressing problems of sustainable development, RCEs are seen as contributors and users of the Global Learning Space (GLS) for Sustainable Development, whose elaboration is seen by UNU as the global output of the DESD. GLS is seen as a resource and an interactive space for the regional actors interested in learning for sustainable development that could be facilitated through ICT and through face-to-face interactions. It will enable regional actors to be informed about sustainability and ESD initiatives beyond their regions. GLS will be, at the same time, a result of knowledge generation at the regional level.

Table 13.2. Regional centres of expertise as of December 2006.

Africa	Ghana	South Africa	
	1. RCE Ghana	3. RCE KwaZulu-Natal	
	Kenya	4. RCE Makana Rural	
	2. RCE Greater Nairobi	Eastern Cape	
Asia-Pacific	China	Japan	Malaysia
	5. RCE Anji	10. RCE Greater Sendai	18. RCE Penang
	6. RCE Beijing	12. RCE Okayama	Pacific region
	India	13. RCE Yokohama	19. RCE Pacific
	7. RCE Guwahati	14. RCE Kitakyushu	Philippines
	8. RCE Lucknow	Republic of Korea	20. RCE Cebu
	9. RCE Pune	15. RCE Tongyeong	Thailand
	Indonesia	16. RCE Incheon	21. RCE Trang
	10. RCE Bogor	Kyrgyz Republic	
		17. RCE Kyrgyz Republic	
Middle East	Jordan		
	22. RCE Jordan		
Latin America	Brazil		
	23. RCE Parana		
Europe	Finland	Rhine- Meuse region	Spain
	24. RCE Helsinki	27. RCE Rhine- Meuse	29. RCE Barcelona
	Germany	Southern North Sea region	Sweden
	25. RCE Hamburg	28. RCE Southern North Sea	30. RCE Skåne
	26. RCE Munich		UK
			31. RCE East Midlands
North America	Canada	USA	
	32. RCE Saskatchewan	35. RCE Grand Rapids	
	33. RCE Toronto		
	34. RCE Sudbury		

GLS is projected as a community of practice where ESD actors would find support for their ideas and inspiration for actions. Access to the inspirational forums outside of their own region would serve as a component frequently recognized as essential for the innovative practices (e.g. Brown *et al.* 2002).

Critical views: clarifications of concepts and pitfalls in practices

The concept of RCE proved to be very appealing to a variety of audiences. One of the stated reasons for such interest is the recognised opportunity for all actors of the region to contribute to the revolutionary aspirations of DESD from their own domains. It is a chance to add to the innovative design of regional learning systems while tackling the all too familiar problems of contemporary educational and training systems that draw the interest of all concerned with challenges of our times. Most of these challenges are easily presented in the documents of UNU and UNESCO and are effortlessly recognised. Highlighted disparities in the quality of education along the levels of the educational column, lack of scientific input into curricula, gaps in communication between educators and other sustainability experts, regionally inadequate educational content and processes attract the attention of regional actors from different fields.

While the RCE strategy calls for collaborative efforts that bring together the knowledge of all regional actors in pursuit of ESD, there is a danger that the RCE strategy might inadvertently reinforce principles and values that are not compatible with the principles of DESD and social learning. Old ways of thinking and acting might be preserved simply because we do not discuss and clarify them enough. Without detailed expatiation, the relative simplicity and appeal of the RCE and ESD concept might serve as a Trojan horse that brings erratic assumptions and, consequentially, sub-optimal practices to the RCE actors and its supporters. This section aims at clarification of the concept that, I hope, will benefit the RCE actors. In particular, I would like to highlight some challenging requirements firstly for knowledge generation and learning in the society and, secondly for collaborative networks.

Defining RCE challenges is important for yet another reason related to the framework of social learning. Tenets of social learning would not be realised unless the process were facilitated in a systematic manner. In contrast to networks run by professional managers and strategists, RCEs and other ESD alliances are mobilised and managed by people who might have little or no experience in management of complex collaborative undertakings. In such cases, there might be a tendency to rely on principles that successfully govern the everyday activities of leading organisations but might not be entirely suitable for more complex networks. Emphasis on critical networking considerations might be helpful for RCE members in their struggle for ESD.

Experts and 'others': what knowledge counts?

As emphasised earlier, different regions are struggling either with the need to respond to continuous change or the need to initiate change for sustainable development. Within the context of DESD such challenges might be better addressed through the provision of life-long learning opportunities for all members of society. This challenge ultimately calls not only upon professional educators or generators of knowledge content typically involved in professional educational services but also a variety of other cognisant actors.

As emphasised by van Dam-Mieras (2005), only five to ten percent of the life-long learning process of an individual takes place within a formal learning setting. Creating an educational system supporting learners outside the formal system demands the engagement of a broader group of stakeholders. A wider variety of topics and different learning techniques relevant to the local context and requirements of learners are essential. Not only learners but also educators, from teachers and university professors to qualified trainers, would require support in professional development.

Demands for innovative solutions in the face of change require a serious transformation of the conventional roles of actors within a learning system. Traditional roles of experts as authorities in knowledge generation, teachers and trainers in knowledge delivery and learners in knowledge reception need to be contested. The common language of knowledge and learning, often present in the area of sustainable development, reinstates the traditional specialisation where knowledge is produced in one domain and consumed in another. The intended evolution towards what is called a knowledge society might not solve the problem. Knowledge and knowledge delivery service become commodities that need to be delivered, often for financial compensation, to the recipients.

Neglecting the question of roles within a knowledge system might lead to the situation where, promoting one theory-in-use based on the norms of collaborative actions and social learning, RCEs would promote practices reinforcing old divisions of labour in the learning communities.

Compartmentalisation of society and the subsequent division of people and organisations as experts and non-experts have yielded massive criticism over the years. Any attempt to recap the debate would be outside the scope of this chapter. I would like, however, to highlight some of the normative statements which I believe to be consistent with the principles of RCE operations.

Expert knowledge is often associated with a particular way of knowledge generation and decision making. However significant the variations in different societies and cultures, experts are often acknowledged as knowledge authorities who know which course of action is the most desirable in which circumstances.

The division of experts as legitimate 'truth holders' on one hand and laymen on the other, is inapplicable to RCE practices. Many members of the communities active in regional processes acquire expertise not by virtue of academic or other degrees but through the experience of dealing with a particular locally significant problem – be it the rights of indigenous people, environment or poverty. In the words of Collins and Evans (2002), they become 'experience-based' experts.

Participation of the experience-based experts in the RCE activities is justified for several reasons. First, they, as well as knowledge experts, are familiar with intricate details of the regional processes. Knowledge of the context makes regional actors, irrespective of their association with the conventional 'expert group', invaluable members of the RCE consortium.

Second, knowledge creation or learning for sustainable development is generally understood as requiring a different approach from learning governed by centralized and expert systems. The need for sustainable development calls for knowledge and learning systems that are more than a rational combination of the best of expert knowledge. The complexity of the system would call, once again, for the engagement of experience-based experts and for the support of the general population. The social learning perspective emphasises that collaboration is essential when questions addressed by the collaborators are of a complex nature or when actors do not have prior experience in dealing with these questions. The convention of experts and scientists expressing the 'truth', while acceptance of the 'layman's' opinion counts only when it is expressed within expert discourse (Wildemeersch 1997/98) should be challenged where 'multilateral control' of the issue is expected.

The third and interrelated reason for discussing the role of expertise is the significance attributed by the ESD and the RCE movement to the scientific community and to institutions of higher education (IHE). Scientific institutions are recognised as knowledge generation entities, while IHE are seen as a link between knowledge generation and the society in two ways – by virtue of their participation in the education of future decision-makers and their contribution to the societal development through outreach and service to society (UNESCO 2004).

Such strategic emphasis should not overshadow the roles of other "*pockets of expertise* among the citizenry" (Collins and Evans 2002), which represent both technical and political domains. In addition to the recognition of the role of the experience-based experts in the regions, one should remember that in the context of RCEs, scientists and other experts would need to make judgements about content and design of the educational programmes, required research and development activities and strategic questions with regards to the RCE network. These questions are based not so much on the "timeless intellectual justification" attributed to the scientific process (Collins and Evans 2002) but on the understanding of regional processes and needs. Experts, in the context of RCEs, would not exclusively play the role of scientists and topical specialists, but, more often that of experience-based experts in public matters. Recognition of such a role would help in eliminating potentially unjustified disparities in the position of different organisations that are members of RCEs.

In the context of RCEs, the frequently assumed prerogative of scientific expertise should be questioned. Its value, however, should not be denied. A similarly critical position should be taken regarding any form of knowledge and knowing, including indigenous knowledge and ways of knowing. None of the single forms of knowledge has enough intellectual authority in addressing questions of sustainability. Knowledge systems informed by multiple perspectives and realised through multiple ways of creating knowledge should be recognised within RCEs.

Actors and structures

In the transition from an old model of decision-making dominated by authoritative relations and compartmentalisation of fields of action to the model of collaborative practices, actors have to look for new roles and new ways of relating to each other. While this statement sounds obvious, our observations over the development of some RCEs show that some concerns are always worthwhile repeating. Universities, schools, private enterprises, NGOs or municipal departments uniting to address ESD goals often come together for the first time. In addition to frequently lacking the experience of working together or working with an ESD issue, they bring along different worldviews, languages, expectations, sense of timing and measures of success. The time that is required for establishing a common ground for such diverse constellations of participants would be much longer than that for homogeneous groups. Various relational problems might emerge. For example, actors with prominent positions, by nature of their experience or prestige, tend to reinforce their status in a new RCE setting and retain control over significant issues (Gray 1985). There might also be actors who would be silenced, co-opted or whose contribution may never find its way to the network choices. Wild and Marshall (1999) noted that in many collaborative ventures engagement of disadvantaged groups had a limited success. Even when these groups became a part of networking there was not a clear connection between their opinions and choices and the network's common priorities.

If RCEs are to become examples of social innovations in the degree of power sharing, the breadth of the actors' engagement, level of public participation, degree of transparency and extent of assessment, the core organisations engaged in the RCE processes should remember some of the essential principles of effective networking.

Membership and relations

The membership composition of the RCE and relations among the members are critical factors for selecting RCE goals and methods of implementation.

It is noted that *homogeneous actors* or actors that have a significant history of dealing jointly with a particular issue might have the advantage of finding quick ways of implementing projects and actions. *Heterogeneous actors*, on the other hand, having different experiences and interpretive frames, might contribute to more creative solutions to the ESD challenges. The development of ideas and their implementation by heterogeneous actors might, however, take longer than that by domain-similar, homogenous ones (Fadeeva and Halme 2001).

Such an observation is strategically important for RCE facilitators and promoters of social learning principles. The invitation of traditional collaborators, whose style is familiar and views are close to those of facilitators, might guarantee rapid formulation of RCE actions. This consideration is applied to the selection of already recognised, or 'safe', actors over controversial or generally uncomfortable ones. Avoidance of organisations raising uncomfortable or politically censored questions, particularly at the stage of agenda formulation, is not an uncommon phenomenon in the network practices. While this strategy might bring about fast recognition of RCE initiatives, the core organisations should remember that maintenance of existent collaborative relations requires significant resources (Burt 1992). In time, growth of RCE actions may not leave time or money for the promised 'future' expansion of membership unless such an expansion is not strategically planned from the very beginning.

The tendency to engage stakeholders partially representing ESD agenda is evident for RCEs. It appears that pioneering organisations within ESD and RCEs are those that are traditionally working with issues of environmental education. As a result, much of the initial, and potentially future, agenda of RCEs might be focused on environmental questions and issues of resource efficiency. Many other questions relevant for the regions, including those of human rights, gender, poverty and health might receive marginal attention. The implications could be serious – the observed divide between the sustainable development community and the community of development specialists may never be breached[22].

Effects of the membership composition for the RCE goals and actions would depend on multiple factors. I would like to highlight one of them, relating to the power of the members, as a prominent one in the existing RCEs. Diversity of membership depends on the power composition of the members. Organisations that have higher coercive power, experience or other forms of legitimacy would have the possibility of dominating the network agenda (DiMaggio and Powell 1991). It is essential then, if the network facilitators are concerned about the contribution of others, to purposefully secure access and contributory space for the less powerful or, potentially competing actors.

To what extent membership characteristics affect creativity and efficiency of networks might be defined by the structural characteristics of RCE networks. A well-established administrative structure (Human and Provan 1997), such as a system of formally defined roles, rules and responsibilities, will help achieve specific tasks of the networks. This system, however, might not be particularly helpful if the RCE is dealing with complex uncertain issues. Such kinds of challenges are better dealt with using what is called an interactive structure – a structure that allows consistent interaction of actors and subsequent emergence of ideas (Human and Provan 1997, Fadeeva and Halme 2001).

RCE: expectations and assessments

The main question of concern to all of the involved parties is how the RCE movement would contribute to the learning systems adequate for sustainable living in the region and, eventually, to the creation of the Global Learning Space (GLS) – a networking space and inspiration for ESD innovations. Like any collaborative network, RCEs run the risk of doing 'business-as-usual' activities under the new umbrella of ESD and giving fresh legitimacy to old activities. Networks and partnerships are known to fall short of expectations. Superior results and innovations expected from networking, their ability to resolve conflicts and optimise resources of achieving goals, provision of additional valuable information (Burt 1992, Dunning 1993,

[22] There are two discourses dealing with capacity building and learning for sustainable development. One of them, represented mainly by the development practitioners, deals with questions of literacy eradication, education for girls and women and, generally, education for development. The second discourse works with questions of environment and sustainable development. Representatives of this movement come primarily from the fields of natural, engineering and management studies. Their current lack of collaboration might undermine the aspirations of DESD.

Håkansson and Shehota 1997, Weber 1998, Clark and Roome 1999, Poncelet 2001) expected from collaborative alliances might never come to fruition. They could be overrun by unclear or differently interpreted goals, lack of credible commitment, lack of trust, insufficient incentives and many other reasons (Weber 1998, Bizer and Julich 1999, Halme and Fadeeva 2000). The challenge is how to ensure that RCE networks become learning networks contributing to the sustainability of the region.

Some of the requirements for defining RCEs guarantee that activities contributing to the development of the regional ESD system are developed, at least to some extent. Highlighting the need for collaboration, the RCE strategy will encourage the RCE initiators to engage other relevant ESD stakeholders. Specification of the research and development might lead to the deepened understanding of the ESD needs of the region and the design of the learning system for sustainable development. Requirement to work with transformative education in all educational spheres could contribute to the introduction of the appropriate educational principles into educational methodology, revision of the existing formal curricula, and design and re-design of training for different regional stakeholders including government, business, media and civil society organisations.

With an appreciation of the areas of RCE contribution, the questions about the RCE role in the progress towards sustainable development and ESD still remain. The primary concerns of the critics, analysts and the RCEs themselves could be expressed with the following questions. Does the RCE deliver identifiable improvements in the social, economic and environmental area? Does it contribute to the creation of a new learning system for sustainable development at the regional and global level?

Principles of action selection and assessment

The aspiration to go beyond 'business-as-usual' practices under the new label of RCE requires that RCEs are guided by a set of principles that support the goals of ESD and the region and help assess the transition towards sustainability.

While having multiple variations, the primary set of guiding principles should be based on several key considerations.

First, RCEs should develop activities with regard to four key elements of RCE activities, i.e. management and governance, networking and partnership, transformative education, research and development.

Second, RCE activities should reflect the four main thrusts of education for sustainable development (ESD) – public awareness and understanding, access to quality education, re-orienting existing education, training programs for all sectors.

Third, bearing in mind that RCEs are seen not only as the major vehicle for translating global ESD agenda into local action but also as co-creators of the Global Learning Space (GLS), RCE activities are expected to address collaborative activities outside RCE's own region.

The above considerations indicate types of *activities* that are necessary for the RCEs to sustain their operations and to implement DESD goals. For the learning goal of RCEs, however, it is important not only to develop activities but also to assess their contribution to the goals of DESD and to the progress in sustainability in the region.

In order to provide a realistic assessment of the RCE's role, the effect of each of the activities should be assessed from the point of view of its *contribution to human well-being, economic development and the impact of human activities on the environment.* In particular, regional strategies focusing on the critical aspects of human and environmental vulnerability and contribution to the natural, human and produced capital should be consulted.

Considerations in RCE activities that address the three dimensions of sustainable development should be complemented by the assessment of the *scale of activities.* Current RCE activities should be compared with the potential maximum scale of activities (e.g. RCE works with two schools out of 10 in the region on re-orienting school curricula) and/or to the past activities of the RCEs.

Reflection on an RCE's regional role might be aided by a relative assessment of RCE actions in terms of their (relative) *novelty.* In it important to remember, however, that the assessment of innovativeness should be conducted in *relation to the local situation.* RCEs could either repeat the routine actions or suggest a novel solution to the problem.

The field of innovation science suggests a number of concepts that could help assess the contributions of RCEs. While it is a difficult challenge, some considerations could be helpful. For example, when assessing factors of change in the socio-technological and socio-cultural systems, attention should be focussed on changes in institutions, management mechanisms and infrastructure. In particular, the changes in the role of actors and qualities of leadership deserve special attention. Depending on the scale of societal changes, whether incremental or radical, actors' roles as active citizens contributing to the RCE networks could vary. Similarly, leadership of RCEs could attend to the activities of building a common vision, strategy and actions aiming at the serious revision of the regional learning system or it could restrict itself to the optimisation process and partial system redesign (Tukker and Bruijn 2002). Active vision-building exercises, e.g. through scenario-building activities, could help to realise current trends, make uncertainties explicit and radically redefine actors' roles and visions.

Finally, RCE activities could be selected and assessed on the basis of how well they could compensate for institutional failures to address a particular issue. For example, if the region experiences a problem with clean water and sanitation or biodiversity and lacks experts in these areas, the RCE could address these challenges by coordinating research and capacity-building activities. If the problems arise from the lack of coordination by responsible organisations, e.g. department of education and department of environment of the regional government, the RCE, through its facilitating role, could bridge the gap.

Conclusions: challenges and the value of the RCE process

One of the demands of our time is for the capacity of individuals and organisations to assemble knowledge and act upon it collectively. It is necessary not only in situations where regions face challenges in reacting to the rapid flow of information, in the market, in social or technological pressures, but also when there is a need to initiate changes or redefine development in accordance with principles of sustainability. Development steered by forces of globalisation or, on the contrary, by insufficient growth, requires learning systems that are able to identify pressures and leverage points, mobilise resources, develop competencies and generate adequate responses.

During the Johannesburg Summit, national governments committed to undertake national actions to address the challenges of ESD. Within the framework of DESD, professional educators and other stakeholders in sustainable development were given an opportunity to legitimately engage with national and regional authorities as well as with other decision-makers to critically examine regional learning systems from the position of sustainable development.

When large forces are at play and political discourses, such as those within DESD, are developed, initiatives born into the process might experience different influences. Legitimacy of practice coming from international plans has unquestionable advantages. It might put an initiative on the 'map', give it access to earlier inaccessible resources and facilitate the building of previously unimaginable initiatives. There are, however, challenges that might be hidden in a large political discourse. Lack of clarification of the conceptual principles might lead to merely symbolic operationalisation of the global agenda. In this chapter, I have presented some of the challenges that RCEs, as a strategy of implementation of the DESD agenda, might encounter. As an innovation of the United Nations University, the RCE concept carries an additional international recognition through its association with another UN agency. A clarification of the principles of their operations helps build effective social learning systems in the regions.

RCEs bring together regional actors who possess both explicit and tacit knowledge of mobilising actions in the regional context. The emphasis of both forms of knowledge is a key in the regional context. While experts representing a particular discipline might possess knowledge of the concepts and principles for a particular field of practice, it is local actors who mediate translation of expert knowledge into regional realities. RCEs, by clarifying the roles of various regional actors, can provide an opportunity for overcoming the traditional institutional divide between experts and non-experts and facilitating social learning.

Actors within an RCE should also be attentive towards the selection of the RCE core members. While homogeneous actors might implement their actions efficiently, heterogeneity of membership could bridge traditional institutional divides and bring more innovative ESD solutions. Optimal development of both administrative and interactive network structures could balance creativity and implementation processes of RCEs. In order for them to become an effective system of learning and action, RCEs are also challenged to consciously attend to the discussion of power dynamics between organizations coming together to form an RCE, the challenges of going beyond customary actions towards more revolutionary innovations, the complexities of embracing the wider spectrum of ESD and many other issues.

RCEs as new collaborative and learning systems: value of a process

RCEs are seen as, primarily, *collaborative networks* pursuing the goals of ESD. This vision makes them, potentially, organisational mechanisms able to deliver not only specific outcomes in the area of ESD but also to be examples of social innovations themselves.

RCEs could be seen as governance mechanisms promoting democratic participation of organisations in regional processes while creating relationships among actors that may not have prior collaborative experiences. One should not have to go far to find targets for potentially fruitful boundary spanning among stakeholders of ESD. Because ESD is concerned with a variety of sustainable development questions, including those of environment, human rights or culture, it requires collaboration between organisational departments that, routinely, do not work together. RCEs that are already in operation (see Table 13.2), work with departments within one government or within one university encouraging them to develop new relations.

Depending on the socio-cultural and political context of a country, the value of the collaborative process developed by RCEs would be different. In some instances the judgement about collaboration should be made on the basis of the quality of their delivery, e.g. revised curricula or research and development activities, while in others the emphasis would be on the contribution to the collaborative and democratic processes of the region. The facilitating role of RCEs would be more important when collaborative ESD alliance is being developed in countries where democratic participation is not a common phenomenon, such as in the countries where economies are in transition or post-apartheid countries (Yanarella and Bartilow 2000). Such an understanding would help organisations associated with RCEs do better justice to the activities of RCEs. Answering the question "Is it the process of collaboration or the educational and research products of RCE that take a priority?" at each stage of RCE development would help avoid disappointment and sustain the interest of a broad range of stakeholders.

References

Bizer, K. and Julich, R. (1999) "Voluntary Agreements – Trick or Treat?", *European Environment*, 9: 59-66.

Brown, H., Vergragt, P., Green, K. and Berchicci, L. (2002) "Learning for Sustainability: Transition through Bounded Sociotechnical Experiment", unpublished paper, GIN2002

Burt, R.S. (1992) *Structural Holes: The Social Structure of Competition*, Massachusetts: Harvard University Press.

Clark, S. and Roome, N. (1999) "Sustainable business: Learning Network as Organisational Assets", *Business Strategy and the Environment*, 8(5): 296-310.

Collins, H.M. and Evans, R. (2002) "The Third Wave of Science Studies: Studies of Expertise and Experience", *Social Studies of Science*, 32(2): 235-296.

DiMaggio, P.J. and Powell, W.W. (1991)" The Iron Cage Revisited: Institutional Isomorphism and Collective Rationality", in P.J. DiMaggio, ed., *The New Institutionalism in Organizational Analysis*, Chicago: University of Chicago Press, pp. 63-82.

Dunning, J.H. (1993). *Multinational Enterprises and Global Economy*, Workingham: Addison-Wesley Punl.

Fadeeva, Z. and Halme, M. (2001) "Establishing and Maintaining Cross-Sectoral Actor Networks: Toward Sustainable Development in European Tourism Sector", in T. Bruijn and T. Bruijn, eds., *Sustainable Development: Partnership and Leadership*, Kluwer, pp. 251-271.

Gray, B. (1985) "Conditions Facilitating Interorganizational Collaboration", *Human Relations*, 38: 911-936.

Håkansson, H. and Shehota, I. (1997) *Developing Relationships in Business Networks*, London: International Thomson Business Press

Halme, M. and Fadeeva, Z. (2000) "Small and Medium-sized Tourism Enterprises in Sustainable Development Networks – Value-added?", *Greener Management International,* 30: 97-113.

Human, S.E. and Provan, K.G. (1997) "An Emergent Theory of Structure and Outcomes in Small-firm Strategic Networks", *Academy of Management Journal,* 40(2): 368-402.

Poncelet, E. (2001) "A Kiss here and a Kiss there": Conflict and Collaboration in Environmental Partnerships, *Environmental Management,* 27(1): 13-25.

Tukker, A. and Bruijn, T. (2002) "Conclusions: The Prospects of Collaboration", in A. Tukker and T. Bruijn, eds., *Partnership and Leadership: Building Alliances for a Sustainable Future,* Kluwer Academic Publishers, pp. 295-314.

UNESCO (2004), *Higher Education Brief for the Decade of Education for Sustainable Development,* available from http://portal.unesco.org/education/en/file_download.php/23490577a0adc755dcb96e737fd16ebdbrief+Higher+Education.pdf.

UNESCO (2005) *Draft International Implementation Scheme for the United Nations Decade of Education for Sustainable Development (2005-2014), Paris, UNESCO,* available from http://unesdoc.unesco.org/images/0014/001403/140372e.pdf.

UNU-IAS (2005) *Mobilising for Education for Sustainable Development: Towards a Global Learning Space based on Regional Centres of Expertise,* Yokohama, UNU-IAS.

van Dam-Mieras, R., with J. Rikers, J. Hermans and P. Martens (2005) "A Regional Centre of Expertise 'Learning for Sustainable Development' in Europe" in Z. Fadeeva and Y. Mochizuki, eds., *Mobilising for Education for Sustainable Development: Towards a Global Learning Space based on Regional Centres of Expertise,* Yokohama: UNU-IAS, pp. 71-78.

Weber, E.P. (1998) "Successful Collaboration: Negotiating Effective Regulations", *Environment,* 40(9): 5-10.

Wild, A. and Marshall, R. (1999) "Participatory Practice in the Context of Local Agenda 21: A Case Study Evaluation of Experience in three English Local Authorities", *Sustainable Development,* 7: 151-162.

Wildemeersch, D. (1997/98) "Paradoxes of Social Learning – Towards a Model for Project Oriented Group Work", *RUCNYT* 3(3).

Wolfe, D.A. and Gertler, M.S. (2002) "Innovation and Social Learning: An Introduction", in M.S. Gertler and D.A. Wolfe, eds., *Innovation and Social Learning: Institutional Adaptation in an Era of Technological Change,* Palgrave Publishers.

Yanarella, E.J. and Bartilow, H. (2000) "Beyond Environmental Moralism amd Policy Incrementalism in the Global Sustainability Debate: Case Studies and Alternative Framework", *Sustainable Development,* 8: 123-134.

Chapter 14

Learning about corporate social responsibility from a sustainable development perspective: a Dutch experiment

Jacqueline Cramer and Anne Loeber

Introduction

Expectations about the responsibilities held by firms for the societies in which they operate are changing. More now than ever, firms are requested to account explicitly for all aspects of their performance, that is to say, not just for their financial results, but also for their social and ecological performance. Openness and transparency are the new keywords. A growing group of companies acknowledges this trend, and actively seeks to take up its 'corporate social responsibility' (CSR) in the pursuit of a sustainable development. Among them are the more than 160 companies who are members of the World Business Council for Sustainable Development (WBSCD).

The WBSCD defines corporate social responsibility as "the commitment of business to contribute to sustainable economic development, working with employees, their families, the local community and society at large to improve their quality of life". Both social and environmental concerns are seen as part of a company's corporate social responsibility (Holme and Watts 2000). The challenge for companies is to find a responsible balance between People ('social well-being'), Planet ('ecological quality') and Profit ('economic prosperity') (Elkington 1997), and to maintain that balance in practice. "What does it really mean for companies to shift their attention from financial to sustainable profit?" they ask.

This practical question led the Dutch National Initiative for Sustainable Development (NIDO) to launch a major programme on the subject, entitled 'From financial to sustainable profit'. The NIDO organisation was active from 1999-2004, and was financed with special funding from the Dutch government. It was set up to facilitate a structural anchoring of 'sustainable' initiatives in society. The objective of the programme 'From financial to sustainable profit' was to actively support companies in dealing practically with the challenges of corporate social responsibility, by initiating and supporting processes of learning (Cramer 2001). The programme focused on the interface between 19 participating companies and their stakeholders, and ran from May 2000 till December 2002 (Cramer 2003). The firms that participated differed greatly in size and type. Both small and medium-sized companies (SMEs) and multinationals were involved, representing a variety of sectors. Furthermore, the firms' representatives varied in terms of institutional position and power. What the companies and their representatives had in common was a general sense of urgency with respect to the need for implementing corporate social responsibility, and some reservoir of experiences with such implementation at the time of joining the NIDO programme.

This chapter discusses the approach to stimulating learning about corporate social responsibility adopted in the NIDO programme, as well as its results[23]. It was found that learning was triggered not only among the company representatives that participated in the programme, but also at the level of the participating companies. Insight was gained as to how to induce learning processes among parties in the corporate sector that may help them to pursue the ambition of a sustainable development. Furthermore, insights were gained into how the NIDO programme could contribute to the development of a legal framework and other structural conditions that were favourable to CSR initiatives.

Triggering learning in practice

In order to stimulate learning processes in corporate social responsibility, NIDO developed a three-track approach in the programme. It focussed on learning processes at the level of the participants of the programme, at company level and at the level of the structural (economic, legal, social) conditions under which a firm operates.

First of all, NIDO organised monthly meetings to exchange experiences among the participating companies, to discuss common problems (e.g. how to make a zero-assessment of a company's performance) and to interact with external stakeholders. The meetings were intended to stimulate learning processes in several ways. Not only could the participants profit from one another's insights and experiences. The setting in which the meetings took place, away from the participants' daily routines and surroundings, and the 'unusual' discussion partners (as the participating firms ranged from chemical industry to banking, their representatives met here with

[23] The authors of this chapter were close witnesses of the programme on corporate social responsibility described here. Jacqueline Cramer, who had been involved in the establishment and design of the NIDO organisation, was the initiator and manager of the programme. Anne Loeber was contracted to evaluate the NIDO endeavour (cf. Loeber 2003a). The data in the research that is described here were collected through document analysis and via participant observation of most of the meetings during the course of the NIDO programme, and through interviews with its participants after it had come to a close.

people that were not their daily sparring partners) placed participants outside their 'comfort zone' and thus stimulated them to take a fresh look at the issues discussed (e.g. how firms could report and externally communicate about their corporate social responsibility performance).

The latter was necessary, as the NIDO programme did not provide clear-cut answers to the questions raised regarding the implementation of corporate social responsibility. Such answers were not readily available. It was the programme management's contention that in every company and in every setting, the question of what entails CSR from a perspective of sustainable development may be answered differently. Moreover, because of the sheer variety in market orientation, competitive position and business context among the participants, clear-cut answers would have hardly fitted the needs and ambitions of the companies involved. Rather, the changes required had to be designed in a process of thinking 'to-and-fro' between environmental or social indicators and ethics, and other more generic conceptualisations of sustainable development on the one hand, and a company's specific characteristics and opportunities, on the other.

The adoption and adjustment of e.g. environmental standards in a company's operation system may be understood as an expression of so-called first-order learning. First-order learning generally results in incremental changes in an organisation's problem-solving strategies; the more fundamental notions, values and assumptions on the basis of which the company operates remain in tact. If these change too, second-order learning takes place.[24]

One might argue that for the pursuit of a sustainable development, second-order learning is imperative. After all, the ambition implies the occurrence of change beyond a mere compliance with formal environmental and social regulations and a mere incorporation of their practical implications within a company's existing operating practices, strategies and standards. Rather, it requires a critical review of a company's values, policy principles and business strategies. Only through a reconsideration of the normative assumptions and background theories that underlie operational practices, can one achieve the shift in focus from mere profit-seeking to corporate social responsibility. This kind of learning, however, is unlikely to occur spontaneously.

[24] The processes described here are in the literature on learning alternatively referred to as 'single loop' and 'double loop learning' respectively (cf. Argyris and Schön 1996). For a discussion of these notions, see Loeber *et al.*, this volume.

The degree of self-reflection that is implied in second-order learning is hard to achieve of one's own accord. In the absence of an impetus to reflect fundamentally on the basic assumptions underlying the present state of affairs, the embedded rules are – often implicitly – factored out of the discussion. A constant questioning of these assumptions interferes with daily routine and would render a working process highly inefficient. Second-order learning may occur only when a person deliberately wishes to reflect on his/her professional practice and others help him or her to take into consideration information and insights that are, in the course of daily routine, easily overlooked or neglected (or dismissed as untrue). The monthly meetings provided a setting in which such self-reflection processes could occur (Loeber 2003b).

Yet, learning by the companies' representatives in the programme, however relevant, is not a sufficient precondition for change. Evidently, a person's ability to convey newly acquired insight and information to his/her company and see it translated into an actual change in the company's course of action critically depends on his/her position, good will, and resources available, as well as on the company's perceptiveness and flexibility. One of the difficulties that participants were likely to meet in practice, for instance, is referred to in the literature as the 'Green Wall' (Shelton 1994). The Green Wall is experienced when environmental managers wish to pursue sustainable development ambitions because they recognise the business advantages it may entail, while their colleague business managers consider the topic in defensive terms, e.g. to keep the company out of the firing zone of NGOs and critical press. The NIDO programme sought to assist in breaking down the 'Green Wall' and to help firms develop an integrated approach to CSR that included a focus on the business opportunities offered by a perspective on sustainable development (Cramer and Loeber 2004).

In order to stimulate the transfer and dissemination of ideas and insights from the programme to the firms involved, NIDO developed a second track to stimulate processes of learning. Each of the companies involved were stimulated to carry out an in-company project on CSR that ran parallel to the NIDO programme. The programme manager assisted the companies' representatives strategically and practically in regard to these projects. She paid visits to every company three to four times to engage senior staff and to keep track of the progress made.

What is said about the individual NIDO participant vis-à-vis the company he or she represented holds equally true for the relationship between the individual firm and its wider societal context: here too, learning may be a necessary, yet not sufficient precondition for change. A firm's willingness and ability to change its mode of operation in the light of newly developed views critically depends, next to its resources and its physical infrastructure, on the legal, economic and societal setting in which it operates. A third track in the NIDO strategy, therefore, was to influence the contextual forces that may affect firms' CSR strategies and initiatives. Such forces are, for instance, the benchmarking of the CSR performance of companies by the financial sector in order to select 'the best in the class' for sustainability indexes, or the activities of NGOs to publicly challenge companies' claims of operating in a 'sustainable' or 'environmentally-sound' manner. Departing from the perspective that what appears a given, limiting condition for one party, is the outcome of purposeful action of others, NIDO set out to connect various developments in society that were supportive to companies implementing CSR. It sought to mutually enforce the programme's dynamics and those of similar initiatives in Dutch society.

Discussion of the programme's learning effects

Learning experiences at group level

In retrospect, all participating company representatives indicated that they felt inspired and encouraged by the programme and notably the monthly meetings, and that they gained knowledge about corporate social responsibility. The understanding described above that corporate social responsibility involves more than a mere procedural exercise in which an additional paragraph is added to existing quality systems was jointly developed in the group in the course of time. The participants grew to see that it requires a new positioning of the company with regard to its (social and natural) environment. In the words of one of the participants: "Corporate social responsibility is not a trick. No, the genie comes out of the bottle. It is not a question of just making accurate records and then getting back to work. That would be a chance lost. Rather, what corporate social responsibility means turns out to be a search process. It is not a cut-and-dried set of starting-points that one can just apply in a vacuum."

What was learnt at an individual level differed largely among the participants, depending on the stage of development of corporate social responsibility within their organisation, and on their personal interests and characteristics. Interestingly, a common denominator in these individual learning processes was that learning more than once occurred as a result of what might look to outsiders as an exchange of rather trivial information. To some participants, access to information on experiences with practical methods and specific procedures for implementing corporate social responsibility was a very valuable aspect of the NIDO programme, even when related to business practices quite different from their own.

Another common trait was that most learning reported involved first order aspects of corporate social responsibility. Often, participants put an emphasis on getting the process of taking corporate social responsibility seriously in their company off the ground by introducing measures that sat well with the company's existing routines and values. Some participants however began to fundamentally question their company's way of thinking and acting, and came to view corporate social responsibility as involving a 'paradigm shift'. Incidences of the second-order learning that occurred as a result are detectable from such utterances as 'the penny dropped' or 'an eye-opener' by which the participants describe the experience.

How such a process of second order learning was triggered by the programme is illustrated by the story of the representative of a multi-utility company involved. According to this participant, the NIDO programme provided him with insight into unexpected opportunities to tackle his company's problems. This learning process was induced notably by the stories on implementing corporate social responsibility from a participant representing a company quite different from his (a multinational carpet company). The latter firm is well-known for the way in which it has integrated sustainable development notions in its core vision and business strategy. The insight stories on how this worked out in practice led him to reflect on the approach his own company had developed to elaborate and embed corporate social responsibility. By consciously contrasting both practices, he came to see that "the topic should take shape in a holistic way". The inference that he drew from this revised opinion of how to approach the issue at the practical level concerned the role of a company's management: "Implementing corporate social responsibility needs to be inspired by *an inclusive vision* of how the various aspects of a company, which takes corporate social responsibility seriously, will connect to and mutually reinforce one another".

With regard to the reasons why the NIDO programme was considered to be conducive to learning, the participants furthermore referred to the atmosphere of trust that was created in the meetings, and to the way in which the meetings were chaired. The facilitating role that the programme manager adopted as a chair enabled the participants to jointly determine the contents of the discussions as well as the course of the programme's process. Characteristically, a meeting ended with the programme manager going over the core elements of a discussion, thereby checking her interpretations with those of the participants. From this summary, she then drew conclusions about which topics required further discussion and suggested a suitable procedure (e.g. inviting a guest speaker to illuminate a topic). While allowing the participants to determine largely the contents and procedures, the programme manager took care to ensure that the way in which she summarised and described a meeting's findings and insights were consistent with the conclusions drawn at previous meetings, and evolved into a coherent understanding of implementing CSR in the face of diversity. This approach to chairing the meetings ensured that the topics elaborated within the programme reflected closely the interests and concerns of the participants, (and thus were sufficiently context-specific to be of relevance to the participating firms), yet at the same time amounted to a more generic understanding of the CSR that would have an appeal to parties outside those immediately involved.[25]

The atmosphere of trust created in this way greatly contributed to the company representatives' willingness to share not only the success stories but also those of failure and frustration. This not only implied that the programme offered an opportunity to learn from others in a different yet comparable situation, but also made participants reflect on their own views and experiences. Furthermore, the impact of the discussions that were held in an open, responsive setting was reinforced by the 'lived' experience of the participants who ran a project in their respective companies.

Learning processes at company level

As was expected from the outset, the company representatives participating in the NIDO programme had a difficult task. They were the intermediaries between the group of people learning from each other, and their own company. The challenge was to transfer the knowledge and experiences they gained in the NIDO group to their own organisation and, preferably, to incite similar (second-order) learning processes there.

As mentioned, the NIDO programme assisted the participants in taking initiatives to involve colleagues in their organisation in the transformation process towards corporate social responsibility. At the outset of the programme, for instance, each company was requested to make a zero-assessment of its performance in terms of corporate social responsibility. To this end, information had to be collected to fill in a so-called Sustainability Score Card. The Card's systematic design was developed by a commercial consultancy agency to assess a company's 'CSR performance'. The Score Card assessment exercise appeared to be a very useful device for triggering processes of in-company learning. It required the active co-operation of various departments in a company. As a result, in all corners of a firm, people became aware of the company's involvement in the NIDO programme, and of the link between the programme, the issue of corporate social responsibility and their

[25] A strategy that proved very successful. Not only did most of the participating firms make major progress on the path towards implementing CSR in the aftermath of the project (cf. Cramer and Van der Heijden 2005). A booklet with the practical insights gained in the programme, which was published shortly after it had come to a close, turned out to be a major success in the management literature on CSR in the Dutch language, heading a top-10 of books in that field for more than a year.

own work. The Score Card exercise succeeded in stimulating internal company discussions as to what the CSR concept meant, and why an organisation should focus on it.

In the next phase of the programme, similar dynamics resulted from the NIDO-related in-company projects. These also promoted involvement and co-operation of different business units and departments within one company. Depending on the particular aspect a participant addressed (e.g. reporting or linking CSR to the current management system in place), specific persons from an organisation were called upon to join in.

The NIDO participants found that translating the abstract concept into down-to-earth activities (e.g. setting up a campaign with schools to promote a healthy lifestyle; cf. Cramer and Van der Heijden 2005) facilitated in-house communication on CSR. Well aware that a company's culture is usually hands-on and orientated towards concrete action, they were thus able to motivate their colleagues to join the effort. 'First-order' information that corresponded well with existing practices, policy statements and standards of a company was easiest to convey. As a result, the changes that were seen to take place in the companies at first glance resulted mainly from first-order learning.

Transferring the fundamental principles that were found to underlie corporate social responsibility proved much more complicated. The cultural shift that was considered necessary was difficult to set in motion. Information on the quality and extent of such a shift that the NIDO participants offered in their respective organisations was often criticised as being 'too soft.' As a result, the company representatives were quite reluctant to address the more fundamental issues. They therefore usually did not bother to convince their company colleagues of the underlying meanings of 'sustainable development' and social responsibility, but instead focused on the practical implications in terms of strategy and return-on-investments.

Still, by adopting such an approach, the more fundamental issues were addressed 'in disguise.' Take, for instance, the participants that focused in their in-company projects on the introduction of new ways of thinking about external communication. By doing so, implicitly, they addressed the underlying question of how the company could and should position itself vis-à-vis society. The effects of such efforts may not be visible immediately in the behaviour of the company, but in the course of time, the mindset of key figures in the organisation might change accordingly. Likewise, the efforts of those who set out to introduce a more people-oriented, tailor-made personnel policy, may help alter more fundamental notions of what 'good management' is.

Learning and processes of change at a structural level

Luckily but not entirely coincidentally, at the time of the programme, the context in which the companies operated was to a large extent favourable to the exploration and adoption of corporate social responsibility from a sustainable development perspective. The NIDO programme was carried out in a period when societal interest in the issue had been steadily growing. Various stakeholder groups were putting pressure on industry to take its responsibility seriously. The government's main advisory body on matters of national and international social and economic policy, the Social and Economic Council (in Dutch: SER), was at the time preparing recommendations on the subject, in close consultation with representatives of employers' organisations and trade unions (SER 2001). The Council's recommendations, which were published a few months after the start of the NIDO programme, triggered stakeholders to formulate more explicitly their own position in the debate. Various stakeholders had already taken up the issue in their policies, but from then on, they intensified their efforts and were much more outspoken in public about the importance of corporate social responsibility. At the time local governments too began to debate their role in supporting industry in their efforts towards corporate social responsibility. The trade unions and employers' organisations also paid more attention to corporate social responsibility. A number of NGOs joined forces and prepared an agreement on what the organisations considered quintessential in propagating CSR.

The NIDO programme was conducted against the backdrop of these developments. The programme manager took on an active role in channelling the insights gained and shared in the programme to third parties concerned with the topic of CSR. She was constantly on the watch for opportunities to combine forces with initiatives taken elsewhere that were inspired by considerations similar to the NIDO endeavour, and which might help 'even out' possible barriers that the companies experienced in implementing CSR.

Vice versa, the programme served as a clearing house of information on the developments in the field of CSR for the participating firms, and allowed them to take advantage of the latest insights. This took shape most concretely in the form of a joint research programme on CSR that was set up by seven universities. The inventory made in the NIDO programme of the knowledge gaps and research needs of the participating firms provided a major input in a series of round table discussions that the NIDO manager set up between a number of academic research groups that, in relative isolation from one another, studied aspects of CSR. The exchange of information among researchers, and between them and the NIDO programme participants eventually resulted in the drafting of a research proposal that envisioned co-operation between various knowledge institutes. This proposal, and the willingness of private partners to participate in the programme, convinced the Ministry of Economic Affairs to such an extent that it agreed to support the research programme financially for an initial period of two years.

The NIDO programme played a similar mediating role in an exchange of information between its participant companies – all involved in the actual practice of implementing CSR – and the Council for Annual Reporting, which at the request of the national government was preparing guidelines for reporting information on corporate environmental and social performance. In this case too, the programme was considered a valuable source of practical information, and the sharing of views was seen as mutually beneficial.

The input that the companies could provide through NIDO to the activities of these and other relevant 'third parties' thus potentially contributed to the development of the economic, legal and other structural conditions that are favourable to their corporate social responsibility initiatives. In turn, because of the exchange of information between them and the various echelons in society, companies had an opportunity to learn about (upcoming) developments that might affect their context in the future. In this way, they were able to assess the relative importance of these developments for their mode of operation, and learn accordingly.

Epilogue

It is obvious that the ambition of pursuing a sustainable development puts demands on the corporate sector. Corporations are pressed to assume responsibility for the social and environmental aspects of their business operations, beyond a mere meeting of regulatory demands. Corporate social responsibility (CSR) offers a useful way for firms to tackle environmental and social problems associated with economic development. The NIDO programme on CSR helped companies shift their attention from focusing solely on their financial performance to including their ecological and social performance, and to help them embed environmental and social considerations into core business systems.

Implementing CSR, the NIDO experience showed, is a complicated and, at times, frustrating process of designing tailor-made solutions that match the specific characteristics, ambition levels and the socio-economic context of a company. The above-reported incidences of learning obviously do not present an exhaustive list of all learning processes that took place with regard to the NIDO programme. While participants learned from and with one another, not all learned to the same extent. What sounded like 'old news' to some could represent entirely new insights for others. Still, in the course of the programme and shortly after its finalisation, the overall impact in terms of learning was considered valuable.

What the experience taught in regard to the process of practically stimulating learning, furthermore, was that a mere exchange of information is not enough to deepen the understanding of corporate social responsibility. The combination of practice and reflection was found to be crucial in inducing processes of learning. An additional practical insight gained was that it pays to translate the abstract notion of 'sustainable development' and a company's 'social responsibility' into down-to-earth, concrete actions (such as setting up an ICT project for people with a handicap or incorporating CSR into a company's quality management system), not only in terms of economic value added and possible higher returns on investments, but also in triggering processes of learning. Helping a company to take CSR seriously through concrete actions may at first glance not appear to help drastically alter standing practices, yet in the longer run, it may help induce processes of second order learning that favour the pursuit of a sustainable development.

In addition, and apparently quintessential to the programme's success, was the open-ended approach to determining what exactly notions such as 'sustainable profit' or a corporation's 'social responsibility' entail. Nowhere in the initial project plans were these topics defined beyond Elkington's general Triple-P (People, Planet, Profit) understanding. This was a deliberate choice. The discussions thus focused on issues that were raised by the companies themselves and related to specific knowledge gaps or problems that they had encountered in implementing CSR. The manager's role of facilitator and mediator enforced this searching, context-specific approach. The learning involved in the programme never involved a 'teaching' of pre-given principles or practical procedures. While the internal consistency of the notions and issues addressed and the insights gained was a recurrent topic of shared concern, essentially, the process *as well as* the programme's contents (and direction of the changes made by the participating companies) were literally unpredictable. It was fascinating to observe how differently the process of implementing corporate social responsibility evolved within the 19 participating companies.

A final remark concerns the people involved. Their personal focus and characteristics largely determined the way in which the topic of CSR was discussed during the meetings, and strongly affected the way in which it was elaborated in their respective in-company projects. Crucial in particular was whether people were personally enthused by the concept, eager to make changes and able to communicate their vision and practical examples in such a way that it inspired others to join in.

This analysis leads to the question, under what conditions the process of implementing CSR from a sustainable development perspective may really take off: at what point does 'the genie come out of the bottle' and why? What exactly triggers a willingness on the part of companies to take their social responsibility seriously? For every firm, it is a different mixture of factors and actors. External pressure (by governments, NGOs, media, consumers) and internal dynamics (e.g. employees' viewpoints and actions), together with the prospect of promising economic opportunities (either in the form of direct financial benefits or more intangible rewards such as a better reputation) apparently in various constellations influence corporate behaviour. Irrespective of the exact trigger, the issue of corporate social responsibility, once taken seriously, is likely to pose problems for companies. Commonly shared are the problems of 'actually getting started', and, subsequently, of how to then get the idea genuinely absorbed into the life-blood of the organisation. That requires learning at all levels: at the level of the individuals in a company, at company level, and indeed between companies and parties in their surroundings (between stakeholders, other companies, governments, knowledge institutes, and so on). Learning, we might say, is the key to dealing with corporate social responsibility. The NIDO programme showed how such learning can be triggered, facilitated and 'exploited' to obtain the best possible effect. Forums for learning such as the NIDO programme may help a growing number of companies and other parties involved to balance People, Planet and Profit in a global environment. The lessons learnt from the Dutch experience may be of help in designing such platforms.[26]

References

Argyris, C. and Schön, D. (1996) *Organisational Learning II; Theory, Method, and Practice*, Reading Massachusetts: Addison-Wesley Publishing Company.

Cramer, J. (2001) "From Financial to Sustainable Profit; Programme Plan", National Initiative for Sustainable Development (NIDO), Leeuwarden.

Cramer, J. (2003) *Learning about Corporate Social Responsibility; The Dutch experience*, Amsterdam: IOS press.

Cramer, J. (2006) *Corporate Social Responsibility: an Action Plan for Business*, Sheffield: Greenleaf.

Cramer, J. and Loeber, A. (2004) "Governance through Learning: Making Corporate Social Responsibility in Dutch Industry Effective From a Sustainable Development Perspective", *Journal of Environmental Policy & Planning*, 6(3/4):1–17.

[26] Several initiatives in this direction have already been taken, for instance, in the shape of a follow-up programme to the NIDO event described here, called 'CSR in an international context' (Cramer 2006). Comparable initiatives that focus on (learning about) transitions and system innovations from a sustainable development perspective are set up under the umbrella of the Dutch Knowledge network on System Innovations (KSI; see http://www.ksinetwork.nl/).

Cramer J. and Van der Heijden, A. (2005) "Sense-making in Business: The Case of Corporate Social Responsibility", in: Jan Jonker and Jacqueline Cramer eds., *Making the Difference; The Dutch National Research Program on Corporate Social Responsibility*, The Hague: Ministry of Economic Affairs, pp. 41-56.

Elkington, J. (1997) *Cannibals with Forks; the Triple Bottom Line of 21st Century Business*, Oxford: Capstone.

Holme, R. and Watts, P. (2000) "Corporate Social Responsibility: Making Good Business Sense", World Business Council for Sustainable Development, Geneva.

Loeber, A. (2003a) *Inbreken in het gangbare. Transitiemanagement in de praktijk: de NIDO-benadering* [Breaking and entering the obvious. Transition management: the NIDO approach]. NIDO, Leeuwarden.

Loeber, A. (2003b) "Learning processes at group level [First and second order reflection among participants in the NIDO programme", in J. Cramer, ed., *From Financial to Sustainable Profit*, Leeuwarden: NIDO.

Shelton, R. (1994) "Hitting the Green Wall: Why Corporate Programs Get Stalled", *Corporate Environmental Strategy*, 2(2): 5–11.

Social and Economic Council (SER) 2001, *Corporate Social Responsibility; A Dutch Approach*, Assen: Van Gorcum.

Chapter 15

Social learning for sustainable development: embracing technical and cultural change as originally inspired by The Natural Step

Hilary Bradbury

This chapter describes and illustrates a learning approach to sustainability that combines principles of technical *and* human-cultural change. Changing behavior is rarely easy. Launching initiatives and maintaining momentum is a great challenge. Uptake of technical insights without attention to cultural-organizational change issues guarantees less than adequate results. This chapter reminds those involved in the work of sustainable development that all change must be implemented by *people,* as individuals, groups, organizations, or societies. Its contribution is therefore to help crystallize a small set of principles from successful, complex change efforts to date. In this chapter actionability of social learning is highlighted.

Introduction

Once upon a time there was an oil company whose business model shifted from one of oil provider to energy provider – the problem was the employees on the oil platforms didn't quite get with the new program. In the same far away place there was also a home furnishings company known for its innovative leasing program that guaranteed a 'zero to landfill' cycle, yet oddly the sales force rarely mentioned it to customers. These companies operated in a country whose population had slowly learned enough about global warming to become very concerned, yet whose President declared the country's unwillingness to join with other societies to address the complex changes needed. In the meantime in the same far away place there were many self described environmentalists whose personal 'ecofootprints' were huge – e.g., the environmental impact of their lifestyle especially the size of their house, car and airplane travel habits – and yet they continued to worry most about *others'* environmental failings.

Perhaps each of us can all too easily locate ourselves in this far away place. Our challenge is that our knowing what to do does not keep pace with our ability to actually make the needed change to move us toward a more sustainable society. Ironically the sustainability community piles on technical insight with little attention to human, behavioral factors that support sustainable change. We find that we have inherited complex human and organizational systems and have grown used to acting the way we have always acted. Unfortunately our collective, everyday actions are increasingly unsustainable; we are learning that we create and reinforce harmful dynamics through our current actions, reward systems and institutions. The urgency of the moment is finding ways to address both immediate crises and contribute to effective, long-term, systemic change.

In keeping with a pragmatist perspective, I suggest that the success of social learning is to be evaluated by *measurable* improvements in moving us toward a more sustainable state, be it through more sustainable products, processes or services. Actionability distinguishes between people knowing about something to being able to produce what they desire, using their knowledge (Bradbury 2007). Social learning for sustainability may therefore be defined as an actionable, participative process that enjoins key stakeholders in generating desired results for a more sustainable level of organization. It may be contrasted with a learning process that focuses on increasing understanding without attending to actionability.

The core question of this chapter then is:

> How can we fundamentally change the ways in which we live together – with all living beings and systems – so that future generations not only survive but thrive?

The chapter is anchored in the story of The Natural Step (TNS) and also looks at a U.S. effort to articulate principles of human systems change that has since been inspired by the process behind TNS. Both highlight the connections between what sustains people internally and personally with their work on sustaining the external, 'natural' environment. In addition, both cases vividly illustrate that it is not enough to develop a 'right solution' to our sustainability challenges; we must also figure out how our individual and organizational-cultural behaviors can be brought into alignment with sustainable development. The chapter shows that there is a range of ways in which this interplay between the internal and external, or personal and professional can occur and that it can deliver results.

The first section describes the work of The Natural Step as a particularly powerful approach to integrating sustainability into organizations. It combines explicit attention to technical issues with implicit attention to human-social concerns and places. Personal engagement and interpersonal dialogue are central to the process. The second section focuses on the articulation of principles for human systems change that are offered particularly as a reminder to those who do not think enough about how to engage people in the changes required for a more sustainable society.

Overview of the early days of The Natural Step in Sweden

The Natural Step in Sweden – more correctly *Det Naturliga Steget* (DNS) – was founded in 1988 as a non-profit educational network. By 1996, it included approximately 10,000 network members. Founded by Karl-Henrik Robèrt, a leading cancer researcher, efforts initially focused on producing a scientific consensus statement about the most pressing environmental issues in Sweden. Approximately fifty of Sweden's senior scientists were involved, offering input and reading the draft statements. Business leaders agreed to underwrite the costs of disseminating the resulting consensus statement, which was produced as a colourful booklet. The booklet was sent to the entire Swedish population, of eight million, in a direct mailing to schools and households. Its reception was celebrated with a televised gala attended by the Swedish King. After its launch, DNS was headquartered in Stockholm with about twenty staffers whose main work consisted of meeting the requests for Natural Step educational presentations that poured in from all sectors of society.

Robèrt's collaboration with fellow scientists John Holmberg and Karl Erik Eriksson, of Gothenborg University, further honed the scientific underpinnings of DNS. In this collaboration the four system conditions for sustainability were articulated using a group learning process that focused on achieving consensus via a 'single version method' among natural scientists. This means that differences of opinion were resolved in a new iteration that was again sent out for review until all involved could agree on the final version. Once it was formulated DNS provided a framework to understand the complex phenomenon of *sustainability* by giving attention to both human-social as well as environmental issues. Once developed, the four system conditions took centre stage as the primary contribution of DNS in helping businesses and other social institutions to strategize about moving toward greater sustainability. Please see Appendix 15.1 with the Natural Step framework.

The clarity and simplicity of the framework with its four system conditions, and the way to apply them to decision-making, had great appeal. Government and professional networks – such as, Doctors, Engineers, Agronomists, Teachers for The Natural Step – formed to see how the framework might apply to their own professional domain. In turn, these networks enriched the insights and application of the DNS framework through the development of professionally-anchored consensus statements. Today, a majority of the Swedish 'states' use the framework to inform their governance. In addition, a number of business networks have been established to apply the system conditions. The multinational furniture manufacturer, IKEA, was the first to have all 30,000 employees exposed to the message of the Natural Step. Scandic Hotels, Electrolux, Swedish McDonalds followed suit. Each company, along with many more to follow, began to use DNS process which drives toward resolution of difference through seeking higher orders of consensus anchored in the four system conditions to shape strategy and help design new products and processes (Bradbury and Clair 1999). Studies independent of DNS (e.g. Meima 1996) have since shown the influence of these networks and their role in effecting change.

The Natural Step also went global. Autonomous Natural Step organizations now exist on all continents. A group of respected North American scientists was convened in February 1997 at the Johnson Foundation's Wingspread facilities to examine the validity of the science behind The Natural Step, in general, and the four system conditions, in particular. After two days of conversation, all those present – some of whom are Nobel Laureates known for their work on environmental issues – summarized their findings in a signed statement agreeing that the principles are based on sound science and provided a "valid approach for addressing the problems [of environmental unsustainability]". They further agreed that the principles are especially useful for the education of non-scientists because of the accessible formulation. Robèrt has since been awarded the prestigious Blue Planet Prize, a type of environmental Nobel Prize, for his efforts to catalyze this momentum. He has written the story of this work which offers additional and interesting detail (Robèrt 2002).

Revitalizing personal aspirations for making a difference

In 1996, the author conducted interviews with over twenty DNS founder-leaders. Robèrt, a highly respected cell scientist and practicing cancer physician, expressed his compunction to break with the technical focus of his field and ask bigger questions. He says that he was not just passionate but 'obsessed' with educating Swedes about environmental issues:

> "When I studied environmental medicine, I was surrounded by scientists who got happy every time they discovered a new toxic pollutant. And later in my career, when I practiced medicine, too many colleagues were more interested in marginal improvements to medicines, rather than in helping people avoid diseases. This was reflected in funding made available for prevention versus medicine improvement – the ratio was 1:10. I became more and more *un*interested in designing ever more elegant medicines. There was a driving force in me, a power and urgency to do something more fundamental."

Robèrt's wife, Rigmor – as famous as her husband for her work in analytical (Jungian) psychology – offered the following insight about his level of personal engagement:

> "I started to understand that something was going on with Kalle [Karl Henrik] as I listened to the dreams he was recounting. He was having amazing, quite elaborate dreams. When men are becoming creative they often have such dreams."

DNS had a founder with a public face, one internally driven to discover and do 'the right thing'. He communicated his sincerity and passion both explicitly and implicitly. Many others signed on to the issues and cause to which he directed attention. His engagement was infectious. He demonstrated the importance of 'emotional intelligence', a capacity to resonate energy and enthusiasm by connecting with people using empathy and self-awareness (Goleman *et al.* 2004).

Hans Dahlberg, then leader of a large insurance company, stepped up as a leading financial sponsor. He pledged a considerable amount of money after his first meeting with Robèrt to be used for promoting DNS' approach to sustainable development. Dahlberg himself had long been concerned with environmental issues. He also wanted his company to regain a leadership position with regard to the environmental agenda given their commitment to customer well-being, which the company defined holistically. Dahlberg was mildly impressed that Robèrt had gotten through his secretary for a meeting, was even more impressed with the scientific credentials and support of scientific colleagues, who are trained to disagree rather than to reach consensus. Most of all, perhaps, he was impressed by Robèrt's capacity for articulating the issues clearly and with passion. Given his strategy to define concern for customer well-being broadly he was happy to find someone as credible as Robèrt to help sell his message. He told me simply, "I was determined that Kalle should succeed".

In supporting Robèrt's work with DNS, Dahlberg was in effect revitalizing his own passion for environmental issues along with revitalizing the reputation and purview of his company. On Dahlberg's request Robèrt was to meet Per Uno Alm for help with business and organizational aspects of DNS development.

Per Uno Alm, called PUA, had worked his way through mainstream Swedish business from a working class background to become a very successful organizational consultant. His practice was oriented to advising large companies, with their paralyzing bureaucratic style, to move beyond the inertia their structure engendered. At first, PUA had no particular interest in issues of environmental sustainability and no real desire to meet Robèrt. He explains how he overcame his initial disinterest during his first meeting with Robèrt:

> "I found myself surprised at how articulate, inspirational and sincere Kalle was. I saw that with Kalle in the public eye there was at least a chance of overcoming the perennial problem of the environmental movement, full as it is with so many well meaning people who just instil such doom and gloom that the ordinary person is just paralyzed. My reluctance lessened, but I would have stipulations. Kalle would have to be the public face of this movement. In effect, he would have to give up his public life for quite some time".

PUA brought his theory of organizational flexibility to DNS and pushed to drive DNS to serve its mission rather than to grow itself as an organization. He explained his strategy:

> "I was willing to work if DNS would set to appealing to large but diverse portions of society. I was willing to see if we could get people to support us, not directly but by themselves getting the idea of what needed to be done inside themselves. From there comes so much energy. There is no limit to it really".

With PUA's commitment to having only a small staff and flat, flexible infrastructure he set the scene for a decentralized network. The network was energized by people internally driven to make a contribution within their own domain of influence and motivated by a personal sense of connection to the issue of sustainability. With PUA a board of business advisors was created. These business leaders were willing to offer their business acumen to help DNS develop strategy. Over and over in interviews I was surprised by the depth, breath and commitment evident in the founding leaders' lives. All were extraordinary in their demonstration of 'walking the talk'. There was the former executive who had acted as CEO of a number of Sweden's largest businesses, who travelled by bus, rather than pollute the air with unnecessary car emissions. A short time before our interview he had been diagnosed with cancer and explained that he was now more determined than ever to fully commit his remaining time to the cause of environmental sustainability.

The stories such as these are numerous and compelling. They vividly demonstrated that what was sown in DNS was a personal commitment to issues of sustainability which revitalized people. The individuals who were organizational leaders, in turn, revitalized their organizations. This revitalization found expression in supporting the DNS learning-oriented networks. Rather than only following a charismatic leader – for one of the first adjectives that people use to describe Robèrt is charismatic – people were following their own deepest commitment, grateful for the direction of resources that Robèrt inspired and fostered.

Revitalizing the capacity of groups to think together

The Natural Step is based on a particular set of normative methods. Specifically, The Natural Step's methods are grounded in a belief that the best way to achieve sustainability – as indicated by the four systems conditions – is through *consensus-building and dialogue*, which in turn helps to *attain one's partners' commitment* to creating a sustainable society. The approach seeks to *avoid cookie-cutter approaches* to implementation by *facilitating TNS and partners to think together* using the same frame of reference. It is therefore possible that the work ends up somewhere different from what might be assumed.

Together these methods allow for a reawakening of an organization's and/or group's capacity to think together. This process expands the group's behaviour repertoire, freeing people up from getting stuck in debate and discussion mode in which new ideas are met with criticism and dismissal before adequate exploration.

What is important is that TNS offers widely divergent people the opportunity to 'get on the same page'. In turn, this method leads to an open system of change efforts designed by partners and *elaborated through network building* across many sectors of society.

It is particularly important to underscore that the success of TNS is evaluated by measurable improvements in products, processes and services, rather than solely in heightened understanding of what is required. Hence the emphasis on getting people on the same page is so that they can coordinate their actions. Constantly allowing for new interpretations gets people away from the realm of coordinated action and into the exchange of sometimes lofty, sometimes petty, but too often unactionable ideas. While the consensus process may be rather unattractive to scholars – who are socialized as to the importance of embracing diverse opinions as a good in itself – it makes for actionability. Actionability distinguishes between people knowing about something to being able to produce what they desire, using their knowledge.

Balancing advocacy and inquiry: yes/and approach

In the business arena specifically, The Natural Step's methods seek to affirm a business person's experiences and concerns through a process of consensus-building and dialogue. Those trained in the science which underpins the work of The Natural Step seek points of agreement between business partners' views and those expressed in the principles of sustainability. This method of conversing is called the 'yes/and' approach. Robèrt himself, after, what he explains were many false starts as a younger, less 'agreeable' man, helped foster the approach. For example, a concerned business executive might say,

> "My organization is totally dependent upon fossil fuels for energy, so the System Conditions are useless to me".

A Natural Step trainer using the 'yes-and' approach, might reply:

> "Yes, you are currently dependent on fossil fuels which are in finite supply and perhaps likely to be regulated due to global climate change concerns. And, yes, this dependence may even make your organization economically unviable over time. Perhaps you might seek alternatives now while there is still time to get an advantage over those who are not concerned with sustainable development".

Action networks: not a cookie cutter approach

As DNS flourished in Sweden professional networks sought to apply Natural Step principles. In keeping with PUA's paradoxical vision of combining a strong public persona as leader with a decentralized organization of energetic people seeking to do 'good work,' these networks experienced the challenge of bringing motivated people together to develop consensus. Once motivated people are gathered together, however, there is no guarantee that they can work or dialogue that well.

Interviews with one of the network leaders who facilitated a consensus process illustrate the learning that took place about the dialogue process itself. She recounts:

> "I slowly realized I was in a group of (well-known) scientists who had been in disagreement for decades. No one was taking the role of discussion leader. Two people chose to leave the group after our first meeting; one because as a researcher he didn't believe scientists should seek consensus. He thought we should always debate. The other believed that bio-technology would solve all our problems, so why discuss".

This opening meeting of a professional group suggested the typical approach to sustainability that exists in many arenas of life. It can be very challenging to actually have a conversation that builds on each other's ideas and generates completely new perspectives on old problems. Some people look to purely technical solutions 'out there' to fix everything. Generally, capacity for conversation and thinking together, rather than mere debate, is low even among (or perhaps especially among?) the most highly educated.

The facilitator of the agronomy consensus process, for example, decided to hold a number of smaller meetings among individuals who disagreed on specific issues, rather than continue convening large sessions. As a result, over nine months, better quality dialogue emerged. This illustrates how difference and diplomacy can coincide as long as the focus remains on actionability. This facilitator asserted:

> "Some people were transformed by the process. The frequent meetings and the feeling of having a common task made it possible. The best outcome was unimaginable at the start. A well-known conventional agronomist (i.e. one who promotes use of chemical fertilizers) went to a well-known organic agronomist (i.e. one willing only to work with the natural processes of pest resistance) and asked for a recommendation to work on an organic farm! It helped that the organic scientist had been very diplomatic in her style within the group".

In retrospect her recommendations for consensus building differ a little from Robèrt's. She suggests the importance of taking disagreements seriously and making them clear, but not focusing on the differences. Moreover, all people involved with the DNS process agree on the importance of respectful interaction, or, as Robèrt puts it, of "not violating the other person's sense of dignity". The revitalization of groups is in large part the revitalization of our capacity to speak with each other in a civil way, in spite of our possibly huge philosophical differences.

Convening consensus on how to approach sustainability

The Natural Step offered a natural experiment for how change that galvanizes sustainable human effort could come about. Reflection on the principles at play occurred after the fact (Bradbury 1998; Bradbury and Mainemelis 2001). What might occur if people who know quite a bit about social change engaged in conscious reflection to inform future practice?

With that in mind, the author led in convening a group of scholars in late 2003, to think together about how change occurs in complex human systems. Including more people in the process is part of giving life to the work. For example, inviting representation from WorldBank and UN leaders led us to hope the knowledge generated during this gathering might begin influencing how international development work takes place. In addition, our intention was also to share our dialogue experience, and specifically our 'call to action' document, with other Organisational Development practitioners to provoke further thinking (and refinements) in ongoing efforts to move organizations and society toward sustainability.

With these aspirations in mind, and a focus on accelerating work on sustainability, in December 2003, a group of social scientists were invited to gather for a couple of days at Case Western Reserve University. Co-convened by the Case/Weatherhead Institute for Sustainable Enterprise and The Natural Step, our goal was to think together about how change happens in complex social systems.

The participants consisted mostly of researchers and academic social scientists as well as eminent change agents and representatives from WorldBank, UN and several 'think tanks'. Understanding ourselves as taking on the task of locating and re-animating conversations that lead to change, we used the concepts of lifeworld and system (Habermas 1984) to conceptualize our efforts.

Conceptualizing 'purposeful' dialogue as a regeneration of 'lifeworlds'

Key in instigating change efforts is fostering new ways to think about the issues involved (i.e., new dominant logics) and creating new inter-personal linkages (i.e., new social structures) to support movement toward the desired change. A social learning process must importantly keep an eye on how the desired change may be differently formulated so as to stay consistent with a vision rather than devolve into mere tactics. The concept of 'lifeworld' (Habermas 1984) is helpful for integrating logic and individual's social worlds together as a way to describe how the human cultural nexus of everyday life is maintained through social relationships and conversations. Bringing attention to this dynamic interplay of interactions and negotiations can allow us reclaim the 'larger than interpersonal world', which can come to be experienced as objective fact and thereby seem much more real and rigid than what we construct among ourselves in our lifeworld. As lifeworld yields to an institutional reality, becoming externalized as objective fact (Berger and Luckman 1966), we have what Habermas simply refers to as '*das System,*' which results in activity in the lifeworld becoming constrained by its predefined logics and structures.

An example might be how our conversations about partnership and coordination at work can become externalized as company policy, which can then lock employees into stringent patterns of relating that can become, ironically, maladaptive. As this example shows, institutional structures can take on a life of their own and the lifeworld comes to support – rather than generate – institutional structures, norms and behaviours. As institutional structures move increasingly toward functionalist rationalization (Habermas 1984), the possibility of changing people's lifeworlds – which undergirds institutions – grows more difficult with the passing of generations. Both Giddens (1984) and Bourdieu (1977) point out how individuals come to deeply embrace, at a precognitive, somatic level, the taken-for-granted norms of the institutions into which they are born.

Change is difficult, but not impossible. And despite the tenacity of institutional structures, these elements are not necessarily enduring. Indeed we have seen significant institutional change occur in just a few generations, such as employee empowerment, concern for diversity, consideration of ethics and a broader circle of stakeholders, to name a few.

In all cases, the level of lifeworld has been the nexus for regeneration, through which new ideas spread in conversations and get enacted in new behaviours. As much as the lifeworld acts as the root of institutional structures, it can also re-root and re-generate these structures.

The work of Organization Development – drawing so much on skills in facilitating cycles of action and reflection, deliberative dialogue, and noticing of inner and outer arcs of attention – is work that re-animates the lifeworld. It draws attention to the ways in which structures are created and recreated in relationship, specifically in partnership with each other. This dialogue effort, therefore, is most accurately understood as an inquiry into how to reanimate the lifeworld in the direction of dialogue and communicative action (Habermas 1984) so as to produce more sustainable organizations.

The dialogue

The December 2003 sustainability and social change conversation was premised on the recognition that systems are difficult to comprehend without inviting in many perspectives, from across specific issue areas. By convening a cross-section of change experts from distinct areas of specialty – ranging from community-based through corporate-focused, and from consideration of legal rights through public health and forestry issues – we focused on accomplishing a joint analytical process of identifying key factors and dynamics around social change within (and across) specific areas of work, exploring shared concepts and frameworks, drawing out the interrelations between focused social change efforts and broader societal shifts, and developing a joint consensus document and statement on the dynamics of social change to be published and broadly disseminated to those interested in social change.

In preparation for the dialogue, we asked each participant share a brief (1-3 page) 'conversation starter' that described their thinking about complex social change and their 'theory of practice'/'theory of change.' The piece was intended to capture:

- Context: What kinds of social change projects/movements have you focused on?
- Themes: What issues do you repeatedly underscore as central in importance to understanding complex social change?
- Legacy: Which core concepts in your work are you most proud to offer to colleagues/the field to help the field better grapple with developing insight and practices related to complex social change?

Specifically, participants were asked to explicitly ground their theory of practice statements in reflections from their own research or fieldwork, responding to questions posed in the dialogue description, including:

- What of what I have learned is necessary for complex social change efforts to occur?
- What are some of the key factors and dynamics around social change?
- What are a few of the core concepts and frameworks that inform my thinking?
- What, if any, gains have been made through interrelations across a range of focused social change efforts?

The responses were compiled and distributed in advance of the meeting to all participants. We asked that each person read these pieces prior to the event. We stressed that because the meeting would be geared toward developing a consensus statement, these conversation starters were to identify the significant overlap (and, to a lesser extent, the important divergences) in thinking that exists among participants.

Moving to consensus

Our dialogue experience was grounded, at the outset, in the appreciative philosophy and methods (Cooperrider 1999) that would be drawn from throughout the dialogue. Facilitators introduced the work of The Natural Step by way of framing the gathering as one of consensus seeking for sustainability. Additional space was created for the group to 'arrive' in the room by opening each session with a check in process (Isaacs 1999) to offer the experience of a shared conversational space from the start. Participants were invited to 'check in' and to share their aspirations for the time together.

To support seeing patterns at a more systemic level of awareness, plenary reflections were captured by a graphic facilitator creating illustrations coded in colour to signify differences in themes (For images see: http://www.naturalstep.org/research/sc_images.php).

Acknowledging group contributions through visualization is a growing way of supporting increased participation, systems thinking, and group memory (Sibbet 2003). Its effectiveness is mirrored in the use of mapping and displays in Participatory Rural Development practices to draw out village stories about how things work (Chambers 1992; Farrington and Martin 1988; Ison and Ampt 1992). Graphic facilitation draws its practices from designing ways of working. Visualizing is a way of conceptually prototyping ideas, seeing new patterns, and remembering greatly increased quantities of information as participants review prior days' charts and informally rescan portions where they may have been less attentive. Because the process is visible and open to change and guidance by the group, there is a high degree of validation of the information, and acceptance of the record as a reference later on.

The dialogue began and proceeded as a cycle of plenary, small group and individual reflective sessions. We took periods in which individuals could write their own reflections or simply meditate. Participants' 'reflective write-ups' were gathered by the conveners. The first evening a small subgroup spent a couple of hours developing a very brief statement on how to approach complex social change that integrated the notes from participants in the plenary dialogue.

The following morning began with all seated in a circle with full view of all other participants. An initial draft of the statement was shared with the group. (Please see appendix for the original statement). The group read and thought about the goal of our statement, the essence of how we wanted it to read, and whom we wanted to read it. Contributions, reactions, suggestions, and questions were offered to further strengthen the document. The discussion led to a desire to rewrite the statement. Everyone wrote up a short piece that was passed to the co-convenors, who then began to put the short pieces together in a longer joint document.

Based on a shared desire to position this joint statement as a call to action, it was suggested that the statement be infused with language of urgency and invigoration, underpinned by the assumption that there are important inequalities and significant difficulties in the world. Mentioning basic human needs that are unmet and environmental degradation was also important to include in our call to action, as was clarifying our intent and voice in writing this document. Members of the group voiced a desire to ground our statement first, in a shared responsibility of the current state of the world, thus not blaming or finger pointing, as well as

Box 15.1. Principles for change in complex human systems.

In creating social change, effective efforts, we...

- address immediate needs while linking them to larger, systemic issues.
  Successful change connects focused efforts with the web of political, economic, cultural, and environmental factors that frame and shape the immediate needs.
- surface discontents, build capacity, and elevate expectations.
  Successful change emerges from dissatisfaction with current conditions, but also celebrates many small victories as well as personal learning, thereby continually building momentum for innovation toward a preferred future.
- raise awareness of how social systems support and resist change.
  Successful change invites people working at multiple levels—individual, organizational, national, international, etc. – to experiment in creating new realities and transforming the forces that maintain the status quo.
- engage diverse people in partnering for positive action.
  Successful change is fuelled by a mix of "un-usual" suspects—from those at the periphery of power to those closer to the centre – in co-producing alternative futures in a context of mutual respect and relationships of trust.
- become the change, innovate with opportunities, and persist.
  Successful change is grounded in personal transformation, encourages experimentation, and eventually evolves the system as a whole.

a positive vision for the change that we are trying to create. Challenging group-think, we remained with our statement, looking for counter intuitive statements, passivity of voice, and fuzzy logic. Re-working the draft, we honed the approach based on our intended audience and ideas on how people would participate in further developing this flexible, living treatise.

Conclusion

From these cases described above we can conclude that social learning for sustainability requires an integration of different ways of knowing ('scientific', 'experienced', 'enacted-in-relationships') that is essentially interdisciplinary and multi-actor. In all efforts to bring sustainability thinking to any human system, the work of systemic change must be simultaneously conceived of as a human as well as technical change. Particularly crucial is to develop a space in which open minded people with similar goals can convene for conversations for new experiments in sustainable living.

Figure 15.1 summarizes the main points of this chapter. The figure shows the relationship of lifeworld to institutional structures. The location of lifeworld inside the institutional structures suggests its ability to act as generative source of change. The primary mode for such change within lifeworlds is dialogue, which convenes networks of change agents in expanding cycles and circles of reflection and action that result in experiments for innovating new and successful sustainable practices.

We summarize the social learning process we have discussed. It is a process that must simultaneously engage technical, experience based and relational ways of knowing. The following are reminders that for sustainability oriented programs to have a higher degree of success:

- Technical requirements of sustainability (scientifically based knowledge) need to be clear.
- The more individuals can connect their personal aspirations for sustainability to the work at hand, the more sustainable will be the work. Such committed individuals are the 'cells' from which the work can develop.
- The convening of likeminded people from multiple domains for collective action requires at least light, skilful facilitation to allow people balance inquiry and advocacy.

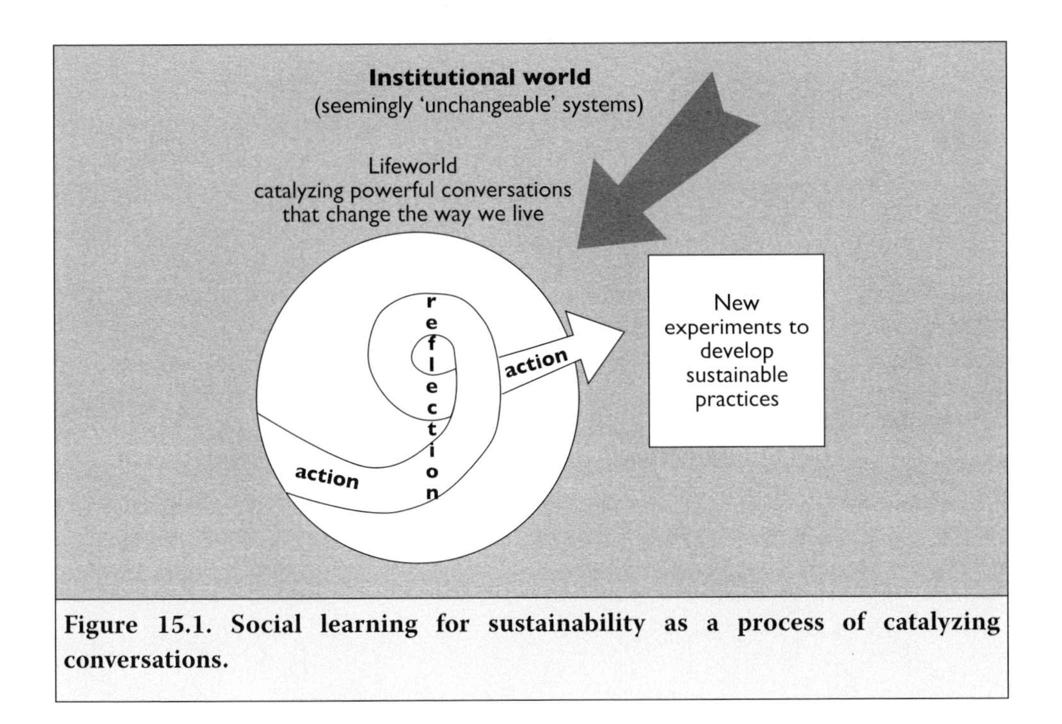

Figure 15.1. Social learning for sustainability as a process of catalyzing conversations.

References

Berger, P.L. and Luckmann, T. (1966) *The Social Construction of Reality: A Treatise in the Sociology of Knowledge*, Anchor Books: Garden City, New York.

Bourdieu, P. (1977) *Outline of Theory of Practice*, Cambridge, UK: Cambridge University Press.

Bradbury, H. (2007) "Actionability and the Question of What Constitutes Good Action Research", in Shani, A.B., Adler, N., Mohrman, S.A., Pasmore, W.A. and Stymne, B., eds., *The Handbook Of Collaborative Research*, London/Thousand Oaks: Sage Publication.

Bradbury, H. (1998) "Learning with the Natural Step: Cooperative Ecological Inquiry Through Cases, Theory and Practice for Sustainable Development", unpublished dissertation, University of Michigan, Ann Arbor, Michigan, USA.

Bradbury, H. and Clair, J. (1999) "Promoting sustainable organizations with Sweden's Natural Step", *Academy of Management Executive,* 13 (4): 63-74.

Bradbury, H. and Mainemelis, C. (2001) "Learning History and Organizational Praxis: Non Traditional Research", *Journal of Management Inquiry,* 10 (4): 340-357.

Chambers, R. (1992) "Rural: Rapid, Relaxed and Participatory", Discussion paper 331, University of Sussex, Institute of Development Studies, Brighton.

Cooperrider, D.L. (1999) "Positive Image, Positive Action: The affirming Basis of Organizing", in Srivastva, S. and D. L. Cooperrider, eds., *Appreciative Management and Leadership,* Euclid, Ohio, USA: Williams Custom Publishing, pp. 91-125

Farrington, J and Martin, A. (1988) "Farmer Participatory Research: A review of Concepts and Recent Fieldwork", *Agricultual Administration and Extension,* 29.

Giddens, A. (1984) *The Constitution of Society*, Berkeley, CA: University of California Press.

Goleman, D., Boyatzis, R. and McKee, A. (2004) Primal Leadership: *Learning to Lead With Emotional Intelligence*, Cambridge, MA: Harvard Business Press.

Habermas, J. (1984) *The Theory of Communicative Action*, Boston: Beacon Press.

Isaacs, W (1999) *Dialogue: The Art of Thinking Together*, New York: Doubleday/Currency.

Ison, R. and P. Ampt (1992) "Rapid Rural Appraisal: A Participatory Problem Formulation Method Relevant to Australian Agriculture", *Agricultural Systems,* 38.

Meima, R. (1996) *The Alpha Case: A Grounded Theory of Corporate Environmental Management in an IT Company. Industry and Environment: Practical Applications of Environmental Management Approaches in Business*, Copenhagen, Denmark: The Aarhus School of Business.

Robèrt, K.H. (2002) *The Natural Step Story: Seeding a Quiet Revolution*, Gabriola Island, BC: New Society Publishers.

Sibbet, D. (2003) "Principles of Facilitation", *The Grove Consultants International.*

Appendix 15.1

TNS Principles

Building on the original scientific consensus statement, the Natural Step's four system conditions for sustainability were articulated by Karl-Henrik Robèrt and John Holmberg. Working with these conditions DNS appeared to tackle the complex phenomenon of *sustainability* that requires attention to environmental issues (see conditions 1-3) *as well as* human-social issues (see condition 4). The Four System Conditions are:

"In the sustainable society, nature is not subject to systematically increasing...

- concentrations of substances extracted from the Earth's crust
- concentrations of substances produced by society
- degradation by physical means
- and, in that society...
- human needs are met worldwide."

An organisation's sustainability objectives are easily linked to these conditions, such as:

"Our ultimate sustainability objectives are to:

- eliminate our contribution to systematic increases in concentrations of substances from the Earth's crust.
- eliminate our contribution to systematic increases in concentrations of substances produced by society.
- eliminate our contribution to systematic physical degradation of nature through over-harvesting, introductions and other forms of modification.
- contribute as much as we can to the meeting of human needs in our society and worldwide, over and above all the substitution and dematerialization measures taken in meeting the first three objectives."

Finally, guidance is offered on how to put each of the sustainability objectives into practice:

- This means substituting certain minerals that are scarce in nature with others that are more abundant, using all mined materials efficiently, and systematically reducing dependence on fossil fuels.
- This means systematically substituting certain persistent and unnatural compounds with ones that are normally abundant or break down more easily in nature, and using all substances produced by society efficiently.
- This means drawing resources only from well-managed eco-systems, systematically pursuing the most productive and efficient use both of those resources and land, and exercising caution in all kinds of modification of nature.
- This means using all of our resources efficiently, fairly and responsibly so that the needs of all people on whom we have an impact, and the future needs of people who are not yet born, stand the best chance of being met.

Chapter 16

Corporate social responsibility: towards a new dialogue?

Peter Lund-Thomsen

Background

The significance of the CSR[27] and development debate is linked to the growth of international production networks in which Northern buyers control a web of suppliers in developing countries. This has led to calls for them not only to be concerned with quality and delivery dates, but also to accept responsibility for working conditions and environmental impacts in developing countries. Simultaneously, leading companies are vulnerable to bad publicity due to the increased significance of global brands and corporate reputation. In addition, the development of global communications technologies have not only enabled corporations to control production activities on a more global scale, but have also resulted in the rapid diffusion of information about working conditions at their suppliers overseas, raising public awareness and paving the way for NGO and trade union campaigns aimed at holding corporations accountable for their actions (Clay 2005, Jenkins 2005).

In this chapter, I will explore the links between CSR, development, and social learning within mainstream business settings. My point of departure is that the CSR discourse has primarily been shaped by business interests and the managerial concern with conceptualising CSR in such a way that sustainable development issues are rendered manageable by corporate decision-makers. In this way, the mainstream business management literature on CSR attempts to turn questions of social, environmental, and economic justice into technical problems that can be solved through a managerial problem-solving approach. Only solutions where corporations can combine profit making with socially responsible behaviour tend to receive serious attention in the literature. The implication is that issues around

[27] I use Blowfield and Frynas (2005, p. 503)'s definition of corporate social responsibility as "an umbrella term for a variety of theories and practices all of which recognize the following: (a) that companies have a responsibility for their impact on society and the natural environment, sometimes beyond legal compliance and the liability of individuals; (b) that companies have a responsibility for the behaviour of others with whom they do business (e.g., within supply chains); and (c) that business needs to manage its relationship with wider society, whether for reasons of commercial viability or to add value to society".

conflict, class struggle, and more radical approaches to citizen participation are sidelined. It is therefore necessary to develop a knowledge base consisting of critical perspectives on CSR and development so that these 'sidelined' issues are brought into the heart of the CSR and development debate. In the second part of the chapter, I argue that new spaces for social learning about CSR and development need to be opened. In particular, it is necessary to initiate a different kind of dialogue where this new knowledge, alternative values, and ways of engaging in CSR and development can be introduced to CSR educators, policy-makers, and present as well as future managers. Finally, I will try to assess the potential and limitations of such an approach to social dialogue with reference to the initial experience of organizing an academic network called the International Research Network on Business, Development, and Society.

The business management literature on CSR

The business management literature on CSR is rapidly expanding. First, an outline of some of the main approaches in the literature is given below. These approaches are then critiqued for their lack of ability to explain the kinds of changes that CSR initiatives can/cannot bring about in the conditions of workers and communities residing adjacent to production sites. Finally, paraphrasing Blowfield and Frynas (2005), I argue that this calls for an alternative critical research agenda[28] on CSR and development, one that focuses on assessing the impacts of CSR initiatives, power and participation in CSR interventions, and finally, one departing from so-called Southern-centred perspectives.

A dominant concern in the business management literature on CSR is related to what constitutes the proper role of business in relation to society. The question is whether business should have any social responsibilities or whether their only social responsibility is profit making (The Economist 2005). A lot of attention is paid to whether it is financially profitable for companies to engage in CSR practices, and how management tools can be devised to help corporate managers solve CSR problems (see e.g. Hopkins 2003, Prahalad and Porter 2003).

Some of the most frequently cited reasons as to why business should or does engage in CSR are:

[28] It could of course be argued that academic research should by definition be critical in nature. However, in the context of business management research, because of the lack of a more in-depth, self-reflective questioning of the basic assumptions about the nature and consequences of CSR for workers and the environment in developing countries, the need for a 'critical' research agenda seems obvious.

- The 'license to operate' argument (Roberts *et al.* 2002); i.e. companies with a poor reputation may be faced with continued criticism and eventual restrictions on aspects of their business operations if they do not alter their behaviour (*ibid.*). Hence, engaging in CSR is a way for companies to secure their license to operate.
- The need for pre-empting legally-binding regulation (Jenkins 2005). This follows from the license to operate argument. Instead of being forced to behave in a particular way by governmental bodies, companies may achieve greater freedom to operate if they voluntarily commit themselves to abide by particular standards; e.g., through environmental standards such as ISO 14000 or social ones such as the SA8000 standard. In this way, voluntary engagement in CSR might reduce regulatory pressure from governmental authorities since companies are seen as taking environmental and social considerations seriously.
- Employee motivation (Roberts *et al.* 2002). This both relates to making the company attractive as a potential employer and retaining employee morale among staff members who may be disillusioned if their employer is routinely singled out in media reports for ignoring social and environmental concerns in its operations. The experience of Shell and Monsanto are classical examples of how employee morale may be undermined by negative publicity.
- Reducing production costs and achieving improvements in working conditions and environmental impact (Porter and Van der Linde 1996). This is one of the so-called win-win situations where investments in cleaner production methods and equipment represent a short-term cost for the company which may be made up for once savings are realized through more efficient use of raw materials, reduction of waste, and recycling of used materials.
- Public relations benefits through stakeholder dialogue (O'Brady 2003, Middlemiss 2003, Freeman and Reed 1983). It is generally believed that companies may improve their image if they are seen as being responsive to public concerns. By engaging its stakeholders in dialogue the company may be able to incorporate some of the concerns of these stakeholders in their business operations, thus minimizing the risk for costly, prolonged conflicts with NGOs and communities with associated negative publicity.
- Managers' personal motives for making a difference (Hemingway and MacLagan 2004). This refers to the personal commitment of company owners or managers who may believe that they have a broader responsibility for securing the welfare of their workers and the communities in which their companies operate.

While evidence is still inconclusive as to whether it financially pays off for companies to be socially responsible, the 'business case approach to CSR' is increasingly being criticized by different authors who claim that financial profitability cannot be an appropriate starting point for evaluating whether companies should engage in socially responsible behaviour (however defined) (see e.g. Frynas 2005, Newell 2005, Lund-Thomsen 2004, 2005). If the incorporation of social, economic, or environmental considerations into business decision-making depends on their financial profitability, "what happens to those issues where such a case cannot be made?" (*ibid.*). It appears as if the role of conflict, class struggle and more radical approaches to citizen participation are largely ignored. As I have argued elsewhere, it is exactly at this point that the limitations of the management-oriented literature become obvious. It ignores the fact that CSR problems are not simply an outcome of management failures but also rooted in international political and economic forces as well as inequalities in the South (Lund-Thomsen 2004). There is clearly a need to develop a critical research agenda that allows for a more in-depth investigation of what CSR initiatives can or cannot achieve in relation to improving conditions of workers/communities in the South (Blowfield and Frynas 2005, Newell 2005, Lund-Thomsen 2005). In the section below, I will try to outline what could be some of the constituent parts of such a research agenda.

Critical Perspectives on CSR in the Developing World[29]

While it is not an exhaustive list, one could imagine that a critical research agenda on CSR could encompass three broadly defined areas: (1) the impact of CSR initiatives; (2) power and participation in CSR; and (3) Southern-centred perspectives.

Impact of CSR initiatives

While numerous arguments have been made about the potential benefits of engaging in CSR initiatives (e.g. Hopkins 2003, Prahalad and Porter 2003), little academic attention has been devoted to generating rigorous, systematic evidence of the social and environmental impacts of such initiatives (Jeppesen 2004, Nelson *et al.* 2005). While some initial efforts are underway in terms of assessing the impact of different types of CSR initiatives[30], a critical research agenda would emphasize the need for creating new ways of systematically assessing the impact of CSR

[29] This section is largely based upon discussions that took place at the Strategic Planning Workshop of the International Research Network on Business, Development, and Society at the Copenhagen Business School on 3-5 September 2005. The International Research Network on Business, Development, and Society consists of: Ana Muro, Anita Chan, Chandra Bushan, Michael Blowfield, J. George Frynas, Peter Newell, Halina Ward, Soeren Jeppesen, Michael E. Nielsen, Rhys Jenkins, David Fig, Maggie Opondo, Marina Prieto-Carron, and myself.

[30] E.g. on industrial clusters, CSR and poverty reduction, see Nadvi and Barrientos 2004; on codes of conduct, see Nelson *et al.* 2005; on labour, see Barrientos 2005; on women workers, see Prieto-Carron 2004; on the use of social auditing; see Opondo and Hale 2005, Barrientos 2005; on industry-wide environmental impact, see Bhushan 2005.

initiatives on issues such as poverty, wages, workers in general, the achievement of the Millennium Development Goals, specific ideas such as the Bottom-of-the-Pyramid[31] notion, and CSR models which are disseminated via business schools and CSR conferences.

Power and participation in CSR

Power and participation are two key issues which also require further exploration in the CSR and development relationship. This is particularly relevant in relation to stakeholder management. The pioneers of stakeholder theory, Freeman and Reed, have argued, "If this task of stakeholder management is done properly, much of the air is let out of critics who argue that the corporation must be democratised in terms of direct increased citizen participation" (1983, p.96). According to Blowfield and Frynas (2005, p. 507), stakeholder management is often portrayed as a process that brings business representatives, non-governmental organizations, and public sector agencies together to address corporate responsibility challenges. Yet, paraphrasing Leeuwis[32] (2000), such approaches tend to overlook the fact that conflict, social struggle, and strategic intent are often part of stakeholder interaction although these factors are perceived as undesirable on normative and theoretical grounds (*ibid.*). Hence, while corporations are increasingly engaging local communities in resettlement schemes and community development projects, questions arise as to whether companies are able to combine their traditional profit-making roles with a new profile as participatory, community development agencies (Frynas 2005). An important concern is whether corporations are sufficiently geared towards taking on community development roles which require their staff to use 'soft', social science skills traditionally used in aid management, while corporations, e.g., mining companies, are often dominated by 'hard science' specialists such as engineers (Szablowski 2002). Or similarly, whether corporations imposing their codes of conduct on supplier factories with abysmal results, can raise labour standards. A critical research agenda will therefore need to emphasize the importance of incorporating underrepresented voices in the formulation of CSR tools and approaches, the use of alternative methodologies such as feminist methods and action research, and the employment of alternative indicators of well-being in relation to addressing some of the existing power imbalances in

---

[31] The idea that businesses can contribute to poverty reduction by selling consumer goods to poor people, increasing their range of choices while reducing prices (see Commission on the Private Sector and Development 2004, Hart and Prahalad 2002). In his book *Capitalism at the Crossroads* (2005), Hart gives a broader definition of the bottom-of-the pyramid notion. He relates it to firms being able to develop disruptive technologies that address society's needs, in a way that is culturally appropriate, environmentally sustainable and economically profitable.

[32] His point is made in the context of participatory approaches used in developing countries, but it can equally well be applied in the context of CSR and development.

the CSR debate. That is, the research needs to be collaborative, with academics and practitioners in both 'developed' and 'developing' countries (Prieto-Carron 2005).

Southern-centred perspectives

As pointed out by Blowfield and Frynas (2005), "we cannot understand how CSR is developing only by examining what has happened; equally important is to explore who and what is being overlooked, taken for granted, ignored, or excluded". A central concern in a critical research agenda on CSR is thus to (1) identify actors that are largely excluded from the CSR debate, and (2) address issues which are ignored[33]. Regarding (1), dominant voices in the CSR debate tend to be Northern businesses, NGOs, governments, and trade unions while women, workers, and communities tend to be underrepresented in the formulation, execution, and evaluation of CSR initiatives (Fox 2004, Prieto-Carron 2004). As far as (2) is concerned, most of the CSR literature overlooks the fact that CSR initiatives have only been implemented to a limited extent so far in developing country contexts. The key challenge may therefore be to ensure company compliance with existing legal frameworks before the potentials and limitations of CSR can be meaningfully debated (Bhushan 2005). In bringing Southern-centered perspectives to the fore, a critical research agenda on CSR thus involves a double-movement. First, a top-down movement that investigates how CSR travels as an idea across borders, how it is operationalised, institutionalised, accommodated and resisted in various developing country settings as well as who benefits/loses from this process. Second, quoting Prieto-Carron (*ibid.*), "a bottom-up approach, from the local to the international with place/space being important in how the localising engages with a globalising process" such as CSR.

CSR in developing countries: towards a new dialogue?

My main concern has so far been to emphasize the need for a critical analysis of the values, norms, and interests underlying the business management literature on CSR. If we accept that a need for a critical investigation of the CSR notion exists, the next question is how we can initiate a dialogue that explores the potential

[33] However, following Blowfield (2004), one can argue that not everything that is ignored can be addressed because there are particular constructs of power, bias, knowledge, etc., embedded in CSR that may be inherently opposed to the norms, values and priorities of many grassroots actors in developing countries. Consequently, either those actors must 'reconstruct' themselves in order to be addressed/recognized, or forever be discounted/ignored (*ibid.*). Hence, the International Research Network on Business, Development, and Society can play an important role in assisting development and business practitioners in understanding what issues cannot be addressed as well as what can be addressed through CSR initiatives.

and limitations of CSR in relation to improving the conditions of workers and communities in developing countries?

Drawing upon a working document produced by the International Institute for Environment and Development in London entitled *Ways of Working for Sustainable Development* (IIED 2004, p. 12), we could say that such a social dialogue can be established by (1) building a knowledge base in terms of research that poses critical questions about the role of business in development, (2) identifying and working with the appropriate actors (CSR target audiences), (3) developing or operating in spaces where this knowledge base and the CSR target audiences are brought together in ways that contribute to better policy making and practice, and that reflect diverse and underrepresented voices. This triangular relationship between knowledge production, working with appropriate CSR target audiences, and linking knowledge and CSR target audiences in conceptual, geographical, or physical spaces is illustrated in Figure 16.1.

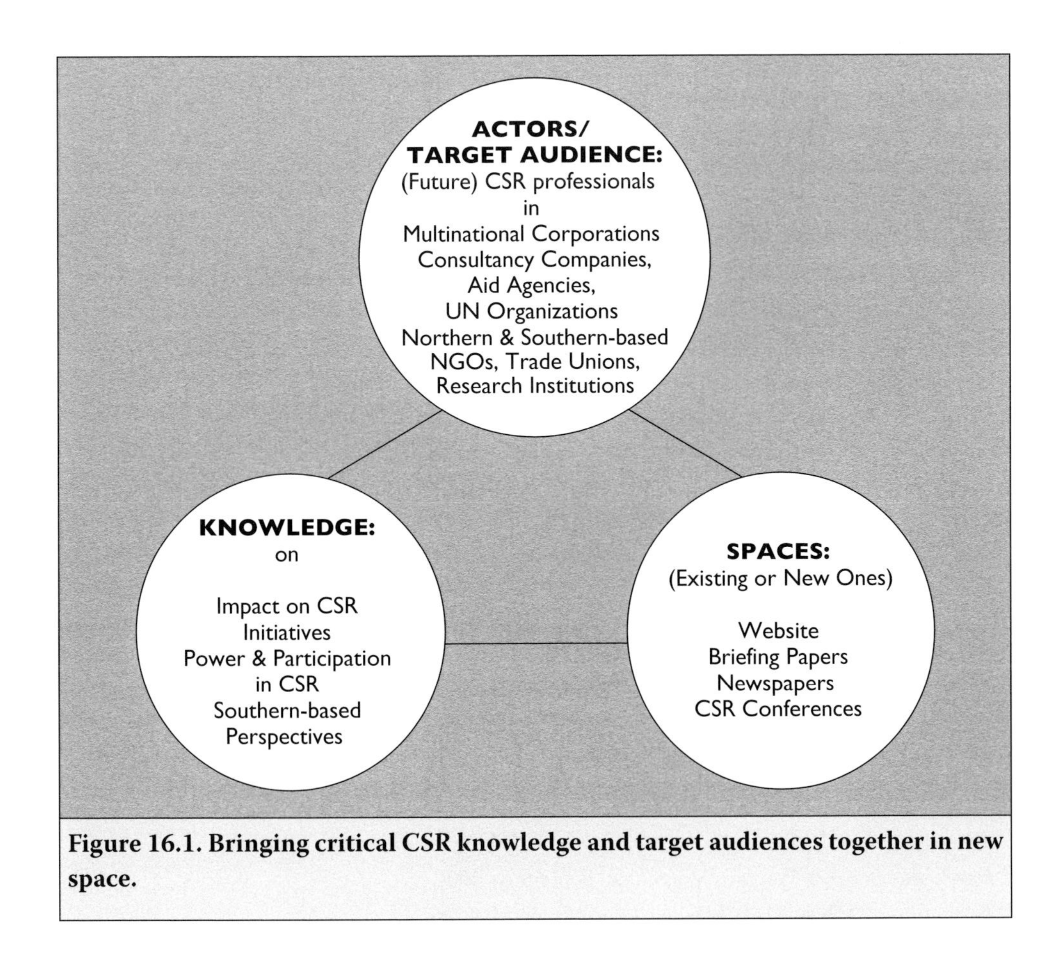

Figure 16.1. Bringing critical CSR knowledge and target audiences together in new space.

We can explain this approach to social dialogue as follows:

- The first step is to construct a knowledge base that poses critical questions about the role of business in relation to CSR and development. Important contributions to the construction of such a knowledge base already exist. For example, the Business for Social Responsibility project[34] of the United Nations Research Institute on Social Development has focused on whether or not transnational corporations (TNCs) and other companies are taking meaningful steps to improve their social and environmental record, particularly in developing countries. Another important contribution was made by a special issue of the journal Development[35] in September 2004, which contained a number of contributions that highlighted the need for a more critical engagement with the CSR and development linkage, while a more recent special issue of the journal International Affairs[36] (May 2005) was specifically dedicated to the theme: "Critical Perspectives on Corporate Social Responsibility in the Developing World". Whereas these contributions constitute initial steps towards a more critical investigation of the CSR and development linkage, I would argue that we need to sustain a long-term and more clearly defined research agenda around themes such as CSR impact assessment, power and participation in CSR, and Southern-centred perspectives.
- The second step is to bring this knowledge base and potential target audiences for critical perspectives on CSR and development together in existing or new spaces that will support a new kind of social dialogue on this topic. These target audiences include a variety of actors who are working on or are likely to work on CSR and development issues on a full-time or a part-time basis (e.g. CSR professionals in multinational corporations, consultancy companies, aid agencies, UN organizations, Northern & Southern-based NGOs, trade unions, and research institutions). However, the future CSR professionals that receive their graduate training in business schools and universities could also constitute potential target audiences. An important consideration is whether spaces for social dialogue already exist or whether they need to be created? On the one hand, the sound bite world of mainstream news reporting on television requires that complex problems are framed in 20-second news spots or one-line headings if we are talking about the printed press. On the other hand, social dialogue requires open-ended processes, willingness to listen to other parties' point of view, engaging with complexity, and time-frames which go beyond twenty-second news spots or one-line newspaper headings. Hence, establishing a meaningful social dialogue may require selective attention to

[34] The project description was accessed at www.unrisd.org on 15 January 2006.

[35] *Development,* 47(2), Palgrave Macmillan, September 2004.

[36] *International Affairs,* 81(3), Blackwell Publishing, May 2005.

which spaces can adequately be used to ensure that a mutual learning process takes place between the researchers and the intended target audiences.

I will now try to assess the potential and limitations of this approach to social dialogue with reference to the initial experience of organizing an academic network called the International Research Network on Business, Development, and Society. An interesting point is that this initiative originated from within the largely mainstream setting of the Copenhagen Business School (CBS). The CBS is one of the largest Business Schools in Northern Europe. The initiative may therefore provide some initial lessons about initiating social dialogue within mainstream business settings.

Phase 1: initial steps towards a new social dialogue

The origin of the International Research Network on Business, Development, and Society can be traced back to a series of conversations on CSR and development that took place between Michael E. Nielsen and the author at the CBS in early 2003. We were both Ph.D. students at the Department of Intercultural Communication and Management at the School. Having read some of the mainstream 'business case for CSR' literature, we felt that the mainstream, business management literature on CSR appeared to be rather narrow in its outlook. In fact, there was a need to develop more academic work that had a critical perspective on CSR and development, challenging some of the propositions found in the mainstream literature. We then explored the idea of arranging a three-day workshop that would bring together some of the world's foremost researchers in the field of CSR and development together in Copenhagen. Initially, we identified relevant researchers via the internet and also used our general knowledge of the field to pinpoint interesting potential participants at the conference. Our main criteria for identifying these researchers included a strong publication record, a critical approach to the CSR discourse, and a commitment to addressing social and environmental justice concerns in relation to CSR and development.

At first it seemed difficult to raise funding for such an event. The primary obstacle was that PhD students did not have sufficient seniority to apply for conference funding with most public or private funding bodies in Denmark. Eventually, the conference was financially supported by the President of the CBS, allowing us as conference organisers to invite a limited number of researchers to Copenhagen to attend this event. Four researchers based in the United Kingdom, one in the United States, and one in South Africa indicated their willingness to attend. Each conference participant was asked to elaborate a paper on the theme of critical perspectives on CSR and development. When the conference was finally held in November 2003, the invited researchers seemed to believe that there was a pressing need for developing a critical research agenda on CSR and development. Whereas the conference was successful in generating a sense of shared purpose amongst the involved researchers, the first day of the conference that was open to public participation was primarily attended by graduate students at the CBS. On the one hand, it was a positive development that graduate business students were also interested in attending a conference that engaged with the CSR and development debate from a more critical viewpoint. On the other hand, the first conference did not build many linkages to other CSR target audiences such as multinational corporations, consultancy companies, aid agencies, UN organizations, Northern & Southern-based NGOs, trade unions, and research institutions.

However, a sense of shared purpose was enough for the hosts of the event and the invited researchers to pursue further work within this area. First, the possibility of publishing a special issue on critical perspectives on CSR and development in an internationally recognized journal was pursued. Several journals were approached, most of which responded positively to the idea. Eventually, International Affairs, an internationally recognized journal hosted by the Royal Institute of International Affairs and published by Blackwell Publishers became the preferred 'space' for publication, primarily because of its diverse readership amongst public policy officials, corporate executives, NGOs, think tanks, and academic research institutions. Before the papers were published in the May 2005 issue of International Affairs, another conference was held with the same researchers at the CBS in August 2004. This time with greater participation from various of the above-mentioned CSR target audiences, and finally a so-called study group was held at the Royal Institute of International Affairs in London in January 2005. The study group format allowed the contributors to the special issue to receive feedback on the papers from a multi-stakeholder group and was an opportunity to discuss the views presented in the different papers in the setting of the United Kingdom for the first time. This process substituted the regular external refereeing process before the papers were finally revised and printed in May 2005.

Social dialogue: lessons learned during phase I

If we try to evaluate the initial process of establishing an international research network that seeks to engage in social dialogue, the following picture emerges. At one level, we could say that it was successful in terms of creating an initial, limited base of 'critical' knowledge on CSR and development. This was presented in the May 2005 special issue of International Affairs. Through a number of conferences and meetings in the United Kingdom and Denmark, it was possible to open some existing, relatively well-established, mainstream 'spaces' (i.e. the CBS, the RIIA, and International Affairs) to initiate a short-term social dialogue around CSR and development. However, other spaces such as the use of a website, briefing papers, pieces in international newspapers/magazines etc. were clearly not embraced yet. At another level, the linkage between this knowledge, target audiences, and relevant spaces was still underdeveloped. Given the potentially large number of target audiences that might be interested in CSR and development across the world, it could be argued that the reach of the social dialogue was so far quite limited. Even if we look at the limited physical 'space' of the CBS where two of the initial conferences took place, the links between the initial pool of knowledge and the potential target audiences could be much more developed. These linkages became stronger, however, as more meetings were held with a broader mix of target audiences attending the conferences.

Perhaps the initial phase could rather be described as a pilot project that allowed the authors of the special issue to assess whether the development of a critical research agenda could generate momentum over a longer period of time. At the same time, in spite of the authors' intention to promote a diversity of viewpoints in the CSR and development debate, it was striking that this initial group consisted of seven males and one female researcher while seven out of the eight researchers were from Europe (the United Kingdom and Denmark) and only one from the South (South Africa). It was therefore decided in January 2005 that the existing group of researchers should be expanded into a broader network with greater Southern participation as well as an improved gender balance. Hence, researchers from Argentina, Australia, India, Kenya, and Spain[37] were invited to take part in the new phase of what became known as the International Research Network on Business, Development, and Society. The subject of critical perspectives on CSR and development seemed important and had sufficiently large number of interested target audiences worldwide to justify a more concerted communication effort that could initiate a new kind of social dialogue around the potential and limitations of CSR in developing country contexts.

Social dialogue: lessons learned during phase 2

The second phase has only recently begun. A strategic planning session was held at the CBS in Denmark in early September 2005 with the aim of rearticulating the research priorities of the network in the light of its recent expansion. This was done by asking new members to prepare presentations on the theme: 'Southern voices in the Global Corporate Social Responsibility Debate' so that their voices and participation would help shape the future research priorities of the network and – over a period of time – create spaces in the CSR debate for viewpoints that were usually not well-represented at mainstream CSR conferences. Along with the May 2005 special issue of International Affairs, these presentations provided input

[37] Four of these were women.

for a brainstorming process whose outcome was the formulation of a network mission statement and a more clearly defined research agenda for the network. As was the case with the previous conferences in the network, one of the three conference days was dedicated to presenting the work of network members to a broader audience. This time the meeting managed to attract a broad variety of target audiences from industry, consultancies, NGOs, trade unions, public sector and UN agencies which made the dialogue more stimulating. It also proved possible to refine the network's critical research agenda[38]. A communication plan has now been developed, focusing on opening further spaces for social dialogue, i.e. the production of policy briefings, a network website, working papers, critical CSR conferences, writing papers in newspapers across the world, etc. In this new phase, the network has already faced a new set of challenges which include:

- Increasing bureaucratisation of the network's activities, including those that relate to dialogue with target audiences. Expansion of the network means that routine tasks take up more time; it becomes necessary to develop management structures to support the growth process of the network, handling logistics, contacting potential funding sources, etc. This obviously takes time away from knowledge production and linking this knowledge directly to target audiences.
- While the International Research Network on Business, Development, and Society is interested in opening existing or new spaces in the CSR and development debate, there is a risk that the Network is unable to initiate a meaningful dialogue with target audiences. In its desire to promote multiple voices, multiple viewpoints and diversity, it may wind up addressing multiple issues, losing focus in its work. Thus, there is clearly a trade-off in social dialogue between being open to diversity of viewpoints/voices and retaining a clearly focused agenda.
- It is difficult to sustain social dialogues without support from funders that prefer two to three-year project proposals with detailed descriptions of purpose, content and activities as well as quantitative and qualitative targets that should be met. Yet the donor preoccupation with impressive looking log frames may contradict the process of social dialogue that should ideally consist of inclusive, open-ended processes where participants learn from one meeting to the next. It may be difficult, if not impossible, to logically plan such open-ended processes over a three-year period. In fact, social dialogue within international research networks requires that social capital is built over time as network members get to know each other better.

[38] Part of which relates to the three themes: CSR impact assessment, participation and power in CSR, and Southern-centered perspectives.

- Many donors are supportive – at least to a certain extent – of the current mainstream conception of CSR and may therefore be hesitant to support initiatives that are not directly tied to corporate interests. At the same time, the Network's focus on conducting independent research into the nature and effects of CSR cannot be tied particularly to corporate concerns, since that would undermine the very raison d'etre of the Network; namely to provide an independent assessment of CSR's contribution to improved conditions of workers and communities in the South.

Conclusion

In this chapter, I argued that the mainstream business management literature on CSR is rather restricted in its outlook. I emphasized the need for initiating a social dialogue on both the potential and the limitations of CSR in relation to development, and stressed that financial profitability cannot constitute the only meaningful basis for judging whether corporations should engage in CSR practices (however defined). My intention is not to say that a critical approach to CSR and development should necessarily be critical of corporations. Rather the point is that neither corporations nor governments, NGOs, trade unions, CSR consultancies or other interested parties in the CSR and development debate can be satisfied with the current state of affairs. At present, the claims made in the name of CSR are not matched by a more critical investigation into the actual effects of CSR in developing countries. Moreover, issues around conflict, class struggle, and more radical approaches to citizen participation are sidelined in CSR teaching, CSR conferences, and most business practice in developing countries. This is why a new type of social dialogue needs to be built around the research findings emerging from a critical investigation of the potential and limitations of CSR in the context of developing countries. However, as this chapter has also shown, social dialogue may be a far-from-perfect, hard-to-predict trial and error process that is gradually improved over time as participants gain more experience. Yet even with the best of intentions, social dialogue activities can be hard to sustain if funding environments are not favourable to the kinds of perspectives being explored through the dialogue. Whereas it may be difficult to influence the overall funding environment in a given country, international research networks need to contact a variety of donors in different countries in order to identify funding sources that are willing to support more critically-oriented work on the role of business in society.

References

Barrientos, S. (2005) "Labour Impact Assessment: Challenges and Opportunities of a Learning Approach", *Journal of International Development,* 17(2): 259-270.

Bhushan, C. (2005) "Southern Voices in the Global Debate on Corporate Social and Environmental Responsibility in the Third World", Conference Paper, Copenhagen Business School, Copenhagen, 5th September, Unpublished.

Blowfield, M.E. (2004) "Ethical Supply Chains in the Cocoa, Coffee, and Tea Industries", *Greener Management International,* 43: 15-24.

Blowfield, M.E. and Frynas, J.G. (2005) "Setting New Agendas: Critical Perspectives on Corporate Social Responsibility in the Developing World", *International Affairs,* 81(3): 499-513.

Clay, J. (2005) *Exploring the Links Between International Business and Poverty Reduction: a Case Study of Unilever in Indonesia,* Oxford: Unilever; Novib Oxfam Netherlands; Oxfam GB.

Commission on the Private Sector & Development (2004) "Unleashing Entrepreneurship", Report to the Secretary General of the United Nations, New York: UNDP.

Fox, T. (2004) "Corporate Social Responsibility and Development: In Quest of an Agenda", *Development,* 47(3): 29-36.

Freeman, R.E. and Reed, D. (1983) "Stockholders and Stakeholders: A New Perspective on Corporate Governance", *California Management Review,* 25(3): 88-106.

Frynas, J.G. (2005) "The False Developmental Promise of Corporate Social Responsibility: Evidence from Multinational Companies", *International Affairs,* 81(3): 581-598.

Hart, S.L. and Prahalad, C.K. (2002) "The Fortune at the Bottom of the Pyramid, *Strategy & Business,* 26: 54-67.

Hart, S.L. (2005) *Capitalism at a Crossroads: The Unlimited Opportunities in Solving the World's Most Difficult Problems,* New Jersey: Wharton School Publishing.

Hemingway, C.A and MacLagan, P.W. (2004) "Managers' Personal Values as Drivers of Corporate Social Responsibility", *Journal of Business Ethics,* 50(1): 33-96.

Hopkins, M. (2003) *The Planetary Bargain: Corporate Social Responsibility Matters,* London: Earthscan Publications.

International Institute of Environment and Development (2004) "Ways of Working for Sustainable Development", London: IIED.

Jenkins, R. (2005) "Globalisation Corporate Social Responsibility, and Poverty", *International Affairs,* 81(3): 525-540.

Jeppesen, S. (2004) "Environmental Practices and Greening Strategies in Small Manufacturing Enterprises in South Africa. A Critical Realist Approach", PhD Dissertation, PhD Series 11-2004, Department of Intercultural Communication and Management, Copenhagen Business School, Copenhagen: Samfundslitteratur.

Leeuwis, C. (2000) "Reconceptualizing Participation for Sustainable Rural Development: Towards Negotiation Approach", *Development and Change,* 31(5): 931-959.

Lund-Thomsen, P. (2004) "Towards A Critical Framework on Corporate Social and Environmental Responsibility in the South: the Case of Pakistan", *Development,* 47(3).

Lund-Thomsen, P. (2005) "Corporate Accountability in South Africa: the Role of Community Mobilizing in Environmental Governance", *International Affairs,* 81(3): 619-634.
Middlemiss, N. (2003), Authentic Not Cosmetic: CSR as Brand Enhancement, *Journal of Brand Management,* 10(4-5): 353-361.
Nadvi, K. and Barrientos, S. (2004) *Industrial Clusters and Poverty Reduction: Towards a Methodology for Poverty and Social Impact Assessment of Cluster Development Initiatives,* Geneva: UNIDO.
Nelson, V., Martin, A. and Ewert, J. (2005) "What Difference Can They Make? Assessing the Impact of Corporate Codes of Practice", *Development in Practice,* 15(3-4): 539-545.
Newell, P. (2005) "Citizenship. Accountability and Community: the Limits of the CSR Agenda", *International Affairs,* 81(3): pp. 541-558.
O'Brady, A.K. (2003) "How to Generate Sustainable Brand Value from Responsibility", *Journal of Brand Management,* 10(4-5).
O'Pondo, M. and Hale, A. (2005) "Humanising the Cut Flower Chain: Confronting the Realities of Flower Production for Workers in Kenya", *Antipode,* 37(2): 301-323.
Porter, M.E. and Van der Linde, C. (1996) "Green & Competitive: Ending the Stalemate", in R. Welford and R. Starkev, eds., *The Earthscan Reader in Business and the Environment,* London: Earthscan.
Prahalad, C.K. and Porter, M.E. (2003) *Harvard Business Review on Corporate Social Responsibility (Harvard Business Review Paperback Series),* Cambridge: Harvard Business School Press.
Prieto-Carron, M. (2004) "Is There Anyone Listening?: Women Workers in Factories in Central America, and Corporate Codes of Conduct", *Development,* 47(3): 101-105.
Prieto-Carron, M. (2005) "Southern Voices in the Global Debate on Corporate Social and Environmental Responsibility in the Third World", Conference Paper, Copenhagen Business School, Copenhagen, 5th September, Unpublished.
Roberts, S., Keeble, J. and Brown, D. (2002) *The Business Case for Corporate Citizenship,* Cambridge: Arthur D. Little.
Szablowski, D. (2002) "Mining, Displacement and the World Bank: A Case Analysis of Compania Minera Antamina's Operations in Peru", *Journal of Business Ethics,* 39(3): 247-273.
The Economist (2005) "A Survey of Corporate Social Responsibility", *The Economist,* 374(8410), January 22– 28.

Chapter 17

Social learning as action inquiry: exploring education for sustainable societies

Paul Hart

Introduction: action research as a social process

Assuming that forms of education directed toward environmental learning and sustainability can constitute legitimate responses to social and environmental challenges, this chapter questions how such responses might be conceived and implemented. The chapter explores conceptions of learning and knowing that serve to broaden bandwidths on what counts as educational experience within visions of a more sustainable world. Recent developments in both learning theory and action inquiry, as they relate to sociocultural practices, serve to trouble meaning systems that educators attach to customary practices. Arguably, it is these taken-for-granted notions about customary practice that must be engaged if the interplay between sustainability-related human thought and actions are to be taken seriously. The idea that action research as a social learning process must be our focus, both in terms of what we shall need to do and how we are to do it, grounds the argument (Scott 2005).

Underpinning notions of action research as a form of social learning is the assumption that knowledge and understanding may be conceptualized beyond formal (i.e. propositional) knowing as socially-situated, practical knowing. Within the field of education, Shulman (1986, 1987) and colleagues have long argued that studies of teaching should be approached by researchers and practitioners together working to understand and represent teachers' pedagogical wisdom. Attention is now paid to the processes by which teachers come to teach and to develop their pedagogy as 'language in use' constructed within communities of practice. This shift toward a more practical epistemology as grounding for teacher research suggests that thinking and knowing can be explored meaningfully in ways that make sense to teachers themselves, that go beyond behaviourist, cognitivist, or even constructivist roots of learning, and that are inclusive of sociocultural dimensions of learning.

Current research on the relationship of thinking and practice incorporates methodological perspectives that consider both professional agency/identity and communities of practice, as methods capable of encompassing both personal *and* social dimensions of learning. Rather than a simple direct individual-centered relationship between language/thought and action, practitioner work is (re)considered as a complex dialectical social relationship with intentionality constituted within complexes of social and cultural discourses.

Conceptualizations of knowing and learning are being challenged by those, like Cochran-Smith and Lytle (1999), who argue that aspects of knowing derived from context-specific social and cultural practices of schools and workplaces are essential in understanding how to improve practice. Strong relationships between tacit, practical and formal ways of knowing are only now being opened for study by research that employs interpretive, critical and other forms of qualitative analysis. Obsession with formal knowledge is being reframed by this research in ways that provide greater freedom for more practice-based considerations of participation in action inquiries. This new emphasis on practice-based learning may give us a better sense of exactly what it is that people learn, how learning occurs, and its relationship to teachers' communities of practice (Wilson and Berne 1999).

According to Elliott (1998), action research, as a frame for communities of practice exemplified in the Environment and School Initiatives (ENSI) Program, has enabled teachers to question some of the assumptions and beliefs that underpin their customary practices. The ENSI model provides a set of curriculum experiments, framed as instances of action research, which allow researchers to examine how young people engage social/environmental issues. Despite resistances one might expect from traditional education systems, responsibility for learning how to engage such issues naturally devolves to school contexts (Elliott 1998, Posch 1994). The kind of teaching and learning strategies engaged in the ENSI process may act to challenge traditional pedagogic cultures.

Elliott (1998) sees the epistemological underpinnings of the ENSI framework rooted in pragmatist theories of knowledge (Wilson 1997), where competence is comprised in terms *both* academic (i.e. cognitive) and practical (i.e. social). That the latter has traditionally been considered as the less privileged of the binary seems to provide a focal point for debate as social and educational researchers prepare the ground for approaches to learning less disconnected from the complexities of practical living and social/environmental issues (Donovan and Bransford 2005). The ENSI process provides an opening for consideration of issues of social learning as epistemologically distinct from individual learning rooted in cognitive psychology and some branches of organization theory (Checkel 2001).

On knowing: intellectual roots of social learning

Debates about learning theory preferences are underpinned by epistemological arguments well beyond the scope of this chapter. However, some consideration of social learning as a way of knowing seems crucial to subsequent critical commentary on the validity of various arguments. Fundamental concerns about how societies and communities can learn to cope with social, economic and environmental change have resulted in new questions about how people know what to do in complex, indeterminate situations. Such questions have spawned research into the social processes of knowing. Although understanding ourselves is crucial in knowing what to do, according to Engel and Salomon (2002), so is understanding how the collective process of knowing governs collective processes of learning (for sustainability). In attempting to conceptualize a bridge between social learning and its origins in views of knowledge, it is easy to see a disconnect between what currently counts as legitimate school practice and what may be needed to serve society's best interests, that is, between academic/behavioural/ cognitive and practical/vocational/social intelligence. While neither dimension of intelligence has proven adequate in itself, broadening the concept of learning may provide openings into how people gain insight and control of the ways in which their actions may affect natural and human domains to ensure a more sustainable future (Lee 1993, Röling and Wagemakers 1998).

According to Greenwood and Levin (2005), learning may be conceived as interactions among three ways of knowing – episteme (common sense), techne (technical knowlege and skills), and phronesis (designing collaborative action). Knowing, as collective and socially distributed, seems to be inherent in such conceptions of knowing-in-action (Schön 1991). These deeper understandings of knowing are linked to collective knowing of phronesis by Flyvbjerg (2001) who argues that techne and phronesis constitute the necessary 'know-how' for social change. Thus episteme, as necessary but not sufficient for social/cultural learning, is accorded no special priority in learning. A lively dissent from dominant perspectives with a decidedly individualist ontology has been generated by scholars who are investigating mutually constitutive and noninstrumentalist bases of social interaction. These scholars emphasize collective learning as a redefinition of interests that can take place during the process of interaction itself (Adler 1997, Checkel 2001, Haas 1990). This work has contributed to a growing realization among educational researchers, including learning psychologists, that the cultural project of knowledge-making must be understood beyond arguments that truth is not only found within the objectivity of science but also in human-social affairs within the intersubjectivity of interpersonal accord. Knowing within this frame is understood to inhere in interactions, that is, to be embodied or enacted in actions

– what one knows (and who one is) are understood in terms of what one does (performativity).

Issues concerning the construction, legitimacy and performativity of various ways of knowing within the educational context are brought into sharp relief within case studies of the ENSI program. As Elliott (1999) says, ENSI agendas transgress many traditional educational boundaries (e.g., across subject specialisms, formal and informal, school and community; teaching and research; knowing and acting; global and local; childhood dependency and adult responsibility). These transgressions are intentional, arising from recognition that adequate educational response to environmental concern requires education that prepares people to participate in shaping the social economic and environmental conditions, local to global. This shift in educational priorities represents a view of learning conceived in terms of multiple ways of knowing (as epistemologically more comprehensive). Within this broader conceptualization of what counts as knowledge, school activities balance learning tasks as relatively passive, systematic and remote or abstract with learning tasks that focus on active generation of knowledge by children and teachers within local communities with real life issues that require critical reflective and thoughtful action (Posch 1999). Thus, the ENSI program raises some fundamental issues about the nature of knowledge and about the nature of learning.

On learning

With the search for general learning principles largely abandoned, recent accounts of learning recognize an inherent, situated and context-bound nature of learning (e.g. Brown and Duguid 2000, Davis *et al.* 2000, Wenger 1998). This shift in perspective has focused research efforts on understanding more of the complex interplay of personal and social learning conditions such as intent participation, cultural emphasis and interaction (Rogoff *et al.* 2003). The notion of learning as something that goes on 'in the head' of an individual in the process of constructing personal meaning based on existing cognitive structures (see Bruner *et al.* 1956) has been challenged by conceptions of learning as a social phenomenon of interactions between learners and a complex set of sociocultural forces (Donovan and Bransford 2005, Bruner 1986, 1990, 1996).

Broadening our understanding of learning has created tensions among learning theorists (Greeno 1997). Arguably, critiques of these perspectives have resulted in better understanding of learning as both personal and social (Anderson *et al.* 2000) but also as a part of larger philosophical positions. Critical and post-modern theorizing has challenged modernist notions of knowledge as direct correspondence with objective reality as well as the idea of a unitary identity (of self), that stands outside power regimes on which they were constructed. Rather, knowing as culturally and historically situated, is understood by recognizing multiple forms and sources of knowing, (i.e. beyond the privileging of science) multiple meanings and interpretations. Within this broadened frame, learning is viewed as a process involving meaning-making as identity-forming in the sense that every person is positioned within the discursive positions that have formed them. Each of us experiences and interprets within our own multiple positionings/subjectivities and within our own social situations. Even within constructivist ideas about learning, although often considered as more individualist than social, theorists such as Checkel (2001) argue that, while it is the individual who constructs meaning, it is never done in isolation of the social context (i.e. influenced by culture, language, politics and history). Therefore, when a person learns, they construct their own knowledge and meaning(s) according to what they already know, within social, historical and linguistic contexts of their meaning. Where such mental constructs may, over time, become that person's reality (worldview), as they see it, such processes can only be understood as multiple, social and troubled.

As an example of the kind of inquiry that supports broadening conceptions of learning in social and practical dimensions, Rogoff *et al.* (2003) contrast participatory forms of learning with traditional formal academic learning based on transmission of abstracted context/information. Their research supports a view of learning through intent participation (i.e. direct observation and 'listening in' that are purposive and anticipatory). They provide examples of the practices by which children are engaged in more 'mature' activities within their cultural communities as legitimate peripheral participants. The incongruencies involve assumptions about the purposes of schooling and learning in respect of mastery of 'valued' knowledge as privileged over processes of critical thinking, social inquiry or problem-solving. School practices must accommodate certain norms of individualism, competition, achievement and independence that will enable students to eventually reproduce and conserve existing social and economic norms and values.

The difficulty, ambiguity, contradiction and cognitive psychological uneasiness engendered in attempts to question dominant discourses of schooling suggest institutional/structural problems involving management strategies and organizational pressures unfamiliar to teachers trained to process students within knowledge-based curricula and limited tempero-spatial organizational and evaluation practices. Contradictions in teachers' own pedagogical ideology may also cause resistances based on personal enculturated beliefs and tacit/unconscious epistemological and pedagogical assumptions about what counts as learning, knowing and educational worth.

Perception is everything. Prevalence of traditional school learning contexts can be attributed to teachers' often unconscious presupposition that expert-derived objective knowledge as the only legitimate type as well as to a culture/discourse of teaching where professional competence is defined by one's mastery of subject matter, by one's pedagogical expertise within traditional structures, and by the status systems conveyed by society. Perceptions of informal or even less formal learning situations as lesser forms of education make it difficult to open the field to create conditions for broader epistemological and ontological considerations about learning. Thus, issues of learning and knowing in curriculum contexts are related both to contextual barriers of assembly-line production models as well as professional hegemony in the institutionalization of dominant models of teaching and learning. Control of the language (i.e., of discourse) at one level becomes power over decision-making resources and practices at another.

Missing or at least understated in the arguments of cognitive psychologists are obvious social and cultural dimensions of learning. Bruner (1999) and others with experience in cultural psychology now argue for forms of learning research that engage anthropological and sociological perspectives. Examples of inquiries by Rogoff *et al.* (2003) are beginning to frame research agendas that employ conceptual lenses beyond assumptions about objectivist inquiry and dominant discourses about what counts as school experience. In examining why we believe that some experiences are more worthwhile than others in terms of their generative potential in changing consciousness, we begin to engage ideas about critical reflection and intersubjectivity that implicate more social and participatory methods. Thus, it is no coincidence that interest in broadening conceptions of learning is occurring at a time when social sciences philosophy and practice is working through the legitimation of broadened perspectives about the epistemological and methodological bases of inquiry.

Perhaps some of the perceptual difficulties both teachers and students have in engaging different participation structures envisioned in programs such as ENSI can only be approached by engaging different inquiry processes as participation/structure issues themselves in need of reconstruction. Perhaps interpretive and critical approaches such as action research, envisioned by Elliott (1999) and Posch (1999), are only now being reconsidered for different reasons. Perhaps it is time to revisit the argument that innovations such as ENSI will be subsumed within larger educational structures unless those dominant epistemological/ontological categories are called to question. Given dramatic changes in those categories within research institutions toward serious considerations of important qualitative dimensions (e.g. emotional, moral, aesthetic), one wonders about how long other educational institutions can sustain particular views of the social order, let alone the environment. How we decide to engage these issues, according to many education researchers (see Reason 1996), may predetermine the outcomes. Perhaps the process should mirror the sustainability processes as participatory and critical.

Action research

In their introduction to the *Handbook of Action Research*, Reason and Bradbury (2001) orient readers to a variety of forms of inquiry they describe as participative, experiential and action-oriented. They distinguish different purposes from traditional academic research, different ways of conceiving knowledge and its relation to practice, different views about quality as a discourse of relational practices and fundamental differences in understanding the nature of inquiry based on a participatory worldview. This shift in worldview represents an ontological stance that subsumes sociocultural perspectives on learning variously described as social learning (connected to ideas of social capital) action learning (connected to ideas of action competence; Fien and Skoien 2002), community learning (Moore and Brooks 2000) and situated learning (Lave and Wenger 2000). As critical research, it also draws attention to choices action researchers can take in raising critically reflexive questions in exemplars of good practices, extending useful conversations about getting the work done amidst the issues and politics of critical inquiry.

It is interesting, when reading about the variety of approaches now characterizing participatory research, to reflect on the ENSI context of the early 1990s when interpretive and critical research methodologies had not gained the degree of legitimacy within the academy that they arguably enjoy today. For example, Posch (1991) positioned action research within the emergence of an extended concept of learning. The ENSI project, as conceived in the background papers, refers to 'active learning' as a process of inquiry applied to practical real-life problems (complex and interdisciplinary) that could result in socially/environmentally important action. Elliott (1991) sees the ENSI project's conception of environmental education as a type of 'practical wisdom' developed through a form of action-based inquiry or an action research process of learning through which students reflect on their experience of living in the environment, identify problems, then develop and test practical solutions. Knowledge acquisition was integrated into this process of enabling students to clarify and resolve problems as real and personal for them. A sense of student initiative, self discipline, team work, open dialogue and co-responsibility comprised a learning process requiring high levels of decision-making and problem-solving skills in both teachers and students.

There is no doubt that ENSI processes challenge existing learning cultures. Controversies developed over the learning process among school staff and community members, student attitudes to knowledge acquisition processes, organizational constraints on teaching/learning processes and assessment criteria. Teachers reported problems of integrating knowledge acquisition with knowledge use which reflected a dichotomy between active (social) and passive (individual) modes of learning. Caught within conflicting social values, education systems attempt to balance their approaches. As Posch (1991) conceded, many cultures do not expect teachers or their students to define and tackle real world issues independently and to monitor themselves. The prevailing culture of teaching and learning in most countries is conservative and static. Today, however, the roots of complexity lie in the conflicting new demands on schools given rapid changes in notions of work, technology and knowledge itself.

In revisiting issues concerning the generation, legitimacy and ownership of knowledge raised within the educational context of the ENSI program, Bonnett (2003) questions the implied apparent faith in rationality. The question is crucial, as rooted in the Habermasian heart of critical theory that grounds participatory action and research. Given that modern rationality is itself not neutral, but embedded with motives (i.e. aspirations to categorize, explain, predict, control, possess and exploit the world), Bonnett (2003) wonders how open is the rationality of teachers, students and community participants in local decision making likely to be? Does ENSI's highly democratic strategy hold the danger of installing antipathetic motives in the name of environmental education? Rather than precluding the value of either disciplinary individual knowing or locally grounded social knowing and action, examining notions of how discourses produce subjectivities can deepen and enrich both the nature of the critical reflective discussions and our consciousness about the nature of our chosen courses of action at pedagogical, curriculum and at philosophical levels. Examining the metaphysics, as Bonnett (2003) uses the term, that orders current social practices in parallel with invitations to both teachers and students to engage valued orientations can provide opportunities for reflection that attempt to penetrate into places and spaces that current thinking on action research seems to demand. As Reason and Bradbury (2001) put it, the wider purpose of action research is to contribute not only to the well being of people but of communities in ways that lead to more equitable and sustainable relationships with the wider ecology of the planet.

What seems to be emerging within a critical ecology or co-evolutionary ontology/epistemology of a participatory worldview is a merging of sociocultural perspectives on learning and the cooperative/critical methodology of action research. Full realization of an Education for Sustainable Development (ESD) paradigm requires an intentional co-evolutionary alliance of a postmodern and an ecological worldview that can be made manifest through a process of intentional social learning. Thus, the largely constructivist/reflectivist view of learning underpinning early ENSI theorizing (e.g. Elliott 1991, Posch 1991) with footings in the literature on critical reflection is now supplanted by feminist and poststructural critiques that ask whether reflection as a learning process ignores the possibility that experience and knowledge are mutually determined within power-laden social processes and cannot be understood outside social meanings (Ellsworth 1997). Perhaps future conceptualizations of action research will trouble reflective processes and begin to work through those breaches between desires, rational thinking, actions and responsibilities.

Agency, identity and learning

Action research, viewed as interactive learning construed as critically reflexive of 'what's behind human actions', implicates notions of identity (self) and agency in linking individual and sociocultural dimensions of this process. The assumption that teachers are active agents who can play meaningful roles in shaping school experiences (Cochran-Smith and Lytle 2004) raises questions about how schooling experiences, as longer term, interactive (complex) processes, work to construct identities (Lave and Wenger 2000). The problem with current analyses of school discourse is in conceiving teacher or student beliefs and attitudes (as reflective of human identities) as essences or unities that are discourse independent. This is the assumption of a person's intention (or tendency) in some unspecified pure form independent of and prior to their action. What needs clarifying in the current context of a 'critical' action research is that identity is not construed as one of those self-evident notions that arise (simply) from one's firsthand, unmediated experience.

If we are serious about developing our notions of action research into a full blown process of social learning, teachers' identities need to be recognized as resulting from unique trajectories through discursive spaces where experiences, constituted as narratives, are never fully formed and are always changing (Sfard and Prusak 2005). Unlike notions of 'personality' and 'character', storied identities are constantly (re)created in interactions between people (e.g. teachers in ENSI conferences) (Holland and Lave 2001, Roth 2004). Thus, unique trajectories of experience and people's own narrativization of these experiences constitute identities. So, as observers of the ENSI people and processes, the focus of our (action) inquiry is not on particular personality or character traits but on the performativity of teachers' experiences in activities. The focus in action research is on teachers' and students' narratives of experience in their activities (see Gutiérrez and Rogoff 2003).

Identity comes through action, that is, through daily activities which are acts of communication collectively shaped (and broadly conceived) to include self dialogue/thinking. The resulting stories, as discursive constructs, are at once accessible and available as well as elusive and changing, as useful fictions, although often considered by participants to be reifying, endorsable and significant. Focusing on identity-as-narrative implicates discourse – as structuring resources that constitute us.

The action researcher's focus is now on *both* actions and stories (as visions of their experiences) rather than the experiences themselves as constituting teachers' and students' identities as environmental educators. Identities can now be studied as discursive counterparts to one's lived experiences. Within this broadened frame for action research, teachers' identities can be viewed in terms of their agency, as both discursive and practical, that is, in terms of their knowledgeability to work toward issues such as sustainability, as part of their working lives. Our interest in rendering their stories and experiences involves what Shilling (1992) calls the search for discursive understanding (i.e. explicit stories) as a window into practical understanding (i.e., implicit stories which are difficult to explain because they are tacit modes of awareness competence). Action research must now consider both of these dimensions of the knowledgeability of individuals in the construction of coherent social interactions.

Although Shilling (1992) equates these ways of knowing/understanding with consciousness, Archer (2000) believes that we also need to account for the idea that consciousness constructs as much as it perceives the world. She argues that embodied understanding/knowing can provide deeper levels of human accounting beyond descriptive, interpretive or even discursive accounts. Thus, as a contested concept, agency needs to account for *how* to understand the preconditions for teacher activity (for example, how teachers are constituted as environmentally predisposed). Here is where the way we choose to construe identity or sense of self becomes the source of argument about whether human relationships can be discursively constituted or whether our continuous sense of self (i.e. self-consciousness) emerges directly from practical activities in the world (independent from language). Whether, as researchers, we see people constituted by practical activity and/or discourse will influence our action research methods. As realists we may see our sense of self emerging directly from our practices/activities, yet, critically, we simply need to question our authority as authors of our own stories and those of others.

Coming to ground: what counts as social learning

Within broadened sociocultural perspectives that now underpin active and ongoing conversations about learning and forms of practitioner inquiry, the challenge to locate sustainability theory within the messy conditions of practice remains. The mediation of teacher learning will continue to regard action research as a species of augmentation rather than a conduit for sustainability education unless these perspectives can also include the pragmatics of interaction and social communication. If the social quality of our learning (i.e. action research) has been neglected, it is because we have failed to conceptualize how to explore active participation, whether in teacher/student activities within school or teacher/researcher encounters.

Sociocultural learning within its social contexts was reflected in the 'networked' learning of coming to understand the operating structures of existing community networks (e.g. how water systems were adapted for use of partially treated water for various industrial/commercial purposes). Complex interplays of scientific background knowledge (e.g. biological systems, chemistry and geography) combined with complex legal and ethical social issues to push learning beyond transmission to interactive processes of negotiating acceptable solutions within the politics of local government. Although the rhetoric of environmental education has long advocated publicly and argued theoretically that learning be conceptualized as sociocultural learning, arguments for attending to social dimensions of learning as a part of a more comprehensive learning theory come from developmental psychology and from mainstream education (Donovan and Bransford 2005). The task remains to assist those (such as ENSI teachers) in researching their action – to find publicly acceptable ways to explore, study and assess learning that includes practical and socio-cultural dimension.

Assessing social learning for a more sustainable world

Individuals learn as they participate by interacting with the community, the tools at hand, as well as the moment's activity (Lave and Wenger 2000). Knowledge comes from interactions, as entwined in doing (Wenger 1998). Understandings are worked out in joint action with others through shared (perhaps partially tacit) understandings of what counts as being, knowing and doing. The process of learning is essentially corporeal, realized through action and perhaps worked out beyond consciousness (as embodied knowing). Knowing or learning is engaged in terms of what make sense in particular situations. Neither knowledge nor context is clearly delineated so no definite boundaries (Sfard 1998), such as participation, distributed cognition and communities of practice, need to be crossed. Such processes are assessed differently, the former by ability measures and the latter with methods and conceptual frameworks from ethnography, phenomenology, narrative and discourse analysis and symbolic interactionism. This shift is important in that teachers (and program evaluators) can recognize students' success in terms of contributions to social groups and community-related actions heretofore not considered as legitimate. Whilst it is a relatively simple matter to test individual cognition, it is quite another to develop narrative accounts where a domain of interactive skills needs to be assessed. Assessing social learning requires evidence across a domain of situation types in which participation involves the kinds of knowing that are of interest in the activity at hand.

Greeno (1997) describes instances of participation, such as gathering information, composing reports and communicating, that can be assessed using methods of qualitative inquiry. However, the concern in social learning is that these skills are viewed and evaluated not simply as abstracted school exercises but as part of students' growth toward mature participation in social communities and development of their identities as responsible, self-directed learners. Thus, the difference between what Sfard (1998) calls two metaphors for learning (i.e. knowledge acquisition and participation) implies different levels of analytical focus. While everything that people do is both individual and social, viewed more broadly, learning through acquisition of individual skills and routine knowledge only becomes important for its contribution to larger social purposes (i.e., the general participation in community sustainability issues).

Consistent with the situative perspective in fields such as medical education, teachers in the ENSI program have rearranged sequences of learning activities to include group problem solving, involving complex problem areas, early on in programs. Although such activity involves individual work, it is meaningfully related to the larger environment-related issues of sustainability. When learning is considered as a trajectory of participation, teachers arrange activities that are somewhat more complicated in the beginning (sometimes as cases, problems or issues) in order for participation to be more personally and socially meaningful (or authentic) (Donovan and Bransford 2005). Such involvement often leads first to more systematic framings and conceptual study within subject area domains as a means to return later to address complex solutions to complex social/environmental problems.

With strong development of qualitative research methods that can be applied to evaluate actual and narrativized social interactions as social/situative learning, practitioners can have convenient assessment processes to make sense of student learning as more broadly conceived. Teachers can themselves participate in action research processes in their efforts to reflectively make sense of their activities and experiences. Both cognitive and social/situative perspectives can be considered as valuable aspects of intellectual performance and learning. Herein lies the major impediment to encorporating social learning in school programs (beyond the ontological concerns expressed earlier). Because we have convenient ways of assessing cognitive aspects of learning, this is what educational systems tend to measure. These assessments are relatively cost effective, easily administered and seem to satisfy a public familiar with mass testing and competitive performance norms and measures.

The hegemony of traditional modes of evaluation carries such weight in people's minds that the shift in thinking required to conceptualize assessment of learning, using unfamiliar qualitative assessments, seems a daunting task. Perhaps, as we develop qualitative 'measures' in social sciences and educational research, teachers might be encouraged to work toward creating conditions for students to engage in participatory action-oriented educational experiences. Only if we are convinced that school learning can be approached more comprehensively and coherently and that we have the will and means to support teaching and assessment of student participation may we have more productive discussions about these changes (Cochran-Smith and Lytle 2004). This will not occur, it seems to me, unless we learn to think about social learning from perspectives beyond modernist conceptions of education.

'Post' critical perspectives

In Fenwick's (2000) terms, relations and practices related to human structural complexities, as well as dimensions of gender, race, class and so forth, determine flows of power and thus the position and ability of any individual to participate meaningfully in particular systems of practice. Learning theories must be considered in terms of power and resistances. This resistance, it seems, implicates the 'critical' in forms of critical action research, for it is sometimes in resistance that people, including teachers, can become open to unexpected, unimagined possibilities for life, work, personal and professional growth and perhaps less vulnerable to those intent on sustaining those 'dominant' discourses and practices which ensure their power.

Post-critical perspectives suggest that certain processes should occur in collaborative action groups that go beyond descriptive and reflective accounts of cases. For example, teachers, as learners, could enable the tracings of their educational situations by directing more of their attention to the politics of the discourses operating. For ENSI teachers, action research involves learning from experiences in environment-related pedagogy, but conditions should be created for 'coming to learn' as critical awareness of the politics of one's situation as well as one's own contradictory investments in their practices and thinking. With this frame, teachers could begin to question dominant relationships in their workplaces and begin to trouble notions of learning and what counts as knowing and as evidence/assessment.

'Post' perspectives are concerned that teachers in collaborative groupings help themselves and others to become more aware of their constituted natures, their role in the structures that power the systems they work in and serve to produce meaning 'for' them.

When viewed as social learning situations, action research groups may help ENSI teachers in subtle but profound ways to acknowledge their multiple subjectivities and to name new subject positions as they learn, through accepted social discourses, to see new categories, perhaps blurring boundaries between existing binaries (e.g., cognitive/social perspectives on learning) and create new approaches, as socially and environmentally sustainable educational experiences. This is not to say that we must not also be cautious of over-zealous cultural critique; all systems may not be inherently manipulative or evil. Some emancipatory efforts of authentic democratic participation should also remain troubled as they are easily co-opted in the name of existing discourses that favour certain knowledge interests over others. Yet, as Davis and Sumera (1997) explain, social/situated perspectives on learning place much greater emphasis on collectivity, co-emergence and mutual affect in action research as possible ways of countering the limitations and negativity of power/resistance-based critical thinking.

References

Adler, E. (1997) "Seizing the middle ground: Constructivism in world politics", *European Journal of International Relation,* 3(3): 319–363.

Anderson, J., Greeno, J., Reder, L. and Simon, H. (2000). "Perspectives on learning, thinking and activity", *Educational Researcher, 29*: 11-13.

Archer, M. (2000) *Being human: The problem of agency,* Cambridge, UK: Cambridge University Press.

Bonnett, M. (2003) "Chapter 10: Issues for environmental education", *Journal of Philosophy of Education,* 37(4): 691–705.

Brown, J. and Duguid, P. (2000) *The social life of information,* Cambridge, MA: Harvard Business School.

Bruner, J. (1986) *Actual minds, possible worlds,* Cambridge, MA: Harvard University Press.

Bruner, J. (1990) *Acts of meaning,* Cambridge, MA: Harvard University Press.

Bruner, J. (1996) *The culture of education,* Cambridge, MA: Harvard University Press.

Bruner, J. (1999) "Postscript: Some reflections on education research", in. E. Lagemann and L. Shulman, eds., *Issues in education research: Problems and possibilities,* San Francisco: Jossey-Bass, pp. 399–409.

Bruner, J., Goodnow, J. and Austin, A. (1956) *A study of thinking,* New York: Wiley.

Checkel, J. (2001) "Why comply? Social learning and European identity change", *International Organization,* 55(3): 553–588.

Cochran-Smith, M. and Lytle, S. (1999) "Relationships of knowledge and practice: Teachers learning in communities", in A. Iran-Nejud and D. Pearson, eds., *Review of Research in Education, 24,* Washington, DC: American Education Research Association.

Cochran-Smith, M. and Lytle, S. (2004) "Practitioner inquiry, knowledge, and university culture", in J. Loughran, M. Hamilton, V. LaBoskey and T. Russell, eds., *International handbook of self-study of teaching and teacher education practices,* Dordecht: Kluwer Academic Publishers, pp. 602–649 .

Davis, B. and Sumera, D. (1997) "Cognition, complexity, and teacher education", *Harvard Educational Review,* 67: 105–125.

Davis, B., Sumara, D. and Luce-Kapler, R. (2000). *Engaging minds: Learning and teaching in a complex world,* Mahwah, NJ: Laurence Erlbaum.

Donovan, S. and Bransford, J. (2005) *How students learn history, mathematics, and science in the classroom,* Washington, DC: National Academies Press.

Elliott, J. (1991) "Environmental education in Europe: Innovation, marginalization or assimilation" in *Environment, schools and active learning,* Paris: OECD, pp. 19-36.

Elliott, J. (1998) *The curriculum experiment: Meeting the challenge of social change,* Buckingham, UK: Open University Press.

Elliott, J. (1999) "Sustainable society and environmental education: Future perspectives and demands for the educational system, environmental education, sustainability and the transformation of schooling", *Cambridge Journal of Education, Special Issue,* 29(3): 325–340.

Ellsworth, E. (1997) *Teaching positions: Difference, pedagogy, and the power of address,* New York: Teachers College Press.

Engel, P. and Salomon, M. (2002) "Cognition, development and governance: Some lessons from knowledge systems research and practice" in C. Leeuwis and R. Pyburn eds., *Wheelbarrows full of frogs: Social learning in rural resource management,* Assen, the Netherlands: van Gorcum, pp. 49–65.

Fien, J. and Skoien, P. (2002) "I'm learning ... How you go about stirring things up – in a consultative manner: Social capital and action competence in two community catchment groups", *Local Environment,* 7(3):269–282.

Fenwick, T. (2000) "Expanding conceptions of experiential learning: A review of the five contemporary perspectives of cognition", *Adult Education Quarterly,* 50(4):243–272.

Flyvbjerg, B. (2001) *Making social science matter: Why social inquiry fails and how it can succeed again,* Cambridge, UK: Cambridge University Press.

Greeno, J. (1997) "Response: On claims that answer the wrong questions", *Educational Researcher,* 26(1): 5–17.

Greenwood, D. and Levin, M. (2005). "Reform of the social sciences and of the universities through action research", in N. Denzin and Y. Lincoln ed., *Handbook of qualitative research* (3rd ed.), Thousand Oaks, CA: Sage, pp. 43–64.

Gutiérrez, C. and Rogoff, B. (2003) "Cultural ways of learning: Individual traits and repertoire of practice", *Educational Researcher,* 32(5): 19–25.

Haas, E. (1990) *When knowledge is power: Three models of change in international organizations,* Berkeley: University of California Press.

Holland, D. and Lave, J. (2001) *History in person: Enduring struggles, contentious practice, intimate identities,* Santa Fe, NM: School of American Research Press.

Lave, J. and Wenger, E. (2000) "Learning and pedagogy in communities of practice, in J. Leach and B. Moon, eds., *Learners and pedagogy,* New York: Paul Chapman.

Lee, K. (1993) *Compass and gyroscope. Integrating science and politics for the environment,* Washington, DC: Island Press.

Moore, A. and Brooks, R. (2000) "Learning communities and community development: Describing the process. Learning Communities", *International Journal of Adult and Vocational Learning,* 1–15.

Posch, P. (1991) "Environment and school initiatives: Background and basic premises of the project" in *Organization for Economic Co-operation and Development: Environment, schools and active learning,* Paris: OECD, pp. 13-18.

Posch, P. (1994) "Networking in environmental education", in M. Pettigrew and B. Somekh, eds., *Evaluation and innovation in environmental education,* Paris: OECD/CERI.

Posch, P. (1999) "The ecologisation of schools and its implications for educational policy, Environmental Education, Sustainability and the Transformation of Schooling", *Cambridge Journal of Education Special Issue,* 29(3): 341–348.

Reason, P. (1996) "Reflections on the purposes of human inquiry", *Qualitative Inquiry,* 2: 58–72.

Reason, P. and Bradbury, H. (2001) "Introduction: Inquiry and participation in search of a world worthy of human aspiration" in P. Reason and H. Bradbury, eds., *Handbook of action research: Participative inquiry and practice,* Thousand Oaks, CA: Sage.

Rogoff, B., Paradise, R., Arauz, R., Correa-Chávez, M. and Angelillo, C. (2003) "Firsthand learning through intent participation", *Annual Review of Psychology,* 54: 175–203.

Röling, N. and Wagemakers, A. (1998) *Faciliatting sustainable agriculture: Participatory learning and adaptive management in times of environmental uncertainty,* Cambridge: Cambridge University Press.

Roth, W.-M. (2004) "Identity as dialectic: Re/making self in urban school", *Mind, Culture, and Activity,* 1(1): 48–69.

Schön, D. (1991) *The reflective turn: Case studies in and on educational practice,* New York: Teachers College Press.

Scott, W. (2005) "ESD: What sort of decade? What sort of learning?", keynote address at the UK launch of the Unesco Decade for ESD, Centre for Research in Education and the Environment, University of Bath, December 13th, 2005.

Sfard, A. (1998) "On two metaphors for learning and the dangers of choosing just one", *Educational Researcher,* 27(2): 4–13.

Sfard, A., and Prusak, A. (2005) "Telling identities: In search of an analytic tool for investigating learning as a culturally shaped activity", *Educational Researcher,* 32(4): 14–22.

Shilling, C. (1992) "Reconceptualising structure and agency in the sociology of education: Structuration theory and schooling", *British Journal of Sociology of Education,* 13(1): 69–87.

Shulman, L. (1986) "Those who understand: Knowledge growth in teaching", *Educational Researcher,* 15(2): 4–14.

Shulman, L. (1987) "Knowledge and teaching: Foundations of the new reform", *Harvard Educational Review,* 57(1): 1–22.

Wenger, E. (1998) *Communities of practice: Learning, meaning and identity,* New York: Cambridge University Press.

Wilson, P. (1997). "Building social capital: A learning agenda for the twenty-first century", *Urban Studies, 34*(5–6): 745–760.

Wilson, S. and Berne, J. (1999) "Teacher learning and the acquisition of professional knowledge: An examination of research on contemporary professional development", in A. Iran-Nejad and P.D. Pearson, eds., *Review of research in education,* Washington, D.C.: American Educational Research Association, pp. 173-209.

Chapter 18

Social learning and resistance: towards contingent agency[39]

Marcia McKenzie

This chapter proposes several modes of resistance that are suggested in three Canadian educational programs with a focus on social and environmental change. Ranging from a grade 12 global education class in a public school in a rural working class community, to a grade 8-10 Montessori mini school in an urban public school, to a non-profit two-year International Baccalaureate (IB) school in a remote residential setting, the programs vary in particular in terms of dominant social class and depth of focus on social and ecological issues. First exploring an understanding of agency as contingent on societal discourses, I then turn to examine how resistance is understood and enacted differently in the three programs. My representation of the data in salient 'portraits of resistance' seeks not to truth-tell, but to question dominant discourses as they affect students' abilities to (un)make themselves in relation to social and ecological issues.

Contingent agency

Rather than simply 'language in use', discourse in Foucauldian terms signals an uncertain world comprised of shifting matrices of power and knowledge through which we are constituted (Foucault 1980). Instantiated by means of practices such as language use, traditions of family and culture, and institutions such as school and media, discourses can be understood as having different degrees of authority, with dominant discourses appearing 'natural' or 'true', denying their own partiality, and supporting and perpetuating existing power relations (Garvey 1997; Pile and Thrift 1995). The discourses dominant in a given time and place constitute the 'subjectivity' of the majority of the people much of time, acting both as, in Foucauldian terminology, 'technologies of power' initiated and enforced by official authorization and as 'technologies of the self', internalized means of self-discipline (Foucault 1982). Rejecting the humanist notion of 'authenticity' in the individual, this suggests instead that subjectivity is fluid and multi-faceted, with its

[39] This chapter is a shortened and otherwise modified version of an article originally published in the Canadian Journal of Education, 29 (1), 199-222, entitled "The (un)making of Canadian students: Three portraits of resistance."

constitution changing in relationship to the relative power of various discourses over contexts and over time.

In contrast to traditional understandings of agency as the capacity for choice and self-determination, this framing indicates limited reflexivity and resistance to processes of discursive constitution. In response, Terry Lovell (2003) suggests recognizing agency as an ensemble performance, with transformative political agency existing in the interstices of interaction between constituted persons. Taking up Judith Butler's (1997) example of the pivotal day in the U.S. civil rights movement when Rosa Parks refused to move to the back of the bus, Lovell suggests it is necessary to look at the cumulative effect of the multiple other resistances that created the conditions for her refusal (not the first by her or others) to become an important 'act of resistance'. The effect of these multiple resistances, including social and political circumstances, point to the possibility that change results from the interaction of multiple discourses, whether at the individual or societal level. Indeed, others have suggested that a high level of interdiscursivity is associated with social change, while a low level signals the reproduction of the established order (Jørgensen and Phillips 2002). Likewise, subjectivity can be viewed as more than a 'sum total of positions in discourse' (Walkerdine 1998), with the opportunity for agency occurring within and amongst discourses, as they bump up against one another – as one discourse enables critiques of others.

This supports the possibility that we do not simply reflect the practices through which we are constituted, but that there is always a possible tension between the discourses available and, as a result, our interpretation and use of them (Søndergaard 2002). Rather than being free from discursive constitution, we may work within that constitution, using alternative discourses to "resist, subvert, and change the discourses themselves" (Davies 2000, p. 67). In this view, agency can be understood as the ongoing process of (un)making ourselves through explorations of our positioning within discourse. Encumbered by constituting discourse, and not at all transparent or outside of power matrices (Applebaum 2004), this understanding of 'contingent agency' is a potential tool as educators work to engage students in their own (un)making in relation to social and ecological issues.

Three portraits of resistance

The aim of discourse analysis is not to uncover an objective reality, but to investigate how we construct objectivity, or sedimented power, through the discursive production of meaning (Jørgensen and Phillips 2002). As such, the analysis of discourse can be viewed as a political intervention that seeks to challenge certain discourses, even as it constitutes or reproduces others. Like all research, discourse analysis itself is unable to avoid constituting the world in particular ways, and

Figure 18.1. Student work on display, Kirkwood Secondary.

thus, also produces 'objectivity'. As Jørgensen and Phillips (2002) suggest, "treating the delimitation of discourses as an analytical exercise entails understanding discourses as objects that the researcher constructs rather than as objects that exist in a delimited form in reality ready to be identified and mapped" (p. 143-144). Validity can then be assessed, not in terms of truth-telling, but in relation to the role the research plays in maintaining or disrupting power relations in society.

Notwithstanding the many conversations left out and the selectivity of the discourses I have chosen to represent here (see also McKenzie 2004), the following three portraits – or perhaps more accurately, caricatures – are intended to provoke inquiry into the ways in which we as students and teachers may understand and enact different modes of resistance in accordance with those discourses which constitute our subjectivities and our schooling.

The three programs are evidently very different in their scope, and in the age groups and populations they serve; but all share a commitment to encouraging socio-ecological activism, and are a result of the hard work of dedicated and resourceful teachers. This research seeks to learn from and contribute to the efforts of these teachers, and not to consider them responsible for more or less promising modes of resistance that should rather be understood as stemming from broader social and cultural narratives and conditions (Van Galen 2004).

Awareness & inactive caring: Hillview

Hillview Secondary School[40] is located in a rural, predominately white, working class community of 5,000 people, about an hour by car from a Canadian urban centre. The teacher, Ms. Meredith Scott, remarkably developed the grade 12 Global Education course as the first of its kind in her school district several years ago. The full-year course is divided into the topic areas of civil disobedience and civil rights,

[40] All places and names of participants have been changed for reasons of confidentiality.

profit and equity, nature and humanity, and development. In addition to class discussion, activities in the course include researching action projects, recycling school cans and bottles, raising money for an orphanage in Asia, volunteering at a soup kitchen for the mentally ill, and hosting guest speakers from organizations such as Amnesty International, Check Your Head, and Canada World Youth. Research participants from this site include 8 female and 3 male students from grades 11 and 12, all of whom were enrolled in the course.

Central to the dominant mode of resistance suggested in the talk of students in the Hillview Global Education course is the perception that their education is, and should be, unbiased – a view that continues to be commonly held and promoted within Canadian secondary schools (Lousley 1999; Kelly and Brandes 2001). This discourse of neutrality is evident in the comments of Angela, who explains:

> I learned a lot about the problems dealing with sweatshops and about cloning, not only with people but with food. And possible solutions for these problems... In this class you get the truth and solid facts about what is going on. Not like the one-sided media (Angela, 18).

An understanding of education as neutral seems to be symptomatic of a broader reliance on a discourse of objective knowing, which makes 'awareness' possible and appears to correspond with a lack of challenging critique of dominant societal discourses. 'Resistance' in the Global Education course tends to involve having one's 'eyes opened', and learning about 'what's going on' in the world, as the following remarks epitomize:

> I've just learned that there are issues and problems that people don't focus on... the States, for example, have so much money... and they would never look at other countries and, and give up pennies for their health care and people are dying and people are getting sick and they've got, we've got medicines in Canada and in the States that, cure some of those diseases and stuff that they have in other countries, but there's no, there's no way of connection... I've just learned so much, um, about, countries that can help, but don't, and just because they're blind – they don't take the time to, to figure out what's going on (Kelsey, 16).

In "teaching students about the world in which they live" (Global Education Course Outline), the course highlights issues that are explored as largely external to the students, and proposes solutions that tend to draw on dominant ethno and anthropocentric discourses, such as Western intervention in 'less developed' countries, globalized economic development, and environmental management (Bowers 1997; Gough 1999).

Ironically, students repeatedly contrast the assumed educational neutrality with an understanding of the popular media as strongly biased, as exemplified in Angela's comments that, "In this class you get the truth and solid facts about what is going on. Not like the one-sided media." Another student explains:

> We've learned that the news is kind of biased and whatever country you're watching in you're going to hear that government's side more than what's actually going on. And I think that's kind of neat, that we found that out. Because you watch the news here and we hear some parts of the war on Iraq, right?, from our news channels. And then you watch American news – it's totally different and I just notice that. Before I thought it was two different things that happened (laughs), and now, it's like the same thing, they just flip it (Corrine, 19).

This 'media is biased' stance, seems to be taken up as part of learning 'what's going on' in the world, although there is little suggestion that students understand why or how they might undertake a more in-depth deconstruction of media. Indeed, students seem to uncritically continue to use mainstream media as their main source of knowledge about the world. This absence of critique was also evident more generally, in marked contrast to the other two research sites, suggesting low interdiscursivity, and minimal reflexivity and agency on the part of students.

> I think for the rest of my life now, I'll be wondering what's going on, looking on the internet and watching CNN more so that I know what's going on. (Corrine, 19)

In addition to discourses around knowing, discourses of subjectivity also appear to be central to students' understandings and enactments of resistance. Adhering to dominant humanist conceptions of the subject, students in the Hillview Global Education course appear to generally understand themselves as somewhat influenced by family and friends, but as primarily autonomous and stable.

> Your friends are the people you hang out with the most, well, other than your family, so they influence the way that you feel about things. I mean, like, *everybody is their own person*, but, if you don't agree with your friends then, I don't know, it causes a lot of conflict. (Corrine, 19)

Holding themselves responsible for their (lack of) achievement and agency, the students in this course have strikingly different aspirations for their lives than students at the other two sites, emphasizing their desire to live 'a steady life'. This position is strongly articulated in the following conversation with Doug:

So, do you think that your experience in the class will, in the long run, affect the way you'll live your life?

Uh, affect it in a good way I would say, maybe help it out and, I would know more about what's going on globally because of it, I guess? Things like that. And being on the field trip to, I'm not too sure, the homeless – that was a good experience, that helped me.

How did it help you?

I don't know, I'm just, never really liked the city very much and going there and seeing how all those people live and stuff like that is just, like, it's an eye opener, for sure.

What does it make you think – did it make you like the city more or less or?

It makes you think of how they got there, and if you want to end up like that, right? Imagining yourself being in that same situation.

It gets you more motivated or?

Yeah...

What things do you think will affect who you are ten years from now?

What will affect me? Probably I will regret my grades in school. I should try better, but I just don't right now. That's one thing I should be doing. If I wanted to get a better job down the road. And, I don't know. That's probably the most important one.

And do you have any specific dreams or goal for the future?

Uh, I'd like to be a personal trainer, but that's just a lot of school work and I'm not very good with school, so – but, just live a steady life and have a family.

Do you have plans for next year?

Uh, I'm just going to get a job and then, after I work here for a bit I want to go the oil rigs. Go to the oil rigs for a couple of years (Doug, 17).

Like many of his classmates, Doug's plans for his future appear inhibited by a sense of lack of agency as he worries about where he might end up and considers his goals for the future. In contrast to the discourse of individual power that is so prevalent at the other two sites, the discourses available to the Hillview students are no doubt bound by their class-specific material realities and life experiences (Jørgensen and Phillips 2002).

Tied to their understanding of social and ecological problems as requiring objective 'awareness' of events happening elsewhere, as well as to perspectives of themselves as autonomous, stable, and lacking agency; students in the Global Education class

Figure 18.2. Class Poster, Hillview.

commonly articulate a second component of their 'resistance', which I have called 'inactive caring'.

Several students indicate that they have 'grown to care' for others 'less fortunate' through the class, and a few talk of wanting to find careers that enable them to help others. However, the caring expressed in students' comments typically does not carry with it a sense of being able to make any substantial change in the world, as suggested in the following remark made by Doug:

> What do you think [Ms. Scott is] wanting to teach you in the Global Ed class, particularly around social issues or environmental issues?
>
> Uh, how the world is and how it runs and problems around the world and things you can do, that you can do personally, obviously you can't change it, but to help it. Things like that (Doug, 17).

> We had that, what's that group, 'Check Your Head'. They came in and they were talking about, um, like, sweatshops and stuff... Like, I know, most of the clothes I'm wearing have been made in sweatshops, but I really don't know where else to buy them from. That makes you feel kinda like there's nothing you can do, like, even when you feel bad about it, it's just like, well, I got to get clothes from somewhere (Shelley, 16).

Another student, Kelsey, articulates a similar notion of caring that is restricted in its ability to effect change. In discussing the possibilities of taking action, she comments that the experience of raising money for an orphanage in Asia was educative in that "it seems like it would be difficult to help them, but actually it's not". However, Kelsey retreats to a position where she wants "not to make the world a better place", but just make "a little bit of a difference or at least put, like, a smile on someone's face that wasn't smiling beforehand". This modest understanding of her potential effect on the world, or resistance to it, is reiterated elsewhere when she states, "I know I'm not going to be able to help [people] dramatically". Through the combined discourses of awareness and inactive caring, Hillview students articulate a limiting portrait of resistance: one that remains strongly influenced by mainstream cultural narratives and suggests some of the difficulties that can be involved in engaging students in deeper levels of reflexivity and activism.

A way of thinking and lifestyle activism: Kirkwood

Kirkwood Secondary School is located in a culturally diverse, lower to middle income community within a Canadian urban centre. Initiated in the early 1990s as a 'school within a school' for grades 8, 9, and 10, the Montessori program has a focus on 'peace education, global issues, and environmental concerns'. Attracting students from beyond Kirkwood each year, the program also regularly draws a number of students from an elementary Montessori program located in the same city. Students are in the Montessori program multi-room space one out of every two days, taking elective courses with 'mainstream' students on alternate days. Within the program, students take Science and Math with Mr. Mansur Karim and Humanities with Ms. Terese Pryde, and also complete 75-100 service hours per year. Typical Montessori program activities include student-run class meetings; action projects; service work with local elementary schools and community organizations; as well as various environmental activities such as school ground naturalization, beach clean up, and nearby habitat restoration. The research participants from this site include 10 female and 6 male students, 6 of whom were graduates of the program, as well as both teachers.

Students in the Kirkwood Secondary School Montessori program commonly take up a discourse of educational neutrality, as did students in the Global Education course. For example, when asked whether the Montessori teachers promoted certain perspectives on the world, one student commented:

> Um, not so much perspectives of, though I guess, in some ways, but, they, they just sort of promote the world. And they don't really say any negative points or positive points. They just explain how the world is and they explain how countries are, and they don't say whether that's good or that's bad, 'cause that's something that we have to learn ourselves (Lara, 14.)

> My viewpoints of what's cool and what's not in grade school was directly from the media, I guess you could say. I think, *now,* I've just changed, *I've realized what's cool and what's not.* Montessori helped a lot. Like kind of learning about child labour and that kind of stuff, questioning the companies and that kind of thing (Daniel, 16).

Unlike at Hillview, however, comments such as those made by three program graduates, suggest a tension between statements of educational neutrality, and acknowledgement of experiences of norming within the Montessori program:

> You said it strengthened your strength, being in the Montessori program?
>
> *Steve:* Well, it strengthened my strength, but only the values I had already. It's not like I developed bad values and then had to change them, I just had, like, sort of, well I already had them, but then because of this I knew they were the right ones.
>
> Which ones?
>
> *Steve:* Not steal, not buy Nike, not whatever.
>
> *Daniel:* The ten commandments.
>
> *Lena:* Yeah. I don't know if it's strengths, as much as morals. It's not really what to do, as much as what not to do (Lena, 17; Daniel, 16; Steve, 16).

Montessori can manipulate [students'] minds, and make them become Montessori (Lena, 17).

The comments of students suggest that while the values of the Montessori program are considered 'right', and therefore perhaps can still be thought of as 'neutral'; in some cases they are experienced as constitutive, or as a form of norming.

While maintaining that the Montessori program does not 'bump up against' mainstream Canadian values 'at all', Ms. Pryde suggests how the Montessori program seeks to enable students to 'resist' the power of the media and related mainstream values: She explains,

> It's no surprise that their life is pop culture and when they put in a CD, when they turn on the TV, when they go see a movie, when they pick up a magazine, they are being targeted as a marketing group. And they are being sold a consumer lifestyle. So, that's totally juxta-positioned to what we're asking them to think about. And it's everywhere, it's pervasive. So, we're really swimming upriver with the kind of power that that has on them (Kirkwood teacher).

How do you think the values that are sort of taught here in the Montessori program – do you think they correspond or conflict with values that are taught by society in general?

> They're quite the same actually, cause I know that Canadian society, they're uh, "We don't want to be part of war", and environmental concerns – little hippie tree hugger country right? (laughs) So that's what Montessori is too (Lara, 14).

This view of students as socio-culturally constituted to some degree, and of media and society as constitutive, contributes to the existence of a discourse of critique in the Montessori program, although a number of discourses such as assumptions of objectivity and educational neutrality seem to generally be beyond the realm of this critique.

[They teach you] freedom of speech, to question authority (laughs), not to challenge it but to question it. Yeah, don't sit back and be spoon fed, you know (Jenny, 16).

The taking up of the combined discourses of socio-cultural constitution and critique appear to translate into students at Kirkwood talking less about 'awareness' of social and ecological issues, and more about a different 'way of thinking' about the world, including in terms of their interactions with media. The Montessori students describe this way of thinking as being quite pervasive and as affecting their actions, including their interactions with peers and family. Lena, a graduate of the Montessori program, self-describes how she took on an 'anti' perspective during her time in the program, which has now shifted back towards a 'middle ground' which is less extreme, but still a different way of understanding the world than the one she started with:

Even the little butterfly flapping its wings I guess, could influence me in some way (Lara, 14).

> Sometime in grade 10, um, it all kind of just, snapped into place. Then I thought I saw a bunch of conspiracies and things, which was I guess the extreme (laughs)... but um, it's kind of like an awakening. It's neat. And then you just get to react to everything differently. Um, I guess it comes about with more knowledge probably, or maybe a deeper kind of knowledge, more critical (Lena, 17).

Another Montessori student, Kim, also talks about developing a different 'way of thinking':

> It just kind of accumulates. Like, from [other students], like, they're kind of the vegetarian spokespeople for Montessori (laughs)... And then we have Off Ramp, which is promoting clean and safe transportation. And we have Evergreen promoting a green school and a green environment. And we just have all these groups, and they just kind of, mesh together, and together it's kind of like a super being, you know, kind of a super global issues/knowledge thing (laughs), and I just think that, everybody's hearing about this, you know, every day at class meeting or whatever, things are brought up (Kim, 15).

Although assuming an underlying discourse of neutrality, the 'way of thinking' described by Kim and Lena seems to go beyond 'awareness' to a deeper, more reflexive kind of knowing, one that causes them to pit certain discourses against each other, challenging their own constitution through media and society more generally, and contributing to their socio-ecological activism.

Connecting to this portrait of resistance is a discourse of agency as 'individual power', a sense of 'freedom' not uncommon in more privileged classes (Dillabough 2004), which is prevalent in the Montessori program and quite distinct from the modest aspirations and lack of agency suggested in the talk of the Global Education students. Contained in this discourse is the notion that students can achieve what they 'set their sights on' if they only work hard enough. What Kirkwood students judge as worthy of striving for commonly seems to match other dominant North American discourses around academic success, social status, and economic achievement. The coupling of 'individual power' with these other unexamined discourses around achievement can be heard in the comments of Kirkwood students, such as those of Kim:

> How would you describe your values? What things are important to you?
>
> Um, most things that are important to me, grades are important to me… but, uh, I'm striving for success in life basically – overall goal. Obviously. Um, and I think grades are a big way of getting there. Grades are getting me up to where I need to be to get into programs for university, for, I want to go to a program in Europe, a boarding school for grade 11 and 12 to earn a baccalaureate… what university or college I attend or law school… I have big goals, but – it lets me strive higher (Kim, 15).

Although indicating a strong sense of agency, or control over her life, Kim also suggests that the 'way of thinking' in the Montessori program extends limited critique to many of dominant cultural narratives, in some cases restricting reflexivity and resistance to particular domains.

The sense of agency, and yet often limited focus of resistance, evident in the Montessori 'way of thinking' goes hand in hand with the discourse of 'lifestyle activism' commonly taken up by students in the program. This approach to 'making a difference' is highlighted in the following discussion with three Montessori graduates:

What do you think the teachers involved in the Montessori program are wanting to teach you during your time here, particularly in relation to social and environmental issues?

Tess: You can make a difference!

Camille: Yeah (laughs). That is the number one lesson they say – like, every little thing counts.

Alix: Be informed. To know what's going on.

Camille: And, and, involve others. Outreach. To your friends, kids. Anything to get out there and get stuff spread, kind of thing...

So, do you believe your experience in the Montessori program will affect your life in the long term?

All three: Totally. Yep.

Why?

Tess: Take shorter showers. The way you eat... Recycling. Just little things. Little things you do that affect the global environment.

Camille: And getting involved. Just, like, even when I'm in grade eleven, and out of the Montessori program, I still want to get involved in workshops and things like that. (Alix, 15; Camille, 16; Tess, 15).

I saw this kid wearing Nike and I said, "Do you know they use sweatshop labour," and he said, "Yeah." So then I asked him, "I'm just wondering why you wear it."(Tess, 16).

As these students explain, the dominant discourse of activism in the Montessori program seems to be one of valuing the many 'little things' that can be done to 'affect the global environment', including staying informed despite media biases, making conscious lifestyle choices, and spreading the word to those around you. However, for some this discourse of lifestyle activism is taken up within an otherwise 'mainstream' life of consumerism and achievement. Kate, for example, suggests that "when I go to buy my house now, I'll probably buy with a low flow toilet", and "When I have money to make the decisions on my eating habits...I can buy organic and shade grown and that kind of thing... I'll make those decisions to, um, eat to save the planet".

The restricted focus of the critique for some students at Kirkwood seems in part to be a result of particular dominant narratives, such as educational neutrality, individual power, and economic achievement, remaining unquestioned, and may also be a function of the reluctance of the Montessori program teachers to contribute to their students feeling 'downtrodden' by focusing on more systemic and challenging forms of activism. One of the Montessori teachers explains:

> I've gotten the sense that they're almost like dogs with their tails between their legs, that there's so much crap and there's so much, you know, because they feel responsible and they want to act responsibly, but there's so *much* to do, and there's so *many* choices and decisions for them to make, and "Gee, I just want to be a kid". They're kids. So they kind of have to balance that with themselves, what can they do, what can't they do, what do they enjoy, what could they change a bit, without feeling downtrodden over it (Kirkwood teacher).

Despite limitations, empowered by a view of socio-cultural constitution and critique, as well as a strong class-based sense of agency, this mode of resistance suggests considerable interdiscursivity and reflexivity, and results in a strong emphasis on lifestyle activism.

Impacting the world and contingent agency: Lawson

Lawson College is located in a remote natural setting half an hour from a Canadian urban centre. This non-profit school has a culturally diverse, predominantly middle class student body of 200 from around the world and operates with a mandate "to promote the cause of international understanding by creating an environment in which students from many countries and cultures are brought together to study and to serve the community". Students come to the school for two years for a grade 12/pre-university International Baccalaureate program of study, which includes a mandatory Theory of Knowledge course, as well as courses in the areas of Languages, Individuals and Societies, Experimental Sciences, Mathematics, and Arts. Students are required to participate in at least three activities per week from the areas of active citizenship, creative expression, humanitarian service, outdoor leadership, and service to the college. Research participants from this site include 6 female and 5 male students, as well as 3 instructors.

Both consciously and unconsciously, through its curriculum and particular environment, Lawson College introduces many of its students to alternative conceptions of knowledge and identity as contingent, thus establishing an important aspect of dominant forms of resistance within the program. Unlike at Hillview and Kirkwood, students at Lawson tend to view knowledge as generally subjective, rather than objective; and understand cultural norms, media, and even their education as biased and potentially alterable, dependent on underlying values and beliefs. This discourse of contingent knowing is included as an important part of the curriculum through the first year course, 'Theory of Knowledge'. One student explains:

Box 18.1. Knowers and sources of knowledge.

How is knowledge gained? What are the sources? To what extent might these vary according to age, education or cultural background? What role does personal experience play in the formation of knowledge claims? To what extent does personal or ideological bias influence our knowledge claims? Does knowledge come from inside or outside? Do we construct reality or do we recognize it? Is knowledge even a 'thing' that resides somewhere?

From "Theory of Knowledge" course website

> That's what TOK [Theory of Knowledge] teaches us to do: think critically about information, and see which one's more likely to be true. It is biased, of course, but all information is biased, but still; the only information that is not biased is say, "I weight 65 kilos", or "I'm 17 years old", that's a neutral statement. But as soon as you're getting involved in, in international politics and points of view, things become really subjective. And the theoretical job of the TOK is that, you inform yourself and decide which one you support, and act based on the information (David, Portugal, 17).

As suggested by David, this approach to knowing includes a strong element of critique.

There is an interesting interplay suggested in the talk of students and teachers at Lawson between discourses of critique and subjective knowing, and a discourse of educational bias. Students generally understand their education as promoting particular perspectives, such as a discourse of media scepticism, which are often quite different from those at home. Yet most students seem to accept and take up the values being advanced by the College, including the emphasis on critique. David explains:

> I mean the media, obviously, we bash the hell out of the media, or the heck out of the media, in terms of popular media (Heidi, Canada, 17).

> Even though they try to be as neutral as possible, there is always bias. Which means they can't produce unbiased statements, and of course there are biases here at Lawson, and they kind of want us to, force us to, think that way ... Even if you think critically there are certain biases

> that the College introduces to you. For example, the word around campus is that, "Don't trust CNN, don't trust a word of what they say". Even if what they're saying is true, I think that a Lawson student will assume that it is false (David, Portugal, 17).

A teacher describes this process of taking up of the 'biases' of the College in the following way:

> This experience of living together in a small global community is something that affects not just what you think, in terms of attitude and background knowledge, but affects who you are, affects the screens through which you see all the world, and we're speaking of knowledge... I think the screens that were developed, the eyes through which you look, are, that they more or less look through – I don't think students look at the world through the eyes after they've left Lawson (Lawson teacher).

I viewed life differently when I first came here. You know, I might have been a bit more racist. But being here has changed me a lot... it opens up a whole load of questions about yourself and about people in general, why we're here, and it's just opened up a whole new world (Adam, England, 18).

As in these examples, the learning students experience at Lawson College is often described as dramatically changing their understanding of the world, or in the words of a Kirkwood student, their 'way of thinking'.

Related to discourses around knowing, are those to do with subjectivity, including the unexamined discourse of individual power. In keeping with the privileged backgrounds and experiences of many of the students at both Lawson and Kirkwood, this discourse is strongly promoted at the College. Violeta articulately outlines this discourse of "I am an individual and I am different and I can do anything":

I think Lawson is within a western model for sure. And I think we westernize students to some extent (Lawson teacher).

> "Whatever you set your sights on", where did you learn that?
>
> It's just that the daily experience of seeing the way you people behave towards each other, the way things function and all the things, it's, it's just how it became engraved in myself... At home we still have, kind of all believe some form of mythic, some kind of the communistic way of thinking... there is still this sort of set

> mould for everything there. Well here it's very much individualistic and the tolerance is valued. "I am an individual and I am different and I can do anything" (Violeta, Bulgaria, 17).

As part of their assumed stance of agency as individual power, as well as through understanding the world as a contingent and shifting object of critique, students at Lawson commonly articulate and enact a resistance writ large, through their desire and efforts to 'impact the world'. For example, Emilia describes the impact her experiences at the College have had on her way of living:

> You have to, to, to *climb the ladder of power* in order to make some big decisions that will impact, that will have a big impact on the world (David, Portugal, 17).

> Lawson has inspired my soul, my spirit, my life, in the way that now I have so many goals, like physical goals but also internal goals, like, as I was saying before, like, converting the educational system in Nicaragua. I don't know, the way you see people, the way you talk to people, but also, the way you live (Emilia, Nicaragua, 17).

While certainly not the case for all, a number of students work between a 'lifestyle activism' approach to socio-ecological change similar to that at Kirkwood and a more outwardly activist stance for effecting change. Heidi, a Lawson College student, describes her own struggle with how to 'help the most':

> I just have one more question – do you have specific dreams or goals for the future?
>
> I thought I did, when I first came here, in terms of wanting to be head of Oxfam or something like that, but, and then as I've come here I've been like, I'm between the lines of just taking care of myself and my immediate area, like you know, having a nice farm and an orphanage of some sort, like very small. I'm torn between that and running Nike so that I can make it so there aren't sweatshops. You know, it's kind of one extreme or the other – how do help the most? And is helping the most important, or do you want quality, quantity. Ahhh!! So I'm torn between that. That will just sort of, time will tell (Heidi, Canada, 17).

While the strong discourse of impacting the world promises much action, it is the less expected discourse of 'contingent agency' which is perhaps more exciting in its possibilities for a deeper reflexivity and more selective resistance to normative discourses of media, society, and education itself.

Rastha and Emilia, two Lawson College students, suggest a sense of contingent agency that works in the spaces of their constituted selves:

> I'm from a very large family… I'm the youngest of them, and there was a lot of pressure on me from other members of my family. And I needed to sort of focus on, "Okay, what do I take from it, and what do I push away from?". And coming away, provided a space for me to sort of reflect on what I want (Rastha, Maldives, 18).

> My experience has made me the way I am. Because you go to so many different experiences and so many different things through the span of your life, and then the way that you react to those, to those experience is the way you are making your own personality, and I would say that's the way. Of course, what informs them? My parents, my culture, my religion, and everything, so, yeah (Emilia, Nicaragua, 17).

Both Rastha and Emilia take up a discourse of socio-cultural constitution in talking about how their previous experiences have exerted pressure on them/made them the way they are.

The possibility of agency within this state of constitution is suggested in their comments that their reactions/reflections are "the way you are making your own personality" Rastha in particular articulates agency as occurring through a process of asking, "Okay, what do I take… and what do I push away from": Agency is suggested to be the working with/against ways of viewing the world (discourses) that have been introduced through various influences. This is an understanding of agency as contingent on previous constitution, but as allowing some degree of resistance to be exerted.

In taking up a discourse of contingent agency, a number of students suggest that at times their sense of agency is overwhelmed by forces of constitution, with students worrying about 'losing' the ways of thinking they have gained at school once they return home. Violeta expresses her concern as follows:

> Do you believe this experience has affected the way you will live your life?
>
> I hope so…It's because again it's true that this is very, very much in a way idealistic, um, but as long as…I've incorporated these ideas in myself I try for them and like fight for them but it depends

> very much on the environment where I go. Because, for example, if I go home, if I am still able to do these things, it will be much harder. And I hope, I dearly hope, that I don't give up with the first failure, because I know if I go home I will have lots of failures with incorporating these ideas but I will try at least. That's maybe, that's what matters, no (Violeta, Bulgaria, 17)?

In realizing the challenges of resisting particular discourses, the students at Lawson indicate a tentative agency that works through a high level of intercultural interdiscursivity to provoke reflexivity and possibilities for working at difficult changes.

Educating for Agency

With a discursive framing, resistance can no longer be understood as replacing wrong with right, but instead must be complicated as something which is never outside of discourse and never proffering a once-and-for-all solution (Lather 1991). In the three research sites, the ways resistance is understood and enacted suggest strong connections to dominant program discourses (e.g. educational neutrality, constitution, critique), dominant societal discourses (e.g. objective knowledge, economic achievement), including discourses more or less available to students with different levels of class privilege based on the sedimentation of early discursive practices and experiences (e.g., critique, individual power).

The 'contingent agency' articulated by students at Lawson College suggests a reflexive response to the interdiscursivity manifest in the shifting between cultural narratives, which is encouraged by an understanding of knowledge as subjective as introduced through the IB curriculum. According to a discursive frame, this state of possible resistance entails engaging in an examination, or an (un)making of one's own discursive constitution, as well as that of one's education, and surrounding culture(s), with the possibility of working within that constitution to effect desirable change. While the 'desires' that drive that change may always rest within discourse, this mode of resistance can be viewed as a more thorough, and always unfinished, probing of their ethical and political implications (Boler 1999). Understanding agency as a matter of positioning within discourse perhaps offers otherwise unavailable opportunities for resistance and change, for (un)making oneself in relation to the dominant discourses of society, and ultimately, for more reflexive and systemic socio-ecological activism.

References

Applebaum, B. (2004) "Social Justice Education: Moral Agency and the Subject of Resistance", *Educational Theory,* 54 (1): 59-72.

Boler, M. (1999) *Feeling Power: Emotions and Education*, New York, NY: Routledge.

Bowers, C.A. (1997) *The Culture of Denial: Why the Environmental Movement Needs a Strategy for Reforming Universities and Public Schools*, Albany, NY: SUNY Press.

Butler, J. (1997) *Excitable Speech: A Politics of the Performative*, London: Routledge.

Davies, B. (2000) *A Body of Writing: 1990-1999,* Walnut Creek, CA: Alta Mira Press.

Dillabough, J.-A. (2004, Sept) "Class, Culture and the "Predicaments of Masculine Domination: Encountering Pierre Bourdieu", *British Journal of Sociology of Education,* 25 (4): 389-506.

Foucault, M. (1980) *Power/Knowledge: Selected Interviews and Other Writings (1972-1977),* New York, NY: Pantheon Books.

Foucault, M. (1982/2003) "The Subject and Power", in P. Rabinow & N. Rose, ed., *The Essential Foucault,* New York, NY: The New Press, pp. 126-144.

Gavey, N. (1997) "Feminist Poststructuralism and Discourse Analysis", in M.M. Gergen, and S.N. Davis, ed., *Toward a New Psychology of Gender,* New York, NY: Routledge, pp. 50-64.

Gough, N. (1999) "Rethinking the Subject: (De)constructing Human Agency in Environmental Education Research", *Environmental Education Research,* 5 (1): 35-48.

Jørgensen, M. and Phillips, L. (2002) *Discourse Analysis as Theory and Method,* London: Sage.

Kelly, D.M. and Brandes, G.M. (2001) "Shifting out of 'Neutral': Beginning Teachers' Struggles with Teaching for Social Justice", *Canadian Journal of Education,* 26 (4): 437-454.

Lather, P. (1991) *Getting Smart: Feminist Research and Pedagogy with/in the Postmodern*, London, England: Routledge.

Lousley, C. (1999) "(De)politicizing the Environment Club: Environmental Discourse and the Culture of Schooling", *Environmental Education Research,* 5 (3): 293-304.

Lovell, T. (2003) "Resisting with Authority: Historical Specificity, Agency and the Performative Self", *Theory, Culture & Society,* 20 (1): 1-17.

McKenzie, M. (2004) *Parrots and Butterflies: Students as the Subjects of Socio-ecological Education*, unpublished Ph.D. dissertation, Faculty of Education, Simon Fraser University, Canada.

Pile, S. and Thrift, N. (1995) Mapping the Subject: Geographies of Cultural Transformation, London: Routledge.

Søndergaard, D.M. (2002). "Poststructuralist Approaches to Empirical Analysis", *Qualitative Studies in Education,* 15 (2): 187-204.

Van Galen, J.A. (2004) "Seeing Classes: Toward a Broadened Research Agenda for Critical Qualitative Researchers", *International Journal of Qualitative Studies in Education,* 17 (5): 663-684.

Walkerdine, V. (1998) *Counting Girls Out: Girls and Mathematics,* London: Falmer Press.

Chapter 19

Sustainability through vicarious learning: reframing consumer education

Sue McGregor

One of the 15 perspectives of the UNESCO Decade for Education for Sustainable Development (DESD) is that there are many spaces for learning (UNESCO 2005). Consumer education can become such a learning site. Designed from the perspective of social learning theory (SLT), educators can help learners practice and adopt perspectives and consumer behaviour which foster sustainable development. This chapter will illustrate how consumer education can be reframed from an SLT/ESD theoretical interface, leading to the development of empowered, responsible global citizens in their consumer role. People feel empowered if they sense inclusiveness, have a voice, are given a chance to participate, are held accountable, have information, and are given opportunities to build capacity and skill sets conducive to social action and change (Nepal Human Development Report 2004).

This chapter makes the case that a particular type of consumer education (Type 4 – Empowerment Approach for Mutual Interest) can lead to sustainable consumer empowerment (McGregor 2005a,b), which then contributes to sustainable development empowerment. After profiling ten basic tenets of SLT, the chapter explores the synergy between ESD and SLT, and then applies this conceptual innovation to reframe consumer education. Because the term student tends to connote learning within the formal education setting, and because SLT assumes that learning happens in and outside the formal education system, this chapter will use the words citizen and learner instead of student.

Social learning theory

SLT posits that people will sometimes act or behave knowing they are *not* going to get an external reward, or any reinforcement. And, some people act because they know their internal thoughts, values, attitudes and beliefs also merit an internal reward (Abbott 2000, Bandura 1977, Heffner 2004). The bottom line is that SLT holds that people can learn vicariously by observing others, in addition to learning by participating in an act personally. The following section provides a primer on ten major SLT concepts, with consumer applications.

Reciprocal determinism

First, SLT embraces the idea that society plays a very large role in the way people think about themselves, the world, how they interact or behave in that world, and how they learn. Bandura (1977) offered the concept of reciprocal determinism to reflect this idea. He proposed that people, their behaviour, and the environment in which they are acting have a three-way effect on each other, and determine human behaviour. This concept is explained in more detail in the second section of the chapter. Reciprocal determinism means that the environment shapes, maintains and constrains people's behaviour. And, it assumes that people are not passive in the process as they create and change their environments by their behaviour (Rimer and Glanz 2005).

From this perspective, learners can appreciate that everyone is connected to everyone and everything else, and that their actions in the marketplace have a profound effect on people living elsewhere, the next generation, and those not yet born. This concept enables consumer educators to concern themselves with constructing learners who are social entities, responsible for others, and the environment (Pelling and High 2005). If Sue's consumer behaviour affects another person, the consequences ultimately affect Sue, too, because everything is interconnected. Seeing themselves in relation to everything else means people can learn to share power for a sustainable future.

Observational learning and modeling

This notion leads to a second SLT concept, observational learning and modeling. People learn from observing the: (1) self-consequences of their behaviour, (2) benefits and consequences of their actions on others, and (3) consequences of the actions of others. Bandura (1977) explained that learners also can learn models for future behaviour through observing the experiences of credible others, and forego any negative repercussions. For example, citizens can plan and produce a sweatshop fashion show (see the Maquila Solidarity Network (MSN) site for full details, http://www.maquilasolidarity.org/tools/campaign/fashionshow.htm). This activity offers people a way to learn about the impact of their consumption decisions without exposing those making the clothing to any further negative consequences.

Listening to, or reading the accounts of, stories of others who have bought goods which were produced without slave or child labour is a way for people to learn to model this behaviour. They do not have to buy the product, and then discover they have harmed others. Learners can learn this way by watching: (1) what parents or other credible role models do after reading about sustainable consumption;

(2) the demonstrations of activists protesting against harmful corporate business practices (e.g. clear cutting forests); (3) videos and documentaries about responsible consumption; and, (4) guest speakers tell their stories of action about global warming or climate change.

Four stages of observational learning

Bandura (1977) took this idea to the next level, proposing four conditions that must be present for learners to successfully model someone else's behaviour. First, SLT holds that, in order for people to reproduce something, they must notice and then pay attention to the model, the person they will be watching. Second, learners must be capable of remembering what they saw or noticed (retention). They must be able to code the information into their long-term memory, so they can re-enact what was noticed. Creating mental images, using their imagination, attaching labels to things, and voicing verbal descriptions of what they watched are mental strategies which help people later recall what they learned.

Third, learners must be emotionally, intellectually and financially capable of replicating the consumer behaviour they observed. For example, they may observe someone buying a Daimler/Chrysler Smart Car™. But, if they do not have the money to pay for it, they will not be able to perform the act of buying the car. On the other hand, people's abilities to perform improve when they imagine themselves performing. So, imagining themselves buying other environmentally friendly products may lead to the purchase of a less expensive yet sustainable alternative, say a bicycle. This way, people learn to be predisposed to certain consumption practices because they watched someone else take action who valued the outcome of sustainability.

The previous instance was an example of watching a live model. Learners can also learn from symbolic modeling by watching television, movies, videos, DVD's, computer games and programs, and other media. The symbols used in media are powerful vehicles of thought which provide people's lives with structure, meaning and continuity, and serve as anchors for their future behaviour (Pajares 2002). Bandura (1977) conducted many studies which proved that children are more violent if they watch too much violence on television. Using this principle, educators can anticipate that showing learners media portraying sustainable consumption behaviour may lead to observational learning, and future sustainable consumer actions.

Finally, learners must *want* to show that they have learned a new way to consume or behave in the marketplace. Learners have to be able to anticipate consequences which will make them want to replicate, imitate or avoid the consumer behaviour they observed. They must have some reason for doing it. This positive reinforcement can come from teachers, peers, parents, friends, the media, and social institutions. It can also come from within the person.

What is in it for me?

What is in it for me? People are more likely to consistently adopt a behaviour they have copied (modeled) if they *expect* this behaviour to serve some function for *them* (Bandura 1977). Educators face a dilemma when trying to convince learners that they will *personally* experience positive outcomes if they learn by watching someone *else* do it. Educators can help learners gain some sense of a moral obligation for the well-being of others and the health of the planet by augmenting learning with global citizenship, moral development, and ethical education (McGregor 2002, 2003a,b, 2006b).

This sounds good in theory. But, getting learners to embrace a global, other-focused perspective is a challenge because of two different kinds of learning. Single loop learning involves being more efficient by learning new activities with increased skills. Double loop learning requires changes to firmly established personal and social value systems, assumptions and ideologies. The dilemma arises because double loop learning, which can lead to a respect for others, cannot happen unless learners question the assumptions guiding their consuming behaviour, and expose this behaviour to public scrutiny (Argyris 1976). This type of learning is much more difficult to achieve because people resist challenging their values if they: (1) want to avoid direct confrontation and public discussion of sensitive issues which might expose them to negative repercussions; (2) have a strong desire to protect or avoid provoking others; and, (3) wish to avoid any public questioning of their personal views (Pelling and High 2005). Although this learning theory is not the same as SLT, it does provide a spotlight on this particular dilemma.

Learning when there is no one to watch

SLT also holds that people do not have to watch anyone to learn; rather, they can 'watch themselves' in different situations, and transpose that learning from one situation to another. Rotter (1954) noticed that people are very capable of transferring what they learned in one situation to similar situations. For example, if people enjoy volunteering for agencies which work to sustain the environment, they may be more likely to apply the sustainability criterion to their shopping behaviour. Rotter's idea would suggest that if people are already involved with social justice, labour, peace or indigenous rights activism in a volunteer situation, they may be more receptive to applying these criteria to a shopping situation (McGregor 2004a,b).

Expectations

Attendant with watching someone do something is the development of expectations about consequences stemming from the observed behaviour (Bandura 1977). Rotter (1954) identified this response as a form of reinforcement which affects a learner's likelihood of performing the particular behaviour in the future. In fact, he suggested that this reinforcement alters the learner's expectations, rather than their actual behaviour. Expectations refer to what people anticipate will happen when they behave in a certain way (Bandura 1977). If hurtful behaviour can be averted by changing expectations, this type of learning is, indeed, a powerful tool for sustainability.

Wals and Heymann (2004) offered four insights into the complex concept of expectations. The first idea is characterization, taken to mean any stereotypical or prejudicial perceptions people have about an issue which affects their potential to reach a positive outcome: "I am only one person. What difference can I make?" "Industry is too big to influence." "It is up to government to protect its citizens so I have no obligation to workers in another country." From these frames (or expectations), people can justify not engaging in sustainable consumption.

Wals and Heymann's (2004) second idea is a process frame, which deals with people's perceptions of their input into a decision process, or their perceptions about the process itself (transparency, inclusiveness, accessibility). For instance, a 'nature is there for the taking' frame has had chilling effects on the earth's biological integrity. On the other hand, a 'we are stewards of the earth' frame has lead some people to believe strongly in sustainability, and act on that belief.

A third type of expectation is related to the outcome, or the preferred solution. If people go into a situation preferring a particular outcome, it can become their sole frame of reference for any acceptable solution to the problem (Wals and Heymann 2004). If consumers believe they play a key role in the economy, and that their interests have to be protected at all costs (the 'I have rights' frame), then any solutions to unsustainable economic development, solutions which place more responsibilities on them, will be rejected because they prefer solutions which protect their right to consume.

Finally, Wals and Heymann (2004) posited that people's perceptions of the source of information will affect the way they respond to an issue. If people are skeptical of the source of the information, they can readily dismiss or discount the information. Conversely, if they hold the information source in high regard, they may accrue the information undue respect. This perception (or expectation) can preclude people from critically analyzing the situation, opting, instead, to accept the status quo. As an example, some people readily accept the word of transnational corporations when their messages are couched in terms of corporate social responsibility. Others are very skeptical of such messages simply because of the source. What is the real intent behind the message?

Citizens need help recognizing their own frames (expectations), and those of others. These expectations are dynamic, not static. While people tend to adopt expectations which they are familiar with, they are also capable of shifting their expectations, over time. Awareness of one's expectations, an openness to figure out how these came to be (deconstruction), and a willingness to create new frames or expectations (reconstruction) is key to sustainablility education.

Behaviourial capability

A model is any pattern that can be observed and used to direct thinking and feeling (Bandura 1989b). SLT assumes that people can see and then copy a particular pattern of actions, model them. The fourth SLT concept, behavioral capabilities, maintains that if learners cannot spot the patterns, methods, strategies, roles, et cetera to be emulated, copied or modeled, they cannot figure out if they already have the ability or skills to perform or model the activity. Without this confirmation, they are less likely to try to learn the behaviour, let alone succeed in the attempt. To ensure that learners follow up after observing 'model behaviour', consumer educators must consciously plan to expose learners to credible role models, and be very clear about which behaviour they want them to emulate. Then, this exposure must be followed up with intentionally planned skill-training and practice related to the new learning.

Self-efficacy

Once people have been exposed to new behaviour, they have to feel motivated to show they have learned it. SLT proposes that this motivational trait is deeply affected by self-efficacy. This is defined as a person's perception of their ability to: perform specific behaviours, cope, visualize achieving a goal, or influence a situation. The higher the self-efficacy, the more likely people are to persevere, and to work harder to learn the behaviour. Also, self-efficacy affects what people will *try* to achieve, because it includes their self-confidence in their ability to successfully perform a specific type of action.

Self-efficacy is a very important sustainability trait because simply receiving praise (rewards) from someone else for sustainable consumption may not be enough if people have not changed inside. People need internal reinforcement. They need to find their inner power, their voice, so they can make a long-term difference

(McGregor 2005a, b). They need to affirm their ability to make a difference. Internal rewards are things people do for themselves to reward their behaviour, and to continually bolster their desire and commitment to change (Rimer and Glanz 2005). People experience these internal rewards as various kinds of attractive feelings, motivations and emotions. The habits and behaviour patterns which people choose to model and adopt are those which are positively reinforced by their internal reward system. Their behaviour and lifestyle are shaped by their goals and valued ends, which are established by their motivations and emotions. If people internalize the notion that shopping without concern for sustainability is 'bad', they can impose penalties on themselves to keep their behaviour in line. These internal rules help people self-regulate their own behaviour.

Self-reflective capability

A learner's predisposition to speculate on their actions is linked with another SLT concept, their self-reflective capability. This deals with people's ability to analyze their own experiences, think about their thought processes, and alter their thinking, ultimately altering their actual behaviour. Self-reflective capability is closely tied to self-efficacy (Bandura 1977). Also, inner reflection will generate a range of emotions, including anger, fear, guilt, grief, and sadness. Bandura recognized that people must develop a range of strategies and tactics to deal with emotional arousal, nervousness or anxiety, if they are to exhibit higher self-efficacy. Most people need help with this process.

Self-regulation

Bandura (1989a, 1993) added self-efficacy and self-regulation to the conceptual repertoire of SLT in the early 90s, and changed the name of the theory to social cognitive theory (SCT). Self-regulation, the seventh SLT concept, is intricately linked to the concept of self-efficacy, and is evident when people have their own ideas of what is appropriate or inappropriate behaviour, and choose to act accordingly. Their inner moral compass kicks in, and guides their external behaviour. Self-regulation, as an SLT concept, conveys the notion that people do not act solely because they have been conditioned to do by society. Instead, it assumes that they can think for themselves before they take action, and that they can apply a moral compass to this intellectual exercise. People can look at their own behaviour, and keep tabs on it. They can compare what they see with an internal or external norm or standard, or both. If people like what they see, they feel a range of positive emotions. If they dislike what they see, they feel a range of negative emotions. Both emotions are forms of reinforcement (Bandura 1977).

For example, people who value the status, sense of safety and personal achievement gained from owning a Sports Utility Vehicle (SUV), will like the mental image created when they own one. They will experience little guilt for the environmental impact of this mode of transportation. People who choose to drive an SUV, while feeling bad because they are contributing to the depletion of the ozone layer, have self-regulated their behaviour, too. They opted to drive the vehicle because their need to display status and financial success is currently stronger than their sustainability value.

Foresightful behaviour

Related to self-regulation is the SLT concept of foresightful behaviour, developed to capture the potential of people to be able to 'see' consequences. People need to be able to represent future events through mental images, pictures and symbols, and then take action based on this forethought. Foresight helps people anticipate consequences, and choose less threatening alternative to people's well-being, and quality of life (Bandura 1977). Using foresightful behaviour in conjunction with the concept of observational learning, we can assume that people can actually learn 'stuff' which can be stored in their memory for future use. So, even though learners do not immediately perform the consumer behaviour patterns for sustainability they have learned, they have the capacity to believe that the outcome of that activity is a valued end, and will be more likely to perform this behaviour in the future because they can envision it.

Self esteem

Closely linked to the process of self-regulation is self-esteem, defined as the degree to which people have a high or low opinion of themselves. This opinion affects their behaviour. Constantly criticizing oneself, because of low esteem stemming from perceived or real failures, sets one up for low self-efficacy. This is the belief that one can make a difference by planning and taking action to achieve a goal. High self-esteem, attained by consistently setting realistic goals and achieving them, predisposes people to have a higher sense of self-efficacy.

The likelihood that people can learn through observational learning is closely tied to self-efficacy, which is closely tied to self-esteem, a person's sense of self-worth. If people do not like themselves, this dislike affects their choices, motivations and well-being, which all influence learning. Indeed, low self-efficacy can lead people to believe that tasks or certain behaviours are harder than they actually are. People operating from this position can become stressed, erratic, and their behaviour can be unpredictable. People with a high sense of self-efficacy are actually encouraged when their learning presents them with challenges and obstacles (Bandura 1986, 1989b,1993).

Imagine learners who are moved by listening to an activist's story of success in getting child labour laws changed or fair trade options for a Majority world community (also referred to as Third World). If these learners have low self-esteem, they may value this experience, but be too intimidated to emulate the process because they think it is too difficult, with too many chances of failure. People with high self-esteem may be fully motivated to embark on a similar task, feeling totally confident that they can beat any odds, overcome any obstacles. Both sets of learners value sustainable human and social development, and they both learned from the modeled behaviour. But, they acted differently because of their sense of self agency and self-efficacy, shaped by their sense of self-worth.

Behavioural potential

Potential is defined as capable of being, but not yet in existence. People's potential to behave in certain ways is measured by the ninth SLT concept, behavioral potential (Rotter 1954). Rotter explained that, in any given situation, there are multiple behaviours in which people can engage. For every behaviour that is possible, there is a likelihood that people will engage in that behaviour. The probability that people will exhibit a particular behavior when they encounter a specific situation depends on how they interpret the same situation, often differently. One person may see buying products made from rainforest wood as a status purchase, while another may be appalled by the cutting of the wood to make the product. This perceptual difference occurs because people have different expectations of the outcome of certain behavior. The former consumer expected peer approval from increased status, and the latter expected further reduction of the well-being of indigenous peoples living in the rainforest.

Also, it is possible that people will over or under-estimate the likelihood of a possible occurrence or outcome (Rotter 1954). Both distortions can potentially be problematic. Buying out-of-season fresh fruits and vegetables may not seem like an unsustainable purchase if people do not appreciate the chemicals, migrant labour, and transportation pollution issues associated with this purchase, an underestimation of consequences. Furthermore, peoples' desire for a particular outcome can vary depending on their respective life experiences (Rotter 1954). Consumers who have watched parents, relatives or neighbours live an environmentally sensitive lifestyle may want to emulate this approach to consumption, if they have learned to value sustainability. They may also wish to avoid this lifestyle, if they continually felt deprived, ridiculed by peers, or stigmatized as not being a normal consumer. Their lived experience greatly affects what they want as an outcome of their consumption behaviour.

In brief, Rotter (1954) proposed that the potential to behave in a certain way is a function of: (1) expectancies (the subjective belief that a given behaviour will lead to a particular outcome), and (2) a desire for that particular outcome. The stronger the belief in an outcome (e.g. sustainability), and the more people want that outcome to happen (they value it), the more likely they are to engage in behaviour which ensures that outcome.

Locus of control

The tenth and final SLT concept to be discussed is locus of control, a term coined by Rotter (1990) to account for people's *general* tendencies to expect to have control over of their actions, and thus the outcomes. While self efficacy refers to people's belief that they are capable of successfully acting out *specific* behaviours, locus of control deals with the degree to which people feel they have control over their *lives*. People with an internal locus of control believe they are in control of their life. People with an external locus of control believe that what happens to them is controlled by others, caused by fate, or by luck. This perceived degree of control is something which affects whether change is self-initiated, or influenced by others.

Some research suggests that 'self as agent' underlies the internal locus of control. This means that people's thoughts control their actions. When people realize this function of their thinking, they can positively bring about changes in their beliefs, motivations, and actions. "The self as agent can consciously or unconsciously direct, select, and regulate the use of all knowledge structures and intellectual processes in support of personal goals, intentions, and choices" (McCombs 1991, p. 6). She asserted that the degree to which people choose to direct their thoughts and energies toward an accomplishment is a function of the realization that *they* are the source of their own agency, and personal control.

The implication for social learning is that a person's self confidence, self esteem and self-agency all have to be fostered, nurtured, and maintained. Also, learners need to see the link between their sense of self and the way they behave in the marketplace. If they see themselves defined by what they own, how much they make and the labels they wear, their sense of self will be fulfilled by materialistic consumer behaviour. If, on the other hand, learners define themselves by their relationships with others and the planet, and with their inner sense of goodness and meaning, they will not need to consume to bolster their self-esteem. Consumer education must involve the examination of beliefs, attitudes, values and meaning systems, if sustainability is to become a desired outcome.

Newer innovations of SLT

In addition to the traditional understanding of SLT set out in the previous text, more recent work will be drawn upon to inform this exercise, work that brings SLT to a higher, more pedagogical and social action level. Scholars are linking the theory to issues of power, diversity, multiple ways of knowing and valuing things, social change, social capital building and social relations, inclusive processes, and empowerment (Wals, pers. comm., November 23 2005).

Gertler and Wolfe (2002) distinguished between the conventional approaches of learning-by-doing, learning-by-using and learning-by-interacting, and the social learning approaches of learning-by-searching and learning-by-learning. The latter two refer to self-monitoring (self-regulation) during the process of modeling the behaviour of others, and involve the ability to shed inappropriate norms and practices, replacing them with those which facilitate the processes of change and adaptation. They also tendered the intriguing notion that the capacity to forget (to unlearn) may be just as important as the capacity to learn.

Newer interpretations of SLT appreciate that, at any given point in time, people have constructed a lifeworld through four frames (norms, values, interests and reality), typically unexamined. Donning their social learning hat, educators would plan curricula on the premise that people are amenable to having their prior learning de-constructed, leading to new frames of reference, awareness and expectations. This mind-opening experience paves the way for reconstruction, leading to a new world view, ideally one that is shared with others (Wals and Heymann 2004, Keen *et al.* 2005).

Consumer education reframed through the ESD/SLT interface

UNESCO holds that sustainability relates to ways of thinking about the world, and forms of personal and social practice, which lead to ethical, empowered, and personally fulfilled individuals. The global vision for DESD is "a world where everyone has the opportunity to benefit from quality education and learn the values, behaviour and lifestyles required for a sustainable future and for positive social transformation" (UNESCO 2005, p. 23).

Indeed, some of the key roles of the DESD are to: (1) foster values, behaviours and lifestyles conducive to a sustainable future; (2) inspire people's belief that they have both the power, and the responsibility, to effect positive change on a global scale; (3) increase people's capacities to transform their visions of society into reality; and, (4) build people's capacity for future-oriented thinking (UNESCO 2005, p. 11).

As discussed earlier, SLT deals with learners' capabilities, expectations, self-power, self-control and self-efficacy, belief systems, motivations, and behaviours (Bandura 1977, 1993, Rotter 1954, 1990). Because ESD requires reorienting basic education to include learning which motivates people to live in a sustainable manner, educators need learning theories which can help them create a 'sustainably aware' citizenry, and work force (McKeown 2002). SLT satisfies this requirement. Also, both SLT and ESD take the conceptual stance that three factors are intertwined. The DESD assumes that the three pillars of social, economic and environment give shape and content to sustainable learning (UNESCO 2005). Social earning theory assumes that people, their behaviour and their environment operate in a three-way relationship during learning (Bandura 1977). The following text describes each of these, and then weaves them together into a new conceptual innovation.

Regarding the three pillars of sustainable development: (1) the society pillar refers to the role social institutions play in change and development, with a focus on full, informed participation in these institutions leading to sustainable development; (2) the economic pillar touches on people's sensitivity to the limits and potential of economic growth (especially consumption), and its impact on the other two pillars; and, (3) the environmental pillar involves people's awareness of the fragility and finiteness of the physical environment, leading to a commitment to favour environmental concerns in social institutions, and economic policy. Clugston (2004) added a fourth pillar to ESD, culture, to reflect the role of values, diversity, knowledge, languages and worldviews associated with sustainability education. Bringing the cultural pillar into the equation opens the door for an appreciation of the impact of a person's actions on 'the other.' It gives educators a lens to help learners gain a sense of the connectedness between themselves and others, which is why sustainability matters in the first place.

SLT posits that: (1) personal factors and cognitive competencies include biological factors, knowledge, expectations, self-perceptions, goals, and attitudes; (2) people's behaviour equates to skills (intellectual and psychomotor), self-efficacy, self-regulation, learned preferences, and practice; and, (3) the environment within which people are acting comprises social norms, access to community, and people's influence on others (their ability to change their own environment). Environment also refers to family members, friends, colleagues (social), and such things as room size, ambient temperature, and furniture arrangement (physical). These three factors (people, behaviour and environment) have a mutual influence on each other, and determine human behaviour (Stone 1998, ETR Associates 2005, Rimer and Glanz 2005). Figure 19.1 illustrates the connections between UNESCO's concept of ESD and SLT.

Within the context of the three pillars of ESD, consumer educators can assume that curriculum has to incorporate concern for the interface between social institutions, the economy, and the natural environment. Specifically, educators have to pay close attention to framing consumption within: (1) the limits of economic growth and the potential of viable alternatives; (2) an expectation for full, informed participation in social institutions; and, (3) a longstanding commitment to the entrenchment of environmental concerns in social institutions and economic policy. Pulling the cultural dimension of sustainable development into the equation (Clugston 2004) means educators must create learning situations which enable citizens to appreciate and respect diversity, shared power, interconnectedness, interrelatedness, and varying value systems and perspectives.

McGregor (2005a,b) argued that critical consumer education can lead to sustainable *consumer* empowerment. To make consumer education for sustainable development really come alive, consumer educators must also foster sustainable *development* empowerment, defined as a situation where people are empowered

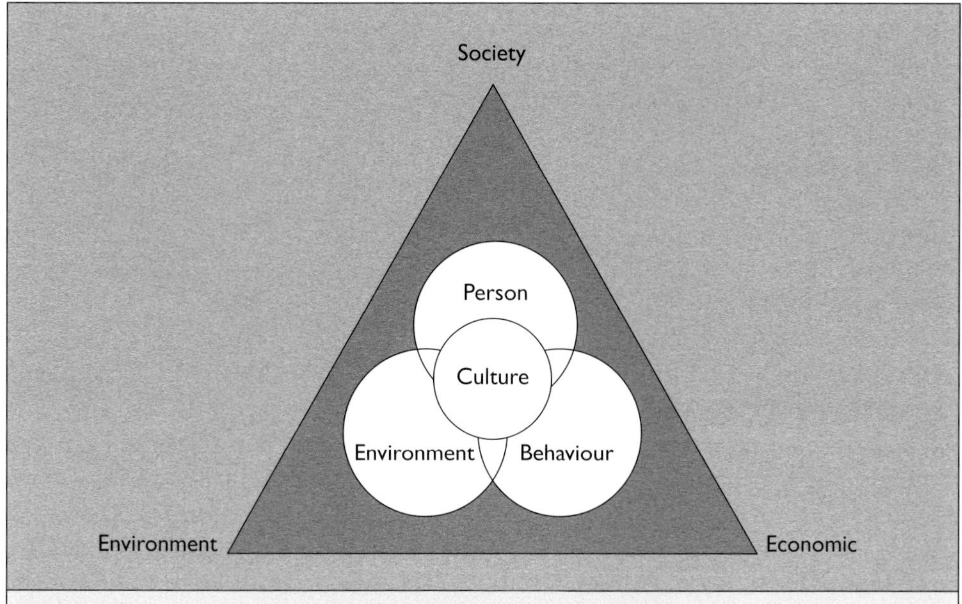

Figure 19.1. Four pillars of the ESD interfacing with three forms of social learning.

in all three spheres that influence development: social, economic and political (Nepal Human Development Report 2004). If people feel empowered in two of these spheres of influence, but not in the third, their empowerment cannot be sustained. If consumers are familiar with economic alternatives and have a deep commitment to environmental issues, but feel they lack a political voice or power, they cannot act from a position of full empowerment to ensure sustainable development. If consumers are familiar with economic alternatives and have been taught advocacy skills, yet have no commitment to sustainability as a social issue, they will not be fully empowered to achieve sustainable development. The preservation of development empowerment is dependent on the type of consumer education provided in schools.

Pedagogical concerns

Exciting new work is being done around the relationship between the way consumer education is taught, and the kind of consumer who is formed. Flowers *et al.* (2001) and Sandlin (2004) offered a typology of three types of consumer education, and three resultant types of consumers. McGregor (2005a, b) recently added a fourth type of consumer education to this typology, the Empowerment Approach for Mutual Interest. This Type 4 consumer education is fully compatible with the integrated ESD/SLT approach to citizen empowerment. From this perspective, consumer educators would teach people using an authentic, critical pedagogy, intending for them to learn how to unveil oppressive power relationships in the global market. Inherent in this approach is critical reflection, a learning strategy which helps people find their inner power, find their inner voice and helps them develop the potential to change the world for the better. They do so by challenging the status quo from a social justice, sustainability, and moral imperative stance. They know that they have a responsibility to help other consumer-citizens to find their voice too, because, once found, they will be transformed, and unable, even unwilling, to consume the same way. They will have evolved toward having a moral conscience in the marketplace (McGregor 2006b).

This other-oriented consumer activity is mediated through moral agency, which becomes a powerful influence on sustainability when coupled with social agency. In a reciprocal relationship, moral standards regulate behaviour, and moral thinking and behaviour are influenced by observation and modeling (Abbott 2000, Bandura 1989a, Rimer and Glanz 2005). In formal education, consumer educators are encouraged to bring an authentic pedagogy to the learning environment. This pedagogy is about getting learners to do *learning work* rather than busy work. It is about engaging learners in big ideas and complex understandings. It includes connectedness (or relevance) which involves helping learners make connections between different aspects of school learning and their past experiences, and the world beyond the classroom. What they learn should have value beyond the classroom (Queensland Department of Education 2001, McGregor 2005a, b, 2006a, b). SLT holds that this out-of-classroom learning can legitimately come from modeling other's behaviour.

Authentic pedagogy also embraces a socially supportive classroom environment, one where learners are able to influence activities, and how they are implemented (democratic pedagogy). It involves a high degree of self-regulation by learners, too. More than making a warm, happy place to be, it is about creating a learning culture which has high expectations of learners, one which encourages them to take risks in learning. Finally, recognition of differences is part of this pedagogy, encompassing inclusivity of non-dominant groups, and positively developing and recognizing differences and group identities (Queensland Department of Education 2001, McGregor 2005a, b, 2006a, b).

Conclusion

Until the world is graced with a tangible majority of people consuming within a sustainable mindset, SLT, integrated with the tenets of ESD, provides educators a theoretical and pedagogical platform from which to approach consumer education for sustainable development and empowerment. They can assume learners are able to: (1) vicariously learn by watching others who exemplify a sustainable lifestyle; (2) draw encouragement and future commitment from others' successes; (3) be taught to critically observe the unsustainable marketplace actions of government, consumer-citizens and business; and, (4) be taught to challenge these practices, leading to increased inner and external transformative change.

SLT, integrated with ESD, lends itself to explaining and influencing the complex behaviour of consumption, because it embraces the notion that observations and attendant thoughts *can* regulate actions (Bandura 1986). Using SLT and ESD as a framework for curricula enables consumer educators to "work to improve their student's emotional states and to [critically examine their] self-beliefs and habits of thinking (personal factors), improve their academic skills and self-regulatory (self-efficacy) practices (behaviour); and alter the school and classroom structures that may work to undermine learner success (environmental factors)" (Pajares 2002, p. 2). This chapter illustrated a way to reframe consumer education such that sustainable consumer empowerment becomes *the* cornerstone of human, social and economic sustainable development, and empowerment.

References

Abbott, L. (2000) *Social Learning Theory*, available at http://teachnet.edb.utexas.edu/~lynda_abbott/Social.html

Argyris, C. (1976) *Increasing Leadership Effectiveness*, New York: Wiley.

Bandura, A. (1977) *Social Learning Theory*, New York: General Learning Press.

Bandura, A. (1986) *Social Foundations of Thought and Action: A Social Cognitive Theory*, Englewood, New Jersey: Prentice Hall.

Bandura, A. (1989a) "Social Cognitive Theory", In R. Vasta, ed., *Annals of Child Development Vol 6*, Greenwich, New Jersey: Jai Press.

Bandura, A. (1989b) "Human agency in social cognitive theory", *American Psychologist*, 44: 1175-1184.

Bandura, A. (1993) "Perceived Self-efficacy in Cognitive Development and Functioning", *Educational Psychologist*, 28(2): 117-148.

Clugston, R. (2004) "The UN Decade of Education for Sustainable Development", *SGI Quarterly*, 38: 2-5. Available at http://www.sgi.org/english/Features/quarterly/0410/feature1.htm

ETR Associates (2005) *"Theories and Approaches: Social Learning Theory"*, available at http://www.etr.org/recapp/theories/slt/Index.htm

Flowers, R., Chodkiewicz, A., Yasukawa, K., McEwen, C., Ng, D., Stanton, N. and Johnston, B. (2001) *What is effective education: A literature review*, Australian Securities and Investments Commission: Sydney. Available at http://www.fido.asic.gov.au/asic/pdflib.nsf/LookupByFileName/EffectConEd_report.pdf/$file/EffectConEd_report.pdf

Gertler, M.S. and Wolfe, D.A., eds. (2002) *Innovation and Social Learning*, New York: Palgrave/Macmillan.

Heffner, C.L. (2004) *"Personality Synopsis: Chapter 8, Learning Theory"*, available at http://allpsych.com/personalitysynopsis/learning.html

Keen, M., Dyball, R. and Brown, V.A., eds. (2005) *Social Learning in Environmental Management: Towards a Sustainable Future*, London: Earthscan.

McCombs, B. (1991) "Metacognition and Motivation in Higher Level Thinking", Paper presented at the *Annual Meeting of the American Educational Research Association*, Chicago, IL.

McGregor, S.L.T. (2002) "Consumer Citizenship: a Pathway to Sustainable Development?" *Keynote at International Conference on Developing Consumer Citizenship*, Hamar, Norway, available at http://www.consultmcgregor.com

McGregor, S.L.T. (2003a) "Globalizing and Humanizing Consumer Education: A New Research Agenda", *Journal of the Home Economics Institute of Australia*, 10(1): 2-9, available at http://www.heia.com.au/heia_graphics/JHEIA101-1.pdf

McGregor, S.L.T. (2003b) "Postmodernism, consumerism and a culture of peace", *Kappa Omicron Nu FORUM*, 13(2), available at http://www.kon.org/archives/forum/13-2/mcgregor.html

McGregor, S.L.T. (2004a) "Workshop on the Challenges of Building a Culture of Peace in a Consumer Society", *Proceedings of the Eastern Family Economic and Resource Management Association Conference*: pp. 82-87. Tampa, Florida. Available at http://www.consultmcgregor.com

McGregor, S.L.T. (2004b) "Consumerism and Peace: Using Postmodernist Thinking to Understand Consumerism from a Peace Perspective", *The Michael Keenan Memorial Lecture*, available at http://www.consultmcgregor.com

McGregor, S.L.T. (2005a) "Sustainable Consumer Empowerment via Critical Consumer Education: a Typology of Consumer Education Approaches", *International Journal of Consumer Studies*, 29(5): 426-436. Available http://www.consultmcgregor.com

McGregor, S.L.T. (2005b) "The Dynamics of Shared Responsibility: Strategies and Initiatives for Participatory Consumerism", *Keynote in the Proceedings of the Consumer Citizen Network conference*, Bratislava, Slovakia. Available at http://www.consultmcgregor.com

McGregor, S.L.T. (2006a) *Transformative Practice*, East Lansing, Michigan: Kappa Omicron Nu.

McGregor, S.L.T. (2006b) "Understanding consumer moral consciousness", *International Journal of Consumer Studies*, 30(2): 164-178. Available at http://www.consultmcgregor.com

McKeown, R. (2002) *"Education for Sustainable Development Toolkit (Version 2)"*, University of Tennessee Energy, Environment and Resources Centre, available at http://www.esdtoolkit.org/esd_toolkit_v2.pdf

Nepal Human Development Report (2004) *"Chapter 1: Empowerment, the Centerpiece of Development"*, available at http://www.undp.org/np/publications/nhdr2004/Chapter1.pdf

Pajares, F. (2002) *"Overview of Social Cognitive Theory and of Self-efficacy"*, available at http://www.des.emory.edu/mfp/eff.html

Pelling, M. and High, C. (2005) *"Social Learning and Adaptation to Climate Change"*, Benfield Hazard Research Centre Disaster Studies Working Paper 11, available at http://www.benfieldhrc.org/disaster_studies/working_papers/workingpaper11.pdf

Queensland Department of Education (2001) *New Basics Project: Productive Pedagogies*, Queensland Department of Education: Brisbane, available from http://education.qld.gov.au/corporate/newbasics/html/pedagogies/pedagog.html

Rimer, B.K. and Glanz, K. (2005) *"Theory at a Glance"*, Washington, DC: US Department of Health and Human Services, available at http://www.cancer.gov/theory/pfd

Rotter, J.B. (1954) *Social Learning and Clinical Psychology*, New York: Prentice Hall.

Rotter, J.B. (1990) "Internal Versus External Control of Reinforcement", *American Psychologist*, 45(4): 489-493.

Sandlin, J.A. (2004) "Consumerism, Consumption and a Critical Consumer Education for Adults", *New Directions for Adult and Continuing Education*, 102: 25-34.

Stone, D. (1998) *"Social Cognitive Theory"*, available from http://hsc.usf.edu/~kmbrown/Social_Cognitive_Theory_Overview.htm

UNESCO (2005) *"United Nations Decade of Education for Sustainable Development"*, available at http://www.unescobkk.org/fileadmin/user_upload/esd/documents/Final_draft_IIS.pdf

Wals, A. and Heymann, F.V. (2004) "Learning on the Edge: Exploring the Change Potential of Conflict in Social Learning for Sustainable Living", in A. Wenden, ed., *Educating for a Culture of Social and Ecological Peace*, pp. 123-145, New York: State University of New York Press.

Chapter 20

Social learning for sustainability in a consumerist society

C.S.A. (Kris) van Koppen

Introduction

Consumption is at the heart of industrialized societies. As a wide range of scholars have demonstrated, the industrial revolution of the 18th and 19th centuries was a revolution in consumption as much as in production. The booming growth of industrial output was made possible by an equally spectacular growth of consumer demand. Much of this demand did not spring from primary needs, but was concerned with luxury products such as clothing, cutlery, and furniture, in other words, with the embellishment of personal appearance, house and garden, and the enhancement of leisure activities. The importance of leisure was also demonstrated by the increasing popularity of books, journals, gardening, sports, and nature recreation. Practices related to consumption – shopping, advertisement, fashion – also took off in this period (McKendrick *et al.* 1983, Campbell 1987). In the two centuries following the industrial revolution, these consumption patterns spread to many other countries, and found their way to large categories of citizens. In the second half of the 20th century, they became dominant patterns in many of the OECD countries.

It is this state of affairs, where patterns of consumption have a major influence on the institutions, discourses, and practices in society that I refer to as 'consumerism'. It is meant as a descriptive term and does not *per se* bear the negative load it often has in environmentalist writing. Rather than arguing against consumerism, I believe that in social learning for sustainability, the consumerist features of modern society should be taken into full account, using them positively where that is possible, and resisting them where necessary. Such a strategy would need to start with an open and thorough analysis of the relationships between consumption and social learning.

Written in an essay style that echoes the inauguration lecture it stems from (Van Koppen 2005), this chapter will present only a rough sketch of such an analysis, starting from sociological theories of consumerism and then exploring the relationships with social learning and education for sustainability. Although most of the empirical examples are based in the Netherlands, this analysis may have a broader theoretical relevance, including most of the highly industrialized countries in the North and, in some respects, parts of the rapidly industrializing countries in the South as well. Another limitation to be mentioned, is that the issue of sustainability is mainly investigated from the environmental, or 'planet' angle. In making this restriction I do not at all mean to suggest that social and economic perspectives on sustainability are less important.

Consumerism in sociology

Status competition, virtual pleasure, and identity

Many sociologists have tried to explain the high level and rapid pace of consumption in modern societies. Most of them agree that it is not simply a matter of satisfying direct practical needs. Obviously, people in the private sphere need products and services for practical use, varying from housing, food and clothes, to means of transport and hobby tools. But this would only account for a limited part of modern consumption repertoires. Many products bought are hardly used, and dumped long before their practical use value has ended. This is obvious for fashion-dependent commodities, but goes for many other products as well. Their value seems to rely on purchase and display, rather than practical use.

The classic sociological explanation is that of Veblen (1994 [1899]), who explained excessive consumption in terms of social emulation or status competition. While for Veblen 'conspicuous consumption' was characteristic for the upper classes, for Duesenberry (1949), who coined the term 'keeping up with the Joneses', it motivated much broader categories of people. Bourdieu (1984) is well-known for exploring the cultural aspect of status competition. His notions of distinction and cultural capital have been widely used to explain the consumption of cultural goods in a status competition perspective.

While admitting that practical needs and status competition have a role to play, Campbell (1987) argues that they are not sufficient for explaining modern consumption. He offers another explanation, based on virtual satisfaction, or 'self-illusionary hedonism' as he names it. Rather than the use of the product, it is the fantasy of what it will bring that gives pleasure to the consumer. Actually purchasing the product is necessary for sustaining the fantasies of consumers; but when the product is acquired, these hedonist fantasies direct themselves to new products. Campbell's theory sheds light on typical consumerist patterns such as shopping for fun, the abundant use of symbolic promises in advertising, and the functionality of fashion.

Another interesting approach to explaining consumption emphasizes the role of consumption in the shaping of personal identity. Against the backdrop of the individualization and increasing reflexivity of modern society, sociologists such as Giddens (1991), Bauman (1988) and Laermans (1991) have pointed out that consumption repertoires of citizens constitute an important vehicle for self-identity and self-representation. As traditional roles – on the basis of gender, profession, religion, social class, or geographic location – relatively lose force, consumption emerges as an important source of identity. Among children and teenagers, the phenomenon of consumption-based identity is highly manifest. The recent marketing book 'BRANDChild' shows how particularly 'tweens' – children between 8 and 14 years old – are strongly influenced by brand marketing (Lindstrom 2004).

Practices

These different sociological approaches to consumption are by no means mutually exclusive. There is little difficulty in considering them as complementary or even mutually reinforcing dynamics of consumption. What they have in common is that they analyze consumption as an action pattern of autonomous individuals, engaged in symbolic imagination and communication. In a recent article consumer sociologist Alan Warde argues for studying consumption from another angle, based on the concept of practices. A practice, in this context, is a domain of routinized 'doings and sayings' that hang together through particular understandings, rules and engagements (Warde 2005). Examples are farming, cooking, managing a firm, or motoring. Consumption, then, is a moment in a practice, rather than a separate, autonomous choice. Much of this practice will be based on convention and routine, and often involve practical consciousness and tacit knowledge instead of deliberate choice. The rewards of consumption are not in consumption as such, but in the practice that is maintained by it. In a way, the practices approach brings us back to the point of departure of social analysis: consumption for satisfying concrete and practical needs. The concept of practices, however, allows for a sociological analysis of such practical needs. Within the context of this practice, other dynamics such as status competition, virtual pleasure, and identity can play a major or minor role. Car driving, for example, can in the same action provide conspicuous display, a fantasy of power, a place of private identity, and actual transport. For these and other reasons, the practices approach emerges as a useful framework for redirecting social theory of consumption to practices of everyday life.

Consumption, power and the public domain

In addition to the above-mentioned efforts to explain consumer behaviour, much social scientific work has been dedicated to analyzing consumption in relation to societal power structures. Over the years, a series of radical authors – starting again with Veblen and including Packard, Galbraith, and Baran and Sweezy – have analyzed consumption resulting from manipulation and coercion by large companies. According to their views, it is big business that drives modern consumption, by manipulating, among others, the social dynamics that are described above. Michael Dawson, a recent exponent of this strand of thought, analyses consumption in the USA as the outcome of corporate marketing, which he characterizes as an "inherently expanding vehicle of class coercion" (Dawson 2005, p. 15).

It is clear that in consumerist societies, multinational corporations, rather than consumers, are emerging as major power containers (Karliner 1997, Beck and Willms 2004). In terms of money and expertise, they have more resources than many of the smaller states, and their operational power vis-à-vis larger states is steadily growing too. Corporate power is manifest in many ways, for instance in the widening gap between incomes of managers-shareholders and average consumers, or in the increased influence of corporations on culture, recreation, and media. However, the power relationship between consumers and corporations in consumer society is different from that between capitalists and labour, or between ruling and underlying classes in other types of societies. As strong as the power of multinationals may be, eventually, it depends on the choices of consumers. In a manner of speaking, it is better to be oppressed by someone who needs you as a customer, than someone who wants you as a worker or a soldier.

If we descend from this rather abstract level of reasoning to practices of consumption, the aspect of consumer freedom is prominent as well. In the shop the consumer is, in a way, sovereign: he or she decides on buying. Even though he departs from consumption models that stress consumer choice, Warde (2005) considers 'discretion' to be characteristic of consumption. The association of consumption and freedom is also manifest from the close relationship between consumptions and leisure (Mommaas 2003). The lion's share of our consumptive practices occurs off-the-job. Consumers clearly experience a strong and continuous stream of advertising and other stimuli, attaching emotions and arguments to products (Leiss *et al.* 1986). But this stream is highly diverse and hardly coercive. It offers a variety of values and models we can accept or reject. The need for consumers to make individual choices, according to Laermans (1991), stimulates *de facto* individual reflection. It is not by coincidence that the age of individualization and reflexivity is also the age of consumerism.

Consumerism has clearly affected the public domain of state and civil society. As Mommaas has shown for the Netherlands, civil society is mainly constituted in the sphere of leisure, and this sphere has been more and more commercialized.

While in the first half of the 20th century broadcasting, sporting and culture were mainly situated in the public domain of citizens, now they are mostly in the domain of private, commercialized consumption. Keywords in this domain are leisure and fun, not education and citizenship (Mommaas 2003). Traditional ties to political parties, trade unions, and religious organizations weaken. Citizens still feel engaged with a wider community, but their loyalties are more impulsive and shifting in time (for the Netherlands, e.g. De Hart 2005). In short, citizens' attitude towards issues of civil society and politics is increasingly moulded on consumerist patterns. While many have deplored the emergence of a consumer-like, calculating attitude towards political issues, it is useful to see the positive side as well. The weakening of traditional loyalties may also result in a more open attitude to other points of view. A more calculating and self-centred attitude towards great narratives may imply a more reflexive and critical stance towards ideologies. Moreover, the mutual influence of consumer and citizen roles not only makes citizens deal with political issues in a consumer-like way. It can also make citizens bring their political views to bear on consumer choices, for instance by boycotting products for political reasons, as was the case in the boycott of French wine as a protest against nuclear testing on the Pacific atoll of Mururoa, 1995. This phenomenon is taken up in recent sociological research under the umbrella of 'political consumerism' (Micheletti 2003, Micheletti *et al.* 2004). Under such circumstances, when the roles of citizen and consumer intertwine, it is plausible to speak of citizen-consumers.

Consumption in environmental sociology

In the 1970s and 1980s, environmental policy was mainly directed to issues of industry, agriculture and infrastructure. Some parts of the environmental movement paid attention to consumption issues, but mostly in a negative way. A better environmental attitude simply meant less consumption. Since the 1990s, however, the importance of consumers as a target group for environmental policy has been fully acknowledged (SER 2003). In the same period, environmental sociology lined up consumerism in its sights. Departing from simple schemes of good and bad environmental attitudes, environmental sociologists have sought for more diversified approaches that not only take account of the risks, but also of the opportunities of consumption for sustainability. Examples can be found in the aforementioned debate on political consumerism, a significant part of which is directed to issues of sustainability.

Consumers can use their buying power as an instrument of environmental protest. As the Brent Spar affair and other similar events have demonstrated, the bundling of individual consumer choices in a broad protest movement can dramatically shift the power balance between companies and consumers. Such boycotts are rare, but the fact they can happen makes powerful corporations keenly aware of the need to legitimize their activities towards citizen-consumers.

Less dramatic, but perhaps even more influential, is the embedding of sustainability aspects into the criteria which consumers regularly use to choose their products. Some products may be avoided, while other products – e.g. eco-label products, local products, refill products – are preferentially bought. In many countries organic products, and other eco-label products, constitute a minor but gradually growing market segment.

Another important aspect of sustainable consumption is the way products are handled and disposed of in consumer practices.

An actual and well-elaborated environmental sociology approach to such aspects of consumption is represented by the model of Spaargaren (2003).

Rooted in Giddens' structuration theory, this model shows that consumer practices are constituted by an interplay of actors (a) and structure (s). Structure is situated in the model as the rules and resources shaping the systems of provision (including shops, utilities, governmental agencies, etc.). Human agency is mediated by lifestyle. The model posits that the interplay between agency and structure occurs through varying social practices. Thereby, it allows for differentiation in these practices, rather than assuming a single 'environmental attitude'. Moreover, it puts practices, rather than individual consumer choice, at the centre of analysis. For

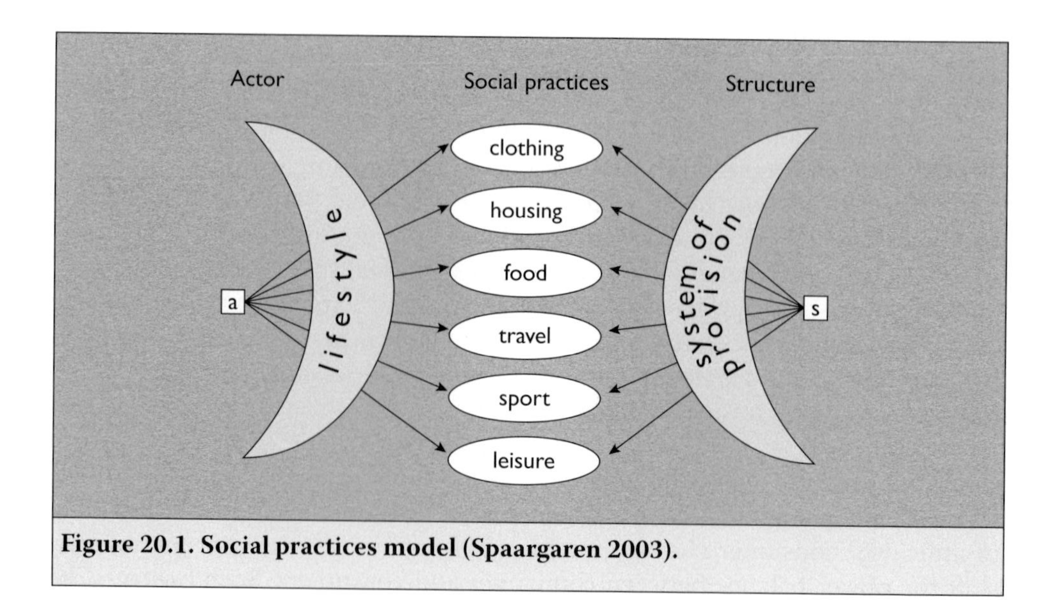

Figure 20.1. Social practices model (Spaargaren 2003).

example, a house owner who decides to buy a solar panel, and install it on the roof, is not just acting out of environmental awareness. A vital condition is that he or she is accustomed to improving and repairing the house (practice of housing). He or she will probably feel that the solar panel fits in well within a favoured image of the house and its residents (lifestyle). The availability of do-it-yourself solar panel kits and options for inserting the panel in the electricity grid are also necessary requirements (systems of provision). As the example may illustrate, and as was argued earlier on theoretical grounds, approaching consumption as an element of practices offers a promising integrative framework for analyzing consumption.

Developed in the context of ecological modernization theory, the model assumes that ecological rationality will emerge both in the rules and resources that drive systems of provision, and in consumer consciousness. Through embedding in practices, ecological rationality will then become institutionalized in consumer behaviour and result in new, environmentally-oriented demands on the systems of provision. However, it is not evident that such processes actually occur, or to put it more accurately, it is uncertain how processes of ecological rationalization relate to and balance with trends in lifestyles that take an unsustainable direction (e.g. Shove 2003). There seems to be a conceptual gap between the idea of ecological rationalization, which suggests a targeted and incremental process of environmental transformation of consumer behaviour, and the concept of lifestyle, which in contrast seems to float easily on consumptive trends, whether ecologically rational or not. Social learning is a concept that may bridge at least part of this gap.

Social learning and consumption

Social learning and education for sustainability

Before embarking on the issue of social learning and consumption, it is useful to better demarcate the concept of 'social learning', which has a wide and highly diverse range of applications. For this chapter, three domains of application are of special relevance. In policy science, social learning is used to refer to lessons learned in policy change, especially when a wider policy community, coalition, or network is involved in policy making (cf. Bennett and Howlett 1992, Pemberton 2000). A related domain where social learning concepts are frequently applied is public participation (e.g. Johnson and Wilson 2000, Stagl 2006, Tippett *et al.* 2005). Many authors in this domain emphasize the confrontations of expert and lay knowledge, and the possibilities of increasing and improving the voice of citizens in policy making. More specifically in the domain of education, social learning is often associated with a constructivist approach to education, focusing on pluralism and confrontation of views and interests as a road to better mutual understanding (e.g. Wildemeersch *et al.* 1997, Wals and Heymann 2004). From these different fields of study, I have distilled three characteristic features of social learning, as it is interpreted in this chapter. First and obviously, social learning is done in and by groups, communities, and networks. Second, social learning is about social issues. More precisely: it concerns matters of common interest and collective action. Sustainability issues typically belong to this category. Third, social learning implies that, even in the middle of conflicts in views and interests, there is an effort at collaboration and consensus.

Contemporary societies vitally depend on such processes of social learning among broad categories of people. Rules of society may be set by business leaders, political specialists, and scientific experts, but under conditions of democracy and market economy, it will be choices of voters, preferences of consumers, and engagements of the citizens that at the end of the day determine which rules will hold, and which not. Social learning means building shared frameworks for communication and action among these constituencies.

Social learning, thus described, is of particular relevance to sustainable development. It is, I would say, the object proper of education for sustainability. Education, here, is defined in a broad sense, encompassing all interventions that are made with an explicit (though not necessarily exclusive) target of creating and facilitating long-term learning trajectories for a more or less specified group of persons (cf. Hovinga 2004).

Relating social learning and education for sustainability to the previous analysis of consumerism, three general observations can be made.

First, with the rise of consumerism, 'consumership' becomes an important field of education. Traditionally, education has been aimed at professional competences and to a less, but still important degree, to citizenship. In a consumerist society, education also needs to address consumption. Citizen-consumers need the competences to shape their consumption repertoire in a way that benefits their own health and well-being as well as that of others and of other generations. Sustainability, therefore, means educating consumers.

Second, education is important in maintaining a public sphere that is capable of regulating and structuring consumptive practices. As was discussed, consumerism puts the public sphere under pressure. As a sustainable society is not possible without regulation from civil society and state government, an independent sphere of citizenship is of crucial importance. The old aim of educating citizens acquires a new and even more urgent actuality under conditions of consumerism.

Third, the trajectories of educating consumers as well as citizens will have to be adapted to a social reality in which the two roles are increasingly part of the same continuum. Practically, this means that modes of formal and informal learning are influenced by consumerism, and have to compete with other modes of consumption. This puts high demands on both the content and marketing of education. Theoretically, it means that the consumer practices model, presented above, also becomes applicable to citizen practices. Practices such as going to school or being a volunteer in an environmental organization could be inserted in the model, as these practices too are subject to the dynamics of lifestyle and systems of provisions.

Rather than theoretically elaborating on these relationships, this chapter will explore them by looking at some examples. Purposefully, they represent widely different contexts of social learning practices. In this diversity, they may illustrate the sorts of demands that consumerism makes upon social learning and education for sustainability.

Child development

Child development is a classical context of education. Much of the efforts concerned with social learning for sustainability have been directed to environmental education at primary and secondary schools. In the Netherlands, as in many other countries, these efforts have resulted in a significant improvement in environmental education in formal education (Sollart 2004). Nonetheless, formal learning for sustainability remains under pressure, and further impulses are needed for continuity and improvement. An important aspect of improvement is to link learning processes with everyday practices, including consumptive practices. For children, much of global society enters their world through consumption, and consumption in its turn is an important means for children to act upon society. The analysis of consumerism suggests that education, rather than telling people that consumption is bad, should be concerned with confronting key symbolic messages of consumption with concrete perception and practices from everyday life. In this way children may learn to look beyond the virtual reality of media and advertising (and many environmental messages as well). And they may be supported in building an identity that relates to their concrete living environment. Many materials and methods for this purpose have been developed by environmental educators, but structural embedding in school education is still lacking.

Child education cannot be dropped at the school door alone. Many schools are already overburdened in their efforts to recover social learning processes neglected in society. Moreover, what happens outside schools is of no less impact on social learning than is formal education. Educational interventions in the out-of-school domain have to target social learning processes that are part of other, non-educational practices. A key issue, in this context, is the facilitation of social learning for sustainability in children's play. For many years environmental educators have made a plea for the establishment of 'wild' neighbourhood areas where children can freely play with plants, small animals, soil, and water. As the everyday world of children in consumer societies has been shifting from outdoors to indoors, it becomes increasingly clear that playing in (semi)natural environments is of major value, not only in terms of building environmental appreciation (Gebhard 2001), but also from an angle of physical skills, self-identity and health (Gezondheidsraad and RMNO 2004). Closely related to the problem of outdoor playing is the development of a children-oriented industry that seduces children into indoor, often screen-bound playing activities. "I like to play indoors better, 'cause that's were all the electrical outlets are" reports a fourth-grader in Louv (2005). This development is related to the virtualization of pleasure that is characteristic of consumerism. Just as food commerce has corroded eating habits, the commercial supply of virtual play and amusement has reached levels of impact on children that actually harm their physical and mental well-being, let alone their appreciation of environment and nature. Interventions in this context therefore should not only be directed to creating playing facilities, but would also imply a fundamental political debate on the impact of television, computer games, and advertising on the development of children, to be followed by appropriate restrictive measures. What is at stake are the learning processes that impact socialization and personal identity formation of children vis-à-vis their human and non-human environment.

Environmental action networks

A second context of social learning for sustainability is environmental action itself, that is, activities of environmental protest and lobbying, awareness raising, and environmental management, carried out by environmental NGOs and volunteer groups. Such activities of environmental organizations and groups are usually evaluated in terms of their effectiveness for environmental protection. Equally important, however, are the social learning processes implied in such actions. Environmental activities in organized settings contribute both to the further development of competences in the civic and political action of activists involved, and to the strengthening and broadening of civil society at large (Dekker 2002). While the latter is vital for raising public concern about sustainable development, the former is also an important factor in educating civil and political leaders of the future. Several nature conservation spokesmen in the Netherlands, for instance, had a history in youth organizations for nature conservation (Gorter 1986).

Under conditions of consumerism, the challenge of stimulating these processes of social learning in civil society becomes even more important and more difficult as well. As we discussed, consumerist society features a shift in civil society from the public to the commercial sphere. It also leads to individualization and a decrease in long-term loyalties, which were characteristic of environmental organization membership in the past. This adds to the importance of social learning as a base for a collective action and public interest awareness, both being indispensable factors in sustainable development. And it increases the need for social networks, to counteract the isolating tendencies of individualization processes. To foster environmental action networks, however, new ways are needed of attracting young people to engage in NGOs and voluntary groups (De Witt 2005). Several organizations in the Netherlands are now experimenting with attractive, short-term projects. The Dutch government is piloting a so-called civil internship (in Dutch: "maatschappelijke stage") offering teenagers the option to work in volunteer organizations (among others, environmental organizations) as part of their school programme. These promising initiatives in formal and informal social learning need further elaboration. Another major factor in education through environmental action networks is funding. It has been a trend in Dutch governmental funding to subsidize innovative projects, selected on the basis of competitive tenders. Such a project-wise and market-based approach, however, has often been to the detriment of existing networks and skills. A better balance and continuity between old and new would be needed to effectively support social learning through environmental action networks.

Public participation

As mentioned before, public participation can fruitfully be analyzed as a process of social learning. A recent Dutch memorandum on Learning for Sustainable Development (Ministerie LNV *et al.* 2003) writes in this context about 'learning arrangements', where stakeholders, citizens and organizations are brought together in concrete situations, and stimulated into a collective learning process. Such learning arrangements should eventually lead to sustainable decision making, and reflexive dialogue about sustainable development. If we compare these words with existing practices of participation, however, the contrast is sharp. Where learning, dialogue, and reflection are central, the participation of citizens is often low. Interest and involvement may grow at the moment that concrete interests are at stake, but then participation frequently ends up in a power struggle, where social learning processes towards consensus and common interests are hard to find. Notwithstanding these problems, education through public participation is much needed, not in the least in issues of sustainability (for Dutch nature protection, e.g. Hermans *et al.* 2004).

Educating the public in participative processes is an ambiguous venture in a consumerist society. On one hand, consumerism makes people more assertive and reflexive, and less dependent on authority and tradition. On the other hand, education does not fit in too well with the short and entertainment-oriented cycle of consumer attention. It may also put people, at least in their own perception, in a position of dependence that contrasts with their sovereignty as customer.

To make 'learning arrangements' of participation work, sophisticated trajectories are necessary, which 'seduce' stakeholders into engaging in the process. Generally speaking, this can be done by coupling learning processes to concrete but gradual steps in decision making, by making good use of social networks, and by safeguarding intensive interaction with experts and authorities. The learning processes targeted have to be elaborated in operational terms, and core concepts need careful translation from one context (e.g. science) to another (e.g. stakeholder practices). In attributing funds to projects, it is important that they are not only assessed in terms of direct results, but also in terms of their impact on long-term learning processes. In spite of the abundant literature on participation, current insights into these processes are at a pioneering stage.

Last but not least, in thinking about education through public participation, consumer roles would deserve more investigation. Up till now, the attention in policy making and research has mostly been directed at the relationship between citizens and government. As the relationship between producers and consumers gains momentum, the question arises as to how this relationship can be part of participatory learning. Initiatives for cooperation between companies, environmental organizations and consumer groups are gradually emerging and it would be worthwhile to approach them from an angle of social learning.

Conclusion

This chapter argues that in a consumerist society, issues of social learning, and thus issues of education, emerge with new vigour and urgency. To address these issues, insights from the sociology of consumption and from social learning studies can be fruitfully combined. On one hand, the social learning approach may offer a perspective for better analyzing environmental transformations in consumer behaviour. On the other hand, the consumer practices approach, which has emerged in consumer sociology, may also be useful in analyzing social learning where citizen roles are concerned, as citizenship is increasingly influenced by consumerism. Against this backdrop, I have sketched some examples of social learning for sustainability in different sets of practices: child development, environmental action networks, and public participation. In all these practices, social learning for sustainability is under pressure from consumerism. To cope with this pressure, multilevel education strategies are needed that partly adapt to and partly resist consumerist patterns. Obviously, much more could be said on each of the examples presented in this chapter. And much more argument would be needed to support its theoretical claims. But if my contribution could be an invitation to further explore such avenues, its main aim would be met.

References

Bauman, Z. (1988) *Freedom*, Philadelphia: Open University Press.

Beck, U. and Willms, J. (2004) *Conversations with Ulrich Beck*, Cambridge: Polity Press.

Bennett, C.J. and Howlett, M. (1992) "The lessons of learning: Reconciling theories of policy learning and policy change", *Policy Sciences*, 25: 275-294.

Bourdieu, P. (1984) *Distinction. A social critique of the judgement of taste*, London: Routlegde and Kegan Paul.

Campbell, C. (1987) *The romantic ethic and the spirit of modern consumerism*, Oxford: Basil Blackwell.

Dawson, M. (2005) *The consumer trap. Big business marketing in American life*, Urbana: University of Illinois Press.

De Hart, J. (2005) *Landelijk verenigd*, Den Haag: Sociaal en Cultureel Planbureau.

Dekker, P. (2002) *De oplossing van de civil society*, Den Haag: Sociaal en Cultureel Planbureau.

De Witt, A. (2005) *Van vervreemding naar verantwoordelijkheid. Over jongeren en natuur*, Den Haag: Ministerie van LNV.

Duesenberry, J. (1949) *Income, saving and the theory of consumer behavior*, Cambridge: Harvard University Press.

Gebhard, U. (2001) *Kind und Natur. Die Bedeutung der Natur für die psychische Entwicklung*, Wiesbaden: Westdeutscher Verlag.

Gezondheidsraad and RMNO (2004) *Natuur en gezondheid. Invloed van natuur op sociaal, psychisch en lichamelijk welbevinden*, Den Haag: Gezondheidsraad en RMNO.

Giddens, A. (1991) *Modernity and self-identity. Self and society in the late modern age*, Cambridge: Polity Press.

Gorter, H.P. (1986). *Ruimte voor natuur*, 's-Graveland: Vereniging tot Behoud van Natuurmonumenten.

Hermans, B., Dekker, J. and van Koppen, C.S.A. (2004) "Natuur van en voor de burger", *Landwerk*, 5: 10-15.

Hovinga, D. (2004) *Zonder bomen geen bos. NME en duurzaamheideducatie*, Utrecht: Universiteit Utrecht.

Johnson, H. and Wilson, G. (2000)" Biting the Bullet: Civil Society, Social Learning and the Transformation of Local Governance", *World Development*, 28(11): 1891-1906.

Karliner, J. (1997) *The corporate planet. Ecology and politics in the age of* globalization, San Francisco: Sierra Club Books.

Laermans, R. (1991) "Van collectief bewustzijn naar individuele reflexiviteit. Media, consumptie, en identiteitscontstructie binnen 'de reflexieve moderniteit'", *Vrijetijd en Samenleving*, 9: 99-118.

Leiss, W., S. Kline & S. Jhally (1986) *Social communication in advertising. Persons, products, and images of well-being*, New York: Methuen.

Lindstrom, M. (2004) *BrandChild*, London: Kogan Page.

Louv, R.(2005) *Last Child in the Woods: Saving Our Children from Nature-Deficit Disorder*, Chapel Hill: Algonquin Books.

McKendrick, N., Brewer, J. and Plumb, J.H. (1983) *The birth of a consumer society*, London: Hutchinson.

Micheletti, M. (2003) *Political virtue and shopping. Individuals, consumerism, and collective action*, New York: Palgrave.

Micheletti, M., Follesdal, A. and Stolle, D. (2004) *Politics, products, and markets. Exploring political consumerism past and present*, New Brunswick: Transaction Publishers.

Ministerie LNV, *et al.* (2003) *Leren voor duurzame ontwikkeling. Van marge naar mainstream*, Amsterdam: Stuurgroep Leren voor Duurzaamheid

Mommaas, H. (2003) *Vrijetijd in een tijdperk van overvloed*, Amsterdam: Dutch University Press.

Pemberton, H. (2000) "Policy networks and policy learning: UK economic policy in the 1960s and 1970s", *Public Administration*, 78(4): 771-792.

SER (2003) *Duurzaamheid vraagt om openheid. Op weg naar een duurzame consumptie*, Den Haag: Sociaal-Economische Raad.

Shove, E. (2003) *Changing human behaviour and lifestyle: a challenge for sustainable consumption?*, Department of Sociology, University of Lancaster. http://www.psi.org.uk/ehb/docs/shove-changinghumanbehaviourandlifestyle-200308.pdf

Sollart, K.M. (2004) *Effectiviteit van het natuur- en milieu-educatiebeleid*, Wageningen: Natuurplanbureau.

Spaargaren, G. (2003) "Sustainable consumption: a theoretical and environmental policy perspective", *Society and Natural Resources*, 16: 687-701.

Stagl, S. (2006) "Multicriteria evaluation and public participation: the case of UK energy policy", *Land Use Policy*, 23: 53–62.

Tippett, J., Searle, B., Pahl-Wostl, C. and Rees, Y. (2005) "Social learning in public participation in river basin management – early findings from HarmoniCOP European case studies", *Environmental Science & Policy*, 8: 287–299.

Van Koppen, C.S.A. (2005) *Zorg voor natuur in de eeuw van de consument*, Utrecht: Oratie Universiteit Utrecht.

Veblen, T. (1994 [1899]) *The theory of the leisure class*, Mineola: Dover Publications.

Wals, A.E.J. and Heymann, F.V. (2004) "Learning on the Edge: exploring the change potential of conflict in social learning for sustainable living", in A.L. Wenden, ed., *Educating for a Culture of Social and Ecological Peace*, New York: State University of New York Press, pp. 123-145.

Warde, A. (2005) "Consumption and theories of practice", *Journal of Consumer Culture*, 5(2): 131-153.

Wildemeersch, D., Jansen, T., Vandenbeeke, J. and Jans, M. (1997) "Paradoxen van sociaal leren", *Sociale Interventie*, 6(4): 198-208.

Part III
Praxis

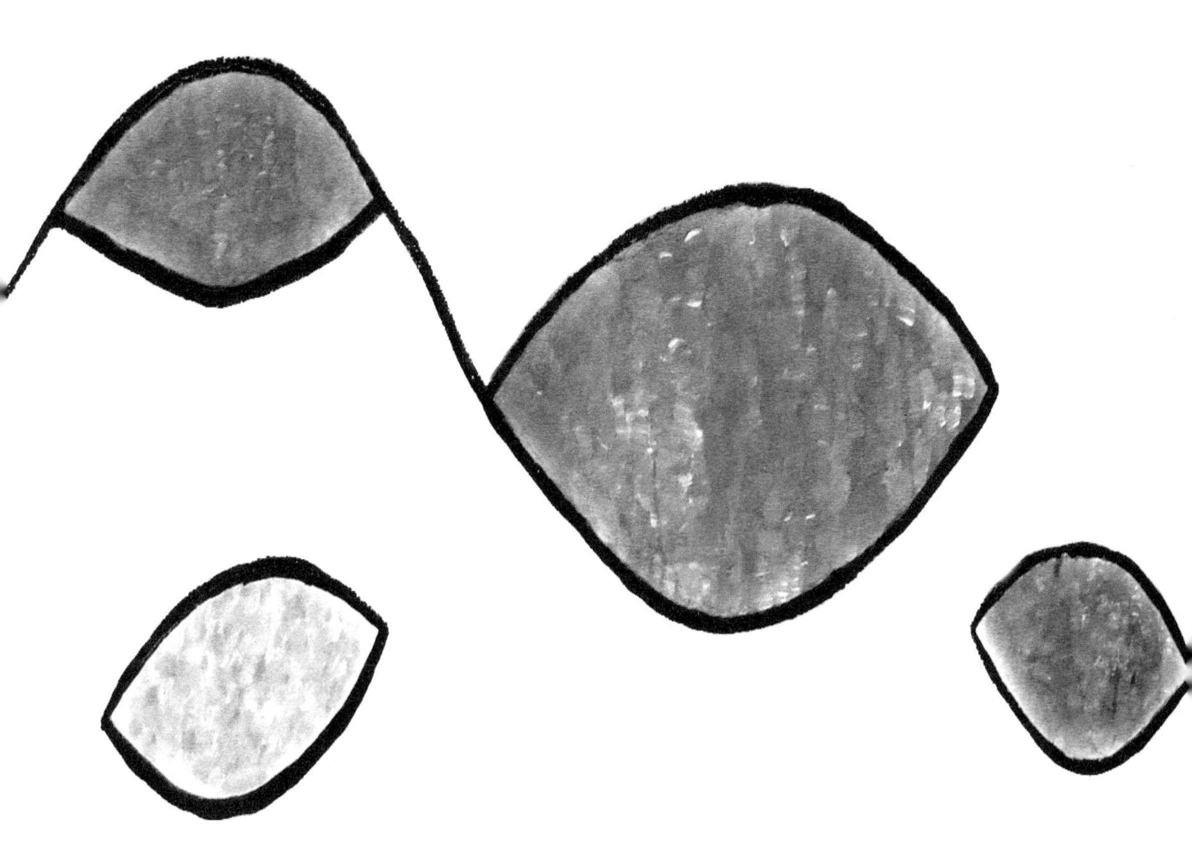

Chapter 21

Partnerships between environmentalists and farmers for sustainable development: a case of Kabukuri-numa and the adjacent rice fields in the town of Tajiri in Northern Japan[41]

Yoko Mochizuki

Introduction

At the Ninth Conference of the Contracting Parties to the Ramsar Convention (Convention on Wetlands of International Importance Especially as Waterfowl Habitat) or the Ramsar COP 9 in 2005, Kabukuri-numa and 259 hectares of surrounding rice fields were designated as a Ramsar site. Kabukuri-numa, located in Miyagi Prefecture of northern Japan, is a 150 hectare freshwater lake that is home to over 200 bird species. One of 1,591 Ramsar sites around the globe, "Kabukuri-numa and the surrounding rice paddies" is the only site which has 'rice paddies' in its official name. This marked a huge triumph for 'waterfowl conservationists' who call for the protection of Kabukuri-numa as the over-wintering habitat for wild geese and 'nature restorationists' who advocate reviving the rice paddy ecosystem.

Back in 1996, a full-scale dredging plan for Kabukuri-numa threatened to destroy the wetland ecosystem upon which greater white-fronted geese *(Anser albifrons)* and other protected or endangered species depend for survival. Not only was the complete dredging plan successfully halted by the efforts of the local NGO

[41] This chapter reports on the partial findings of ongoing research at the Regional Centre of Expertise on Education for Sustainable Development (RCE), which is proposed by the United Nations University as its contribution to the United Nations Decade of Education for Sustainable Development. The description of partnerships between environmentalists and farmers in this chapter is based on data collected between October 2004 and February 2006 as part of the author's research on RCE Greater Sendai during her Postdoctoral Fellowship at the United Nations University-Institute of Advanced Studies (UNU-IAS). The author visited the Town of Tajiri four times (February, August, and November 2005 and February 2006), and conducted formal and informal interviews with local stakeholders. Data collection was partially funded by a collaborative research project on Education for Sustainable Development (ESD) between Yokohama National University and UNU-IAS. As a result of ongoing consolidation of municipalities in Japan, the Town of Tajiri became part of the newly formed City of Osaki in March 2006. For more information on RCE in general and RCE Greater Sendai in particular, see Fadeeva (this volume) and Mochizuki (2005), respectively.

Japanese Association for Wild Geese Protection (JAWGP) and other stakeholders, a citizens' movement to conserve Kabukuri-numa evolved into a participatory programme for engaging with the community for the preservation of biodiversity – both in natural wetlands and rice paddies – and sustainable agriculture. In 1998, an experimental programme to utilise post-harvest flooded rice fields as feeding and resting grounds for wintering waterfowl (ducks, swans and wild geese) was initiated in the Town of Tajiri, where Kabukuri-numa is located. Overcoming the initial antagonism between those who called for the protection of wild geese and rice farmers who viewed waterfowl primarily as a rice-eating pest, the Town of Tajiri is aspiring to promote both environmental and economic agendas at the local level.

Background

Ninety years ago, natural wetlands in Miyagi Prefecture contained 40 marshes, now mostly converted to anthropogenic uses, notably wet-rice agriculture. Out of 40 marshes, 31 were completely drained away and lost, and six large ones, including Izunuma-Uchinuma and Kabukuri-numa, were partially drained and decreased in size (Kurechi 2004). Miyagi Prefecture is one of six prefectures in the Tohoku (literally North East) region. Kabukuri-numa is located about 8 kilometres from Izunuma-Uchinuma, a 387-hectare wetland designated as the second largest Ramsar site in Japan in 1985. Today Tohoku is the highest yielding rice region in Japan, occupying about a quarter of the nation's rice-field acreage. Miyagi Prefecture is especially known as a major production area of popular high-end rice varieties such as *Sasanishiki* and *Hitomebore.*

Greater white-fronted geese (*Anser albifrons*) travel from their northern nesting grounds in Arctic Russia's tundra area to their southern wintering grounds in Japan. The autumn rice harvest is completed largely before migratory birds arrive at the Izunuma-Uchinuma and Kabukuri-numa wetland area. Since freshly harvested rice is moist, however, the grain must be dried prior to threshing. Before combine harvesters became common usage, ducks and geese ate harvested rice while it was being dried in the sunshine, and in the evening and at night, nocturnally active ducks pecked at harvested rice. When ducks arrive right before the harvest, they come to rice fields as the sun goes down and suck at the golden ears of rice plants. This has made rice farmers loathe ducks and geese altogether as '*kamoko*,' which literally means 'petty ducks' in the local dialect but is used to refer to wild geese as well. Grain depredation by ducks and geese made them a major pest for rice farmers. Even after the diffusion of combine harvesters which reduced actual feeding damage, farmers' hard feelings against ducks and geese have remained to date.

Today, the Ramsar sites Izunuma-Uchinuma and Kabukuri-numa are estimated to harbour over 90 per cent of more than 100,000 white-fronted geese wintering in Japan. Several factors contributed to a dramatic rise in the size of the white-fronted geese colony in northern Miyagi. First, there has been a remarkable increase in the number of wild geese in Japan, partly due to the official designation of the white-fronted goose as Japan's protected species in 1971 and the ensuing population rebound. Before the legal protection took effect, the number of white-fronted geese wintering in Japan decreased to just a few thousands. Over 20 years between 1979 and 1999, the number increased by 5.6 times to approximately 46,000 in 1999 (Shimada 2002). Secondly, there are fewer wetlands that can serve as their habitats in the country. Compared to swans and ducks, geese are very cautious and sensitive creatures that cannot survive without rich wetland environments, and only 1 per cent of Japan's wetlands where waterfowl roost today are suitable as goose habitat (Kurechi 2004).

The dramatic increase and concentration of white-fronted geese in these two Ramsar sites attests to more than the effectiveness of the legal protection of the goose, the environmental quality of these marshes, or the lack of alternative sites. The increase reflects a larger trend of the substantial population growth of Arctic breeding geese which forage in agricultural habitats during the non-breeding season (Jefferies *et al.* 2003). The population concentration is linked to the use of agricultural food sources, coupled with a change in the harvesting machine utilised in rice fields in northern Japan. According to Shimada's (2002) study on the daily activity pattern and habitat use of the wild geese around Izunuma-Uchinuma between 1996 and 1999, the amount of rice grains left in the fields by combine harvesters was 8.7 times the amount left by reapers. Between 1980 and 2000, combine harvesters replaced reapers in the area under study, and this has presumably contributed to increasing the amount of 'waste rice' – grain escaping collection by harvesters – which serves as valuable food for wild geese.

The concentration of waterfowl at specific habitats is considered problematic since it makes those birds more susceptible to epidemics of avian influenza and poultry cholera. In 1996, when the River Management Department of Miyagi Prefecture announced a full-scale dredging plan for Kabukuri-numa, waterfowl conservationists were alarmed that the plan would worsen the problem of over-concentration of migratory birds in Izunuma-Uchinuma. In response to this announcement, the Japanese Association for Wild Geese Protection (JAWGP) started devoting much of its energy to the conservation of Kabukuri-numa (JAWAN 1999).

Innovation for a sustainable future: Winter-Flooded Rice Fields

The birth of Winter-Flooded Rice Fields

The innovation of 'Winter-Flooded Rice Fields' (hereafter WFRF) was born out of synergies created between different groups who engaged in action at the local level arising from different challenges. For waterfowl conservationists, the driving force for action came from the discrepancy between the dredging plan for Kabukuri-numa and their goal of conservation and 'wise use' of wetlands. For local farmers, the challenge was to increase agronomic efficiency in rice. Market prices of rice have been steadily falling since 1995, when Japan freed its rice market from government control. A small number of farmers in Tohoku were searching for an alternative rice farming method which could help them earn more income with less labour, sometimes drawing upon a 'non-tillage method' proposed by Nobuo Iwasawa (see Iwasawa 2003) and at times a low-cost, labour-saving organic method developed by the NGO Minkan Inasaku Laboratory[42].

When JAWGP embarked on a self-assigned mission to stop Miyagi Prefecture's full-scale dredging plan for Kabukuri-numa in 1996, major tensions arose between the environmental NGO which called for the conservation of Kabukuri-numa and administrative authorities who insisted on the necessity of dredging Kabukuri-numa for the purpose of water control. As well as lobbying local and national politicians and having them address their concerns about the proposed dredging plan, JAWGP was seeking to mobilise public opinion. In his efforts to engage with the community for the conservation of Kabukuri-numa, the President of JAWGP came to learn that farming households growing rice in Shiratori District, 50-hectare state-owned land adjacent to Kabukuri-numa, were being put in a difficult situation by the government's request to return their rented farmlands as an integral part of the proposed flood control plan as well as of a national policy of reducing rice acreage to alleviate the problem of rice overproduction. While Shiratori District had suffered from repeated floods and many farmers had abandoned farming in the District, there were also many who were disinclined to give up their rented rice fields.

The JAWGP President approached the representative of a rice growers' association in Shiratori District and persuaded him to cooperate with JAWGP. This led to

[42] Nobuo Iwasawa organises the Japanese Association for the Diffusion of Non-Tillage Farming *(Nippon Fukōki Saibai Fukyu-kai).* Minkan Inasaku Laboratory (*Minkan Insaku Kenkyusho*), which literally means a private institute to research rice farming, is headed by Mitsukuni Inaba. Both Iwasawa and Inaba participated as speakers in the First WFRF Symposium held on 24 February 2002 in the Town of Tajiri.

collaboration between JAWGP and local residents for creating a multi-stakeholder committee to organise nature observation tours and flora and fauna surveys of Kabukuri-numa. In May 1996, around the same time as the Ministry of Construction and Miyagi Prefecture expressed their views that there was no need for full-scale dredging of Kabukuri-numa, the First 'Kabukuri-numa Expedition' was held with more than 40 participants ranging from experts on aquatic plants and animals, water quality and civil geotechnology to environmentalists, local farmers, government officials at different levels, and politicians including members of municipal and prefectural assemblies and Parliament.

A series of Kabukuri-numa Expeditions contributed to raising awareness of the rich biodiversity of Kabukuri-numa, leading to the creation of transparent and participatory processes to address the environmental and technological challenges of conserving Kabukuri-numa. In February 1997, the River Management Department of Miyagi Prefecture convened a multi-sectoral roundtable committee – including local authorities, local farmers, NGOs, research scientists and other stakeholders – to draft a basic plan for managing Kabukuri-numa as a floodwater reservoir and a natural wetland. The creation of a platform for multi-stakeholder dialogue served to transform Shiratori District's chronic flood problem into a means to enhance the ecological functions of Kabukuri-numa. In November 1997, with the approval of the Tajiri Town Assembly, farmers abandoned farming in Shiratori District and returned the district to the state so that it could be restored to a natural wetland. By allowing rainwater to collect, Shiratori District soon came to act as habitats for diverse species, including wild geese. Today, Kabukuri Wetlands Club, a local NGO which grew out of the Executive Committee of Kabukuri-numa Expeditions, makes recommendations and proposals to administrative authorities for conservation and wise use of Kabukuri-numa[43]. The Club is also commissioned by Miyagi Prefecture to manage Shiratori District, where it carries out environmental education programmes in cooperation with the Prefecture and Miyagi University of Education.

The dramatic success of Shiratori District in enhancing biodiversity and serving as a waterfowl habitat prompted waterfowl conservationists to try out the idea of winter flooding of rice fields in order to diversify waterfowl habitats and risks associated with the concentration of wild geese in Izunuma-Uchinuma and Kabukuri-numa. At this point, many of the alleged benefits of WFRF described below were largely unknown. In 1998, an experimental programme to utilise 3 hectares of post-harvest flooded rice fields for the benefit of wintering waterfowl

[43] Kabukuri Wetlands Club *(Kabukuri Numakko Club)* was established in 1998 and was granted the status of an incorporated non-profit organisation in 2000. See Kabukuri Web Link at www2.odn.ne.jp/kgwa/kabukuri/j/ for useful websites on winter-flooded rice fields and Kabukuri-numa.

was initiated with the cooperation of a local farmer adopting a non-tillage method in the Town of Tajiri. In the following year, four farmers came to cooperate with this experiment, and the flooded acreage doubled.

Meanwhile, there were emerging efforts to restore the soundness of the rice paddy ecosystem, including NGO-driven initiatives to develop organic rice farming techniques by enhancing the conservation value of rice paddies as replacement habitat for wetland flora and fauna. In 2001, the Ministry of Agriculture, Forestry and Fisheries and the Ministry of the Environment jointly initiated an annual national 'Rice Paddy Fauna Survey' *(Tanbo no Ikimono Chōsa)*. The survey is carried out in cooperation with interested citizens, and some school teachers in the Town of Tajiri came to be actively involved in it. The 2004 national survey identified 98 fish species and 17 frog species in the rice paddies and associated waterways, including 17 red data list fish species and 1 red data list frog species (Ministry of Agriculture, Forestry and Fisheries, 2005). Coupled with the preliminary observation that WFRF bring back various insects, fishes, snails and frogs that have largely disappeared from paddies drained over winter, the survey has contributed to broadening and refining the concept of WFRF and mobilising more actors for WFRF.

Based on experimental WFRF undertakings by individual farmers between 1998 and 2003, waterfowl conservationists and nature restorationists were able to develop some WFRF 'good practices'. Nevertheless, the flooded acreage was still very limited. There was a definite ceiling to the amount of water an individual farmer could take from irrigation and drainage canals during winter since farmers do not have a right to draw water from rivers – known as a water right – during the non-agricultural season. In order to promote WFRF, the Town of Tajiri decided to choose a model district to implement WFRF on a larger scale and to coordinate with the authorities concerned to facilitate winter flooding in the chosen district (Takahashi 2004). In December 2004, a three-year project to implement WFRF in Shinpō District, facing the southern part of Kabukuri-numa, was initiated. In the 2004-2005 season, twelve farming households who grow rice in the District participated in the project and flooded their fields which totalled 20 hectares.

Possible Benefits of 'Winter-Flooded Rice Fields': a 'win-win' solution?

WFRF are considered to be a viable strategy for addressing different environmental, ecological and agricultural challenges. After harvest, farmers leave stubble in rice fields, which increases the 'waste rice' for waterfowl. They subsequently flood harvested fields, without tilling them, and immerse the standing stubble with pumped water. From an environmental perspective, WFRF are managed as temporary wetlands sustaining a rich biodiversity outside Kabukuri-numa. Research has shown that wet-rice agriculture is capable of providing habitats for diverse wetland fauna (Lawler 2001, Sprague 2004), and WFRF are thought to enhance the conservation value of rice paddies[44]. In addition, WFRF potentially contribute to conserving and purifying water. In WFRF, it has been observed that the algae grow and multiply rapidly. Winter-flooding nourishes the water table, while the entire paddy field filled with algae may function like a water purifying plant. Since the WFRF method minimises the need for intermittent flooding, it conserves water. Moreover, since it does not use agricultural chemicals, no pollutants and nutrients are released into rivers.

By means of promoting biodiversity in the fields, WFRF may increase the production capacity of rice fields. WFRF dramatically increase the number of tubifexes or tubificid worms (JAWGP 2004, 2005). WFRF replace chemical fertilizers, herbicides and pesticides by nurturing tubificid worms, chironomidae larva and other microorganisms that form the basis of food chains in the rice paddy ecosystem[45]. WFRF support insect pests' natural enemies such as frogs and spiders. Not only do tubificid worms contribute to pest management, they also play an important role in soil fertility and weed control[46]. Tubificid worms excrete

[44] WFRF offer breeding grounds for resident birds, feeding grounds for passage birds, and feeding and resting grounds for migratory birds. In other parts of Japan, there have been efforts to use WFRF to restore the population of resident birds such as ibises and storks. For example, Japanese crested ibises in Sado City, Niigata Prefecture, and white storks in Toyonaka City, Hyogo Prefecture, have been observed to feed on small fish and insects in WFRF.

[45] In addition, Ito (2006) offers a tentative explanation of the soil enrichment mechanism of WFRF from a pedological (soil science) perspective. It has been observed that paddy water of WFRF turns reddish. Ito speculates that this is caused partly by floating soil particles due to the activities of proliferating tubificid worms and other microorganisms in the soil and partly by the drastic increase in photosynthesis bacteria which have a red pigment. Depending on the oxygen amount in WFRF, photosynthesis bacteria may contribute to daytime nitrogen fixation, transforming atmospheric nitrogen into nutrient nitrogen.

[46] In the WFRF method currently practised in Shinpō District, poultry manure, rice bran and crushed soy beans are often used as organic fertilizers. Past research on WFRF in the Town of Tajiri reports that WFRF contained more nitrogen, phosphorus and potassium (three major nutrients in fertilizers) in soil than drained paddies did (Iwabuchi 2002). Whereas some are quick to conclude that bird droppings are turned into natural fertilizers, others are more careful about assuming a direct link between bird droppings and soil fertility. While waterfowl which forage in flooded fields leave droppings in the fields, it is unclear whether there is a threshold amount of droppings which has a significant impact on the amount of soil phosphorus. This needs further investigation, since many farmers are discouraged from adopting the technology of WFRF on the grounds that their rice fields are too far away from Kabukuri-numa (or Izunuma-Uchinuma) and birds would not come to their fields. With regard to weed control, Ōhata and Yamamoto's (1999) study on WFRF in Kaga City, Ishikawa Prefecture, reports on the weed control effects of winter flooding. Research on WFRF in the Sacramento Valley rice straw decomposition also confirms the weed control effects of foraging waterfowl in WFRF (Bird *et al.* 2000, van Groenigen *et al.* 2003). Ito (2006) suggests that turbid-reddish paddy water of WFRF may help prevent weed seeds from germinating by reducing the amount of light that reaches the soil's surface.

their faeces on the mud surface to be mixed with fungi, forming layers of nutrient-rich fine soil, called '*torotoro* (creamy) layers', which can sometimes push the weed seeds deep into the ground, preventing them from germinating. Furthermore, geese eat weeds, and ducks and swans eat weed seeds[47].

From an agronomic perspective, WFRF can make the dream of profitable organic farming with lower input come true. As fields are flooded without being tilled, it can decrease labour input in autumn. Rice producers may save money in production costs since the WFRF method cuts the costs not only of tilling but also of spraying pesticides and herbicides. By participating in the WFRF Project, farmers usually incur a loss in production by 20-30 per cent, but rice harvest from WFRF may sell at a much higher price than regular rice grown in more conventional methods with agricultural chemicals. Consumer demand for safe organic rice is expected to continue to be high, but there is a need to differentiate WFRF rice from regular organic rice which is becoming more common in the market. One of the keys to making WFRF a successful endeavour for sustainable development is to find a niche market where people are willing to pay a premium not only for organic produce but also for a 'vision' of sustainable society.

Success factors of social learning in Tajiri

In many parts of the world, waterfowl and rice farmers are longstanding competitors for lands that were once wetlands and have now been replaced with rice production. The Town of Tajiri made a departure from the old line of thinking that waterfowl and rice farming interests could not be harmonised. Today Tajiri may demonstrate a case of an optimal multi-sectoral use of wetlands. The 1997 decision to restore abandoned rice fields to a natural wetland contributed to turning Kabukuri-numa into a more stable habitat for wild geese, leading to a substantial increase in the number of migratory birds recorded at the site. This further led to winter-flooding of rice fields for the mutual benefit of waterfowl and organic farmers. In July 2004, the Town of Tajiri was designated by the Ministry of the Environment as one of thirteen model districts to promote eco-tourism in Japan, and tours are organised to watch large flocks of wild geese taking to the wing at dawn, soaring through the skies to their daytime feeding areas.

What factors fostered social learning towards a more sustainable future in Tajiri? While there is no denying that a series of coincidences and accidental discoveries led to the birth of the innovation of WFRF, the combination of bottom-up and

[47] *Aigamo*, a crossbreed between wild and domesticated ducks, eat insect pests as well as seeds and weed seedlings. An integrated organic rice-duck farming method known as the '*Aigamo* Method' uses *aigamo* ducks for pest and weed control in rice paddies (see Furuno 2001, Ho 1999).

top-down approaches, visionary leadership provided by environmentalists, and the development of trust between environmentalists and farmers were the key to refining WFRF as a collective solution for differing problems at stake. The space provided here does not permit discussion of personal and professional learning that took place on the part of pioneer practitioners of WFRF since 1998. In this section, I limit my discussion to some important factors that contributed to the collective practice of WFRF in the Town of Tajiri.

Roles of local municipal government

After Shiratori District was restored to a wetland in 1997, Kabukuri-numa, a small marsh virtually unknown in the community back then, became an unquestionably important habitat for wild geese in Japan. Having set the peaceful coexistence of migratory birds and agriculture as its goal, the Town of Tajiri played a major role in enabling collaboration for WFRF by providing the official mechanisms for farmer support. Given some farmers' passionate hatred of rice-eating '*kamoko*', the Town of Tajiri first needed to demonstrate that the protection of wild geese would not hurt local farmers. On 20 December 1999, Tajiri Town Assembly enacted an ordinance to compensate farmers for crop damage caused by waterfowl in Kabukuri-numa. This ordinance required Tajiri Township to offer financial compensation for crop damage that was not covered by other sources. The Town Mayor himself convened a committee to examine compensation eligibility and discuss measures to prevent waterfowl crop damage every year. Furthermore, in April 2004, the Town decided to start a direct payment system for rice-growers who converted to organic method and flooded their harvested fields[48]. In addition, Tajiri Township Government committed 5 million yen for improving WFRF cultivation techniques and carrying out rice paddy fauna surveys for three years between 2004 and 2007 (Takahashi 2004).

Prior to the designation of Kabukuri-numa and the surrounding rice fields as a Ramsar site, there was concern in the local community that the designation would harm local agriculture. A traditional line of thinking that wild geese and rice farmers cannot equitably share the Town of Tajiri is still persistent; many farmers feared that the designation would elevate the already high status of the white-fronted goose as a protected species and adversely affect their livelihood. Instead

[48] Under the Town of Tajiri's direct payment system, cooperating farmers receive 10,000 yen per 1/10 hectare of flooded paddy fields. The Town of Tajiri defines WFRF as rice fields that are flooded with a water depth of at least 5 centimetres over 60 days between November and the end of February and cultivated without agrichemicals and chemical fertilisers. Currently WFRF in Shinpō District are not necessarily combined with Iwasawa's (2003) "strictly no-till" method partly due to Shinpō farmers' lack of access to rice planters specifically designed for the hard, non-tilled land.

of denying the almost 'sacred' status of the goose, the Town of Tajiri decided to capitalise on it. This is not to suggest that the Town is dreaming of using eco-tourism as a lure for tourist yens. In fact, there is neither an observatory nor a visitor centre for bird watching. Rather, the Town is trying to sell the idea that this is a special place – not just one whose products are safe and organic but one that supports an annual pilgrimage of wild geese which is nearly 4,000 kilometres long. The Town is using WFRF to raise the value of local agriculture, as a way to reinvent the identity of the town and revitalize the rural community. The Town of Tajiri is even described as a town 'chosen' by wild geese in an information magazine published by the Regional Development Department of Miyagi Prefecture (Miyagi Prefecture 2004).

Creating a common vision: beyond the protection of wild geese

While JAWGP's visionary leadership was crucial in transforming local people's perception of the wild goose from a rice-eating pest into an indicator of environmental quality, scenic resources, and even potentially a saviour of rural communities, one of the keys to social learning in Tajiri was internal flexibility and change on the part of JAWGP itself. Over time JAWGP's mission came to include not only the well-being of migratory birds but also cultural and socio-economic perspectives that are needed for ensuring the well-being of rice growers. JAWGP's vision now resonates with the movement to revive *Satoyama*, "a traditional agricultural landscape or compound ecosystem" (Washitani 2001, p. 120). In June 2004, JAWGP submitted a proposal for creating "biodiversity-rich rural environments through WFRF" to the Japanese government as inputs to agricultural policy reform (JAWGP 2004). By promoting the WFRF scheme, JAWGP is campaigning for the reassessment of rice paddies not simply as farmland but also as a feeding ground for waterfowl and a man-made wetland that sustains rich biodiversity (see Figure 21.1) and supporting rice growers as deliverers of food and environmental services.

JAWGP made a critical contribution to creating a common vision of WFRF by disseminating the results of experimental undertakings of WFRF through its website, participation in academic conferences and other meetings all over Japan, and holding of a series of symposia, seminars and workshops on WFRF[49]. Not

[49] A series of 'WFRF Symposia', which were held five times between February 2002 and February 2006, achieved successful outcomes not only in terms of creating a platform for multi-stakeholder dialogue but also of turnout and local media coverage of these events. For example, the Fourth WFRF Symposium was held on 4-5 December 2005, and this two-day event attracted more than 500 participants. The Fifth WFRF Symposia was held at the Tajiri Town Ramsar Festival in February 2006. See /www.jgoose.jp/wfrf/ for detailed information.

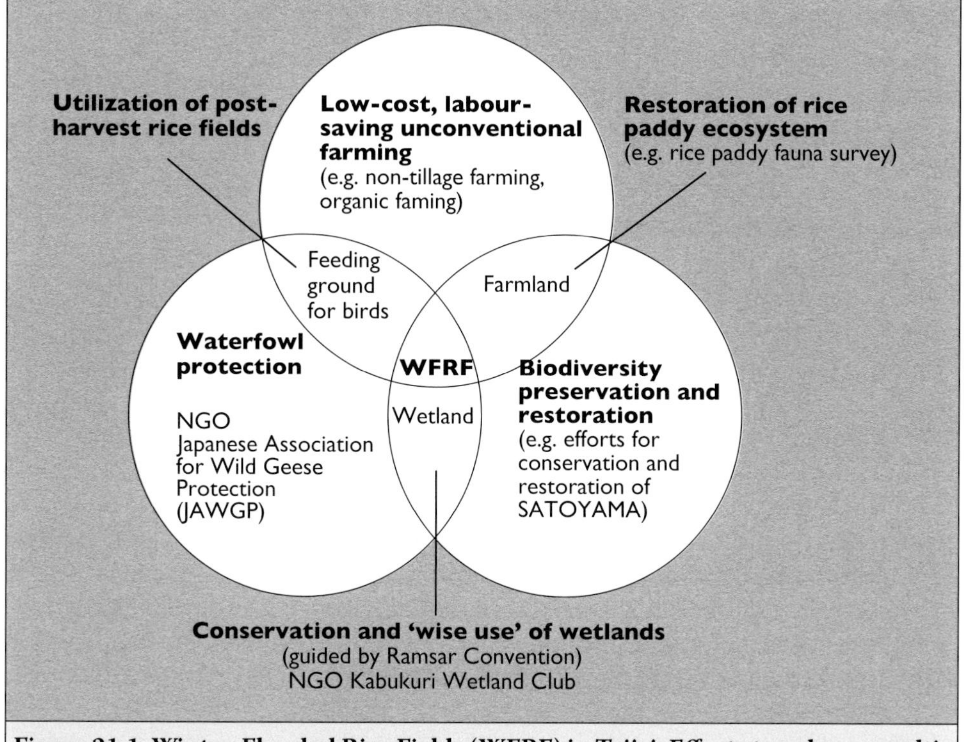

Figure 21.1. Winter-Flooded Rice Fields (WFRF) in Tajiri: Efforts to enhance multi-functionality of rice paddies.

only did these public forums play a significant role in bringing the relevant actors together and creating a space for exchanging information and identifying symbiotic relationships between waterfowl conservationists and social entrepreneurs working to save Japanese agriculture, they also contributed to raising the public profile of WFRF and thus persuading important groups and sectors – including state actors in environmental policy making, distinguished university professors, and the media – to throw their weight behind WFRF.

Enhancing the respectability and credibility of WFRF

Environmental activities in general are not welcomed by local farmers because they fear that such activities will regulate and restrict their agricultural practices. Moreover, WFRF combined with a non-tillage method runs directly counter to the current norm of well-drained post-harvest rice fields coupled with the conventional tilling method. Given the nature of WFRF, which challenges agricultural common sense, environmentalists have tired to legitimize WFRF as an 'authentic' practice by rationalising WFRF in scientific terms and referring to winter flooding practices from different places and different times. In addition to endeavouring to establish a model case of WFRF themselves, environmentalists working on WFRF were ready and eager to learn from "best practices" abroad and traditional agricultural practices.

Legitimising WFRF through research

One of the ongoing efforts to legitimise WFRF is to collect empirical data to validate claims about the benefits of WFRF in collaboration with research scientists. The long-term agronomic and ecological consequences of combining winter flooding and in-season flooding for rice cultivation are largely unknown, and interdisciplinary research is currently in progress to examine WFRF from various angles (see Figure 21.2). The research is designed to address the concrete regional challenges of protecting waterfowl, conserving wetlands, and practising sustainable agriculture. Based on the findings of this comprehensive research, agricultural and environmental education materials and programmes will be developed. In addition, the research aims at clarifying the multifunctionality of WFRF (see Figure 21.1) and developing a model of environmentally-friendly wet-rice agriculture.

Finding global allies

When a member of JAWGP visited Valencia, Spain, to attend the Ramsar COP 8 in November 2002, he noticed that acres of rice fields were flooded as far as he could see. He then learned that there is a traditional farming method to flood all post-harvest rice fields from 1 November until 31 January in Mediterranean Spain. He further learned that this method is called *Perellona* after the village of Perello, where the method originated, and that *Perellona* is practised in the Ramsar sites Ebro Delta, Catalonia, a major rice-growing area in the Mediterranean, and La Albufera, marine marshes in Valencia. While the original purpose of *Perellona* seems to have been to prevent soil salinization, it also serves to provide habitats for wild birds such as purple herons, black-winged stilts and flamingos. The Spanish Ornithological Society or Sociedad Española de Ornitología-SEO/Birdlife (Birdlife in Spain) initiated a European Union (EU)-funded LIFE project in the Ebro Delta in 1997 and demonstrated that organic farming could benefit farmers, wildlife and the environment (Ibàñez and Ripoll 2006)[50]. Today the organic rice grown in the Ebro Delta is sold through Riet Vell, S.A., a company established in May 2003 by SEO/Birdlife for producing and marketing organic rice, as well as online

[50] The original, longer text of Ibàñez and Ripoll (2006), single-authored by Carles Ibàñez, can be downloaded from www.livinglakes.org/images/ebrodelta.pdf.

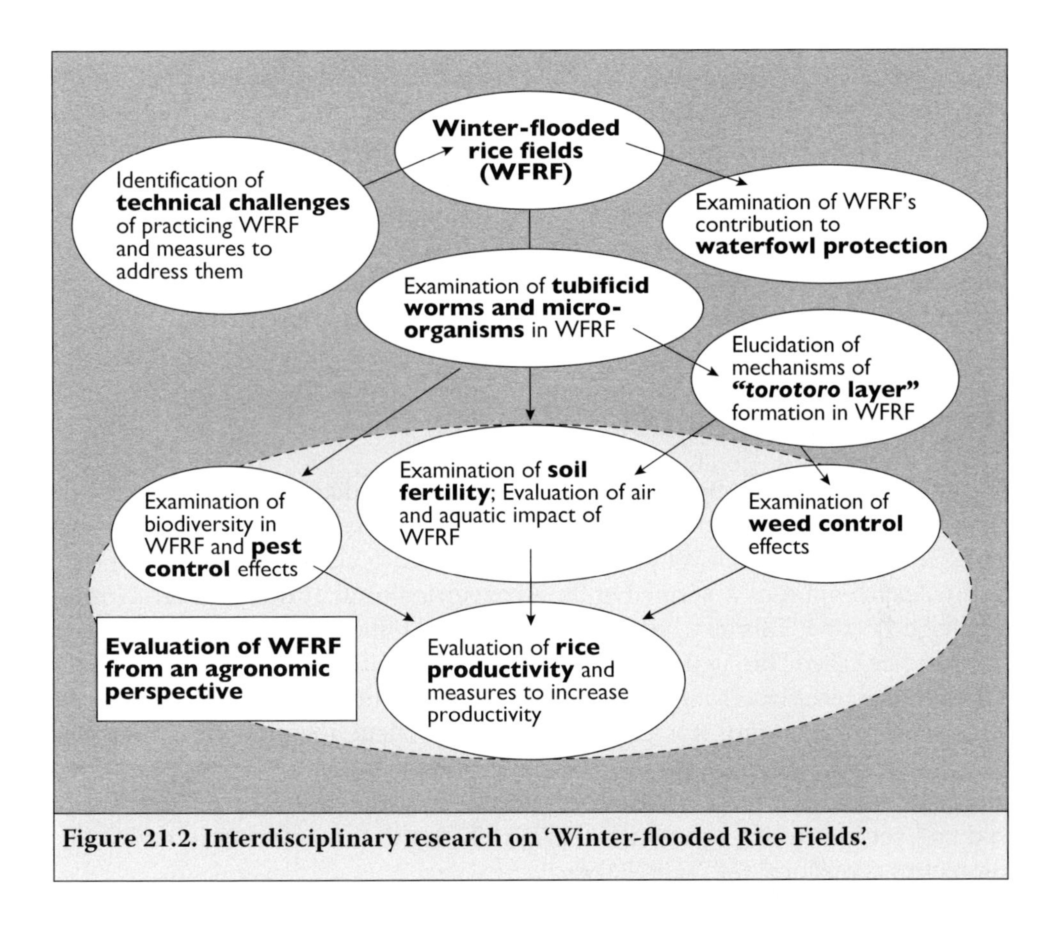

Figure 21.2. Interdisciplinary research on 'Winter-flooded Rice Fields'.

on the website of the Royal Society for the Protection of Birds-RSPB (Birdlife in the UK), Europe's largest wildlife conservation charity which has over one million members.

The knowledge of *Perellona* and the work of SEO/Birdlife gave the WFRF Project confidence that it was heading in the right direction. When JAWGP started experimenting with winter flooding in 1998, it was not aware of similar practices in other parts of Japan or abroad. By discovering similarities between WFRF in Tajiri and a farming method with 150 years of tradition in Spain, the Project embarked on a long-term process of transforming an experimental project that had started out as a localised, improvisational practice into a new regime of sustainable agriculture. Drawing on the Korean Government's direct payment system which was initiated in 2001 to give farmers an economic incentive for sustainable rice paddy agriculture, the WFRF community has paid a lot of attention to a direct payment system as a feasible solution for saving Japanese rice farmers within the constraints of WTO regulations. At the Ramsar COP 9 held in Uganda, JAWGP,

Japan Wetlands Action Network (JAWAN) and a Korean NGO co-organized a side event on rice paddies in the Asian Monsoon climate region. The Ramsar COP 10 will be held in Korea in 2008. Seizing this opportunity and building on the groundbreaking resolution on 'Agriculture, wetlands and water resource management' passed at the Ramsar COP8 (Resolution VIII.34), which called for "concerted efforts...to achieve a mutually beneficial balance between agriculture and the conservation and sustainable use of wetlands", the WFRF Project is planning to promote the idea of WFRF as a new model of sustainable wet-rice agriculture in Monsoon Asia.

Reviving indigenous knowledge

Fuyu-Mizu-Tanbo community does not assume that indigenous ways of knowing have less validity or epistemological sophistication than modern ways of knowing. JAWGP sees the origin of WFRF in the traditional agricultural practice of flooding rice fields in winter as described in the Aizu Agricultural Textbook (Aizu Nōsho) of the Edo Period. This text, written more than 300 hundred years ago, encourages the practice called 'Ta-Fuyu-Mizu' (which literally means 'Paddies-Winter-Water). One scholar speculates that 'Ta-Fuyu-Mizu' was practiced as a farming technique for over 200 years until it was abolished by the Meiji government (Haraikawa 2004). In the brochure on WFRF, JAWGP defines its mission in terms of reviving the traditional knowledge of 'Ta-Fuyu-Mizu': "Winter-flooded Rice Fields are an old and yet innovative agricultural skill. ... Today we are working to revive this agricultural wisdom" (JAWGP 2005).

Seeing a direct link between 'Ta-Fuyu-Mizu' of 300 years ago and today's WFRF, however, may be to overinterpret 'Ta-Fuyu-Mizu'. Whereas WFRF as practiced today have been originally proposed to make no-till farming more effective, the concept of 'no-till farming', let alone that of 'environmentally-sound agriculture', is not likely to have existed in the Edo Period. While it may be an interpretation reflecting the agenda of environmentalists to establish WFRF as a universal model, references to the traditional practice of 'Ta-Fuyu-Mizu' give WFRF legitimacy as indigenous knowledge.

Engendering a sense of ownership of the project

While environmentalists have perfected the logic behind promoting WFRF, their work has been far from limited to conceptual work or lobbying activities. The WFRF Project owes much of its success to a deep emotional relationship and a sense of ownership of the Project among participants. Environmentalists provided WFRF farmers with constant encouragement and reinforcement to continue with winter flooding and created ample opportunities for the farmers to interact with various experts and receive technical advice from them. Environmentalists and the town officials worked in the rice fields together with farmers. Environmentalists, farmers and local authorities built a history of working together, eating and drinking together, and discussing the future of rural communities in Japan. Main actors of the project interviewed stressed the importance of egalitarian personal relationships in building the WFRF Project. The recognition by both environmentalists and farmers of their reciprocal relationships was also crucial in fostering a sense of ownership of the WFRF Project. Environmentalists needed the cooperation of farmers to implement WFRF, and the farmers needed the help of the environmentalists to add value to the WFRF rice and sell it at a higher price.

In addition to the development of trust, the conscious creation of catchy phrases rooted in Japanese culture served to generate common feelings among the WFRF community and enabled the attribution of shared meanings and significance to the collective practice of WFRF. When JAWGP returned to Japan from the Ramsar COP 9 in 2002, inspired by 'Perellona', a need was felt to coin a term to describe WFRF in plain Japanese language. Back then, the Japanese word for WFRF '*Tōki Tansui Suiden*' was agricultural jargon and could not be easily understood by a lay person. The coined term 'Fuyu-Mizu-Tanbo' is similar to 'Ta-Fuyu-Mizu' and sounds like classical Japanese, yet it is easier to pronounce and helps a lay person to visualize WFRF. A Japanese song entitled 'The Theme of Fuyu-Mizu-Tanbo' was written and composed by a government official who is a self-designated Save-Rural-Village singer and has connections with the WFRF community, and the CD was released in 2006. The WFRF community now often uses the coined term '*Inochi Mandala*' ('Inochi' means living things in Japanese and 'Mandala' is a geometric design symbolizing the universe in Buddhism) to describe 'biodiversity'. In present-day Japan, where the merits and disadvantages of rapid economic development are strongly felt, creating a sustainable future is closely linked to creating an alternative future which resonates with the 'good old days' before Japan rose to become an economic superpower (Mochizuki 2005). Not only have these catchy phrases served to foster a sense of ownership among the participants of the WFRF project, they have also helped the project gain instant credibility among certain circles, including consumers who are willing to pay a premium for 'slow food' and the vision of a 'nostalgic future'.

Challenges and the way forward

While the case of Tajiri is an inspirational story of communal self-determination and empowerment, the hard part still lies ahead. Given the incompatibility between WFRF and the modern regime of intensive agriculture that has been promoted by agricultural administration and chemical and pharmaceutical agribusinesses for decades, participation in the WFRF project puts farmers in a difficult position in their own community. WFRF farmers constitute a tiny minority – less than 1 per cent – of 1,655 farming households in the Town of Tajiri (Takahashi 2004). Even in the model district of Shinpō District, only about 20 per cent of 58 farming households are practising WFRF. These 20 per cent of farmers drain their fields at the end of February in order to allow the majority of farmers practising a conventional method to till their lands using tractors. Draining water in early spring, when the temperature just starts rising, deters the *torotoro* layer formation and substantially reduces the alleged weed-control effects of WFRF, thereby making the WFRF method 1.5 to 2 times as labour intensive as a conventional method. Will the global agenda, science, and national policies help farmers overcome these challenges and continue to foster social learning based on partnerships between environmentalists and farmers?

Interactions of the global and the local: universalising ambitions and indigenising missions

It has been a while since international organisations came to emphasise the importance of the local and individual. The universalist model of development no longer has much currency. UNESCO states that a global vision of education for sustainable development "will find expression in varied socio-cultural contexts – where 'positive societal transformation' will be articulated in different ways" (UNESCO 2004, p. 23). The Fuyu-Mizu-Tanbo project may demonstrate an ideal case where local problems and issues of immediate, visible and tangible concern are tightly wired to global sustainability principles. While manifesting a romanticized wish to return to the pre-modern past, the Fuyu-Mizu-Tanbo project is also explicit about its ambitions to universalise WFRF as a model of sustainable wet-rice agriculture.

Aiming to revive 'Ta-Fuyu-Mizu' or 'Satoyama' does not mean 'going back' to subsistence farming before Japan came into contact with the 'West'. Rather, it means rejecting a unitary 'modern' condition that is unsustainable and developing an alternative model that can offer a counterpoint to economic globalisation processes and the WTO regime. Given that agricultural production accounted for only 1.2 per cent of GDP in 2003 (FPCJ 2005, p. 80), it is no longer a viable policy option for Japan to try to protect the 'narrow' interests of the agricultural sector through subsidies, quotas, tariffs and other forms of 'economic nationalism'. The Fuyu-Mizu-Tanbo community instead promotes a form of 'environmental nationalism' that allows environmentalists to idealize 'traditional' Japanese (and Asian) wet-rice agriculture as *the* sustainable (agri)culture and campaign for protecting rice farmers as deliverers of environmental services and guardians of aesthetically pleasing traditional rural landscapes – a symbol of Japanese culture. While such 'global-local', 'economy-culture' dichotomies may serve to promote sustainable agriculture in the Japanese context, it is important not to get trapped in the simplistic global-local binaries that can undermine international cooperation for the global agenda of sustainable development.

Roles of scientific research and 'expertise' in social learning

Whereas modern technologies of intensive agriculture are designed by scientists and transferred to farmers, the technology of WFRF was born out of human imagination and ingenuity and collaborative learning between farmers and environmentalists. There is no denying that WFRF are a positive outcome of social learning. As one of the informants put it, environmentalists and farmers who first experimented with WFRF were 'semi-professionals' who were trying to make a difference. The problems and solutions to them were defined by people living and working in Tajiri themselves, not by 'certified experts' or 'formally accredited scientists and technologists' (Collins and Evans 2002, p. 237). The Fuyu-Mizu-Tanbo project demonstrates a model case where 'pockets of expertise' of non-certified experts were combined with expertise of scientists for an optimal result (Collins and Evans 2002). Scientific research can reduce the risks associated with alternative production technology, and it will also serve to correct false beliefs about the benefits of WFRF and identify challenges to establish WFRF as a model of sustainable agriculture[51].

As the WFRF project becomes more than an isolated local initiative and its objectives become lofty, however, scientific research has deprived some pioneering WFRF practitioners of a sense of ownership of the project. With the involvement of research scientists, farmers have been put in a place where they are 'instructed' what to do, rather than able to act on their own. While researchers share their own experiences and knowledge with communities and add to the local knowledge base, they also have the power to discredit some of the results of social learning. Furthermore, the design of academic research – though interdisciplinary – reintroduced traditional disciplinary specialisation to the WFRF community (see Figure 21.2). Ideally, scientists are involved in the project to contribute to "technical decision making" in the public domain, not so much as authoritative scientists who have access to the 'Truth' but as 'experts' who help people make a

[51] For example, JAWGP (2005) suggested that WFRF contribute to preventing global warming. Indeed, in the case of California, straw decomposition in flooded fields has been found to be more environmentally friendly than straw burning, which emits carbon dioxide, or straw incorporation into soil, which produces methane (Wong 2003). The alleged role of WFRF in deterring the emission of greenhouse gases, however, is being seriously challenged by Ito's (2006) report that WFRF in Shinpō District actually emitted more methane than drained fields. Ito (2006) suggests that winter-flooding, when coupled with a non-tillage method, may deter methane emission, and emphasizes a need to develop a technology that would reduce methane emission levels.

'better' decision (Collins and Evans 2002). How the project unfolds in the coming years will offer a good case to reflect on the roles of non-certified and certified 'experts' in social learning for a more sustainable world.

Sustaining equitable learning partnerships

While the existence of 'dissenting' farmers who do not welcome the Ramsar designation is no secret in the media coverage and the environmentalists' narrative of WFRF, subtle tensions between environmentalists and 'cooperating' farmers are left largely unaddressed. The story is usually about visionary environmentalists and 'cooperating' farmers who have been 'enlightened' and 'empowered' by those visionaries, but it is not the case that the environmentalist position is automatically accepted by 'cooperating' farmers. For many environmentalists, agrichemicals and the many modern technologies which freed farmers from intensive, painstaking labour represent the very 'irrationality' of modern Japan, while such technologies do not necessarily represent irrationality for farmers, including those participating in the project. The recent WFRF discourse emphasises that WFRF bring back the 'joy' of farming and help farmers regain confidence in their profession, viewing modern technologies as having robbed farmers of what they should do rather than as having freed them from drudgery. This discourse, although meant to empower rural communities, runs the risk of rejecting present-day farmers' expertise in farming altogether, denying farmers' rights to benefit from the convenience of modern technologies and thus placing unreasonable demands on aged farmers. The majority of farmers in Japan – 57 per cent in 2004 – are aged 65 years or older (FPCJ 2005, p. 80), and they are not ready to go back to the old days of long hours of weeding.

When asked what would be their next goal, just a few weeks after the Ramsar designation, one of the 'cooperating' farmers interviewed replied, albeit jokingly: "A paradise of wild geese". Indeed, the Ramsar designation might well save the geese, but it is not likely to save rural communities from economic decline caused by youth out-migration, the aging of the farming population and a downward spiral of rice prices. Japan has no specific national legislation corresponding to the Ramsar Convention such as a comprehensive national wetland law, and there is no guarantee that, by the time the Town of Tajiri completes its official WFRF project in 2007, the Japanese government will have developed agri-environmental policies and measures for sustainable agriculture to support undertakings like WFRF. Environmentalists are highly aware of this problem, and both JAWGP and Kabukuri Wetlands Club point out the urgent need to develop official mechanisms for farmer support. There are also uncertainties about market development for organic rice production. In the face of these challenges, there is an increasing need for maintaining and enhancing "equitable learning partnerships between the combined expertise of communities, professions and governments" which enabled social learning for WFRF in the first place (Keen *et al.* 2005, p. 6).

References

Bird, J.A., Pettygrove, G.S. and Eadie, J.M. (2000) "The Impact of Waterfowl Foraging on the Decomposition of Rice Straw: Mutual Benefits for Rice Growers and Waterfowl", *Journal of Applied Ecology,* 37: 728–41.

Collins, H.M. and Evans, R. (2002) "The Third Wave of Science Studies: Studies of Expertise and Experience", *Social Studies of Science,* 32(2): 235-296.

FPCJ (Foreign Press Center Japan) (2005). Facts and Figures of Japan, http://www.fpcj.jp/e/mres/publication/ff/index.html

Furuno, T. (2001) *The Power of Duck: Integrated Rice and Duck Farming,* Tasmania, Australia: Tagari Publications of the Permaculture Institute.

Haraikawa, N. (2004) *Fuyu-Mizu-Tanbo no Rekishi (1): Aizu Nōsho no Ta-Fuyu-Mizu (The History of Fuyu-Mizu-Tanbo (1): Ta-Fuyu-Mizu in Aizu Agricultural Textbook),* http://park17.wakwak.com/~suhi25/03material/06history_tafuyumizu/01tafuyumizu1.htm

Ho, M. (1999) "One Bird – Ten Thousand Treasures: How the Duck in the Paddy Fields Can Feed the World", *The Ecologist,* 29(6): 339-340.

Ibàñez, C. and Ripoll, I. (2006). "Integrated Management in the Spa of the Ebro Delta: Implications of Rice Cultivation for Birds", keynote speech delivered at Tajiri Town Ramsar Festival, Tajiri, Miyagi Prefecture, Japan, 3-5 February, Proceeding of Tajiri Town Ramsar Festival, pp. 47-52.

Ito, T. (2006) "Dai-san Bunka-kai: Tanoshii Komedukuri to Ikimono no Nigiwai (Working Group 3: Fun Rice Farming and Rice Biodiversity)", proceedings of Tajiri Town Ramsar Festival, Tajiri, Miyagi Prefecture, Japan, 3-5 February 2006, pp. 92-94.

Iwabuchi, S. (2002) "Gan/Hakuchō-rui no Tōki Tansui Suiden to Kanden de no Riyō-hō to Kōdō no Hikaku (Case Study: Comparison of Waterfowl's Habitat Use of and Observed Activities in Winter-Flooded Rice Fields and Drained Rice Fields)", Paper presented at the Fourth Annual Meeting of Japan Ornithologist Group for ASNEAF (JOGA), Suiden Nōgyō to Gan/Kamo-rui: Tairitsu kara Kyōsei Sono Torigaku-teki Senryaku (Paddy Agriculture and Waterfowl: From Conflict to Coexistence, Ornithological Strategies), Nippon University, Tokyo, 14 September, http://www.jawgp.org/anet/jg007b.htm

Iwasawa, N. (2003) *Fukōki de Yomigaeru,* Tokyo: Soushin-sha.

JAWAN (Japan Wetlands Action Network) (1999) *Case Study: Kabukuri-numa. Wetland Restoration Project by Local NGO.* From JAWAN's Report to the 7th Conference of the Contracting Parties to the Ramsar Convention, Costa Rica, May 1999, http://www2.odn.ne.jp/kgwa/kabukuri/kabukuri_e/jawan_rp.htm

JAWGP (Japanese Association for Wild Geese Protection) (2004) "Ikimono Yutaka na Nōson Kankyō no Sōzō ni Mukete: Tōki Tansui Suiden, Miyagi kara no Teian", proposal submitted to the Ministry of Agriculture, Forestry and Fisheries, Japan on 21 June 2004.

JAWGP (2005) *Winter-flooded Rice Fields: Fuyumizutanbo, Environmentally Sound Rice Fields,* brochure of JAWGP, Miyagi, Japan: Tohoku Regional Nature Protection Center for the Ministry of the Environment of Japan.

Jefferies, R.L., Rockwell, R.F. and Abraham, K.F. (2003) "The Embarrassment of Riches: Agricultural Food Subsidies, High Goose Numbers, and Loss of Arctic Wetlands – A Continuing Saga", *Environmental Reviews,* 11: 193-232.

Keen, M., Brown, V.A. and Dyball, R. (2005). *Social Learning in Environmental Management: Towards a Sustainable Future*, Sterling, VA, USA: Earthscan.

Kurechi, M. (2004) "Mizu no Michi o Kari ga Wataru", keynote speech delivered at Forum: Minna de Tsunagu Mori, Kawa, Umi 2004, Iwate Prefectural University, Iwate, Japan, 25 July, http://www.pref.iwate.jp/~hp0208/02mizu/forum2004kichou.pdf

Lawler, S.P. (2001) "Rice Fields as Temporary Wetlands: A Review", *Israel Journal of Zoology*, 47: 513-528.

Ministry of Agriculture, Forestry and Fisheries, Japan (2005) "Nōrinsuisansho to Kankyōsho no Renkei ni Yoru 'Tanbo no Ikimono Chōsa 2004' no Kekka ni Tsuite (Results of the National 'Rice Paddy Animal Survey 2004' Collaboratively Implemented by the Ministry of Agriculture, Forestry and Fisheries and the Ministry of the Environment)", Press Release on 25 March, http://www.maff.go.jp/www/press/cont2/20050325press_3.htm

Miyagi Prefecture (Regional Development Department, Planning Division) (2004) "Gan ga Tobi Yume Hirogaru Machi", *Hustle*, 13, http://www.pref.miyagi.jo/tisin/hustle/hustle_13/feature13

Mochizuki, Y. (2005) Articulating a Global Vision in Local Terms: A Case Study of the Regional Centre of Expertise on Education for Sustainable Development in the Greater Sendai Area of Japan, Working Paper No. 139, United Nations University-Institute of Advanced Studies, Yokohama, Japan.

Ōhata, K. and Yamamoto, H. (1999) "Aretsudukete ita Kyūkō-den ni Tanbo Fukkatsu: Gan/Kamo wo Yobiyose Nōson Amenity wo Tsukuridasu (Restoring Paddy Fields in Desolate Idle Farmland: Bringing in Waterfowl and Creating Rural Amenity)", *Nogyō Gijutsu Taikei*, 8, Addendum 21.

Shimada, T. (2002) "Daily Activity Pattern and Habitat Use of Greater White-fronted Geese Wintering in Japan: Factors of the Population Increase", *Waterbirds*, 25(3): 371-377.

Sprague, D.S. (2004) "Research and Policy on Agri-Biodiversity", unpublished paper presented at the FAO International Symposium: Harmonious Coexistence of Agriculture and Biodiversity, United Nations University, Tokyo, 20 October.

Takahashi, N. (2004) "Miyagi-ken Tajiri-cho no Fuyu-Mizu-Tanbo (Tōki Tansui Suiden): Chiiki-zukuri to Know-how (Fuyu-Mizu-Tanbo in the Town of Tajiri, Miyagi Prefecture: Community Building and Its Know-how)", speech delivered at the Symposium Ikimono Chōsa no Igi to Tenkai Jōkyō: Shizen-Kyōsei-Gata Nōgyo/Chiiki o Motomete, Chiba, Japan, 21 November, http://www.jgoose.jp/tanbo/pdf/0411takahashi.pdf

UNESCO (2004) *Draft International Implementation Scheme for the United Nations Decade of Education for Sustainable Development (2005-2014)*, Paris: UNESCO.

Van Groenigen, J.W., Burns, E.G., Eadie, J.M., Horwath, W.R. and van Kessel, C. (2003) "Effects of Foraging Waterfowl in Winter Flooded Rice Fields on Weed Stress and Residue Decomposition", *Agriculture Ecosystems & Environment*, 95(1): 289-296.

Washitani, I. (2001) "Traditional Sustainable Ecosystem 'SATOYAMA' and Biodiversity Crisis in Japan: Conservation Ecological Perspective", *Global Environmental Research*, 5(2): 119-133.

Wong, A. (2003) "Comparative Emission of Methane from Different Rice Straw Management Practices in California: A Statewide Perspective", *Journal of Sustainable Agriculture*, 22: 79-91.

Chapter 22

Social learning in the STRAW project

Michael K. Stone and Zenobia Barlow

On a crisp January morning, Paul Martin is hosting a special group of visitors at his dairy in southern Sonoma County, California. The ten miles from Petaluma (a city of about 50,000) to Martin's ranch, and the ten from there to the coast, traverse rolling grassy hills, dotted with stands of oak, bay and buckeye. This is dairy and sheep-ranching country. These hills know only two colours, golden brown and green. Since it's winter, the land is emerald.

Past Two Rock Presbyterian Church, large cardboard 'STRAW' signs mark the ranch's driveway, next to an open structure sheltering 10- or 12-foot-high stacks of hay bales. Laurette Rogers, director of STRAW (Students and Teachers Restoring a Watershed), is standing in the driveway. "Listen to the meadowlarks!" she exclaims. "I don't recall ever seeing so many here" (unless otherwise noted, quotations of Laurette Rogers are from L. Rogers, January 2001, personal communication).

From where she's standing, Stemple Creek's route through the pastureland is easy to trace by the lines of willows, interspersed with oaks, extending several feet on either side of the creek. The foliage is high and thick at the east end of the property, where STRAW did its first planting in 1993. Farther west, where the students will be planting today, it thins considerably. (In the U.S., the term 'students' is used at all educational levels, from kindergarten through graduate school. Throughout this chapter, 'students' usually applies to elementary school children.) "When we came for our first planting", Rogers says, "I didn't realise that that was the creek. It looked more like a drainage ditch".

The day's workers, nine- and ten-year-olds from Lagunitas and Wade Thomas Schools, arrive. A line of sedans, station wagons, and SUVs, driven by parents, pulls in. About forty kids pile out and run to climb the hay bales. "Off, right now", yells Rogers. "We've been doing these projects for years without any injuries, and we're not going to have the first one today". Later she confides, "When I'm in the classroom, I'm very mellow. Out here, I get intense". Her carefulness is one reason Paul Martin trusts STRAW on his property.

Rogers directs the students' eyes to the lush growth in the original planting. "See those trees? The sprigs you're planting today will be that tall by the time you're in high school". The students pull calf-high wellies over their shoes, and line up for work gloves. They're divided into groups of four, each accompanied by a teacher or parent. Each team is issued a heavy digging bar, about six feet long and an inch in diameter, with one pointed end. After a final reminder, "Last chance to use the portable toilet", students, parents, and teachers trek across a muddy field to the creek. They're led by Boone Vale, a staffer from Prunuske Chatham, Inc., a design and construction firm that specialises in restoration and is overseeing today's restoration. Staff members from Prunuske Chatham and STRAW have already been out to the worksite, to lay temporary board bridges across the creek and double-check that Paul Martin's electric fences are turned off. On the other side of a barbed wire fence, a herd of Holsteins turns its full attention to the noisy newcomers.

The creek is three or four feet wide, a few inches deep, down two-foot embankments. The Prunuske Chatham workers have placed flags at the places they chose for planting the willows. Boone Vale shows the students how to use the digging bars, three or four people at a time, pounding them into the ground, wiggling them around, pounding again, until they've dug a narrow hole a couple of feet deep. He hands out three-foot-long willow sprigs, a half-inch in diameter, cut from trees on the property. He shows the students how to tell which end is 'up', how to plant them and tamp down the earth. Recent rains have left the ground soft, making digging and planting easier. The children invent songs and chants to accompany themselves as they take turns with the digging bars. They work for about ninety minutes, break for lunch, then get back to work. By the time they leave, they've planted more than 300 willow sprigs.

Students and Teachers Restoring a Watershed involves 3,000 students yearly in habitat restoration in the San Francisco Bay Area. The programme, the product of fourteen years of social learning work, depends on the cooperation of students, teachers, administrators, ranchers, for-profit businesses, philanthropic foundations, other nongovernmental organisations, and governmental agencies. Not all of the members of that network are natural allies; some of them are frequently inclined to regard each other with suspicion. Through a process that may be characterised as 'multi-stakeholder social learning', this nature conservation project has incorporated a disparate collection of participants, with diverse purposes, goals, and values, into a network working together for sustainability. This case also demonstrates ways that even young children can be helped to acquire attitudes and habits that will prepare them to engage in additional social learning processes in the future.

Our ideas about sustainability and social learning are strongly influenced by the work of physicist and systems thinker Fritjof Capra, who is a cofounder and chair of the board of directors of the Center for Ecoliteracy one of STRAW's earliest supporters, where we both serve. He writes,

> "The term 'sustainable' has recently been so overused, and so often misused, that it is important to state clearly how we understand it at the Center for Ecoliteracy .... A sustainable community is usually defined as 'one that is able to satisfy its needs and aspirations without diminishing the needs of future generations'. This is an important moral exhortation. It reminds us of our responsibility to pass on to our children and grandchildren a world with as many opportunities as the one we inherited. However, this definition does not tell us anything about how to build a sustainable community. We need an operational definition of ecological sustainability.
>
> The key to such an operational definition, and the good news for anyone committed to sustainability is the realisation that we do not need to invent sustainable communities from scratch. We can learn from societies that have sustained themselves for centuries. We can also model human societies after nature's ecosystems, which *are* sustainable communities of plants, animals, and micro-organisms. Since the outstanding characteristic of the biosphere is its inherent ability to sustain life, a sustainable human community must be designed in such a manner that its ways of life, technologies, and social institutions honor, support, and cooperate with nature's inherent ability to sustain life" (Capra 2005, p. xiii).

To do this, says Capra, we need to teach our children the 'fundamental facts of life': for example, "that matter cycles continually through the web of life; that the energy driving the ecological cycles flows from the sun; that diversity assures resilience; that one species' waste is another species' food; that life, from its beginning more than three billion years ago, did not take over the planet by combat, but by networking" (Capra 2004, p. 8).

These notions, especially diversity and networking, are also basic to our understanding of social learning, which we see as a process by which diverse individuals and organisations collaborate to define and act on issues of sustainability that jointly affect them. Because we, at the Center for Ecoliteracy, understand sustainability to be a property of networks, rather than of individuals, we recognise that solving problems in an enduring way requires that people addressing the problem develop networks of cooperation and conversation, even when they bring differing understandings and perspectives to those conversations.

We are also influenced by Capra's description of emergence, derived from recent work in the theory of living systems, as a process in which new forms occur at points of instability in social and environmental systems (Capra 2002). Instability, even conflict, can engender creativity where there is leadership capable of helping participants develop and maintain relationships, devise solutions, and create new structures. The project that became STRAW began with one fourth-grade class. The project was possible because the faculty at the school where that class was located had spent a year confronting internal differences and creating a plan for project-based whole-school reform. Then, out of what might have been a recipe for instability – nine-year-old 'environmentalists' attempting to save a species endangered by practices on economically marginal ranches – emerged an unpredicted structure, a network involving more and more parties which had not previously worked together. Through the course of this process, the children learnt and practiced some of the skills needed to initiate and maintain the conversations that make social learning possible.

The origins of STRAW

STRAW's origins lie in 1992 at Brookside School in suburban Marin County, about 25 miles north of San Francisco, where Laurette Rogers taught fourth grade (in the U.S. most fourth-graders are about nine years old).

She had showed her class a National Geographic film on rainforest destruction. "It was filled with haunting music and pictures of chain saws", recalls Aaron Mihaly, a fourth-grader in 1992 who graduated from Harvard University in 2005 (A. Mihaly, February 2001, personal communication). A depressing discussion about endangered species followed, until one student raised his hand. "But what can we do?" "I looked into his eyes", says Rogers, "and somehow I just couldn't give him a pat answer about letter writing and making donations".

Because the teachers and principal of Brookside School had recently made a commitment to environmental project-based learning, Rogers had the flexibility to propose to her class that they choose and design a project around which to organise lessons. The process that eventuated in this commitment to project-based learning required a year of sometimes-difficult debate and negotiation during which the members of the Brookside School faculty had to resolve significant differences. Like many other social learning processes, this one succeeded because its director, Brookside principal Sandy Neumann, was a strong leader with enough self-confidence to allow faculty members to express disagreements and to confront differences until they felt that they, and not an outside authority, were responsible for making the ultimate decisions.

Laurette Rogers consulted with Meryl Sundove, a trainer for a now-defunct California State Adopt-A-Species programme. Rogers suggested some criteria: she wanted the species to be local, and she wanted it to be obscure, to counter the bias toward beautiful and charismatic species being the most worth saving. Sundove suggested a trout, a salmon, and the California freshwater shrimp, *Syncaris pacifica* (about the size of a child's little finger), now found only in fifteen creeks in Marin, Sonoma, and Napa Counties. The students voted for the shrimp, but weren't that enthusiastic. Neither the students nor she expected to develop much affection for the shrimp, Rogers reports.

In retrospect, "the shrimp were perfect", says Aaron Mihaly. "We weren't joining someone else's campaign to save a distant cuddly animal. No one had ever heard of them, so we had to use our creativity to interest other people. They fit our image of ourselves...we were just a little fourth-grade class. If we didn't work on them, no one else was going to".

Meryl Sundove offered Rogers a key strategy: "Pick any species. Go into depth about its life. Find out all about it, and you'll fall in love with it". Out of that love, she said, and not out of a transitory feeling of obligation, would come the resolve to do whatever was necessary (including making compromises) to save the shrimp.

The class did fall in love. They found that the shrimp are beautiful, almost transparent creatures. The males are up to 1-1/2 inches long, the females up to 2-1/2 inches long, with rust-coloured spots. They've been in local creeks since the time of the dinosaurs (a fact the fourth-graders loved). They are the creeks' rubbish collectors, feeding on dead and decaying plant material. Because they are terrible swimmers, they must cling to riparian roots in order not to be washed away.

Rogers learnt an important lesson the first year. Many people assume that nine- and ten-year-olds are not good candidates for social learning projects because they are impatient, and need to see immediate payoffs. But her students worked for six months on the shrimp before they ever saw one. (When they did, "There was this big, 'Ahhh'. You'd think they had seen a movie star".) They kept focused even after learning it would probably take fifty to a hundred years for the restorations to have a significant impact on the shrimp's habitat. They talked about taking their grandchildren to see their work, and telling them, "We did that", she says.

Rogers adhered to another principle with applications for social learning: before entering potentially controversial territory, get your facts in order. She refused to predigest material for her students. She gave them original scientific papers on the shrimp; each fourth-grader was responsible for understanding and accurately reporting the most important information from one to two pages of a paper, including figuring out the scientific jargon. Students analysed the data for each of the fifteen creeks where the shrimp live. They worked in class two hours a week, but frequently put in more time on weekends or after finishing other lessons. Other classroom lessons kept coming back to the shrimp – shrimp drawings during art lessons; shrimp poems, songs, and fairy tales during language arts sessions.

The students learnt that the shrimp are threatened primarily because of habitat destruction around the streams where they live. Dairy, beef, and sheep ranches are the agricultural mainstays of west Marin and Sonoma Counties. In former years, agricultural agents advised dairy farmers to build their pastures near creeks to water their stock. Now, the students discovered, the shrimp habitats were pressured by the damming of creeks, petroleum and chemical runoff, manure in the water, and sedimentation from soil erosion caused by stock trampling the creek banks and grazing the foliage that could otherwise stabilise the soil. It wasn't just cows, though. It was also off-road vehicles, and dumping of rubbish, and damage by potato farmers. And it wasn't just shrimp that were affected. They turned out to be one strand of a web that includes trees, grasses, aquatic insects, songbirds, creeks, estuaries, and the entire San Francisco Bay. The students began to understand the 'shrimp problem' as a watershed problem.

They learnt that Native Americans used to eat the shrimp, which are now so rare that no one, including scientific researchers, can even touch one without a permit. They also saw how the story of 'their' shrimp was repeated over and over again for other endangered species. (The only other known Syncaris species, *Syncaris pasadenae*, became extinct when a football stadium, the Rose Bowl, was built over its entire habitat in the early 1920s.)

The class chose to focus on Stemple Creek, one of the most deteriorated, which flows from the hills of Petaluma, through about ten miles of cattle ranches, before passing into the Estero de San Antonio. They made presentations to meetings of the local Resource Conservation District and the Stemple Creek/Estero de San Antonio Watershed Program. They determined that the best way to help the shrimp was to exclude livestock from the creek and stabilise its banks by planting fast-growing willows, which would also provide root systems, on which the shrimp could anchor themselves, and shade to cool the water.

The students could, perhaps, have chosen a more confrontational strategy, perhaps demanding government action on the basis of laws such as the Endangered Species Act. Most ranchers are, of course, wary of government imposition on their operations. As in other instances of social learning, they are much more likely to agree to a project if they believe that their agreement came of their own volition rather than by government mandate. The students chose to go to ranchers with an offer to do the plantings themselves if permitted to do so. First, though, they needed to reframe ranchers' perceptions of the likely consequences of allowing them on their property.

Liza Prunuske, cofounder of the design and construction firm Prunuske Chatham, identified a potentially interested rancher, Paul Martin. He was concerned about erosion, and wanted to improve his pasturage, but he also remembered the Valley quail he had grown up with, and hoped to see them again on his land. However, he didn't know if he wanted a lot of fourth-graders running around on his property, and was concerned by the prospect of the students' inspiring environmentalists to descend on him and try to dictate how he could run his business. As he tells the story, "I wasn't sure what they were up to. Then Laurette told me that she had told her students to imagine what it would be like if someone came into your bedroom and said, 'From now on, you can't get anything out of your closet – none of the toys, clothes, or anything'. You can imagine the kids saying, 'But that's our property, what do you mean?' Then Laurette told the kids that's how unfair it would be if they went to the rancher and started telling him what to do. After I heard that story I knew it would be all right, and we started working together" (P. Martin, February 2001, personal communication).

Martin, now coordinator of environmental services for the Western United Dairymen, had another goal. Just as he had to adjust his assumptions about what environmentally motivated students might do, and had to be willing to take a risk, he wanted 'citified people' to know what his life was like. When the class came to his ranch, he brought out milk and ice cream, and reminded the students where they had come from. He helped them understand the economic pressures on family farmers, workdays that begin at 2:00 a.m., and why ranchers sometimes don't have the time or money for restoration work that they would like to do. "See that man?" he once asked a group of students who had come to a ranchers' meeting. "He'll be eating beans tonight. Five nights a week, that's all he can afford". "The ranchers have taught us so much", says Rogers. Recognising and honouring the ranchers as teachers, rather than as learners who needed to be 'taught' how to protect the shrimp on their property, helped create a dynamic which made mutual learning possible and gave the ranchers additional incentives to remain engaged in the process.

In March of 1993, the class did its first planting on the Martin ranch. Martin had already fenced off part of the creek, to keep the cattle from returning and undoing the work. The class planted willows and blackberries along the creek banks. "In our area, you get more bang for the buck with willows than anything else", says Rogers. "Students can see results. In four months, the sprigs they plant will have branches three to four feet long. In two years, they'll look like little trees. They stabilise the soil. They provide shade to cool the water and reduce evaporation. Birds nest in them, and bring in seeds of other trees like alders and oaks".

Students have returned to Stemple Creek every year since. The first plantings are now a tall, dense growth that blocks sight of the creek. Five years after the first plantings, the Valley quail, which Martin remembered from his childhood, came back. Songbirds are nesting in the trees. And, to everyone's surprise, California freshwater shrimp – which were not expected to re-establish themselves for decades – had migrated downstream by 1999 and begun clinging to the roots of willows planted by students six years earlier. It is many years too early to know whether the shrimp will establish long-term residence at the restored sites, multiply, and eventually be rescued from their endangered status, but the results from the first few years are encouraging.

Nested systems

The Brookside fourth-graders' project didn't end with doing a planting. The natural ecology of shrimp, cattle, willows, and streams overlapped with the social ecology. For instance, an important ecological process identified by Fritjof Capra is 'nested systems': classrooms exist within schools, which in turn are parts of school districts, overseen by state boards of education. They also reside within towns and cities, which are nested within counties, Congressional districts, states, nations, and ultimately within global economies and politics. The children's concern for the shrimp involved them in agricultural economics, politics, and conflict resolution. The fourth-graders wrote letters to government officials and testified at hearings before local government bodies and Congressional committee.

These experiences were sometimes discouraging. The students and their teacher had to try to negotiate a route through polarised situations. They arrived at one hearing about endangered species to find the 'environmentalists' on one side, wearing green shirts, opposed by ranchers on the other side, wearing blue hats. They attempted to reframe the discussion: when the environmentalists offered them green shirts, they said they would wear them only if they also wore the blue hats. They learnt that efforts at reframing are not always successful, though. Adults are not always civil (some of the adults booed comments by the nine-year-olds) and politicians are sometimes more interested in grandstanding for the audience than in seeking solutions to problems.

Rogers reports that some parents were unhappy when their children were confronted by unpleasant reactions or were reminded at a young age about environmental problems and the difficulties of resolving those problems. It helped to remember that the children had become involved originally because they couldn't avoid being aware of problems such as endangered species. Among the most important things students learnt, she writes, "were that life is complex and that understanding different perspectives is required for peaceful living. Life is not all black and white. It is not enough to say, 'OK, let's all be nice and help the shrimp'" (Rogers 1996, p. 30).

It's also important, she believes, to keep communication flowing, even when faced with disagreement. At one point, she invited Dennis Bowker, director of the Huichica Creek Land Stewardship and the Napa Resource Conservation District, to address the class. Bowker had recently won a National Wetlands Award from the Environmental Law Institute, which had cited his work in bringing business people and government regulators together in order to work jointly on issues. He told the students that a key to problem solving with people with different positions is "Never let the CAT (Confrontation, Accusation, Threat) out of the bag". He also gave them a motto, which could be taken as a social learning watchword: "There are to be no compromises, just elegant solutions" (quoted in Rogers 1996, p. 11).

The children put some of their lessons in diplomacy to work when they heard that the city of Santa Rosa wanted to build a wastewater dam at the headwaters of Stemple Creek. To understand this plan, they hoped to invite the engineer for the proposed dam to speak to the class. Rogers talked with him first, and was told that he was suspicious of an invitation from a schoolteacher from Marin County, which had a reputation for environmentalist leanings. He figured that the class had already made up their minds, and that meeting with them would be confrontational. One of the fourth-graders called him several times; he finally agreed to come, he said, because of the gracious, polite way she spoke to him on the telephone. By listening to him, the students realised that assuming the dam would be bad and they didn't want it was an oversimplification. The dam would put ten acres of old bay forest under water, but it would also allow residents to recycle their water, for the good of people and environment. They saw, once more, the need to reframe assumptions that there was a 'pro-environment' and an 'anti-environment' side to deciding whether to build the dam. The students also learnt, Rogers wrote later, "how to be polite and diplomatic and still ask the tough questions like 'Will the dam's water leaching into the creek affect the shrimp and other creek species in a negative way?'" (Rogers 1996, p. 27).

Learning from setbacks

The students addressed educational conferences, sold 'Shrimp Club' T-shirts, arranged media coverage, painted a gigantic mural featuring a six-foot-long shrimp at the local ferry terminal. They won Anheuser-Busch's 'A Pledge and a Promise' award as the environmental project of the year for 1993, and increased the $32,500 they had won from the prize into a total of $100,000 for shrimp protection, all of it raised by the students. In each of these activities, they learnt lessons from setbacks as well as successes. One aspect of social learning is accepting mistakes as opportunities for learning. While acknowledging her responsibility for her students' safety (for example, when using potentially dangerous tools), Laurette Rogers understood the importance of avoiding the temptation to intervene in order to protect them from making any mistakes. When a student forgot to notify the media about an event, no reporters showed up, and the event did not garner the expected publicity. The students needed to learn both that mistakes have consequences and that they can provide feedback for doing something better the next time.

Rogers estimates that 90 to 95 percent of the plantings since 1993 have survived. Some have not. In ecological terms, Fritjof Capra notes that the complex relationships among members of both natural and social communities are nonlinear, and that even well-intentioned change agents can never completely control systems change. These failures also became learning opportunities. In one instance, people trampled a planting in a location with public access. Rogers felt that she needed to share the bad news with the students, who had worked hard on the project. They were disappointed, but then they brainstormed, and decided to post signs on a near-by fence, asking users of the property to respect the fragile plantings. Another site was vandalized, to the great discouragement of the students. This setback called for a different remedy: redoing the restoration, then disguising it so that it would not be a temptation for mean-spirited vandals. But occasionally no satisfactory remedy could be found. One rancher persisted in allowing horses to run through the creek, destroying the restoration work. After many efforts to work with that rancher, the program was forced to discontinue its relationship with him, a reminder of the need sometimes to acknowledge failure and shift efforts.

Sustainability: a network property

The Shrimp Project gained an ally when the Center for Ecoliteracy, then a new foundation, became a sponsor. "The Shrimp Project was an ideal model of an integrated curriculum", says Fritjof Capra. "Lessons were organised around an issue kids were passionate about. They developed ecological values out of first-hand experience. They got excited about shrimp, which led them to learn about the problems caused by cows. They had to take into account the ranchers' ideas. To write letters to City Hall, they had to learn to spell well" (Unless otherwise noted, quotations of Fritjof Capra are from F. Capra, January 2001, personal communication).

The Shrimp Project continued, on one or two ranches a year, until 1998. By then, the strategy of beginning with one sympathetic rancher rather than trying to meet the concerns of all initially, paid off. The students demonstrated their trustworthiness.

The programme showed results by meeting ranchers' needs and desires. Projects like this are rare on private land. They couldn't happen without cooperation by ranchers, who offer access to their property and contribute their own labour. Ranchers bear the cost, or must find funding, for installing and maintaining fences to keep their herds away after plantings. "This is our land", Marin rancher Al Poncia says. "We want to maintain it too" (A. Poncia, February 2001, personal communication). Says Rogers, "You can't help the shrimp without helping the ranchers". More ranchers began approaching Prunuske Chatham, requesting students and projects. By then, Rogers had left Brookside School. Ruth Hicks, who had taken over the Shrimp Project, told her, "We need to expand this thing. We need to go to scale". The project became a network, an important moment in its ecological development and an affirmation of the premise that sustainability is a network property.

The Brookside students' networking paid off in unexpected ways. Grant Davis is executive director of The Bay Institute (TBI). The Institute was founded in 1981 to promote work from the then-novel perspective of seeing the entire Bay-Delta ecosystem (which covers 40 percent of California) as a single, independent watershed. TBI uses scientific research and advocacy on behalf of protecting that watershed. In 1998 TBI had just begun working with local schools. But five years earlier, when Davis was on Congresswoman Lynn Woolsey's staff, Shrimp Project students at Brookside School called to invite the Congresswoman and him to events. He remembered those calls – "How often do you get a call from a fourth-grader?" – and offered a base for expanding the Shrimp Project. The Center for Ecoliteracy stepped in with additional support. STRAW was born, initially as a joint project of The Bay Institute and the Center for Ecoliteracy. Laurette Rogers became its director.

The network began to attract other members. The Marin County Stormwater Pollution and Prevention Program (MCSTOPPP) is a joint effort of Marin County's cities, towns, and unincorporated areas to prevent stormwater pollution and enhance creek and wetlands quality. It now plays the same role in the STRAW network that Prunuske Chatham does for rural restoration: it serves as liaison with property 'owners' (e.g. parks, schools, and open space districts); plans projects; identifies and prepares sites; orients and oversees students; provides plant materials, equipment, and follow-up maintenance. Half of STRAW's projects are now urban.

Urban projects have the added advantage of close proximity to students' neighbourhoods and schools (students can walk to half of them from their classrooms). Liz Lewis, director of MCSTOPPP says, "It's important for students to see they're caring for their own neighbourhoods. They'll think twice next time about throwing rubbish in the storm drain. It's also important that they learn where their water comes from, that water doesn't magically get treated on its way to them, how it must be filtered, that it's habitat for native animals, and that the health of the creeks affects the health of the people living near them" (L. Lewis, February 2001, personal communication).

Teachers are critical links in the STRAW network. STRAW charges teachers nothing. It requires only a commitment to do a watershed project, attendance at 'Watershed Week' during the summer (on teachers' own time) and at two dinners and a culminating activity where participants present their projects. The Watershed Week and dinners are partly orientation and training, partly inspiration, partly chances for teachers to share with each other. "When we started the Center for Ecoliteracy", says Fritjof Capra, "we thought we would be helping teachers design educational curricula. We didn't realise that so much of our work would be building personal relationships among teachers". Rogers tells programme veterans, "Even if you already know how to do the programme, we want you there to help the others".

Maintaining these relationships is one way that the STRAW network addresses one of the issues facing any social learning project – sustaining participants' motivation. Respecting all the networks' members, including creating events for the purpose of demonstrating that respect, is another prime strategy. These events become a place to honour teachers for working above and beyond what their jobs require. Says Sandy Neumann, Laurette Rogers's former principal, and now a consultant to STRAW, "We find the most respectful place we can (e.g. a beautiful site on the edge of the Bay), we get the best food we can, we give the teachers lots of time to walk by the water, we ask them what they want" (S. Neumann, January 2001, personal communication).

She says the programme really works when it enters the culture of the school. Teachers come and go, but the principal provides continuity. It's very difficult for a teacher to take the risks that teaching in a different way requires unless she or he has a supportive principal. So STRAW also sponsors events to give its principals recognition and opportunities to share experiences.

STRAW requires one parent or teacher for every four students on a project. The 4:1 ratio is partly a safety precaution, but it also draws parents intimately into their children's education and reaffirms the importance of the projects in students' eyes. "Parent involvement is key", said Bill Bryant, the father of a Wade Thomas fifth-grader participating in the restoration at Paul Martin's ranch. "We talk with the kids on the way out and back. We participate in fieldwork with them. When it's time to fund-raise for the Parent-Teacher Association, we're already committed" (B. Bryant, January 2001, personal communication).

The network also relies on the expertise and advice of private consultants such as Prunuske Chatham and governmental agencies such as MCSTOPPP – not to replace social learning with pronouncements from 'Big Brother', but to supply good information on which participants can make their own decisions. From the start, one of Rogers's watchwords has been, "Use good science".

"We discovered one time that we were actually pulling out native grasses in order to plant willows", says Rogers. At the same time, the mantra of many environmentalists, 'native good, non-native bad', proved to be another assumption in need of reframing. "Sometimes the non-native blackberries are holding the bank together", says Jennifer Allen, Southern Sonoma County Resource Conservation District watershed coordinator. "Until you've stabilised the bank, you can't start pulling them out" (J. Allen, February 2001, personal communication). The right action also needs the right timing. "We were going to pull out a bunch of non-natives one year", says Rogers. "We called Melissa Pitkin, who was then the education coordinator at the Point Reyes Bird Observatory. She said, 'This is the wrong time. The birds are just starting to nest in them'".

Public agencies are also vitally linked to the network, especially agencies that encourage social learning by providing resources not accompanied by coercion. The Marin County and Southern Sonoma County Resource Conservation Districts (RCDs) are special districts of the state of California. Because they have no regulatory authority, participation by landowners is voluntary. The RCDs offer technical assistance with soil, water, vegetation, and wildlife conservation. They sit down with ranchers, often around a kitchen table, to ask, 'What do you see as problems?' A primary role is helping to secure funding from public and private sources for expenses such as fencing, water troughs, and cattle crossings. Grants often require matching funds and/or labour provided by the landowner; STRAW can sometimes count as part of the match.

Lessons from shrimp

The Shrimp Project, and its evolution into the STRAW network, yields a number of lessons that we believe can be useful to educators and practitioners of social learning for sustainability. Among them: Sustainability is a network phenomenon. New forms can emerge from points of instability, and even conflict, in social and environmental systems, when a vital network of conversation, free flow of information, and mutuality is maintained. Maintaining these networks requires an attitude of respect, and often reframing assumptions 'about either/or'; 'us versus them'; 'black and white'; 'we are the teachers, you are the learners' positions. Social learning can be facilitated by leaders with enough self-confidence and trust in the process to allow participants to express disagreements, confront differences, and take responsibility for their own decisions.

Hands-on work on issues of consequence can lead to openness to solutions that theoretical speculation would not predict. Where Big Brother intrusions are likely to be resisted, people can make good decisions when given latitude and respect.

Even young children can participate meaningfully in social learning. They are capable of committing to long-term projects without promises of immediate success. Even when they believe passionately about something, they are have the capacity to practice attitudes that will equip them as lifelong participants in social learning: "Never let the CAT (Confrontation, Accusation, Threat) out of the bag"; "There are to be no compromises, just elegant solutions".

Systems change cannot always be controlled. Willingness to take risks is a prerequisite for the emergence of new forms. "It's not enough to say 'OK, let's all be nice and help the shrimp'". Mistakes will be made and setbacks will occur, but they can be treated as feedback loops for improving future action.

"The Shrimp Project was like a pebble thrown into the water", Laurette Rogers writes at the end of *The California Shrimp Project*, her book on the project. "It did many things we did not know it would do. It touched many people we did not know it would touch" (Rogers 1996, p. 35). As for the students, then-fourth grader Megan summed up her work on the project, "I think this project changed everything we thought we could do. I always thought kids meant nothing. I really enjoyed doing this, it was fun and I felt like our class just knew exactly what to do. I feel that it did show me that kids can make a difference in the world, and we are not just little dots" (quoted in Rogers 1996, p. 32).

References

Capra, F. (2002) *The Hidden Connections*, New York: Random House.

Capra, F. (2004) "Landscapes of Learning", *Resurgence*, 226: 8–9.

Capra, F. (2005) "Preface: How Nature Sustains the Web of Life", in M.K. Stone and Z. Barlow, eds., *Ecological Literacy: Educating Our Children for a Sustainable World*, San Francisco: Sierra Club Books, pp. xiii–xv.

Rogers, L. (1996) *The California Shrimp Project: An Example of Environmental Project-Based Learning*, Berkeley, California: Heyday Books.

Chapter 23

Social learning in situations of competing claims on water use

Janice Jiggins, Niels Röling and Erik van Slobbe

Water resource management constitutes a resource dilemma

Water managers must be able to deal with competing claims on the use of water, i.e. they must be able to manage resource dilemmas, as well as the purely technical aspects of their work. Resource dilemmas arise when water is a *common pool resource* where use is conditioned by (1) *sub-tractability* – the use by one subtracts benefits from others; (2) *high transaction costs* incurred by excluding individuals or groups from using the resource; and (3) *high risk* of degrading water quality, including threats to: groundwater recharge, natural water purification, water retention capacity and stability of flow (Ostrom 1990, 2005). *Multiple stakeholders* make different claims on the resource, from recreational fishing, to abstraction, from use as drinking water to a medium for carrying off effluent (Steins 1999). There is *interdependence* in that stakeholders can realise their own objectives only through the actions of and in agreement with others. There is *controversy*: stakeholders hold strong but divergent values and perceptions about what is at stake. The issues around what is at stake arise from multiple causes and have multiple effects, with different expressions in space and time, and the irreducible value dimensions of 'the problem' cannot easily be measured or modelled; scientific data cannot resolve these (Funtowicz and Ravetz 1993). And there remains therefore an irreducible *uncertainty*: in complex situations, surprises are to be expected.

Such dilemmas ask for forms of governance that go beyond hierarchy and market. In the 'tragedy of the commons', Hardin (1968) had shown that, based on the assumptions of rational choice, it is in the interest of individuals to destroy common pool resources. Ostrom (1990) subsequently showed, however, that human communities everywhere throughout history have managed to maintain their common pool resources by creating institutions (interactive mechanisms, agreed rules, surveillance and sanctions) that limit access and off-take and regulate 'rational choice'.

Hydrological systems, such as the catchments and river basins in the European Union, are multiple-scale and multiple-use common pool resources that often span several countries. However, their ecological status is the outcome of the collective impact of the activities of millions of stakeholders acting locally in very diverse contexts. The challenge is to devise governance mechanisms for the sustainable management and use of water at multiple scales, including the largest scale, the catchment, which can include such systems as the Rhine or the Danube and their tributaries.

At present, the boundaries of existing governance structures, be they communities, townships, districts, provinces, countries, or indeed the Union itself, do not coincide with the boundaries of hydrological systems. It is a sign of our times, that governance of integral bundles of natural resources and ecological services has become a necessity (Röling 1994). The question is how to design governance systems that work for such complex bundles of natural resources and ecological services as river basins.

This chapter is about social learning as a response to the challenge to find more adequate forms of governance of water resources in the European context. *Social learning* is treated as an interactive process of shared, experiential learning, amplified by facilitated communication and dialogue across spatially and hierarchically differentiated scales of interaction (SLIM 2004a-c). It is based on the authors' involvement in a European six-country study, SLIM in support of the European Union's Water Framework Directive (WFD). The chapter presents the WFD as an attempt to manage complex resource dilemmas sustainably, examines the implications for governance mechanisms, and then focuses on the facilitation of social learning as an approach to the coordination of human behaviour that supplements more familiar forms of resource governance. The chapter ends by drawing out the implications for knowledge processes, and by offering some guidelines for social learning.

The Water Framework Directive

The Water Framework Directive (WFD) passed into European law in 2000 as Directive 2000/60/EC. It aims to provide a common, legally enforceable framework for the governance of Europe's surface and groundwater and protected areas under River Basin Management Plans (SLIM 2004a). While it codifies and simplifies much of the previous tangle of legislation, it also marks a departure from previous ways of thinking about water resources. In fact, one might say that it recognises for the first time that water management in European conditions has become the site of multiple resource dilemmas. The history of water management in every European country of course offers many examples of conflict, as different resource users have asserted competing claims on the same unit of water, and a rich variety of conflict resolution processes and water governance institutions have emerged in response.

However, it would be true to say that prior to the introduction of the WFD water resources in the modern era were managed largely for their use functions, for a variety of economic, utilitarian, and recreational end uses. Management regimes were informed most often, or most powerfully, by the expertise of specialist disciplines, such as hydrology, irrigation, and water engineering, and confined within rather narrowly demarcated sectors and interests.

The WFD's cognitive dimension (or, way of thinking about water) is radical in that it reifies the status of water as a living entity that has the force to affect the human habitat (Ollivier 2004). As a 'life milieu' it is recognised as a resource fundamental to the potential for life, including human life. Furthermore, water as an 'ecological agent' (in the form of river systems) is positioned as acting on terrestrial ecosystems, and thereby as capable of shaping and constraining human opportunity within the water catchments that construct the living landscape. The WFD's normative dimension thus requires that river systems should achieve water of 'good ecological status', within a given time frame (2015). This quality requirement provoked considerable debate during the negotiations leading up to the WFD and, as the WFD has been introduced into national legislation, it continues to be seen by some as a merely utilitarian obligation to make efforts to achieve such a status. However, the procedural dimension of the WFD demands adoption of indicators, standards, and monitoring capacity that can demonstrate improvement, and that are scientifically-based, unambiguous, and quantitative. The WFD also specifies explicit sanctions for failing to achieve the ecological quality standards formulated in the river basin management plans.

The difficulty for those trying to bring the intentions and provisions of the WFD into effect rests on the fact that the designation and boundaries of river systems and of catchments, and of their desirable states, are purely human choices. The definitions of ecological standards cannot be derived purely from science. The framework thus does not so much resolve the dilemmas, let alone remove them, as cast them in another form. In so doing, it brings new kinds of expertise (such as nature management and ecology) into management decisions, and brings conventional water authorities face to face with other organisations that for the first time have been given a formal stake under the WFD in the systemic management of water.

The additional requirement under the WFD for citizen involvement in implementation adds further complexities. Article 14 calls for "active involvement of all interested parties in the implementation of the Directive .... The success of this Directive relies on close cooperation and coherent action at Community, Member States and local level, as well as on information, consultation and involvement of the public, including users ... to ensure the participation of the general public ... in the establishment and updating of river basin management plans". This demands a new approach to water governance as well as to the knowledge processes that inform these activities.

The implications for resource governance

The renewed interest in water governance and knowledge management moves the focus away from purely technical matters and towards institutions – a 'messy' concept that often lacks precise definition and in common usage is frequently conflated with 'organisations'. A useful distinction, that we follow here, has been made between (Eaton *et al.* 2005):

- The *institutional environment* (framework): the set of constraints that conditions economic interaction (rules of the game); and
- *Institutional arrangements*: groups of individuals bound by some common purpose to achieve objectives (organisations, governance structures, or coordination mechanisms): state, market and network or community.

Table 23.1 illustrates these coordination mechanisms and the nature of the rules that govern them. These coordination mechanisms represent different ways of arranging human affairs. They also represent three internally coherent 'frames', or paradigms within which people think about ways by which people manage to engage in concerted action. Many observers have empirically validated the distinction between these three frames (e.g. Habermas 1984, Douglas and Ney 1998, Powell 1991, Bowles and Gintis 2002). *Hierarchy* emphasises instrumental rationality and power, and the state, policy, political will, etc, as the most important conditions for getting things done. The frame assumes that it takes some central authority to ensure that desired action is taken, be it through some project, measure, policy, or regulation. We are talking of 'the big stick'. *Market* is the dominant frame. It assumes that self-interested individuals pursuing their individual preferences will, collectively, lead to an optimal outcome for society (methodological individualism). It therefore emphasises market liberalisation as the panacea for society's ills. Given the domination of neo-liberal, if not neo-conservative economics, as the paradigm for designing the future of society, most of us find it difficult to think outside this box. The break-through made by new institutional economics is the realisation that markets are based on man-made institutions, such as money, banks, etc., that sometimes took centuries to develop and become the dominant institutions in modern society (North 1990). *Networks* follow a logic of their own. The third coordination mechanism draws attention to the fact that, in addition to power and competition among self-interested individuals, people are able to collaborate, engage in reciprocal agreements, and realise concerted action on the basis of shared learning, negotiation and other interactive processes. It is typical

Table 23.1. Coordination Mechanisms (based on Powell 1991, Röling 2002).

Descriptor	State, hierarchy	Market	Network, community
Operational dynamic	Law, regulation, hierarchy	Invisible hand	Interdependence, social learning, agreement
Intervention	Policy implementation	Liberating market forces	Process facilitation
Mechanisms	Cause-and-effect Social engineering Technical measures Force	Rational choice driven by individual preferences	Co-creation of knowledge Reciprocity Cooperation Negotiation
Outcomes	Directed action	Aggregation of individual choices	Concerted action
Human energy	Power	Financial capital	Social capital
Implications for resource governance	Expert determination of ecological imperatives Ecological imperatives translated into law	Internalisation of social & ecological costs shown to be possible within market competitive pricing regimes	Ecological interests become key stakeholders Ecological literacy becomes widespread Non-expert values & experience incorporated in ecological standards
Risks	Non-compliance	Continued externalisation of social & ecological costs	Non-equivalence of negotiated outcomes to ecological imperatives

of our society that we have continued to emphasise hierarchy and market and have tended to neglect the third coordination mechanism (Douglas and Ney 1998, Thompson *et al.* 1990).

Yet strategies based on the instrumentalism of hierarchical thinking and on the methodological individualism of market thinking, alone or in combination, do not hold much promise for effective governance of the kind of complex resource dilemmas that we are dealing with in this chapter. Such resource dilemmas often do require regulation ('big stick behind the door') that signals that the default position of 'no change' is no longer acceptable. Yet in the characteristic situations outlined at the beginning of this chapter, regulation alone typically gives rise to widespread non-compliance, and the related costs of surveillance, enforcement, legal procedures – all of which in turn can lead to bad publicity, political embarrassment, and even civil unrest. The introduction of appropriate incentives (e.g. 'right prices') has also proven useful but price mechanisms have not been able to deal with the problem of the externalisation of social and environmental costs in competitive market conditions. Neither of these two mechanisms recognise ecological entities as 'having agency'. This point is often ignored or dismissed as naïve by economists and engineers but it may be illustrated with reference to a rice scheme in northern Japan that was required to share water with a drinking water company. The development of the farming system and clever engineering eventually meant that the water leaving the scheme became cleaner than the water entering it and as a by-product the red dragon fly, a cultural symbol with deep resonance in the life of the community, returned to the rice fields. Would the technological and agronomic choices have been the same, if the research team had begun by aiming to restore a rice habitat favourable to the red dragon fly? (Jiggins 2002). An optimal mix for coordinating action requires paying attention to the third mechanism as well.

The strategic and operational aspects of the third mechanism are by no means self-evident. Interest in them has risen exponentially in recent years as the ecological predicament, driven by human activity and purposes, asserts itself ever more forcefully. The intransigence of poverty in developing countries has led to the emergence the world over of participatory approaches to environmental management (Pretty *et al.* 1995). Resource dilemmas have also given rise to the elaboration of adaptive management and social learning (e.g. Gunderson *et al.* 1995, Röling 2002), to the facilitation of platform processes (e.g. Röling and Jiggins 1998), and to multi-stakeholder approaches (Ramirez 2005). But in other arenas as well, the need for the third type of coordination mechanism has become evident. Interactive approaches similar to social learning, modelled on 'Communities of Practice (Wenger 1998), abound in business and public administration environments, as evidenced, for example, in the work of the members of MOPAN – Multi-Organisational Partnerships Alliances and Networks[52].

In the remainder of this chapter, we will focus on social learning as essential to effective governance of the complex resource dilemmas embodied in water basin management. The focus means that we take an operational stance in elaborating on what the 'third column' entails. We see this focus as complementary to ongoing work on polycentric governance (Ostrom 2005, p. 281), which is identifying

[52] More information is available on the MOPAN website: http://www.mop-a-net.

the advantages of decentralised local resource management regimes, in various combinations with expert and hierarchical coordination mechanisms.

The contribution of social learning to water resource management

On the basis of the SLIM research we have found it useful to think about learning as constituted in the experience and practice of a subject capable of communication and action. We also make a distinction between the 'single loop learning' that occurs when people modify their point of view on an issue and their practice, asking 'how can I do what I do in a different way?', and the 'double loop learning' that occurs when people ask 'why am I doing what I am doing, and how could it be done differently?' i.e. they modify the rationale underlying their perceptions and actions. It is the experience of the SLIM partners that double loop learning can be designed as a deliberately organised process to achieve social-scale effects by interaction among stakeholders.

Social learning thus can be seen as a transformative process in resource governance, which allows stakeholders to engage in concerted actions that lead to more sustainable resource use. By 'more sustainable resource use' we mean both that the stakeholders perceive the situation to have improved and also that this improvement can be objectively measured against standard indicators and threshold values. However, one of the outcomes of the kinds of social learning experiences documented by SLIM researchers is the realisation among stakeholders that perceptions of what constitutes improvement, and choices about which standards and values to privilege, are in the end the outcomes of institutional arrangements.

On the basis of the SLIM research (SLIM 2004b), the key elements may be described as usually arising from a history of crisis and conflict, marked by a convergence of understanding arising from the co-creation of knowledge, representing a change in relationships from individualism and competition to interdependence and collaboration, and based on the creation, on multiple scales, of social spaces and arenas for interaction and learning activities. The process is punctuated by the emergence of routines, procedures and institutions that can be re-stabilised in structural relationships, and facilitated by deliberate strategic interventions (such as creating the social spaces for interaction). The whole is placed in context by framework conditions and predicated upon the willingness of public authorities to transfer responsibility for achieving public aims to the area-based interaction of stakeholders.

Social learning as a governance mechanism is clearly not a universal panacea – it is apt for situations in which the resource dilemmas described at the start of the chapter are evident. Given proper facilitation, institutional support, and a conducive policy environment, it is best seen as a powerful complement to other governance mechanisms rather than as a stand-alone process. The SLIM research studies have documented its potential to achieve wide and significant impacts that become evident in material effects such as increased amount of groundwater saved, the improved ecological status of a salmon river, a higher potential for multifunctional land use, or a restored water retention capacity. These technical effects can be viewed as the outcome of transformations in understanding, behaviour, and relationships that lead to concerted actions.

What makes the learning process a 'social' process is the deliberate effort to facilitate concerted action among stakeholders and stakeholder organisations (SLIM 2004b). The process can be observed principally along three axes:

- the convergence of goals, criteria and knowledge, reflecting more accurate understanding of mutual expectations, and the building of relations of respect and trust;
- the process of co-creating the knowledge needed to understand the context of interdependence, the issues, and practices;
- a change in behaviours, norms and procedures arising from shared actions, such as empirical enquiries, physical experiments, joint fact-finding, and participatory analysis.

Concerted action is not brought about just by people talking together, by idealistic effort to develop common visions, or by expert planning – it requires facilitation (Box 23.1).

Facilitation has emerged as a key form of expert support in resource management, i.e. as a professional activity capable of promoting social learning processes among diverse individuals and organisations in a position of interdependence (SLIM 2004b). Five kinds of facilitation tools proved useful in the context of the SLIM research studies: (1) various participatory mapping and diagramming techniques; (2) information and communication technologies, such as cameras and satellite photography and Geographic Information Systems; (3) performance arts, as a means of engaging a wide range of actors in the enactment of the dilemmas experienced, and in the sharing of new 'stories' about how these might be solved; (4) metaphors, serving as a means to mutually question the language used by stakeholders and to explore the diversity of (spoken and unspoken) meanings; and (5) material objects that could serve as the focus of discovery of new options for action.

Box 23.1. Examples of facilitation tools to support social learning.

1. Systems diagrams were used by the U.K. SLIM team to facilitate interactions among the different organisational members of the steering group of the Tweed Forum, a body set up as a cross-border platform for the management of the Tweed river system.
2. Disposable cameras were handed out to stakeholders implicated in the interactions between nitrate pollution and landscape management in the Sierra de Conti, in the Marche region of Italy, to record landscape elements they desired, and those that they disliked. The processed snapshots were discussed and analysed at a stakeholder workshop, giving rise to a richer mutual recognition of the interdependence of activities that lead to the pollution, and creative ideas about ways forward.
3. SLIM researchers and project partners in the Netherlands held a 'mock parliamentary debate', facilitated by a media personality, on the water conservation project. The DVD of the event has been widely shown and used to promote understanding of the social learning approach.
4. Examples of metaphors used by the U.K. and French SLIM researchers to stimulate reflection included: 'rolling out', 'platforms', 'tools', and the language of the WFD. Metaphorical exploration also proved helpful in deepening mutual understanding among the SLIM country teams.

Figure 23.1 illustrates point (5). It shows a small weir, installed in a location chosen through dialogue among neighbouring farmers and water board staff, being manipulated by a farmer. The basic idea is to retain winter rainfall longer in field ditches, thus to reduce the need for drawing down groundwater through overhead sprinkler irrigation in the summer time. Because every field and farm is different, a farmer must learn what amount of water suits his own needs best but, because water *moves*, he needs to coordinate the location of his weir, and his management of it, with others. In the course of discussing optimal sites and best practices, and the effects of the weirs, farmers develop a different understanding of how water moves through a hydrological profile and its ecological effects at different scales, while water officers learn what water means to farmers and how they perceive the risks of surface flooding or summer drought.

Figure 23.1 shows an example of so-called 'socio-technical' objects that serve to define the 'boundary' of what is at stake (Carlile 2002). They function as a 'shared context' that 'sits in the middle' of the discourse among participants. The object serves over time to re-define stake-holding in ways that take account of the diverse views and experience of the participants, and to develop a new understanding of who could be considered a stakeholder. Other examples of 'socio-technical objects'

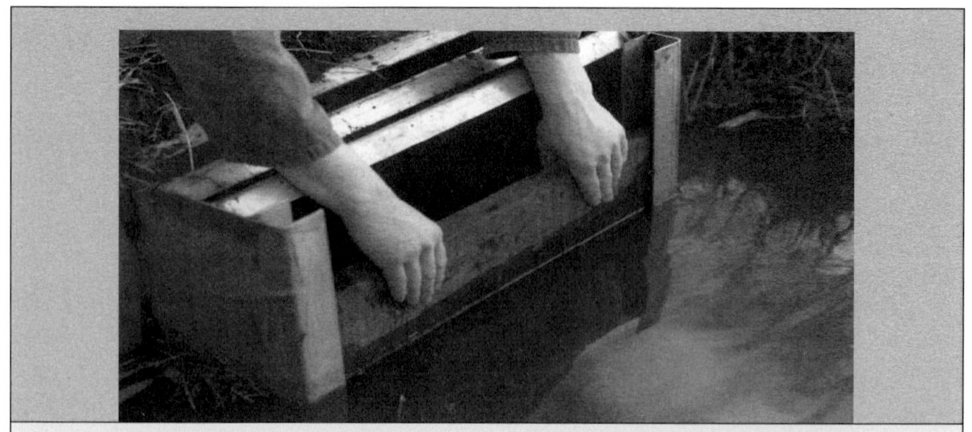

Figure 23.1. A farmer manipulating a weir in his field ditch (ZLTO 2003, Tilburg).

studied under the SLIM project included a local breed of cow, the Maraîchine, in the Atlantic marshlands of north-west France, and a catchment management plan developed for the Tweed river on the Scotland-England border.

Implications of social learning for knowledge processes

The implications of bringing social learning into play as a governance mechanism and as a way to move action and achievement forward in the presence of water resource dilemmas, are both bold and subtle. Three implications are addressed here: in terms of expertise; the co-creation of knowledge; and the institutionalisation of knowledge.

In terms of expertise, the WFD requires that individuals from hitherto largely separated disciplines be brought together in constructive dialogue – something easier to plan in the abstract than accomplish in reality. The process requires not merely the aggregation of hitherto separated information and data but the negotiation of different ways of thinking about the world, the transgression of intellectual boundaries, and measures to contain or resolve the claims to power of different traditions of thought and of the organisations within which these are embedded. The effort can be both threatening and intellectually rewarding for those concerned – but it also creates a new difficulty. For the danger is that, as the emerging expertise becomes stabilised in the indicators and standards required by the WFD, it comes into confrontation with the knowledge and expertise of the millions of water users and managers who are operating in widely distributed localities, and who are embedded in quite other institutions and ways of inter-acting with water (an effect documented by Steyaert (2004) also in the implementation of Natura 2000 legislation in France). If social learning is to build the potential to bring about the desired kinds of concerted action among networks of individuals situated in different ecological contexts, organisations, and hierarchical levels, then the question of who is considered knowledgeable comes sharply into focus.

The co-creation of knowledge becomes necessary wherever expert knowledge is inadequate to inform the ferociously place-dependent and time-specific decision making necessary in ecological management. Co-creating knowledge between experts and lay persons becomes possible on the basis of an understanding that knowledge is constituted in action – in the case of water, typically actions around the manipulation of real-world objects situated in a human activity relevant to the water issue at stake, rather than those of the laboratory, software programme, or experiment field.

Social learning, as described and instanced in this chapter, in theory leads institutions to embed the knowledge outcomes of the processes in which they participate in (at least) three institutional dimensions: organisational arrangements, norms and values, and behaviours. The presumption is that by participating in the creation of knowledge, such changes may be the more easily brought about. It is our experience that this does not happen automatically but can be brought about by careful design of the social spaces in which social learning occurs.

Box 23.2. Creating the social spaces for learning.

The project for Water Conservation in the Benelux Middle Area involved two Dutch and two Belgian provinces, two Dutch water board unions and two Belgian provincial water agencies, three farmers organisations, and a Belgian drinking water company. Each partner contributed members to a series of working committees, which had authority to commit the allocated budget but also the responsibility to ensure progress against agreed activities. The committees met every two months throughout the three years of the project, with the venue rotating amongst the partner organisations' offices in order to optimise the learning potential. Each partner also was involved in the various interactive learning activities conducted in the field. The outcomes of the learning processes and the co-creation of knowledge thus directly and progressively informed the committee's work and decisions.
The learning impact was optimised by the organisation of action-focussed meetings in natural settings (kitchen table, barns, farm visits), at times convenient to the diversity of participants. Others were organised for the wider public (open days, study tours, evening 'show and tell" events), and for a wider diversity of organisations (Open Space meeting, symposia).

»

Evaluations studies showed not only significant impacts in technical terms (installed capacity over 140,000 ha. for saving 4 million cu.m. water p.a.), and in adoption (3,500 farmers adopted one or more measures; 1,000 approx. began to develop water conservation plans). Other institutional changes included permanent arrangements for joint planning among the partners, new or more inclusive professional networks, development of capacity for and competence in facilitation among field level water managers, as well as significant and widespread changes in understanding of the complex interactions between ecological and hydrological processes and human activity. The organisational partners subsequently have negotiated two successor projects, which have brought nature management organisations also within their ambit.
Source: Jiggins and Röling 2004.

Conclusions

The need for social learning was first raised by ecologists (Gunderson *et al.* 1995) on the basis of their understanding of the incompatibility between the linear growth of economic systems and the cyclical nature of ecological processes. Social learning straddles the interface between natural science and social science, between the management of natural systems and the institutional and other human processes that allow people to engage in concerted action. Ecological sustainability can usefully be seen as an emergent property of stakeholder interaction, rather than merely as a technical property of ecosystems.

Conway (1994) suggests that policy in today's world needs to promote action in four dimensions: economic growth, equity, sustainability and stability. These dimensions are not additive; the mix that emerges in any context most commonly arises from processes of trade-off and negotiation. Social learning to ensure sustainable use of natural resources and ecological services represents a different kind of effort, in recognition that ecological goods and services do not lend themselves to management by the trading of interests.

The nature of social learning is 'ferociously local', given the place-dependency and time-specificity of ecological relationships, and it can be exquisitely sensitive to the outcomes of human interference in ecosystems. On the other hand, local human and ecological relationships are nested in higher scale interactions. Social learning processes must be able to span multiple scales, and thus must be designed to allow shared learning to occur in multiple social spaces. Social learning as a deliberate social choice and as a mechanism for resource governance presupposes a willingness to invest in interaction in multiple spaces and at multiple scales, as well as a readiness to set in place conducive framework conditions. These conditions may well call into play supportive market and regulatory measures.

More research is needed to understand the combinations of governance mechanisms that have been found to be useful in particular resource contexts. In the meantime we single out the following 'guidelines for social learning towards a more sustainable world':

- Investment in social learning requires investment in interaction. This means up-front funding, instead of paying once harm has been done for the costs of litigation over non-compliance with regulations and for remedial measures.
- Social learning brings individual and organisational actors at higher scales, whose decisions influence interactions at the local level, into a new relationship with local level actors.
- Social spaces need to be designed in which these encounters with the other can take place, and which form the *locus* of informed decision-making, responsibility for co-creating knowledge through action learning, and accountability for outcomes.
- Social learning processes benefit more from working around material objects than from spending endless hours on trying to develop shared visions in the abstract.
- Social learning is easier when it occurs in multiple spaces and multiple scales than when it is expected to take place in large commissions of officials who each represent their sector or discipline.

Acknowledgements

This chapter draws on research funded by the European Commission (DG Research, 5th Framework Programme for research and technological development 1998-2002), called Social Learning for the Integrated Management and Sustainable Use of Water at the Catchment Scale (SLIM), contract No EVKI-CT-2000-00064 SLIM. It draws especially on Policy Briefing No. 6 'The Role of Learning Processes in Integrated Catchment Management and the Sustainable Use of Water' (http://slim.open.ac.uk). The institutional partners were: INRA/SAD, France; Universitá delle Marche in Ancona, Italy; Swedish University of Agricultural Sciences, Uppsala; Wageningen University and ARCADIS in the Netherlands; and the Open University, U.K. The researchers worked with project partners on a total of 14 case studies located in England, Wales, Scotland, the Marche region of Italy, the Atlantic wetlands in France, the Benelux Middle Area in Belgium and the Netherlands, and two other sites in the Netherlands.

References

Bowles, S. and Gintis, H. (2002) "Social Capital and Community Governance", *Economic Journal*, 112 (483): F419-F437.

Carlile, P.R. (2002) "A pragmatic view of knowledge and boundaries: boundary objects in new products development", *Organisational Science*, 13(4): 442-455. http://cipd.mit.edu/documents/workingpaper_files/EEL_pragmaticview.pdf

Conway, G.R. (1994) "Sustainability in agricultural development: trade-offs between productivity, stability and equitability", *Journal for Farming Systems Research-Extension*, 4(2): 1-14.

Douglas, M. and Ney, S. (1998) "Missing Persons. A Critique of the Social Sciences", New York: Russell Sage.

Eaton, D., Meijerink, G. and de Jager, A. (2005) "Institutions and Agricultural Innovation in Sub-Saharan Africa", paper for international Workshop on Convergence of Sciences, held at Cape Coast (Ghana), Elmina Resort, The Hague: Landbouw Economisch Instituut, October 25-28, 2005.

Funtowicz, S.O. and Ravetz, J.R. (1993) "Science for the post-normal age" *Futures*, 25(7): 739-755.

Gunderson, L.H., Holling, C.S. and Light, S.S. (1995) "Barriers and Bridges to the Renewal of Ecosystems and Instituitions", New York: Colombia Press.

Habermas, J. (1984) *The Theory of Communicative Action. Vol. 1: Reason and the Rationalisation of Society*, Boston: Beacon Press.

Hardin, G. (1968) "The tragedy of the commons", *Science*, 162: 1243-1248.

Jiggins, J. (2002) "Participatory Approaches: Goals, Processes, and Organisation" in S. Sato, J. S. Caldwell, and A. Sato, ed., *Agricultural Innovation for Sustainability. Characteristics, Constraints and Potential Contributions of Participatory Approaches*, Association of Agriculture and Forestry Statistics, Tokyo, pp 70-81.

Jiggins, J. and Röling, N. (2004) "Water Conservation in the Benelux Middle Area", SLIM Case Study Monograph No. 2., http://slim.open.ac.uk. Research reports in English available at: http://waterconservering.nl

North, D.C. (1990) *Institutions, Institutional Change and Economic Performance*, New York: Cambridge University Press.

Ollivier, G. (2004) "An analytical understanding of the Water Framework Directive questioning its potential to enable sustainable management of water", SLIM Case Study Monograph No. 9. http://slim.open.ac.uk

Ostrom, E. (1990) *Governing the Commons. The Evolution of Institutions for Collective Action*, New York: Cambridge University Press.

Ostrom, E. (2005) *Understanding Institutional Diversity*, Princeton: Princeton University Press.

Powell, W. (1991) "Neither Market nor Hierarchy: Network Forms of Organisation", in G. Thompson, J. Frances, R. Levavcic and J. Mitchell, eds., *Markets and Hierarchies and Networks: The Co-ordination of Social Life*, London: Sage, pp 256-277.

Pretty, J, Guijt, I., Thompson, J. and Scoones, I. (1995) *A Trainer's Guide for Participatory Learning and Action*, London: IIED.

Ramirez, R. (2005) "Stakeholder Analysis and Conflict Management. An Overview", in *Concept, Ottowa*: IDRC, Books on Line. http://www.idrc.ca/en/ev-2791-201-1-do_TOPIC.html

Röling, N. (1994) "Platforms for decision making about eco-systems", in Fresco, L.O., Stroosnijder, L., Bouma, J. and Van Keulen, H., eds., *Future of the Land: Mobilising and Integrating Knowledge for Land Use Options,* Chicester: John Wiley and Sons, pp. 386-393.

Röling, N. (2002) "Beyond the Aggregation of Individual Preferences. Moving from multiple to distributed cognition in resource dilemmas", in Leeuwis, C. and R. Pyburn, eds., *Wheelbarrows Full of Frogs. Social Learning in Natural Resource Management,* Assen: Koninklijke Van Gorcum, pp. 25-28.

Röling, N. and Jiggins, J. (1998) "The ecological knowledge system", in N. Röling and A. Wagemakers, eds., *Facilitating Sustainable Agriculture. Participatory Learning and Adaptive Management in Times of Environmental Uncertainty,* Cambridge: Cambridge University Press, pp. 283-307.

SLIM (2004a) Ecological Constraints in Sustainable Management of Natural Resources, *Policy Briefing No. 1,* May. http://slim.open.ac.uk

SLIM (2004b) Facilitation in Policy Processes: Developing New Professional Skills, *Policy Briefing No. 4,* May. http://slim.open.ac.uk

SLIM (2004c) The Role of Learning Processes in Integrated Catchment Management and the Sustainable Use of Water, *Policy Briefing No. 6,* May. http://slim.open.ac.uk

Steins, N.A. (1999) "All Hands on Deck. An Interactive Perspective on Complex Common-Pool Resource Management Base on Case Studies in Coastal Waters of the Isle of Wight (UK), Connemara (Ireland) and the Dutch Wadden Sea", Published Doctoral Dissertation, Wageningen University, the Netherlands.

Steyaert, P. (2004) "Natura 2000: from consultation to concerted action for natural resource management in the Atlantic coastal wetlands", SLIM Case Study Monograph No. 7. http://slim.open.ac.uk

Thompson, M., Ellis, R. and Wildavsky, A. (1990) *Cultural Theory,* Boulder (Co.): Westview.

Wenger, E. (1998) *Communities of Practice: Learning, Meaning and Identity,* Cambridge: Cambridge University Press.

Chapter 24

Exploring learning interactions arising in school-in-community contexts of socio-ecological risk

Rob O'Donoghue, Heila Lotz-Sisitka, Robert Asafo-Adjei, Lutho Kota and Nosipho Hanisi

Introduction

Today, few educators would dispute that learning arises in diverse socio-cultural contexts of meaning-making interaction. As such, learning can strengthen social relationships across school and community and has the potential to develop as reflexive praxis in response to environment and health risks in a local context. These processes of 'social learning' have recently appeared as a new 'category' for thinking about human meaning-making interactions.

It is difficult to conceive of any human learning interactions that are not social processes of engaged meaning making either by learners as social agents in context or from the point of view of what is learned relating to social life in a world of interdependent living-things. Given the complexity of contemporary sustainability questions and an arising ambivalence in modernist notions of knowledge transfer, we note how educators are usefully using this somewhat ambivalent category for probing socio-cultural perspectives on how we see and approach learning interactions for environment and sustainability education. In foregrounding a critical perspective, we signal a cautious approach to a popularising of the term 'social learning' as a 'renaming' that provides a more coherent perspective for research and reflection on social processes of meaning making and change.

The chapter reports on three case studies of meaning-making interaction around indigenous environmental knowledge in South African school curriculum settings. The perspective on learning in our schools is currently more open to a plurality of ways of knowing and to an engagement with questions of environment and sustainability. The cases each report social interactions around intergenerational ways of knowing in local community and school curriculum settings, situated processes of reflexive learning interaction around tensions, discontinuities and risk in local context.

Here, the move away from a curriculum conveying modernity's definitive categories of apparent certainty through communication / knowledge transfer to create awareness of environmental concerns, is reflected in a recognition that learning arises amidst continuities, tensions and risks in the 'real' fabric of social life. Bauman (1991) and Harraway (1997) have both commented on the prevalence of ambivalence / uncertainties in modern society, and in fact Bauman (2000) has theorized that modernity was a project to reduce ambivalence, but has consistently failed to do so. Beck (1992, 1999, Beck and Beck-Gernsheim 2002) who has theorized discontinuities of emergent risk in his Risk Society, argues that the associated axes of tension involve both knowledge and unawareness (that which is known, and that which is not known and cannot be known, remaining, so to speak, 'under the radar screen'). These perspectives have implications for theories of education to foster change through awareness-raising induction into the cultural capital (social orientation and knowledge) necessary for better lifestyle choices and more sustainable livelihoods in local contexts of risk.

As environment and sustainability perspectives have become increasingly pluralist, recognizing socio-cultural worldviews as diverse and different, even to the individual level, the emerging social constructivist notions of human learning have not provided an adequate account of either a robust and enduring social habitus or limits of reflexive human agency. In her recent work on structure and agency, Archer (1996, 2002) argues that social change is morphogenetic in nature. She theorises these processes of social change ontologically (being in the world), noting a primacy of practice in shared meaning-making interactions. She argues that all social action is contingent upon socio-cultural, historical and structural conditioning (not all of which is rationally knowable), and that social action thus arises within socio-historical and socio-cultural context, in agential actions that are both deliberative and reflexive. Would such a perspective be labelled 'social learning', or quite simply human learning interactions in a socio-cultural setting? In her view social action is not reducible to the socio-historical / cultural, and it is through critical meaning-making within developing socio-cultural context (reflexive engagement) that 'new learning' and change arises, leading to structural elaborations and cultural changes (note that Archer does not elide structure and culture).

Drawing on these orientating perspectives, we explore the mobilizing of indigenous knowledge as 'social' learning processes. In so doing we attempt to take account of habitus, emergent socio-cultural perspective, risk and reflexivity to begin to contemplate learning interactions towards a more sustainable socio-ecological orientation in a developing context.

School-community partnerships have recently been identified as significant in enhancing teacher agency for changing pedagogical practices and in improving school environments in response to environmental issues and risks (Lotz-Sisitka

2005, Lotz-Sisitka and Timmermans 2005). Here we seek to probe these widening concerns through examining some of the findings of teachers working with a plurality of indigenous ways of knowing and learning area (subject) propositions in school-community interactions around questions of social justice, environmental health and sustainable living. We do this by asking the following questions:

- How do educational responses to risk shape moral imperatives and ethical orientation in social contexts? Here we seek greater insight into the engagement of environment and sustainability concerns in local contexts.
- What local deliberative processes arise in response to risk? Here we probe learning interactions in local socio-historical / cultural contexts.
- What steering choices and processes of change emerge in situated learning interactions such as this? Here we probe reflexive agency and lifestyle choices.

These questions are examined across small-scale case studies of local work with indigenous knowledge in three contexts of the Eastern Cape Province in South Africa. Each case was developed by a teacher-researcher working with learners on a shared imperative to enhance local learning interactions towards a more meaningful and reflexive engagement with the socio-ecological and human livelihood concerns. Here, the interest that we had in common was more meaningful and effective engagement with risk in socio-historical context.

The subtle but significant shift from environmental education imperatives that set out to create awareness and to foster change in others, to a concern for local engagement in learning interactions around emergent concerns ('social learning'), has accompanied our work to illuminate some of the intricacies in reflexive processes of social change. The contextual praxis (interplay between practice and theory) in each case, similarly provided the teacher researchers and those involved with insights into the context of their activities. Here the interplay of shared history and context reflects an intermeshing of:

- Engagement with socio-historical context (who / where).
- Emergent local imperatives (why).
- Deliberative research / learning activities (what / how).
- Reflexive consideration of possible change (for what).

These emergent and intermeshed processes are used to report the three discrete cases. Our purpose is to enable readers, including ourselves, to clarify 'social processes of reflexive learning interaction' at the school-in-community interface.

Case 1: from imifino to umfuno

Socio-historical context

Case 1 is the work of Robert Asafo Adjei (2004) a teacher in a rural area, in a small apartheid-created township high school, near Queenstown in the Eastern Cape. Robert is responsible for Agricultural Science.

The township and its local area are on marginal land, and the area is in a rain shadow belt that is prone to drought. Pressure on the land is high, due to large numbers of livestock being kept without adequate access to pastures, leading to land degradation and a loss of agricultural productivity. The community is primarily dependent on government pension grants which are provided to the elderly (US$100 per month) and child support grants provided for young mothers without work (US$25 per month), and poverty levels are high with families of up to 10 people living off one or two such grants. The area has a history of colonial intrusion, and through separatist development policies of the apartheid state, it was proclaimed a 'black homeland'. Learners have been disenfranchised by Bantu Education policies which led to poor quality educational provision in most black schools in South Africa.

Learners at Robert's school are hard pressed to find work after completing their schooling, and Robert's concern was for a lack of local 'relevance' in the Agricultural Science curriculum that he teaches, as his observations indicate that learners are not able to 'use' what they learn in school to sustain local livelihoods, or to find work or participate in agricultural activities in the surrounding farming community.

The Agricultural Science syllabus that he has been working with for the past 20 years at this school is 'outdated'. In 2005 the national Department of Education devised a new National Curriculum Statement, and has defined an outcomes-based approach to Agricultural Science which requires engagement with indigenous knowledge concerns, a stronger praxis focus and greater learner participation. In response to the socio-cultural and socio-ecological context outlined above, Robert chose to research the processes of mobilizing indigenous knowledge in the context of a changing Agricultural Science curriculum. To do this, he identified imifino (the 'looking after' of wild vegetable plants in a garden), as being a potentially significant Agricultural Science activity. Traditionally, the vegetable garden at the school has only been concerned with *umfuno* (vegetables that are planted and cultivated).

Risk and the moral imperative

In this case Robert recognized the risks in the school-community, constituted by the history of marginalization and colonial intrusion, marginal land, regular drought, poverty, and increased pressure on the land with resulting land degradation. Poor agricultural practices were also identified as a factor increasing risk. The quality of education provided for the learners, and their ability to participate meaningfully in society after leaving school was also a key risk factor identified by Robert in his contextual profiling work. The outdated nature of the Agricultural Science curriculum, and its pedagogy were also seen as contributing to the 'state of play' evident in socio-cultural context of the school-community. Robert's feeling 'for' the learners motivated his work to establish a more relevant curriculum, which would be responsive to school-community context, and in particular to food security questions, as *imifino* has a high nutritional value, and Robert wanted to encourage his learners to care for wild edible plants, and encourage their use in a community context where food security and health was a key concern.

Educative deliberations amongst learners, community and teacher

In his work, Robert firstly asked learners to identify those indigenous vegetable plants that they were familiar with. He then discussed the range of plants identified by the learners, and asked them to decide on a few plants that were most widely used in the community. These were *utyuthu (Amaranthus spp.), imbikicane (Chenopodium album)* and *ihlaba (Sonchus olearaceus).* Following this, he worked with the learners to interview community members to find out what was known about the three plants. A synthesis of these insights helped the learners to plan how to 'look after' the plants in the school vegetable garden, and they engaged in practical experiments and observations over a six-month period.

During this work, Robert and the learners continued to discuss the plants with various elderly community members, finding out what was known about the plants and how they were being, and had been used in the community. Findings from this work indicated that there were various social myths about the plants. For example male members of the community did not want to be seen to eat *imifino* as traditionally it was said to reduce their strength. Despite this, they found that some members of the community were eating the plants, in response to food security issues. Few community members were aware of the 'scientific' information on the nutritional value of the plants. They also discovered that modernization had influenced the use of the plants as a food source in the school community, as people perceived *imifino* to be wild plants only eaten by those who were too poor to buy conventional vegetables at the market, signifying the influence of cultural change on social belief systems.

Resulting steering choices and change

This study illustrated that Agricultural Science, when mobilizing indigenous knowledge in school-community contexts, cannot restrict itself to conventional disciplinary boundaries. If Robert and his learners had done this, they would simply have focused on the planting experiments in the school garden. Through deeper socio-cultural engagements, they were able to establish that changing *imifino* (wild vegetables) to *umfuno* (cultivated or 'looked after' vegetables) would require socio-cultural understandings *as well as* scientific learning (i.e. the technical aspects of growing food). Resulting from this school-community engagement, Robert has now defined a curriculum module on *'From imifino to umfuno'* which engages learners in a broader understanding of local agricultural knowledge. This work has been published by Robert in a national textbook for schools, and may, in future, be utilized by more schools as the South African curriculum changes begin to play out in other socio-cultural contexts.

Case 2: Umqombothi life sciences and lifestyle choices

Socio-historical context

Case 2 is the work of Nosipho Hanisi (2006), a science teacher in Grahamstown. The area has a history of early trek Boer and British colonial intrusion (1820 Settlers) and the separate development policies of the apartheid state. The legacy of Bantu Education policies associated with this is apparent in low literacy and large class sizes with poor facilities and provisioning in most township schools.

There has been a recent period of rapid urban expansion in this small university town in the Eastern Cape. The urbanization coincided with increasing freedom of movement that accompanying the demise of the pass laws of the apartheid state. A far more significant process, however, was rapid change and a loss of livelihood on commercial farmlands. Here most of the forced migrants were farm workers displaced by a shift from agriculture to game farming and tourism that accompanied the expansion of the international tourist market as the affluent sought sunshine, wildlife and wild places for recreation in Africa.

Over decades of working on commercial farms and migrations from rural areas to urban townships, many people of Xhosa and Fingo (amaHlubi) ancestry have lost touch with their rural origins and the cultural practices associated with rural communal life. Many of the youth in particular still see themselves as Xhosa but of the Eastern Cape and Makana (Grahamstown) without strong links of identity with the rural homestead areas of their ancestors.

Today up to seventy percent of people living in the townships and squatter settlements in Grahamstown are unemployed and a high proportion are not actively seeking employment as there are no jobs and their background and skills are in farm labour. The unemployed in the township community are primarily dependent on government old age pension grants (US$100 per month) and child support grants provided for young mothers without work (US$25 per month). Poverty levels are high with one employed person supporting 6-8 others and families of up to 10 people living off state grants.

The number of people 'affected by and infected with' HIV-AIDS is high so attending funerals is a regular weekend commitment for many. Some of the elderly former farm workers tend small garden plots and many of the unemployed sell home brewed beer. Alcohol abuse and teenage pregnancy are a problem amongst learners in many of the schools.

Risk and the moral imperative

Nosipho noted that many members of the community, associated with the High School at which she teaches science, had lost knowledge associated with indigenous fermented foods, notably *umqombothi (sorghum beer)*. She had been intrigued by a science experiment that her group participated in during science education activities the previous year. Not only was she fascinated by the science of fermentation but noted how her students did not relate the insights on nutrition to Xhosa cultural practices.

The school calendar has an annual cultural day when parents and learners celebrate their Xhosa heritage in song dance and foods. Although this event is well supported, many of the learners and their families have 'lost' the practices and ways of knowing associated with rural cultural life. Nosipho is concerned that, with this, there has been a loss of respect for and understanding of cultural heritage with Xhosa becoming a performative culture of song, dance and dress for occasional celebrations and visiting tourists.

Educative deliberations amongst learners, community and teacher

Nosipho thus set out to engage community members in her lessons by inviting them in to give a demonstration of the making and cultural significance of *umqombothi* (sorghum beer). Her main professional motive was, however, to make sense of the demands of the new curriculum that indigenous knowledge be included in science lessons. The shift from a Christian National Education instructional ideology of the apartheid state to Outcomes Based Education has brought radical changes in school, classroom and learning management. Most teachers have not fully grasped and have not been able to keep up with the changing demands of the new system. As a Master of Education research student at the university, Nosipho developed a research programme to engage parents and pupils in a learning activity through which she hoped to develop a better understanding of imperatives to include indigenous knowledge, cultural heritage as well as environment and sustainability concerns in her teaching.

Nosipho supported the learners to share all they knew about *umqombothi* and alcoholic fermentation. The students then took on the role of researchers when attending a demonstration conducted by parents. Nosipho moderated the learning process by conducting focus group discussions with all involved. She noted that the students struggled to apply and make sense of the concept of alcoholic fermentation but had a good grasp of the making of *umqombothi* and its cultural significance. Being responsible for teaching the concepts of science she developed activities to support learners to link the two. Effective use of the concept of alcoholic fermentation opened a realization of the intergenerational knowledge of the Xhosa and questions related to the problems of alcoholism in the community.

Resulting steering choices and change

Here alcoholic fermentation as a practice of Xhosa cultural significance was picked up by the learners and contrasted with alcohol abuse in the school and local community context. How colonial interpretations of Nguni culture and the religious beliefs of Christians had served to marginalise and foster a widening urban rejection of isiXhosa cultural practices related to fermented foods was also deliberated by the learners. In their learning and discussion they developed new insights and respect for Xhosa fermentation practices that bring out the food value and nutrition in the grain. Nosipho now intends to write up a learning programme unit for the new curriculum. This will be developed to replicate the social processes of engaged meaning-making with members of a local community as well as working with the curriculum concepts.

Case 3: Amarewu: consumer studies and nutritious food for healthy living

Socio-historical context

Case 3 is the work of Lutho Kota (2006), a teacher working at Nosizwe High School in Pakamisa Township near King Williams Town. The area developed as a township following forced removals from white farms and people being resettled on small-holdings in the former Ciskei Homeland of the apartheid state. During this period there were few economic opportunities and people were poor, being forced to live by farming smaller parcels of land.

The consolidation of the apartheid homeland system led to financial incentives that encouraged factories to locate near homeland sources of labour. The nearby industrial area of Zwelitsha expanded and attracted many job seekers. This led to the development of shack settlements in low-lying areas that were flooded in 1979. In the 1980s people were settled on higher ground that became known as Pakamisa (to lift up), upliftment both in terms of altitude and the prospect of skills training and better job opportunities.

Nosizwe school draws about half of its learners from Pakamisa and others from nearby rural villages. One of these villages, Cliff location, was a farm some years back but it has now developed into a township. Many of the people here still keep their Xhosa cultural practices and value this heritage. When it was a farm, people still had a subsistence living but now there has been a population increase owing to rapid rural-to-urban migration with the demise of the apartheid state. Today there is little farming other than the keeping of a few cattle and goats as people have moved from a self-reliant rural way of life to a consumer lifestyle in urban areas.

Today there are high levels of poverty and unemployment with the collapse of industries that had been subsidized and supported by the apartheid state. DaGama Textiles is still a major employer in the area but it has been hard hit by the recent decline in the South African textile industry. It now mainly prints imported fabrics and there have been many retrenchments over the last decade. Pensions and social grants now sustain most families and one employed person supports an extended family of 6-8 unemployed adults. There are high levels of teenage pregnancy and crime, the former having increased rapidly with the introduction of child support grants a few years back. HIV-AIDS is prevalent with many grand parents left to take care of orphaned children. The extended family culture of the Xhosa is still sufficiently intact in the area that the orphaned children are taken up into the community when the old people pass away.

Risk and the moral imperative

As an young African professional, Lutho feels driven to uncover and recover much of the wisdom in indigenous ways of knowing that enabled the Xhosa to thrive as a pre-colonial democratic order prior to the imperial occupation and conquest of the region. She has a passionate interest in health and nutrition with so many community members being affected by poverty and infected with HIV-AIDS. In her research she found a widening use of amaRewu in the care of people suffering from HIV-AIDS. AmaRewu was found to stabilise their digestive system and they commented about its use in the old days to give energy for work in the fields. She was struck by the creative ingenuity of amaRewu being made from leftover maize porridge from the morning meal. Historically the remaining porridge was used to make a refreshing energy drink that was available for the rest of the day and was mainly consumed when everyone was tired and there was still work to be done in the fields. Her initial research pointed to how the making of amaRewu was now seen as a practice of the poor and for the sick, with most of the people of King Williams Town / Bisho now not drinking it at all or preferring to buy it in cartons from the supermarket.

Educative deliberations amongst learners, community and teacher

Lutho chose to work with women in the local community to provide learners the opportunity to uncover their indigenous heritage and to illustrate changing patterns of production and consumption in modern times. The curriculum also specified the need to eat healthy foods and have a balanced diet.

The community members came to demonstrate the making of amaRewu with the learners observing and asking questions. An interview with the women who did the demonstration revealed how the sharing of the starter ferment (*umlunuso*) was a way of fostering a feeling of community, an ethos of mutual care in the community when nobody would go to bed hungry.

Focus group discussions were used to track what the learners found significant and to identify what aspects of the practice they valued. This research was used to develop a learning activity on sustainable food choices. The learners researched the food preferences of the local people, notably whether they made and consumed homemade or preferred to buy commercial amaRewu that is now available in differing fruit flavours in the local shops. The learners found that people still valued the home made over the commercial. The youth involved knew of the cultural significance of amaRewu but most had never tasted it. Many of the older people still drink it but now prefer to buy amaRewu as they do not have time to prepare it.

Resulting steering choices and change

Community members, school learners and teacher became co-engaged in probing the health and community benefits of amaRewu as a fermented energy drink. Learners began to see the value and some proposed to begin making and selling it as a small business opportunity with tangible health benefits to the community. The teachers began to consider what other cultural foods could be brought into the Consumer Studies curriculum. The demonstration did not make the clear links to specific aspects of nutrition but it was clear that digestive health and energy-giving was the main reason for valuing amaRewu. Learning activities were developed to draw out health science information on carbohydrates for energy, vitamin B for health and enzymes for digestion. This contributed to a valuing of and a respect for indigenous food preparation and nutrition.

What we are learning about social learning interactions

A local socio-historical and generative character is apparent in an arising moral imperative that reflects concerns for poverty relief, health and quality of life in contexts of degradation and risk that are reflected in each of the case accounts. Notable is how the participants were able to bring a plurality of experiences and perspectives into learning interactions around the focal concern being addressed. Here knowledge and experience capital emerged in the deliberative practice of those involved. Deliberation arose in context through a responsive imperative to a shared sense of risk in socio-historical / cultural context, and was then carried into wider mediated interactions to clarify and to steer developing ideas. Emergent steering insights came with re-appropriating associations within the sustaining social logic of practices of the past and propositions emerging as practical possibilities and informative insights to support these. The latter was particularly evident and points to a common error of opposing indigenous and scientific knowledge in academic discourse rather than relating both to the realities of best practice and tolerance of diversity.

Contrary to the expectation of significant differences between indigenous (local cultural know-how) and scientific propositions (curriculum concepts), the cases were notable for how these corresponded and were complementary in developing learning interactions. Here much of the knowledge embedded in the focus of local intergenerational practice ('how-to') resonated with the more explanatory capital ('why') provided by scientific concepts in the curriculum. This intermeshing of how(s) and why(s) had the effect of fostering a situated understanding of and respect for indigenous practices. The emergent synergies that resonated with local realities thus shaped enhanced agency and a sense of renewed respect for and pride in the value of the cultural heritage being explored.

Here what was notable to us is a distinction between opposition as assumed differences within plural worldviews and socio-symbolic learning processes where propositions are opposed within a differentiating of a more 'object-adequate' orientation (Elias, 1989) in a meaning-making journey of learning interactions. A conflating of dialectic processes that are commonly experienced in learning interactions amongst different ways of seeing things with notions of differences in cultural world views, appears to have given rise to the popular opposing of indigenous and scientific propositions in contemporary academic discourse. The simple issue here is that the indigenous and scientific that are commonly opposed do not refer to each other in learning interactions but to socio-ecological realities in/of the world. In these cases it is thus not a matter of 'opposing world views' but 'views of the world' the multilingual pluralism of which are differentiated in reflexive meaning making interaction around a shared concern for healthier and more sustainable lifestyle choices[53]. Suffice it to say that in these cases, an expectation of opposing knowledge systems (indigenous and scientific) was not apparent.

Most notable in all cases was the reflexive and generative emergence of human agency and social change. It should however be noted that the cases examined are school-in-community research initiatives in local socio-ecological context. This gives the cases a particular character as processes of situated multilingual social learning interaction. Besides a circumventing the constructivist trap of assumed difference in cultural and individual world view, the most striking insight was how generative social engagement is at once reflexive and enabling of agency whilst strengthening community and cultural identity in a widening modern world of and at risk. The focus on community learning processes involving a culturally situated pluralism deployed in reflexive deliberation around local health and environment concerns, has provided a useful window on reflexive praxis in school-in-community contexts of risk.

The cases reflect the importance of time being given to the gathering and assembling of diverse historical and local information so as to circumvent the assumption that locals know the context in which they are working and living together. We do have social and local knowledge with which to start reflexive work but this needs to be informed by wider histories and diverse ways of approaching and seeing things. Most notable here is taking developing tensions and disputes into close and detailed examination in local realities, especially through the engaged mediation of these matters in Mother Tongue, notably practical focus group activities that are, simultaneously and with due mutual respect, reflected in other languages so that outside participants are also able to contribute their

[53] The colonial and institutional blind spot at issue here is the appropriation and mediating interpretation of intergenerational knowledge capital in/of local socio-ecological context by outside research institutions. Here, illustratively put, the local is mediated by the outside against the institutional view on/of reality. A dominance of modernist processes such as these that have shaped a school curriculum delivering comparative and illustratively mediated propositions rather than the mobilizing of pluralist knowledge capital, has effectively disenfranchised indigenous cultures and disabled local meaning making engagement within generative and reflexive processes such as those evident in the case studies. An elaboration of these socio-historical processes of appropriation and marginalisation must remain beyond the scope of this chapter.

stories and experiences related to a concern. Here we are finding it more useful for Mother Tongue speakers to work with simultaneous translation and deliberative mediation, for example, than for those engaging the concern to be making a point in one language and then be restating this in another language so that all present are part of the deliberations. The former allows for a social mediation in context that recognises cultural pluralism. The social is thus both local and part of a wider socio-cultural milieu of democratic pluralism that we share. And it is at the interface of social processes such as these that the necessary social sensitivities and deliberative respect seem to emerge as key social processes that can make the 'social' in learning a useful qualifier when contemplating meaning-making interactions around pressing concerns related to environment, health and sustainability.

References

Archer, M. (1996) *Culture and Agency: The place of culture in social theory*. Cambridge: University of Cambridge Press.

Archer, M. (2002) "Realism and the problem of agency" *Journal of Critical Realism*. 5(1): 11-20.

Asafo Adjei, R. (2004) "From *imifino* to *umfuno*: A case study foregrounding indigenous agricultural knowledge in school-based curriculum development", Unpublished M.Ed study, Department of Education, Rhodes University, Grahamstown.

Bauman, Z. (1991) *Modernity and Ambivalence*, Cambridge: Polity Press.

Bauman, Z. (2000) *Liquid Modernity*, Cambridge: Polity Press.

Beck, U. (1992) *Risk Society: Towards a New Modernity*, London: Sage.

Beck, U. (1999) *World Risk Society*, Cambridge: Polity Press.

Beck, U. and Beck-Gernsheim, E. (2002) *Individualization*, London: Sage.

Elias, N. (1989) *Involvement and Detachment*, London: Blackwell.

Hanisi, N. (2006) "Nguni Fermented Foods: Mobilising indigenous knowledge in the Life Sciences", Unpublished M.Ed study, Department of Education, Rhodes University, Grahamstown.

Harraway, D. (1997) *Modest Witness @ Second Millenium Female Man Meets Onco Mouse: Feminism and Technoscience*, New York: Routledge.

Kota, L. (2006) "Local Food Choices and Nutrition: A case study of amaRewu in the Consumer Studies Curriculum", Unpublished M.Ed study, Department of Education, Rhodes University, Grahamstown.

Lotz-Sisitka, H. (2005) "Engaging ambivalence: A critical examination of a rights-based environmental education discourse in South Africa's new national curriculum statement", paper presented at the 3rd World Environmental Education Congress, Turino, Italy, 3-7 October 2005.

Lotz-Sisitka, H. and Timmermans, I. (2005) "Exploring the practical adequacy of the human rights, social justice, inclusivity and healthy environment principle statement in South Africa's national curriculum", paper presented at the Kenton at Mpekweni Conference, 27-29 October 2005.

Chapter 25

Professional ignorance and unprofessional experts: experiences of how small-scale vanilla farmers in Uganda learn to produce for export

Paul Kibwika

Introduction

Falling world market prices for agricultural commodities are likely to sustain poverty in developing countries that rely on traditional agricultural exports. In Uganda, small-scale agriculture employs over 80% of the population (GoU 2000), but the proportion directly depending on agriculture for a livelihood is even higher in rural communities (Abdalla and Egesa 2004). With shocks due to price falls in agricultural commodities, producers are likely to be poorer even relative to the country's average income (Page and Hewitt 2001). Dependence on traditional exports such as coffee, cotton and tobacco for income only worsen the poverty situation among small-scale farmers as Bahiigwa *et al.* (2005) clearly illustrate:

> "while the producer prices as a ratio of World prices had steadily increased from 12% in 1987 to 79% in 1998, these positive developments for coffee producers were subsequently undermined by declining prices, with the 1999-2000 price falling to 36% of what it was in 1994-95".

Recently, several African countries have gone into production of non-traditional fruits and vegetables to diversify their exports and increase hard cash earnings (Singh 2002). In Uganda, small-scale farmers are exploring new opportunities for niche market crops such as vanilla and cardamom as alternatives to traditional cash crops. Tamale and Namuwoza (2004) emphasize Uganda's comparative advantage on vanilla production that: "Uganda is the only country on the mainland of African continent, which grows vanilla and it is the only one in the world, which harvests vanilla twice a year".

The shift to new crops is accompanied by demand for new knowledge and technologies, but research and extension for many reasons (including a continued focus on traditional crops in training, institutional priorities and bureaucracy) are unable to adjust quickly enough to offer the support needed. This opens new windows of opportunity for farmers to engage in social learning processes to generate their own knowledge and technologies. This chapter describes a case of how small-scale farmers self-organised to learn and share knowledge and innovations on vanilla without the intervention of research and extension. It illustrates endogenous phenomena of social learning driven by farmers, which challenges professional relevance to such processes. In agriculture, the process of social learning requires that farmers become experts, instead of users of other specialists' wisdom and technologies (King and Jiggins 2002) what then is the role of research and extension? This case highlights 'new' or additional functions for research and extension that warrant a new breed of professionals.

The case study is based on comprehensive interviews with 31 vanilla farmers in Ntenjeru sub-county, Mukono district, to understand how they came to know the crop and how they learn about it. Ntenjeru was the pioneer sub-county for vanilla production in Uganda and therefore ideal for understanding farmer learning mechanisms. To trace these mechanisms, farmers were selected in categories based on the period when they took up vanilla growing, i.e. before 1990; 1990-1995; and 1996-2000+. But first, I will provide a brief history of vanilla in Uganda, followed by a description of the circumstances and platforms for learning. Based on these, I suggest functions through which research and extension can enhance social learning before making conclusions.

History of vanilla in Uganda

Background

Vanilla planifolia, a fruit of the Orchid is not an indigenous crop to Uganda. It is a native of Mexico and Central America now grown in parts of the tropics including Madagascar, Indonesia, Reunion, Seychelles, Comoro Islands and Uganda (Tamale and Namuwoza 2004). Its fruits (beans) are harvested before they are fully ripe, fermented and cured as flavour and spice for food and pharmaceutical industries (Purseglove 1972).

Pioneer farmers who well know its history say vanilla was introduced in Uganda during the colonial period by British farmers as far back as 1940s. Salama estate farm in Ntenjeru sub-county in Mukono district was one of three farms owned by British farmers where vanilla was grown. It was exclusively protected as "white" farmer crop and to ensure that it remained so, the Ugandans employed on the estate as labourers were routinely checked before leaving the farm to ensure that none escaped with planting materials (vines). This type of control aroused the curiosity of some labourers who stealthily manoeuvred to take away some vines. Secretly they planted it in the middle of coffee gardens for fear of losing their jobs (or possible arrest) if the British farmers found out that they were growing vanilla. In the late 1960s, however, the British farmers experimented with an outgrower scheme with a few farmers in the neighbouring Kooja parish as a strategy to increase production. Kooja parish later became the pioneer and nucleus for vanilla production in Uganda.

When dictatorship creates opportunity for small-scale farmers

The 'economic war' declared by the military regime of Idi Amin (1971-79) made the economic and political environment unfavourable for foreigners; so the British farmers left and abandoned their farms around 1972. Their departure halted commercial vanilla production due to difficulties of marketing. Those who had vanilla only used it locally to spice tea and local brew. The most vibrant economic activity then was illegal cross-border trade or 'smuggling' in which many youths engaged directly or as brokers.

In 1980 some business men deployed brokers in Mukono district to search for vanilla allegedly to be used to conceal drugs like marijuana which they trafficked abroad. Kooja parish was the target since it was the place suspected to have vanilla. The price offered (USh. 300/= per kg) by brokers was very attractive compared to other crops which stimulated interest in growing vanilla. Due to demand, the price more than doubled to USh. 800/= per kg by 1983 (Bank of Uganda 2004). Many farmers obtained planting material (vines) from the former Salama estate where vanilla was growing wild then. At the time, only former labourers at Salama estate, some of whom were also the experimental outgrower farmers had knowledge of vanilla production.

What is unique about vanilla?

The uniqueness of vanilla production practices compared to other crops in the farming system is what incited my curiosity to understand how farmers learn about it. Its successful production therefore was not just adaptation of indigenous knowledge to a new crop but involved generating new knowledge and practices, yet all this happened amongst farmers without the intervention of research and extension. Vanilla is peculiar in that:

- It requires shading of two thirds to one half of normal sunshine (ADC/IDEA Project 2000). For this reason, vanilla is interplanted with banana and coffee; however, through experience, farmers have identified characteristics of appropriate shading trees. Small leaf trees like *Glyiricidia* are preferred to broad leaf trees like mangoes. Small leaves decompose faster, allow better water

infiltration and reduce incidence of soil related fungal diseases compared to broad leaves.
- The climbing vines are staked and looped to control plant height and ease pollination. At the point of contact with the ground, the looped vines are buried to increase root establishment for higher nutrient uptake. A local shrub, *Kirowa* that is commonly used as boundary landmark has been found to be the most appropriate stake because it establishes well even under dry conditions, provides shade and is strong and flexible enough to support the weight of a big cluster of vines.
- Being a surface feeder, weeding with hand hoe is minimised as it would affect the root system; at the same time roots have to be protected from the heat of the sun especially during the dry season. In this regard, a variety of weed management options are experimented with.
- Flowering is naturally induced by the dry season, however this is not adequate. Inducement is enhanced by cutting some of the looped vines (pruning) to stress the plant.
- Pollination is done by hand. Unlike other crops, vanilla is not naturally pollinated by wind or insects, the flower is opened and the male (anthers) and female (stigma) parts are joined physically. Timing is important here as the flower is viable only for 12 hours, according to the farmers. This is probably the most scientifically technical practice that farmers can be expected to do successfully.

These unique practices stimulated farmer exploration and experimentation in search for effective ways to produce vanilla. Non-involvement of research and extension to provide knowledge and technological support enhanced farmer experience sharing and learning from one another. This self-directed learning turned farmers into 'unprofessional experts' while the researchers and extensionists kept away from the learning process to retain their 'professional ignorance' about vanilla. I call them *unprofessional experts* because they are self-made experts. *Professional ignorance* depicts the rather arrogant notion that what is not taught during career training is less valued knowledge, an attitude that delineates professionals out of important social learning processes. Often this is defended with self-professed "national" priorities that may have little relevance to farmers' real needs and aspirations. This case illustrates processes of social learning that rapidly spread vanilla as an export crop in Uganda. By 2002, vanilla was grown in over 18 districts (Tamale and Namuwoza 2004).

Conflict and competition in social learning

Conflict induces social learning

While farmers exchanged materials and knowledge, real self-organisation for learning was triggered by conflict (Dewey 1922, Eshuis and Stuiver 2005, Heymann and Wals 2002). Two major conflicts influenced social learning: first was the conflict between brokers and farmers, and second that between the Ministry of Agriculture, Animal Industry and Fisheries (MAAIF) and farmers.

Until 1990, farmers did not have direct contact with vanilla exporters; they dealt with brokers who bought the produce at less than the actual price and sometimes purchased on credit but often defaulted on payments. The challenge for farmers was how to deal with dishonest brokers. By sharing the challenge, the idea of forming a farmer cooperative society emerged. This idea came as advice from a visiting son of one of the farmers in 1988. In January 1989, Kooja Vanilla and Fruits Growers Association (KVFGA) was started to promote the marketing of vanilla and other fruits. Using their subscription fees, they advertised on radio advising the vanilla exporters to deal with KVFGA directly. Soon, two exporters visited them to verify their existence and to clarify the quality of vanilla they wished to buy. As an association, they negotiated the price and devised a system for bulk sale of vanilla in two neighbouring parishes of Mpunge and Nsanja.

In 1990, a prominent business man, Agha Sekalala who had links with an American firm, McCormicks Ltd that procures vanilla contacted KVFGA with the aim of promoting vanilla production for export business. He provided credit to farmers on the understanding that they would supply him with all their produce. For technical support, Sekalala in conjunction with McCormicks Ltd secured support from USAID through a project on Investment in Developing Export Agriculture (IDEA project) to hire an expert, Steve Caiger to work with farmers for about three years. Having had experience in different climatic conditions, his expertise was not directly transferable. Instead, he engaged with contact farmers in a learning process to generate context-specific knowledge and practices (cf. Eshuis and Stuiver 2005) which he compiled into a production manual (also translated into the local language). For language translations, he was assisted by the area extension worker (an agricultural assistant then). Basically, Caiger's main function was to scale up learning through experimentation and sharing of experiences.

Shared knowledge flows in the community verbally through informal networks. It is therefore not surprising that only two farmers still kept the manual but more as a souvenir than as a source of information. Other than marketing, KVFGA became a network for learning. It secured funds from Sekalala and the IDEA project to support a regular radio programme, called *vanilla buggaga* (vanilla is wealth), to increase awareness and disseminate knowledge about vanilla. The thirty-minute programme was presented once a week by a farmer, John Nviiri, who had vast experience (from production to primary processing), with the British farmers. Note here that the name of the programme portrayed the shared goal of wealth creation, which also inspired learning. The resultant awareness created an overwhelming demand (from within and outside Mukono district) for planting material. The vanilla farmers then reaped the rewards from selling vanilla beans and vines. But beyond wealth creation, sustainability was in contention. This was the basis for the second conflict.

The conflict between MAAIF and farmers was due to difference in perceptions of the impact of vanilla on soil fertility and whether it was sustainable. Our institutions and mechanisms of governance seem increasingly archaic (Woodhill 2003); they usually respond with pessimism to unfamiliar circumstances to conceal their inabilities. MAAIF discouraged farmers from growing vanilla on the pretext that it would lead to rapid soil degradation. But farmers saw it as an opportunity and it was difficult to dissuade them from growing vanilla anyway. On the contrary farmers argue that the vanilla practices, namely, planting shade trees, minimal weeding, non-use of chemicals, guaranteed a more sustainable environment than other crops. Amidst such conflict, it would therefore be naïve for farmers to expect support from research and extension which falls directly under MAAIF. This strengthened farmer interdependence to learn through their own initiatives.

Remember, KVFGA was an emergent property of the conflict between farmers and brokers. In this case, conflict becomes beneficial in social learning when it is turned into a shared challenge for which solutions are jointly sought. Articulation of a shared challenge too is a social phenomenon that is anchored in a common goal, which in this case was 'being wealthy'. Harnessing conflict into opportunity for joint learning is easier when stakeholders pursue complementary objectives than when they compete.

Competition as barrier to social learning

Because social learning is a move from multiple to collective and or distributed cognition (Röling 2002); competition can be counterproductive. Röling emphasizes that parties involved in social learning must develop overlapping or at least complementary – goals, insights, interests and starting point. In other words, it is only when the process offers sustainable mutual benefit that the parties will engage to learn together for better living.

As vanilla demand increased, it attracted many buyers competing to raise export volumes; in the process, quality was compromised. Competition eroded cohesion in KVFGA and its collective bargaining power was diminished as individual farmers struggled to supply to highest price offers, creating an opportunity for brokers again. Coinciding with the disastrous storms in Madagascar, the world-leading vanilla producer sky-rocketed the price reaching over USh. 100,000/= per kg in 2004, making vanilla the farmers' 'gold'. At that price, envy, jealousy and a mentality of quick gain manifested themselves in rampant theft of vanilla, forcing farmers to hire gunmen to guard their gardens at night. This created suspicion and mistrust amongst farmers reducing their interaction to learn from each other. Only farmers with a high level of social trust could visit each others' garden.

Consequently the radio programme stopped as the buyers could not cooperate to invest in educational programmes. Social learning is a cooperative process and for it to be sustained in a competitive environment, appropriate levels of competition and cooperation have to be clearly defined. Resumption of production in Madagascar, coupled with neglect of quality have recently plunged prices below USh. 1,500/= per Kg. The 'gold' has simply melted away. If, for example, the level of competition had been conceived to be between producer countries, the buyers would probably have cooperated to keep the Ugandan quality high to sustain a good price on the world market.

Learning mechanisms

The capacity for farmers to innovate, and share knowledge and rationale for doing so was best articulated by a farmer:

> "We as farmers, when we face a problem or opportunity, we become creative, we explore, discover and share this amongst ourselves. For example, I was the first one in this parish to demonstrate that you can transplant a mature vanilla plant but I also learnt it from my friend in another parish. When there is a good price, you can become creative in many things because everyone wants to get more from what they have." (Nsonera)

Exchange of knowledge and experiences from experimentation were largely based on individual interactions. The motivation for experimentation was clearly stated by a farmer who started growing vanilla at the age of 15:

> "I was motivated to discover more by producing more but I also strategically located my garden by the roadside to make it an example/ demonstration for others to learn from. I do my own research to be

> outstanding and because of this, I have hosted many farmers from other districts. They come to learn and I also get orders to supply them with planting materials". (Kiyaga)

It is not the case that farmers learn from all farmers in the village. Information flows through a relatively sparse social network (Conley and Udry 2001) based on interpersonal relationships. In this work, I describe social relationship and inclusiveness as fundamental values that underpin continued social learning.

Interpersonal relationship as the vehicle for social learning

Learning is an interactive process that takes place on a platform for exchanging knowledge and experiences. Here a platform is defined simply as *a forum for interaction to learn, negotiate and/or resolve a conflict* (cf. Röling 2002, p. 39). In learning about vanilla, four platforms were prominent: source of planting materials; radio programme; farmer experimentation and exchange visits; and informal sharing and conversation (See Figure 25.1). Throughout, interpersonal relationships based on friendship and trust were a key factor that I will now elaborate on:

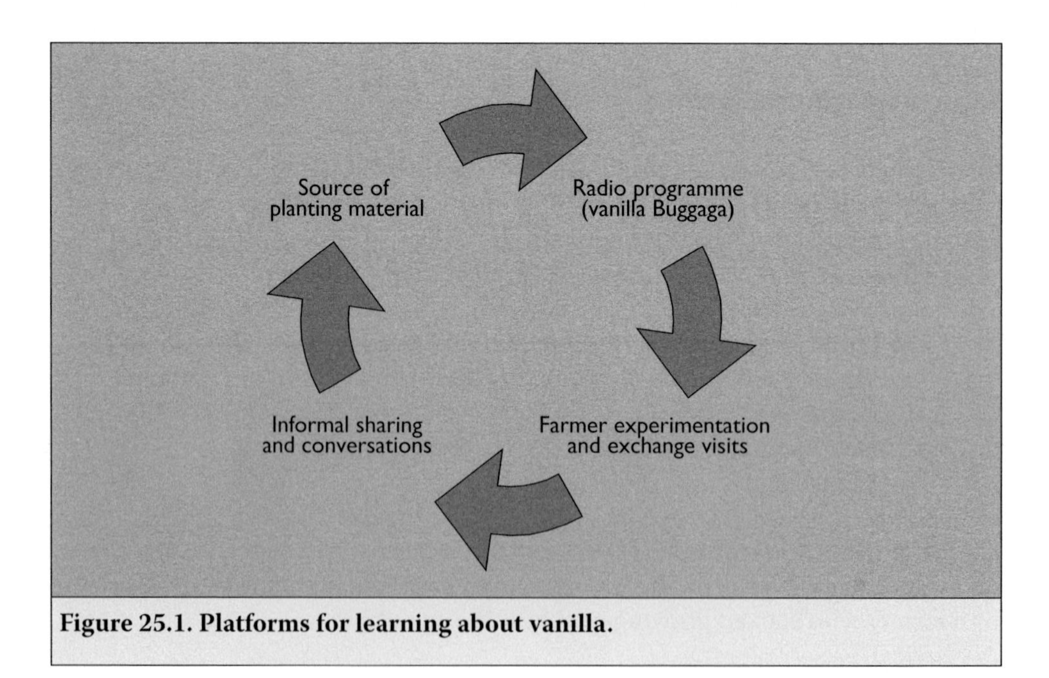

Figure 25.1. Platforms for learning about vanilla.

Source of planting material

Source of planting materials was a platform for exchange of knowledge and experiences about vanilla. All the farmers involved in this study obtained their first planting materials from other farmers they had contact with, and often their first batch of materials were offered free – an indication of a friendly relationship. The market created a passionate environment for unreserved sharing of personal experiences.

Radio programme

All farmers acknowledge they learnt a lot from the radio programme presented by an experienced farmer who also used the experiences of other farmers to explain constraints and possible solutions. Farmers tended to listen to the programme in small groups and after the programme they discussed the content with respect to their own experiences. The choice of who to listen with was based on the social relationship between them, and the discussions were also a means of mobilizing social energy to experiment.

Farmer experimentation and exchange visits

The absence of 'blue-print' recommendations from research and extension provided the space and freedom for farmers to experiment widely, for example, on spacing, weed management options, appropriate shading trees and different ways of inducing flowering. Exchange visits were a platform for sharing successes and failures from their experimentation. Since these visits were informal, they were again based on interpersonal connections. As illustrated by Kiyaga's quote above, successful experiments were also learning sites for farmers from within and outside the area.

Informal sharing

Vanilla, the "gold", became the subject of everyday conversation at all social forums. These conversations permitted exchange of experiences, identification of best practices and creating linkages for learning. The pride of recognition for innovation was an incentive for enlarging social linkages to share individual discoveries with others. Such pride is apparent in Nsonera's quote above.

Inclusive shared learning

The core value of participation is inclusiveness of all stakeholders in whatever will affect them. I look at inclusiveness beyond just involvement to include building platforms for intergenerational exchange. This case presents an intriguing example of gender inclusiveness and more so the intergenerational exchanges in the learning process. In discussing inclusiveness, the emphasis is on the learning process and equity issues are beyond this discussion.

Gender inclusiveness

Based on interactions with farmers, there was no indication that the women were any less knowledgeable about vanilla than men and vice-versa. What is rather more apparent is the recognition of specific expertise for men, women and children, allowing for a distributed knowledge system. It is commonly said, for example, that women are better than men at pollination but children are even better. This, however, should not be misunderstood for child employment. It is a situation where children contextualise their education and translate it into life skills. This is essential for sustainability of rural life as many young people drop out of school early without any gainful skills (Kibwika and Tibezinda 1998).

Contrary to gender related studies that cash crops are dominated by men, here the entire family tended to work together as a social unit. It was also common that women and in many cases children had control over some plots from which they derived income to meet their individual needs. It is probably this kind of arrangement that facilitated complementary use of knowledge and expertise within the family.

Intergenerational exchange

Due to limited experience, children are rarely recognized as key players in community knowledge system and for that reason they are often left out of adult learning processes. This poses a serious concern for intergenerational sustainability. In this case, children are recognized for their expertise in pollination, which can be attributed to their knowledge of science acquired in school. Their understanding of the flower morphology puts them at an advantage to carry out more successful pollination. But what is interesting is how this knowledge/expertise is solicited from the children in a learning context as one farmer explained:

> "I have learnt to pollinate from children. I used to invite children from homes that grow vanilla to come and pollinate for me. I would give them sugarcane in return and they would demonstrate to me as I watched carefully. This is how I learnt". (Luwalira)

These intergenerational platforms for learning also integrate local and 'scientific' knowledge that enhances adaptation and therefore sustainability of social learning processes. Intergenerational exchange is enhanced by targeting schools for social learning as an old lady and former contact farmer reported:

> "While promoting vanilla we set-up demonstrations at schools to train the children so that they could train their parents. The children are much easier to train and they grasp the principles much more easily than adults". (Lusulire)

The schools then are not only institutions for teaching children but rather institutions for community learning.

Functions for research and extension in social learning

I have described a social learning situation where farmers generate relevant knowledge and technologies without intervention from research and extension, but that does not mean that there is no place for research and extension. The challenges of social learning call for more professional engagement than before but in a way that is different from the 'expert' prescriptive approaches. Fundamentally, what needs to be done is to redefine the roles and functions of research and extension within a social learning context otherwise their relevance is contested. I use experiences from this case to highlight some of their core roles and functions to support sustainable social learning processes:

Markets and market information brokerage

Learning is motivated by its economic and or social value to the learners in a sustainable environment. It has a cost in terms of time and effort; it is therefore unthinkable that farmers like other stakeholders will engage in learning processes for the sake of it. In this case, the major motivation for learning was an attractive market initially linked by brokers as highlighted above. Therefore, the catalytic role for research and extension on the social learning landscape is that of market and market information brokerage. This requires an understanding of market dynamics – demand and price trends, quality standards, business linkages as well as environmental requirements to sustain the business.

For example, there were very wide price fluctuations between 1992 and 2005 mainly due to Madagascar's supply to the World market. Most spectacular was the price rise to a climax of USh. 100,000/= per kg in 2004/5 when Madagascar's supply was affected by the disastrous storms; however, its recent recovery plunged the prices to below USh. 2,000/= in late 2005. Surprisingly, nearly all farmers interviewed did not have enough understanding of the reasons for price fluctuations to make informed strategic decisions. To survive in global competition, farmers need to be more proactive than reactive to the market opportunities; for this, small-scale farmers must be organised.

Farmer organizational development

One of the great appeals of market systems is their self-organising nature (Woodhill 2002). Survival of small-scale farmers lies in strong farmer organizations that allow pooling of resources and products to access reliable markets, increase bargaining power, and strengthen their demand for services and rights. KVFGA was a good start in this direction but it was overwhelmed by the interests of competing market agents.

Through strong organisations, farmers can collectively pursue a common goal, avoid manipulation, and demand services from various providers including government agencies and politicians in a coordinated manner. With regard to learning, farmer organisations are forums for expanding social networks for learning, but who is responsible for organising farmers? Private business agencies are profit driven; it would be unrealistic to expect them to invest in organizing farmers especially where they do not have a monopoly of service. Organising farmers therefore is a prime responsibility of extension and researchers as a public service.

Facilitating joint learning and promoting innovations

Effective social learning as an interactive dialogue and decision making does not just happen but needs to be consciously and proactively facilitated (Woodhill 2003). Facilitating such interactive learning processes requires social skills to build confidence in each other; being open to sharing knowledge and experiences for mutually agreed goals; collectively defining beneficial levels for cooperation and competition; and develop shared values and guidelines for engagement. Moreover, innovations are encouraged through continuous experimentation and sharing, which is possible when there is mutual recognition of each other's contributions. Given their supportive mandate, training and relatively neutral interest in such platforms, researchers and extensionists are best placed to facilitate them. But to do so, they need new skills, mindsets and power relations.

Facilitating multi-stakeholder dialogues

Indeed not all stakeholders would engage in learning processes; some would engage with farmers for specific services that may promote social learning. Aside from learning, there is a need to conduct dialogue for services and conflict resolution, e.g. market services, input supply, infrastructure, and communication services. All these require negotiations and lobbying. As in the case of joint learning, such negotiation processes are characterised by tension arising from conflicting interests and facilitation is needed to amicably resolve such tensions.

Developing entrepreneurial skills and attitudes

One of the major challenges of small-scale farming in Africa is how to develop the entrepreneurial spirit and attitude among farmers who apparently treat farming as a way of life. Success or failure is often attributed to good or bad luck respectively rather than to deliberate or poor planning. This attitude only creates self pity and a feeling that there is not much one can do to change their situation. On the contrary learning is enhanced by confronting challenges with deliberate actions/plans that are continuously reviewed in light of clearly defined targets. The drive to achieve targets is derived from an entrepreneurial attitude, which provides the impetus for adaptive confrontation of challenges for better living.

The culture of saving and investment is not part of the normal thinking among small holder farmers in Uganda. The common argument is that they have nothing to save, but even when they get it (like in the case of vanilla), they still save nothing. During the vanilla boom, farmers earned money they never ever imagined in their lives, but this was taken for granted and they resorted to luxurious lifestyles rather than investing in ventures that would guarantee regular income. The drastic price fall may push many to poverty levels they experienced several years back. The 'gold' has simply melted in their hands. This situation is analogous to environment, which is often taken for granted and consumed luxuriously without consciousness for its sustainability – it too can melt away in our hands. Entrepreneurial attitude is a mental shift from living by the day to ensuring daily living with a consciousness for sustainability including the not so obvious environmental aspects. Professionals who work with farmers have to champion this mental shift.

Conclusions

The traditional linear model that suggests that innovations are developed by scientists, disseminated by extension and put into practice by users has proved its ineffectiveness (Kline and Rosenberg 1986, Rip 1995, Woodhill 2002). For sustainable livelihood in agriculture, social learning is inevitable; given the right incentive, farmers can innovate and learn through their social networks. But as more stakeholders engage with farmers, the situation becomes more complex than they can handle to systematically continue learning. Research and extension have the social obligation to enhance social learning by playing a facilitative role and bring to consciousness environmental sustainability. However, their professional training is still much skewed towards the linear model characterised by hierarchical power relations. As a result, research and extension practitioners react rather defensively when confronted with situations that require joint learning and discovery because they are supposed to be the 'experts' who provide answers. Clearly, the defensiveness is a cover-up for an inability to productively engage in such complex processes. In agriculture, the process of social learning requires that farmers become experts, instead of users of other specialists' wisdom and technologies (King and Jiggins 2002).

Based on this case, I argue that a sustainable livelihood is not something that can be offered to people. It is the adaptive capacity of people to respond to challenges with creativity in a solution-oriented manner. Social learning therefore contributes to sustaining the adaptive capacity to cope in changing environment by providing the impetus and confidence to take collective action. Aside from the knowledge and skills exchange, social learning also has to address embedded moral values that instil social responsibility for sustainable living.

If research and extension are to meaningfully contribute to social learning in agriculture, their roles and functions have to be re-examined. They have to take on functions such as: brokerage; organisational development; facilitating joint learning and multi-stakeholder dialogues; and developing entrepreneurial skills and attitudes. This means that for each of these, key competences have to be identified and integrated in the curricula for agricultural professionals. Attendant to this are mindsets oriented towards the recognition of different bodies of knowledge and expertise that can be shared. In short, a new breed of professionals is needed to advance social learning in agriculture.

References

Abdalla, Y.A. and Egesa, K.A. (2004) "Trade and Growth in Agriculture: A Case Study of Uganda's Export Potential within the Evolving Multilateral Trading Regime", Bank of Uganda Working Paper, WP/05/01.

ADC/IDEA project (2000) "ADC Commercialisation Bulletin No. 1".

Bahiigwa, G., Rigby, D. and Woodhouse, P. (2005) "Right Target, Wrong Mechanism? Agricultural Modernization and Poverty Reduction in Uganda", *World Development,* 33(3): 481-496.

Bank of Uganda (2004) *Annual Reports for the Period 1984 – 2003*, Uganda.

Conley, T. and Udry, C. (2001) "Social Learning Through Networks: The Adoption of New Agricultural Technologies in Ghana", *American Journal of Agricultural Economics*, 83(3): 668-673.

Dewey, J. (1922) *Human Nature and Conduct: An Introduction to Social Psychology,* London: Allen & Unwin.

Eshuis, J. and Stuiver, M. (2005) "Learning in Context through Conflict and Alignment: Farmers and Scientists in Search of Sustainable Agriculture", *Agriculture and Human Values,* 22: 137-148.

GoU. (2000) *Plan for Modernisation of Agriculture: Government Strategy and Operational Framework for Eradicating Poverty in Uganda*, MAAIF and MFPED.

Heymann, F. and Wals, A.E.J. (2002) "Cultivating Conflict and Pluralism through Dialogical Deconstruction", in Leeuwis, C., Pyburn, R., ed., *Wheelbarrows Full of Frogs. Social Learning in Rural Resource Management*, Assen: Royal Van Gorcum, pp. 25-47.

Kibwika, P. and Tibezinda, J.P. (1998). "Participation of Youth in Agriculture in Iganga District, Uganda." *MUARIK Bulletin*, 1: 1-5.

King, C. and Jiggins, J. (2002) "A Systematic Model and Theory for Facilitating Social Learning", in Leeuwis, C. and Pyburn, R., eds., *Wheelbarrows Full of Frogs. Social Learning in Rural Resource Management*, Assen: Royal Van Gorcum, pp. 25-47.

Kline, S.J. and Rosenberg, N. (1986) "An Overview of Innovation", in Landau, R. and Rosenberg, N., eds., *The Positive Sum Strategy: Harnessing Technology for Economic Growth*, Washington: National Academic Press, pp. 275-305.

Page, S. and Hewitt, A. (2001) *World Commodity Prices: Still a Problem for Developing Countries?* London: Overseas Development Institute.

Purseglove, J.W. (1972) *Tropical Crops: Monocotyledons*, Longman Group Ltd.

Rip, A. (1995) "Introduction of New Technology: Making Use of Recent Insights from Sociology and Economics of Technology", *Technology Analysis and Strategic Management*, 7: 417-431.

Röling, N.G. (2002) "Beyond the Aggregation of Individual Preferences. Moving from Multiple to Distributed Cognition in Resource Dilemmas", in Leeuwis, C. and Pyburn, R., eds., *Wheelbarrows Full of Frogs. Social Learning in Rural Resource Management*, Assen: Royal Van Gorcum, pp. 25-47.

Singh, B.P. (2002) "Nontraditional Crop Production in Africa for Export", in Janick, J. and Whipkey, A., eds., *Trends in New Crops and New Uses*, Alexandria, VA: ASHS Press, pp. 86-92.

Tamale, I.. and Namuwoza, C. (2004) "Addressing the Challenge of Providing Technological Optons that Respond to Demands and Market Opportunities for Vanilla in Uganda: The Experience of Taimex (U) Ltd", *Uganda Journal of Agricultural Sciences*, 9: 776-770.

Woodhill, J. (2002) "Sustainability, Social Learning and the Democratic Imperative: Lessons from the Australian Landcare Movement", in Leeuwis, C. and Pyburn, R., eds., *Wheelbarrows Full of Frogs. Social Learning in Rural Resource Management*, Assen: Royal Van Gorcum, pp. 317-331.

Woodhill, J.A. (2003) "Dialogue and Transboundary Water Resouces Management: Towards a Framework for Facilitating Social Learning", in Timmerman, J.G. and Langaas, S. ed., *Environmental Information in European Transboundary Water Management*, IWA Publishing.

Chapter 26

Multi-level social learning around local seed in Andean Ecuador

Marleen Willemsen, Julio Beingolea Ochoa and Conny Almekinders

Introduction

Seeds as public and private goods

For more than 800 million people living in the more marginal and heterogeneous areas, food security and poverty continue to be a daily challenge. They have hardly benefited from modern agricultural technologies. Moreover, in many instances, they have become excluded from natural resources they need for a sustainable livelihood. Instead they seem to carry the main share of the negative externalities (side effects) of modern agriculture and reduced access to natural areas. Pretty (2002) describes this in his book *Agri-culture* as a process in which people become 'disconnected' from land and nature. Genetic erosion for example, is associated with a loss of knowledge and culture: the disconnection affects food production as well as the social fabric of communities, thereby affecting the sustainability of livelihoods.

Apart from seed being a private good with direct benefits for the farming and rural communities, agriculture and nature also have functions with a value for mankind in general. Agriculture and nature are sanctuaries of genetic resources, provide water storage and filtering capacity, stocking carbon oxide and landscapes for recreational purposes. They thereby represent public goods and services (Weiskopf *et al.*, 2003). Through these functions they also directly and indirectly benefit local communities. However, the benefit for those who actually manage and protect them, i.e. rural people and farmers, is less than the value they represent.

In this chapter we present the process and experiences around the initiation of a project in three Andean provinces of Ecuador. The process aimed to make farmers more aware of the importance of seeds for their 'agri-culture'. The awareness-raising was only the beginning of the project and meant to provide a space for farmers to identify desirable actions in regard to their seeds. The NGOs and local organisations involved aimed for a project that dealt with seeds and food security and was to be designed in a participatory way. In the first part of the chapter we elaborate on the issues of use of seeds, genetic erosion and sustainability in the

Ecuadorian communities in the Andes and how we feel this initiative fits with the concept of learning for sustainability. Thereafter we present the experiences of the actors in the learning process at various levels and cycles. Finally, we reflect on the development of the participation in the learning process over time.

Seeds, social learning and participation

Sustainable agriculture seeks to make the best use of nature's goods and services while taking into account the needs of future generations (Pretty 2002). Redesigning agricultural practices into more sustainable forms of production, (re)connection to land and nature, for example through community-supported agriculture, farmer groups and slow food systems, requires reconstruction of the values of agricultural products, the seeds and the land. This reconstruction involves revisiting assumptions and perceptions. We can call this (un-)learning about nature. This involves, among other things, building relationships of trust with different users and user groups and looking for shared norms and rules about practicing agriculture.

This reconstructing and (un-)learning through interaction with other actors we call in this chapter social learning. It implies that participants in the process (re-)frame the way they think about the use of seeds, soils and other natural resources. The 'seeds' play an important role in this process because of their agri-cultural value and their symbolic representation of life (Posey 1999, Pretty 2002).

The first step in the process of social learning we describe is the exploration of current meanings and frames of the various actors, i.e. different community members, local organisations and NGO partners.

In this process of learning for sustainability, we can find different forms of participation by the different actors, changes in the form of participation and multiple levels of learning. Different classifications of 'participation' exist. Pretty *et al.* (1995) define seven forms of participation: Passive Participation, Participation in Information Giving, Participation by Consultation, Participation for Material Incentives, Functional Participation, Interactive Participation and Self-Mobilization, of which the first can be placed lowest on a 'ladder of participation', and the last on the highest level. Wals and Heymann (2004, Figure 26.1) consider four types of participatory relations.

Wals and Heymann (2004) define four main types of participation, represented in the quadrants of the diagram (Figure 26.1). They depend on the level at which participants are involved in the design process (active vs. passive) and the level at which there is space for their ideas (pre-determined vs. self-determined). A

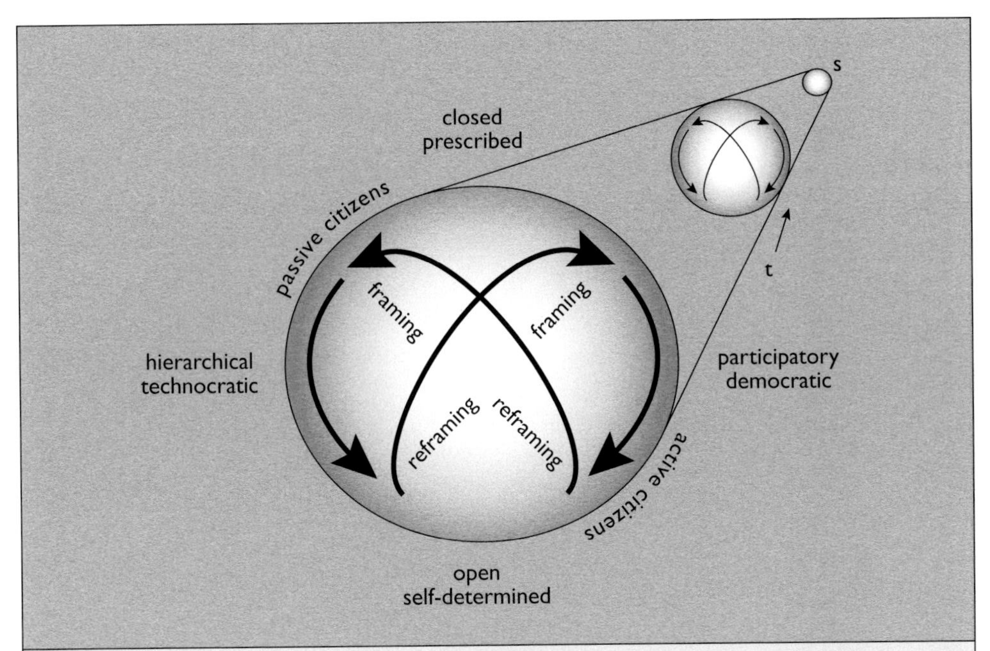

Figure 26.1. Participation, re-framing and change towards sustainability (Wals and Heymann 2004).

combination of these levels in a quadrant leads to four main forms of change towards sustainability (S). In the most extreme cases the processes lead to 'instrumental sustainability' (passive and pre-determined) and 'emancipatory sustainability' (active and self-determined), from which the latter would be most ideal as 'sustainable living agreed upon by all'. In practice, processes can take place partly in one quadrant and partly in another, shifting back and forth in time (t).

Based on Kolb (1984, Figure 26.2) we consider four phases in the 'learning': experiencing, reflecting, conceptualisation and planning. All learning cycles consist of an experience-, reflection-, conceptualisation- and planning phase (Kolb, 1984). *Experiencing* or *immersing oneself in the 'doing'* of a task is the first stage in which the individual, team or organization simply carries out the task assigned. *Reflecting* involves stepping back from task involvement and *reviewing what has been done* and experienced. *Conceptualizing* involves *interpreting the events* that have been noticed and *understanding the relationships* among them. *Planning* enables taking the new understanding and translating it into *predictions* about what is likely to happen next or *what actions should be taken* to refine the way the task is handled.

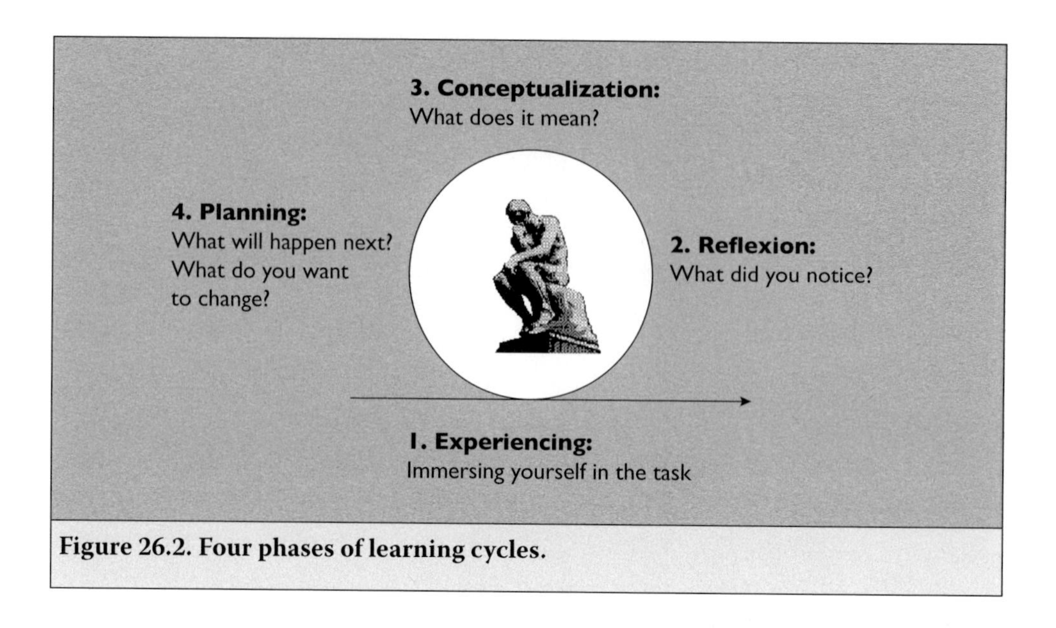

Figure 26.2. Four phases of learning cycles.

Together these phases form a learning cycle. In most change processes, combinations of learning cycles can be distinguished. Depending on the participating actors and the relationships between them, these learning cycles may be associated with different levels of learning. We have used these learning cycles to design and reflect on the initial phase of the project.

The use and loss of seeds in the Ecuadorian Andes

In Chimborazo, Bolívar and Cotopaxi, three highland provinces in Ecuador, genetic erosion has become a generally accepted fact. In these provinces, many small-scale farmers produce for subsistence and the local market. Crops like quínoa, mashua, melloco, beans and peas have lost importance and are only grown on a small scale. They have largely been replaced by modern varieties of maize and potato. Farmers say this change is a consequence of a limited demand and low prices in the market for the traditional Andean crops as compared to maize and potato. Farmers do not refer to the ecological effects of such changes. The overall trend in literature indicates that reduced on-farm crop diversity increases risks of damage from nematodes, insects, night frost or hailstorms. Also, modern varieties of potato and maize tend to be less robust: they do not perform well under adverse conditions and usually need more fertilizer than the traditional Andean crops and the local varieties. Furthermore, farmers seem not to be aware of the nutritive values of the different crops. Replacement of mixed crops with legumes and Andean crops like quínoa with mono-crops of maize and potatoes leads in practice to a less varied diet, usually with fewer proteins and vitamins. Often a mono-crop is grown for marketing and 'fideos' – which contain only carbohydrates and therefore have a lower nutritional value – are bought with the income instead. Children in particular suffer from malnutrition (FAO 1986). It is also shown that the loss of genetic material coincides with the loss of knowledge and cultural identity (Posey 1999). When certain crops or varieties are not grown anymore, young people do not learn about them and the cultural history related to them disappears as well. As a result, food security and the possibility to live autonomously and with dignity are at risk. The process of losing biodiversity can therefore be seen as a consequence of the 'disconnection' between farmers and their crops.

The Non-Governmental Organisation (NGO) World Neighbours, in coordination with two other NGOs, SwissAid and IIRR (International Institute for Rural Reconstruction), wanting to change this trend that threatened the sustainability of the communities, developed a proposal for setting up a process of learning through interactive exploration with the final aim to strengthen Informal Seed Systems in a way that would be meaningful to farmers. They would do this together with the Local Action Agencies (LAA's) that would be responsible for the implementation of the project on the ground. A Technical Advisory Committee with members from Wageningen University, National Agrarian Research Institute INIAP and the Universidad Inter-cultural de Pueblos Indígenas y Autóctonas del Ecuador was formed to support the future project. The project partners successfully applied for funds from the McKnight Foundation.

The assumption underlying the initiative is that farmers are currently not sufficiently aware of the role of seeds and their diversity. As a consequence, farmers' decisions on what to grow are based on their perceived notion of the opportunities for commercialisation of the crop products. They do not or hardly consider the wider importance of their crops and seeds and the variety of roles they play in their livelihoods. Supporting the farmers in developing their awareness (through the facilitation of learning) is expected to lead to farmers redefining the value of local seeds and empower them to take action to recover and/or maintain them. Activities that increase local production, exchange of seeds and seed diversity, are considered to strengthen the local or informal seed system (Almekinders *et al.* 1994). Strong local seed systems allow farmers to use and maintain a level of crop genetic diversity that positively contributes to their livelihood. For the farmers this may mean a return to traditional practices – possibly in a modified and better-adapted form, based on deeper cultural values, and re-connecting them to their local agro-ecological reality (Pretty 2002)

Learning around seeds

Four learning levels

The communities, LAA members and project partners went through a series of nested *learning cycles* (Figure 26.3), which in combination make up the overall learning process. In the various learning cycles we distinguish four different *levels* (Figure 26.3). The first level is the community level. It refers in the first place to the learning in relation to seed problems and takes place in, among and with the community. Learning by community members and the facilitation team members took place around the history of the communities, the actual unsustainable situation of seeds, and visions of the future of agriculture. The second level of learning refers to the level of the facilitation team. It consisted of three cycles (three different facilitation teams) with its respective sub-cycles. In every province, a team with LAA members was set up and prepared itself to implement workshops in the communities. At this level, the facilitation-team learned practical lessons about the process: organising and facilitating workshops, improving the community level learning and the team-learning itself. The third level refers to the NGO level. The overall process and structure of this phase of the project, i.e. the process that aimed at involving the communities in the next step, i.e. the participatory project design, was coordinated by the three NGO's. They were learning about designing a project with the involvement of different actors. The individual learning level can be seen as a fourth level of learning. All participants learned in their own way during the process while some were probably more reflective than others. In Figure 26.3 we focus on the personal learning of one of the authors of this chapter, Marleen. She played a key role in the setting up of the teams, the facilitation and the learning at all three other levels. Her reflection was the objective of her MSc study. Her thesis and the preparation of this chapter form part of this reflection.

Fifteen learning cycles

A short explanation is given of the total of 15 learning cycles (Figure 26.3) that together make up the total learning process:

- Cycle 1 was a personal learning cycle for Marleen. In the early phase of the project she met the three NGO's and the three Local Action Agencies (LAA's) with whom she would implement the process. With the LAA's she formed facilitation teams. They visited communities to learn about local practices.
- Cycles 2, 3, 4 and 5 refer to learning cycles in the province of Chimborazo. The first three cycles relate to the learning in and from the communities in three different locations. Cycle 5 refers to the 'collective' workshop in which team

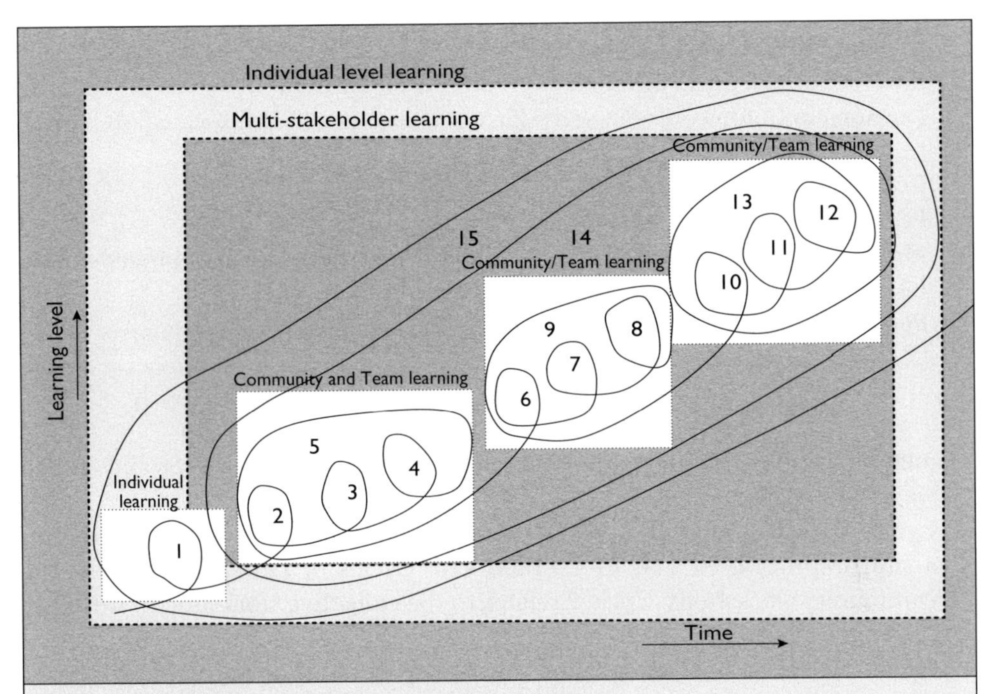

Figure 26.3. Multiple-level learning of the process.

1. **Personal learning cycle** Learning from field visits and meetings with LAA's and Coordination Committee.
2. **Learning in Chimborazo** Community: Pueblo Viejo, Team DEIB-CH: field visit, school activity, community workshop and evaluation.
3. **Learning in Chimborazo** Community: Gramapamba, Team DEIB-CH: field visit, school activity,community workshop and evaluation.
4. **Learning in Chimborazo** Community: Paquibug, Team DEIB-CH: field visit, school activity,community workshop and evaluation.
5. **Collective learning** Workshop Chimborazo, Team DEIB-CH. Participants : delegates from the 3 communities of Chimborazo.
6. **Learning in Bolívar** Community: Queseras, Team CEMOPLAF: field visit, school activity, community workshop,evaluatio
7. **Learning in Bolívar** Community: Tolapungo, Team CEMOPLAF: field visit, school activity,community workshop, evaluation.
8. **Learning in Bolívar** Community: La Cria, Team CEMOPLAF: field visit, school activity, community workshop, evaluation.
9. **Collective learning** Workshop Bolívar, Team CEMOPLAF/FUNPRODIP Participants: delegates from the 3 communities of Bolívar.
10. **Learning in Cotopaxi** Community: Compania BajaTeam DIPEIB-C/ CEMOPLAF: field visit, school activity, community workshop, evaluation.

»

11. **Learning in Cotopaxi** Community: Wakra Wasi Team: DIPEIB-C/ CEMOPLAF: field visit, school activity, community workshop, evaluation.
12. **Learning in Cotopaxi** Community: Planchaloma, Team: DIPEIB-C/ CEMOPLAF: field visit, school activity, community workshop, evaluation.
13. **Collective learning** Workshop Cotopaxi, Team DIPEIB-C/ CEMOPLAF. Participants: delegates from the 3 communities in Cotopaxi.
14. **Multi-stakeholder learning** Team: NGO's Participants from communities and LAA's of the 3 provinces.
15. **Personal learning cycle** Documentation of the process.

members and members of the three communities came together and shared views among each other.

- Cycles 6, 7, 8 and 9 took place in the province of Bolivar and represent a similar process to that in Chimborazo. Cycles 6, 7 and 8 represent the community workshops. Cycle 9 relates to the collective workshop of the three communities.
- Cycles 10, 11, 12 and 13 represent the same process once more, implemented in the province of Cotopaxi.
- Cycle 14 refers to a final 'umbrella' workshop, with members of the three NGO's, the LAA's and all involved communities.
- Finally, cycle 15, represents Marleen's personal process of reflection on the research while documenting and analysing the process with some more distance.

The whole process took place in early 2005 over a period of three months.

Description of the process

First learning cycle: initiating the process

At the start there was not much of a design. It was obvious to the project partners that somehow the local agencies and the communities had to be brought 'on board' so that they would participate in the definition of the project. All agreed that this was best done through a series of workshops. Marleen, who had shown interest in further designing and facilitating this phase as part of her MSc research, felt she should start with a general 'getting-to-know-the-people'. She first met with the partners from the three NGO's involved, and exchanged ideas on the organisation of the series of workshops, activities that could be carried out in the project, and possible local partners, communities and schools to work with. On the basis of these meetings the project partners agreed she should develop a plan.

She decided to first meet with the various LAA members that would be part of the facilitation teams in the three provinces. Together with these members she visited a number of communities they could possibly work with. These visits gave the potential facilitation teams opportunities to share experiences.

The communities reacted very open and positively. They confirmed that particular seeds and animals were not used anymore and they identified the consequences they suffered. Many farmers reacted enthusiastically to the idea of a seed-conservation project. Also schoolteachers reacted with interest and suggested that a project could be combined with the vegetable cropping in school gardens, which served to teach the pupils about organic farming. But apart from the enthusiasm, concerns were raised as well: some earlier projects had failed because false promises had been made by other NGO's. From discussions it became obvious that all were aware that without serious commitment workshops would be a waste of time and money.

Marleen felt she had gained the trust of the LAA representatives. This was confirmed by a firmly expressed willingness of the LAA representatives to co-organise workshops in the communities in which they would further explore the communities' perspectives on seeds they used. The LAA members agreed with her that apparently an initial basis of reciprocal trust and interest had been developed with the communities and that this provided a good foundation for future collaboration.

Learning in Chimborazo

Three community workshops (cycles 2, 3 and 4)

The facilitation team (members of the Dirección de Educación Intercultural Bilingüe de Chimborazo and Marleen) planned visits and workshops in three different communities. Technicians, community educators and community promoters were invited for a four-day program with essentially four elements:

- Introduction of the research to the community leaders.
- Activity at the school. A format was developed on what and how to talk with the children about seeds in their community.
- Workshop with the adult community members. Based on the earlier visits, the team defined the objective and methods of this workshop: (1) to jointly explore the communities' history, (2) the future vision and an analysis of the community; and (3) a *SWOT* analysis (analysis of strengths, weaknesses, opportunities and threats).

- Evaluation of the events by a facilitation team and some community members. The outcomes that refer to the community's knowledge on the use and loss of their seeds are presented in Box 26.1.

Box 26.1. The Chimborazo communities' knowledge and awareness concerning their use and loss of seeds.

Children of fourth and fifth grade defined *seed* as 'that what produces, grows, reproduces and dies'. It gives fruit and seeds, cane, leaves, herbs and flowers. It also gives medicine. They associated seed with germination and seed selection. Seeds are necessary for sowing, harvesting, cooking and eating, milling, peeling, to make puree, to dry and save, to feed the animals and to sell on the market. Most important seeds for children were maize, potato, beans, barley, *oca* (a potato-like turnip), and peas. Problems with seeds were: illnesses, fungi, drought, worms, birds and hailstorm. They did not know if and what seeds the community lost in the past years.
The adults in the three Chimborazo communities identified the same characteristics of seeds as the children. Contrary to the children, they were able to mention a number of species lost. They mentioned papa Uvilla (*Uvilla* potato) and *mashua* (another potato-like turnip) as seeds that were lost from their community. Other seeds which were slowly disappearing are the traditional grain *quínoa*, *melloco* (turnip), papa Chaucha and papa Leona (potato varieties). Also peas, beans, barley and wheat are not produced in quantity anymore.

Visions and SWOT
The community members said that in the future they'd wish to improve their Andean seeds and animals like cows, goats and guinea pigs. They would like to have vegetable gardens, improve education and getting a health care centre. In order to do so, they wanted to conserve their medicinal plants, improve seed selection practices. They wanted to improve their quality of life through integrated farming. They also wanted to train health-care promoters, utilise organic fertilizer, take care of good nutrition, and to (re-)forest the hill sides.
The communities identified strengths, weaknesses, opportunities and threats in four areas: agriculture, organisation, education and health. The following points emerged:

- Strengths
  - In agriculture: collective and private parcels, own seeds, plants, animals and water.
  - In organisation: committees, shops, mills, nurseries, electricity, participation, knowledge, support in the form of ideas and economic, force, interest, compromise, formation of leaders, decision-making, complying promises, punctuality.
  - In education: schools, professors and students, school gardens. »

  - In health: medicinal plants and knowledge.
- Weaknesses:
  - In agriculture: lack of knowledge, lost knowledge about the agro-ecological solar calendar and seed selection, overuse of agro-chemicals.
  - In organisation: de-organisation, lack of knowledge, confidence, financial resources, active participation, interest, morality, complying with promises, morality, valuing our own things, leadership and co-ordination, annual plans and policy.
- Opportunities:
  - In agriculture: sowing of pastures, vegetables, Andean products, forestation, make use of more organic fertilizers, use windbreakers and other known techniques for soil conservation.
  - In organisation: supporting institutions like NGO's, municipalities or the provincial government to start new projects. Observing, dialogue, researching, and learning. Make use of alphabetisation.
- Threats:
  - In agriculture: natural phenomena (plant diseases and insects, hailstorms, night frosts and changing climate) and economic threats (unstable markets and migration).
  - In organisation: criticism of others in the community, egoism of leaders, ignoring of rules, disinterest.
  - In education: teacher strikes, migration of youth, low quality of education.
  - In health: over-use of chemicals in agriculture, environmental pollution, and no toilets.

Collective workshop (cycle 5)

In a collective workshop, the representative community members of the three Chimborazo communities discovered they shared similar histories and visions. Together participants discussed the problem of losing seeds, its root causes, its impact on community living and how to address it. They formulated the following vision:

> "Capacity building of community promoters and educators in the management of integrated farms, sustainable agriculture, health and education to improve the quality of life of the communities' families."

Through a brainstorm on how to realise the vision, suggestions emerged to experiment with seeds, set up seed banks, exchange seeds and improve soil management with organic fertilizers.

Reflection

At the end of the workshop the participants from the communities said they had appreciated the workshop and confirmed their faith in the future project. It was the first time, a farmer from Pueblo Viejo said, that an institution wanted to involve them in the design of a new project. The Chimborazo facilitation team felt good about the relationship they had built up with the community. In a short time the team had developed quite effective methods to collect information on the use of seeds, the peoples' perspectives and their engagement (Box 26.2). The degree of self-reflection of the community members had especially surprised the team. Communities were apparently quite aware of the process of losing seeds and the role seeds played in their livelihood. The team felt that this was an important motivation for them to follow through and engage with these communities in a seed project that responded to the needs and vision of the communities.

Box 26.2. The Bolivar communities' knowledge and awareness concerning their use and loss of seeds.

The situation in the communities in relation to seed systems was quite comparable to that in the Chimborazo communities, only the types of seed they lost varied (white carrot, *mashua*, *quínoa*, potato's Santa Rosa and Ovaleña, lentil 'putsa' and linseed). Like in Chimborazo, they were also losing barley, beans, *melloco*, wheat, peas and *oca*. They now had fewer animals because there was less pasture land.

Vision and SWOT

There was also similarity in the future visions of the community members. They wanted their agriculture to be more sustainable. They felt, however, that they lacked knowledge and wanted to learn about the management of integrated farms, organic fertilizers and soil conservation. Like in Chimborazo, the communities in Bolivar see the natural environment as their biggest challenge. In addition, economic problems and unstable markets make the use of sustainable practices problematic.

Learning in Bolívar

Three community workshops (cycles 6, 7 and 8)

In the province of Bolívar Marleen started with a new team of members from the Centro Médico de Orientación y Planificación Familiar (CEMOPLAF) and farmers of member organisation *Wiñay Kawsay*. They applied a similar approach in this province: three local-community visits in which the team would discuss 'seeds' with the school children, hold workshops with adults and carry out evaluations. The team incorporated the lessons from the Chimborazo workshops. We felt this and the strong participation of the LAA in Bolívar resulted in improvements in the process as compared to Chimborazo: 'smooth' workshops and the collection of more information from the communities. We learned that the communities' knowledge and awareness of their seeds, their use and loss had quite some similarity to the situation in the Chimborazo communities (Box 26.2).

Collective workshop (cycle 9)

In the collective workshop of the three communities in Bolívar, Freddy, a local farmer who successfully cultivated and commercialised thirteen different potato varieties, played an important role. He challenged the community members to act on their collective future vision "to recuperate seeds and plots in an organic way, in integrated farms." He obviously impressed and stimulated fellow farmers.

Reflection

From the various responses we can conclude that also in the province of Bolívar, the process had been successful. The communities were enthusiastic and showed awareness of the need to control the loss of their seeds. The team members of LAA's mentioned that they had learned about the use of different workshop tools and had gained confidence in the success of the project and the participants.

Learning in Cotopaxi

Three community workshops (cycles 10, 11 and 12)

In the province of Cotopaxi, Marleen formed a facilitation team with two members of Dirección Provincial de Educación Intercultural Bilingüe de Cotopaxi (DIPEIB-C). Because the team members were teachers, they were very interested in the possibilities to start a project with the schools. For linkage between the schools and the community, the team invited Centro Médico de Orientación y Planificación Familiar (CEMOPLAF) to co-organise the workshops. This organization has many direct contacts in the community. The collaboration of the two organisations gave the workshops new input and another dimension.

The outcomes of the workshops were comparable to the ones in the other provinces. One difference was that schoolteachers and students in particular were very enthusiastic; community members mentioned that they had little time to participate in workshops.

Collective learning (cycle 13)

As in the other provinces the collective workshop in Cotopaxi was a success. In this workshop, adults as well as children were participating. This gave the discussions an interesting dimension and made the adults aware of the fact that their children had a good understanding of the unsustainable situation in their communities and also had ideas about how to change it. The facilitation team had fruitful conversations and again a basis of trust among community members and with NGO members had grown.

Final learning cycles

Multi-stakeholder learning (cycle 14)

Finally, after the workshops in the provinces, a two-day multi-stakeholder workshop was organised with participation of the NGO's World Neighbors, SwissAid and IIRR, the involved Local Action Agencies DEIB-CH, CEMOPLAF, Wiñay Kawsay and DIPEIB-C; and members of the nine involved communities. The participants were introduced to each other and jointly made a 'tree-analysis' of the causes of the seed problem. The shared problem was defined as 'losing control of our seeds'. It was not always easy for the participants to distinguish causes and consequences of the problem. People also tended to see natural phenomena and markets as a problem, rather than the lack of organisation and interest.

From the discussion the participants concluded that a change process needed to start with a change of attitude: from placing oneself in a victim role to an active role of researching one's own situation to find solutions for problems. This formed the basis of the action plan to be made together.

Communities committed themselves to sharing the outcomes of the tree-analysis with their community members and documenting the feedback and ideas of the community. All LAA's took responsibility for continuing the workshops with their communities. Workshops for analysis of opportunities for the communities to act on the identified seed problem were planned for the three provinces. All communities and organisations were to select two members to participate in a workshop on networking, exchange and documentation of information. Finally, a follow-up of the multi-stakeholder workshop was planned.

Marleen's reflection (cycle 15)

This learning cycle consisted of the documentation of the whole process, using the learning cycle of Kolb. This was an example of an individual learning cycle for Marleen.

Multiple level learning

During the process, participants were learning at one or more of the levels described in the previous paragraphs and visualised in Figure 26.3. Community members mostly learned at the 'community level'. The facilitation team members learned at their level of workshop organisation, implementation and facilitation, but they also learned at community level about the importance of seeds and livelihoods. The Coordination Committee members learned at the community level, the facilitation-team level and the NGO level. This learning at the different levels implies that participants of different groups learned from each other. The facilitation teams, for example, learned from the analysis the communities made of their situation and the Coordination Committee learned from the workshops organized by the facilitation teams and from the reflections of the communities. The facilitation teams learned from the expertise of the Coordination Committee, and so on. The exchange of experiences between the different levels made learning more interesting and deeper. It made the learning process a *social interaction* in which framing and reframing of concepts related to seeds and agriculture on the different levels played a key-role. Social learning – learning that occurred with different stakeholders, in a setting in which people searched for solutions to the actual problem of seed erosion they experienced – helped to create a collective basis to start a project on Informal Seed Systems.

Lessons from the various levels are the following:

- Community level learning: learning about seeds and sustainability
  Community members became more aware of the fact they were losing seeds and the significance that those seeds had in their life. The collective workshops revealed to community members that this is a problem shared with other communities. The facilitation team became better informed about the seeds being used and lost. They were surprised about the level of seed-related knowledge of community members and children. This has helped Marleen to re-frame her picture about seed as a public good: it is necessary to understand

how seeds have a private, i.e. direct, agricultural value to farmers in their daily life to be able to understand better the discussions around conservation of seeds and biodiversity.

- Facilitation-team level learning: learning about involving communities
  The workshops have been an effective way to collect good information on the seeds fairly quickly. Learning about the organisation of workshops and team working was a valuable process and made the team stronger. A long list of lessons about organising and facilitating workshops was developed during the process (Willemsen 2005). The teams realised that organising workshops were a pleasant way to interact and build up a trusting relationship with the communities.
- NGO-level learning: learning about participatory development of an initiative
  The workshops have also been an appropriate way to involve actors in developing ideas for a project. To the members of World Neighbours, IIRR and SwissAid, the experiences of the three teams showed how an 'open' and 'participatory' project design could be planned and implemented. For the three facilitation teams involved the workshop methodologies and tools were an interesting learning experience.
  The facilitation teams learned that the process, in particular the joint analysis of the problem, had contributed to the awareness of the community members regarding genetic erosion. Farmers know about the disappearance of particular crops and seeds, but this is apparently not enough to make people act. The transformation of knowing into an increased awareness indeed resulted in increased participation and willingness of people to take action in order to gain more control over the use of their seeds, which corresponds with the theory of levels of conscience of Freire (1971). Freire defined four levels of awareness: magic consciousness (a belief that everything that happens is in the hands of a higher power), ingenuous consciousness (awareness of the power relationships that exist in the world but accepting these as they are and conforming to them), critical consciousness (understanding the power relationships in the world, not accepting these but trying to change them) and fanatic consciousness (convincing others about the situation that we live in, sometimes by manipulating others). According to Paolo Freire the *critical level* would be the highest level of the four. When people learn and become aware of their own situation of power relationships their level of consciousness can change. Here we also see the connection between Paolo Freire's theory and Pretty's ladder of participation: when people arrive at the critical level of consciousness, they can also move upwards on Pretty's ladder of participation.
  It is important for communities to become aware of their situation and even more about possibilities for change, says Freire. If not, awareness can lead to hopelessness and apathy rather than to collective action. It is still an open question

for Marleen and the co-authors of this chapter to what extent facilitation is an indispensable ingredient for translating awareness into action.

- Individual learning: the learning cycle of Kolb
For the facilitation teams who guided the learning process in the communities, Kolb's theory on four phases in the learning cycle (Figure 26.2) has been very helpful. The theory of the different phases also provided the facilitating teams with a design format for their own learning. Going through the workshop cycles, the reflection moments (the second phase in the cycle of Kolb) helped to find the positive and negative aspects of the new experiences. The lessons learned allowed the team to improve the next cycle. Lessons from one province were taken by Marleen to integrate them into the process for the next province. Kolb's theory also helped to structure the process and the documentation of experiences afterwards.

Reframing the seed issues: learning for sustainability

Following Wals and Heymann (Figure 26.1), the approach to the project development could be called 'participatory', but it was in a way also pre-determined. Although the information on the use of seeds and genetic erosion came from the communities (participation by consultation), via local agencies, the initiative for the project came from NGO's and a donor organization. The intention was to involve communities and LAA's in the following phase in the design of the project through an interactive participatory approach. In order words, they wanted to move from the left upper quadrant (Figure 26.1) to the lower right one. The initial phase of the project, therefore, has become an experiment in how to move from one to another type of participation in the learning for sustainability.

The changes in the participation, i.e. the move of one type to another type, like in Wals and Heymann (2004), is not given much attention by Pretty and others (1995). A lesson from the process described in this chapter is that participation is not fixed or static but 'dynamic'; it can change in the course of the process. How people start to participate is not necessarily the same as the way they participate in a project later on. Also, different people can be acting on different levels of participation in the same process. In this initial phase of the project we aimed at increasing the participation or, in the words of Pretty *et al.* (1995), make people move up the ladder-of-participation. For the Coordination Committee and Marleen it was relevant to question how an environment could be created in which people want to participate, and how to increase the participation. The experiences from this initial phase of the project showed the facilitation teams that the building up of trust is a condition for people to reflect on value issues such as they play a role in the use and loss of seeds.

The reflections have yielded increased awareness for all participants. It has generated a clearer and more explicit frame than that in existence before the workshops. The sharing with others consolidated, enriched or changed the frames. It will be interesting to learn how the framing of the various stakeholders and the participation will change during the continuation of the project.

Acknowledgements

The authors thank in the first place the communities for their hospitality, openness and sharing of their experiences. World Neighbours has been most supportive in facilitating the work with the other actors and creating the possibility for Marleen to become involved and to learn. Finally, the financial support of the McKnight Foundation to this phase of the project is gratefully acknowledged.

References

Almekinders, C.J.M., Louwaars, N.P. and de Bruijn, G.H. (1994) "Local seed systems and their importance for an improved seed supply in developing countries", *Euphytica,* 78: 208-216.

FAO (1986) "Informe final – Reunión sobre cultivos andinos subexplotados de valor nutricional", Oficina Regional para América Lantina y el Caribe, Santiago.

Freire, P. (1971) *Pedagogy of the oppressed*, Harmondsworth: Penguin.

Kolb, D.A. (1984) Experiential learning: Experience as the source of learning and *development*, Englewood Cliffs, NJ: Prentice Hall.

Posey, D.A. (1999) *Cultural and Spiritual Values of Biodiversity. A Complementary Contribution to the Global Biodiversity Assessment*, IT Publications/ INEP, p. 731.

Pretty, J. (2002) *Agri-culture. Reconnecting people, land and nature*, London: Earthscan.

Pretty, J., Guijt, I. & Scoones, I. (1995) *A trainers guide for participatory learning and action*, London: IIED.

Wals, A.E.J. and Heymann, F.V. (2004) "Learning on the edge: Exploring the change potential of conflict in social learning for sustainable living", in A.L. Wenden, ed., *Educating for a culture of Social and Ecological Peace*, New York: State University of New York Press.

Weiskopf, B., Almekinders, C. and von Lossau, A. (2003) "Incentive measures for on farm conservation", in CIP-UPWARD, *Conservation and sustainable use of agricultural biodiversity: a sourcebook*, International Potato Center, Users' Perspectives with Agricultural Research and Development, Los Baños, Laguna, Philippines, pp. 578-589.

Willemsen, M. (2005) *Learning on different levels in the process of participatory project design. Sembrando la Semilla para un Proyecto Participativo*, MSc Thesis EDU/TAD, Wageningen University.

Chapter 27

Learning and living with the Earth Charter

Michael C. Slaby, Brandon P. Hollingshead and Peter Blaze Corcoran

> We must join together to bring forth a sustainable global society founded on respect for nature, universal human rights, economic justice, and a culture of peace. Towards this end, it is imperative that we, the peoples of Earth, declare our responsibility to one another, to the greater community of life, and to future generations. (Earth Charter Preamble)

Introduction

We take the concept of social learning as particularly apt for the processes at work in the community of commitment and practice of the Earth Charter Youth Initiative (ECYI). The ECYI is a small, dynamic group of young Earth Charter advocates from around the world.

The Earth Charter itself is a declaration of fundamental principles for building a just, sustainable, and peaceful world (Appendix 27.1). We see it as an inclusive ethical vision seeking to inspire in all peoples a new sense of global interdependence and shared responsibility for the human family and the larger living world. The Earth Charter is a statement of ethics developed through a larger social learning process of worldwide participation by many tens of thousands of stakeholders in meetings and online discourse over thirteen years. We describe the Earth Charter as a people's treaty and a living document. The Earth Charter Steering Committee decided at its meeting in Amsterdam, in November 2005, to ask the Earth Charter community to suggest changes and improvements in the text itself for future consideration.

Founded in 2000, the Earth Charter Youth Initiative includes several hundred young leaders in some forty countries. The mission of the ECYI is to encourage young people to bring alive the values of justice, sustainability, and peace as they are articulated in the Earth Charter, and to effect positive changes by using the Earth Charter as an ethical guideline.

We believe that the work and structure of the Earth Charter Youth Initiative can be effectively examined through the lens of social learning. Vandenabeele and Wildemeersch (1998) write that social learning is a collaborative reframing process involving multiple interest groups or stakeholders, and is located in the multitude of actions, experiences, interactions, and social situations of everyday life. In bringing together youth across geographic, cultural, and religious boundaries, the ECYI engages young people in just such a collaborative learning process. By encouraging local stakeholders to decide on their own ways of translating the global values and vision of the Earth Charter to the specific contexts of their local communities, they are encouraged to take ownership of the document.

The ECYI strives for a kind of ripple effect – such as after someone throws a stone into a lake, each ripple creating another one until the whole lake vibrates. The more stones that are thrown into the water, the more the ripples can transform into waves – waves of healing water to wash the wounds of our fragmented societies and our tortured planet (Slaby 2005). Grassroots youth activities around the world create a substantive process. In this way, the Earth Charter Youth Initiative demonstrates social learning that is "collaborative and collective, i.e. 'learning communities' or groups of stakeholders...working together to probe, discuss, and test various insights and solutions to environmental problems, and forming networks to promote continuous interaction and communication" (Krasny and Lee 2002).

History and provenance of the Earth Charter

The saga of the writing of the Earth Charter began in 1987 in the lead-up to the 1992 Rio United Nations Conference on Environment and Development and covered thirteen years in all. In the preparatory process, many governments disagreed with the idea of an ethical commitment, so efforts within the United Nations structure were ended. It was also felt that there could be an imposition of an environmental agenda on the global south by northern nations.

Therefore, in 1994, a civil society initiative was launched to advance the development of a people's charter of ethical principles for sustainability. Under the leadership of Maurice F. Strong and Mikhail Gorbachev, the Earth Council was established. The Earth Charter Commission with worldwide membership was formed in 1997. An International Secretariat was established, and a formal drafting process took place from 1997 to 2000.

From a social learning perspective, the process that was used to draft the Earth Charter is quite noteworthy. We believe it to be the most open and participatory collaboration ever used in preparing a global document. Hearings were held throughout the world. Both the Drafting Committee and the Earth Charter Commission, which approved the final wording, were large and widely representative of a diversity of regions of the world, races and ethnicities, faith traditions, and backgrounds. Tens of thousands of additional stakeholders participated in regional, national, and internet-based Earth Charter drafting conferences.

These efforts proved extremely valuable in highlighting areas of consensus, as well as areas of conflict, in relation to the structuring and phrasing of early drafts of the Earth Charter (Earth Charter Commission 2000, Vilela and Corcoran 2005).[54] One such example is from the Inuit Circumpolar Conference, which became deeply involved in the debate concerning the wording of the Earth Charter text; in particular with regard to Principle 7 of Benchmark Draft II, "Treat all living beings with compassion". This was because of the interpretation of the word 'compassion'. Compassion for animals is a very important notion in many religious traditions, but it was unacceptable among the indigenous hunting cultures as related to animals. The Inuit and Hindu stakeholders represented "totally different cognitive agents with multiple perspectives" (Röling 2002). After significant deliberation, the notion of 'respect and consideration' in relation to animals was accepted by all. In terms of social learning theory, we believe this example represents what Niels Röling (2002) has called 'distributed cognition', or 'different but complimentary contributions' to the Earth Charter drafting process.

The last thirty years of international dialogue on environment and development have produced an array of international declarations and paths of action that, together, articulate the international community's understanding of sustainable development. The unique contribution of the Earth Charter is that it provided an integration of the values and principles contained in these documents – and was then refined by a decade-long civil society consultation process where a multiplicity of vantage points and perspectives were included.

This provided credibility and a framework for making clear the ethical vision for sustainable development. The Earth Charter also has the capacity to deal with several of the other problems of social learning in education for sustainable development. It can bring meaning to the multiplicity of viewpoints, diversity of ecologies, and complexity of cultures in which learning for sustainable development must take place.

We know such a vision takes form in a specific cultural setting and at a particular scale. In reality, then, challenges in learning for sustainable development emerge in local interpretation and local implementation approaches. We think it is useful

[54] This story is as told by Mirian Vilela, Executive Director of the Earth Charter International Secretariat, and Peter Blaze Corcoran in *Building Consensus on Shared Values* in Corcoran (2005).

to have the context of an overarching ethical framework to make sense of the diversities and multiplicities of practice.

Social learning within the Earth Charter Youth Initiative

In this chapter, we reflect upon the learning that takes places within the Earth Charter Youth Initiative at two levels: in the small local youth groups that comprise the Earth Charter youth network, and in the international virtual community of electronic communication. At both levels, the initiated processes of social learning are transformative to individuals and communities alike. Social learning has been described as "collective and collaborative learning that links biophysical to the social, cultural and political spheres, the local to the global arena, and action to reflection" (Finger and Verlaan 1995). In this sense, we believe that by engaging with the Earth Charter, young people not only learn about the different aspects of sustainable development, but experience themselves as active promoters of change and contributors to global discourse on sustainability.

Earth Charter community level youth activity includes strictly local efforts inspired by the Earth Charter's call to act collectively and locally towards the common goal of fostering sustainable development. In particular, Subprinciple 12.c of the Earth Charter says, "Honor and support the young people of our communities, enabling them to fulfill their essential role in creating sustainable societies". Initiatives of local Earth Charter Youth Groups, as they are known, such as those in The Philippines to campaign locally against the introduction of genetically modified organisms, are analyzed below.

Initiatives on a larger national scale, such as the Earth Charter Youth Group in Sierra Leone, are then described. This group uses the tenets of a specific principle of the Earth Charter, Principle 16, to reconcile ex-combatants of the horrible decade-long civil war with their local communities. Although the geographies, technologies, and socio-political contexts of Earth Charter activists vary from nation to nation, common aspects of social learning are found in the projects of the ECYI network. In these cases, stakeholders – Earth Charter youth – demonstrate interactive problem-solving, conflict resolution, shared learning, convergence of goals, and concerted action as they promote the ethical values and principles of the Earth Charter: respect and care for the community of life; ecological integrity; social and economic justice; and democracy, non-violence, and peace.

Locally in The Philippines

The Eco Trekkers Society, Inc. (ETSI) of Negros Occidental, The Philippines, was created as a college organization at the Technical University of The Philippines

– Visayas and solely as an adventure club.[55] At the time of inception, members' interests focused on mountaineering and outdoors activities. However, in response to timber poaching and illegal forest activities encountered while hiking on the island of Negros Occidental, as well as genetically engineered 'Bt-Corn' introduced throughout The Philippines, the group widened its focus to include environmental awareness and education.

The shift in focus from recreation and social outings to environmental and educational activism demonstrates a key tenet of social learning: stakeholders learning and responding through observation and interaction with their social and environmental context. In 2002, ETSI formally registered as a non-governmental youth organization with the goal to "advance the role of (Filipino) youth and actively involve them in the protection and promotion of sustainable development" (Yap 2005). In 2003, the group registered as an official Earth Charter Youth Group. The group has endorsed the Earth Charter as its guiding principles and ethical framework.

By basing their local activities on the ethical principles of the Earth Charter, Eco Trekkers expresses its solidarity with youth organizations who are working on similar issues in other parts of the world. However, the process of locally adapting the Earth Charter is not a linear one. The Earth Charter is not used as an 'ultimate truth' that has to be strictly applied to any given situation or local context. Rather, we see it as a stimulus to reflect upon the aims and values that guide our collective behavior. This questioning of guiding motivations makes room for manifold processes of social learning.

Maarleveld and Dangbegnon have written that social learning includes "learning by individuals through interaction with their social context ...learning pertaining to social issues ...and learning that results in recognizable social entities" (1999). Against a backdrop of youth unemployment and environmental degradation, Eco Trekkers' main focus is organizing youth from a grassroots level, where poverty and lack of education often hinders young people from actively participating in finding solutions to issues that affect them. In particular, the Eco Trekkers Earth Charter Youth Group embraces the environmental and ecological aspects of the Earth Charter.

One main facet of Eco Trekkers outreach and education programs is to campaign locally against genetically modified organisms (GMOs). It is based on Earth Charter Subprinciple 5.d, "Control and eradicate non-native or genetically

[55] This story is as told by youth activist Khyn P. Yap in *Using the Earth Charter in Local Campaigns Against Genetically Modified Organisms* in Corcoran (2005).

modified organisms harmful to native species and the environment, and prevent introduction of such harmful organisms". While other activist groups lobby for the government's rejection of GMOs, the young people of ETSI work at a local level to educate consumers in their community about the threat that GMOs pose to health and environment. They do this by organizing small group discussions in schools, hosting monthly public forums, distributing flyers that list GMO foods sold in local markets, and displaying public exhibits that warn about the risks of introducing engineered crops. This dialogue engages Earth Charter Youth within and among local communities. Against this background, discussions on the Earth Charter principle on GMOs broadened the participants' perspectives and revealed the inextricable interconnection of The Philippines' environmental and economic challenges.

By using the Earth Charter Youth Group identity, ETSI demonstrates its connectedness to other youth around the world who promote the vision of the Earth Charter in their local communities. In addition, Eco Trekkers sees participation in the Earth Charter Youth Initiative as a way to strengthen its efforts of disseminating the Earth Charter within its community, within its networks of outdoor recreation and adventuring, and within its networks of the wider youth sustainability movement. Khyn Yap, President of Eco Trekkers Society, Inc., has written

> "Our involvement in the Earth Charter Youth Initiative makes us realize that we are not alone in striving to make our world a better place, but that there are individuals like us who persistently pursue peace and sustainable development. We became aware that small local actions, like...campaigns against harmful genetically modified organisms, make a big impact nationally and internationally. And if we all acted together, who says that we cannot build a sustainable and peaceful world?" (Yap 2005)

The worldwide global dialogue taking place among the larger group of the Earth Charter Youth Initiative, to which Yap alludes, is analyzed below.

Nationally in Sierra Leone

In some parts of the world, several Earth Charter Youth Groups are formed within a single nation, as in the case of Sierra Leone. Youth activists across the country have united as the Earth Charter Youth Group-Sierra Leone (ECYG-SL).[56] Where

[56] This story is as told by youth activist Sylvanus Murray in *Using the Earth Charter with Ex-Combatants in Sierra Leone* in Corcoran (2005).

Eco Trekkers works toward sustainability through the Earth Charter's call for environmental activism, the Sierra Leone group responds to issues of social and economic justice, as well as democracy, non-violence, and peace. Sierra Leone currently faces the challenge of social reconstruction after enduring a civil war that concluded in 2002. The bulk of the warring factions were comprised largely of young people. They took instructions from their older commanders to loot, burn, rape, and kill. Thus, the issue of reconciling ex-combatants with their local communities is imperative in the current post-war situation.

Working as a committee of representatives from umbrella youth organizations and community-based organizations, Earth Charter Youth Group-Sierra Leone uses the Earth Charter to "support mutual understanding, solidarity, and cooperation" (Earth Charter Subprinciple 16.a) between ex-combatants and their communities. This work is done through community meetings and sensitization sessions for young men who are being disarmed, demobilized, and reintegrated into society. These sessions gave former soldiers the chance to repent for the deeds of cruelty that they were forced to commit during the times of civil war, and they were welcomed by the victimized community. In other instances, the dialogues hinge on the economic problems that were among the root causes of the conflict, and lead to collective efforts to generate new forms of income for the members of the community.

ECYG-SL empowers ex-combatants to participate in projects that develop and safeguard communities, with a special emphasis on peacemaking, tolerance, ecological, and cultural components. This work relates to social learning in that it allows for "interaction in which all who feel the need are free and have equal chances to express their views, and that they do so in an understandable, legitimate, and truthful manner" (Maarleveld and Dangbegnon 1999). This interaction is particularly important in sensitization and tolerance-building. It provides a framework for creating a safe environment in which different views are shared in a truthful, understanding, and non-violent manner. The skills of peaceful conflict resolution and conflict transformation are prerequisites for social learning, as only on their basis can opposing views and opinions be regarded as beneficial and enriching to one's own perspective.

Members of the Sierra Leone youth group use the Earth Charter as a comprehensive strategy to prevent violent conflict and to manage and resolve environmental conflict disputes. One example of action comes from the Firestone Community, where Earth Charter youth activists established a home-garbage collection program that created employment opportunities for young people, helped to create sanitary conditions, and reduced the number of violent conflicts over pollution in the local river and water supply.

Sylvanus Murray, coordinator of the Earth Charter Youth Group in Sierra Leone, sees the links of environmental protection, economic development, and peace espoused in the Earth Charter. He writes "In search of a new vision that promotes economic stability, respect for all forms of life, good governance, human rights, and democracy, the youths in our country have found the Earth Charter as a guiding document. ... [T]hese issues must be addressed in an integrated approach, as it is outlined in the Earth Charter" (Murray 2005).

Social learning and the Earth Charter Youth Initiative in cyberspace

Social learning theory recognizes the existence of collective learning goals and the need for creating the right conditions for stimulating the learning of individuals. It can be viewed as an intentionally-created, purposeful learning process that hinges on the presence of the 'other' or others (Wals and Heymann 2004). Since 2002, the Internet has been the primary means of communication through which ideas and experiences are shared among the widely-scattered Earth Charter Youth Initiative. We see it as giving a human 'face' to the abstraction of interconnectedness. E-mail communication links activists from different cultural and economic circumstances. In this process of linking action, diversity, and global thinking, much learning takes places. It is a challenging but hopeful process which creates shared participation and engagement. This zeal and commitment contributes an informed perspective to the larger Earth Charter Initiative.

As an example of this electronic communication, the Earth Charter Youth Initiative has discussed the difficulties of locally adapting the Earth Charter principles. A Chinese member of the network shared experiences of how difficult it is to clear the ground for the Earth Charter's approach to propagating ethical principles in the cultural context of her country. As a reaction, Canadian ECYI members with Chinese background shared their insights of how they involved their peers in meaningful dialogues and activities on the Earth Charter. This Chinese-Canadian interaction resulted in a partnership between the two organizations.

The Earth Charter Youth Initiative benefits from its loose and creative structure, linking youth around the world who share the ethical vision of the Earth Charter and strive to make it a reality for their local and national communities. These youth have developed a remarkable range of ideas for bringing the Earth Charter into action and spreading its message among their peers: from the Armenian summer camps focusing on environmental issues and distributing children's versions of the Earth Charter in three different languages; to the Costa Rican Earth Charter Concerts; to the pupils and students on the Balearic Islands who drafted their own local Earth Charters and lobbied their school-boards for endorsements.

By facilitating the sharing of local and international youth experiences and successful projects using the Earth Charter, the ECYI serves as a global 'high five' for young activists around the world to acknowledge and support each other beyond geographical, cultural and religious boundaries. The groups and individuals who are linked through the internet get a sense that they are not alone in their efforts. This connection empowers them to stay active and helps to recharge their enthusiasm.

Through the Earth Charter Youth Initiative, young people from the Western world are brought into direct exchange with young activists from Africa, the Middle East, Latin America, and Asia. Activists share the vision of creating a more sustainable world, but translate the principles of the Earth Charter into different culture-specific programs and projects. In entering into the cross-cultural online communication, participants learn from diverse approaches to Earth Charter principles. A sense of solidarity is strengthened through transnational collaboration and global North-South partnerships.

Recently, such youth perspective has been crucial to the global activists' meeting celebrating five years of the Earth Charter, which was attended by many from the ECYI. They also contributed significantly to a new book published for the occasion, *The Earth Charter in Action: Toward a Sustainable World* (Corcoran 2005). Here, they represent about one out of five contributors.

Conclusion

We hope the authentic participation of youth in the Earth Charter movement can be a model; there is a great need globally to integrate the knowledge and perspectives of youth into society. This is the aim of learning and living with the Earth Charter – personal empowerment in order to participate actively in sustainable development.

In addition to the significant learning at the individual level, important social learning is enabled by the structure of the Earth Charter Youth Initiative. This social learning takes place in locally-organized projects and in nationally-organized efforts. Learning from these initiatives enriches the global virtual community of electronic exchange. This organic process represents the kind of social learning that informs and strengthens the work at all levels.

Since its launch, the Earth Charter Youth Initiative has confirmed youth's interest in establishing a sound, ethical foundation for the emerging global society. This takes place with no significant financial assistance, but rather by nurturing the enthusiasm of dedicated young people ready to spend their free time striving to make the world a better place. Young people, who are directly and adversely affected by economic globalisation, have realised how essential it is to find holistic solutions to these challenges. They have demonstrated that the Earth Charter plants seeds of hope in people's hearts. Youth who are learning and living with the Earth Charter can use the tool of social learning to reflect critically upon their work.

References

Corcoran, P.B., ed. (2005) *The Earth Charter in Action: Toward a Sustainable World,* Amsterdam: Royal Tropical Institute (KIT) Publishers.

Earth Charter Commission (2000) *The Earth Charter.* San Jose, Costa Rica. Available online at http://www.earthcharter.org

Finger, M. and Verlaan, P. (1995) "Learning Our Way Out: A Conceptual Framework for Social-Environmental Learning", *World Development,* 23(3): 503-513.

Krasny, M. and Lee, S. (2002) "Social Learning as an Approach to Environmental Education: Lessons from a Program Focusing on non-Indigenous, Invasive Species", *Environmental Education Research,* 8(2): 101-119.

Maarleveld, M. and Dangbegnon, C. (1999) "Managing Natural Resources: A Social Learning Perspective", *Agriculture and Human Values,* 16: 267-280.

Murray, S. (2005) "Using the Earth Charter with ex-combatants in Sierra Leone", in P.B. Corcoran, ed., *The Earth Charter in Action: Toward a Sustainable World,* Amsterdam: Royal Tropical Institute (KIT) Publishers, pp. 151-152.

Röling, N. (2002) "Beyond the Aggregation of Individual Preferences: Moving from Multiple to Distributed Cognition in Resource Dilemmas", in C. Leeuwis and R. Pyburn, eds., *Wheelbarrows Full of Frogs: Social Learning in Rural Resource Management,* Assan, The Netherlands: Koninklijke Van Gorcum, pp. 25-48.

Slaby, M.C. (2005) "Making Ripples of Change: The Hopes of the Earth Charter Youth Initiative", in P.B. Corcoran, ed., *The Earth Charter in Action: Toward a Sustainable World,* Amsterdam: Royal Tropical Institute (KIT) Publishers, pp. 113-114.

Vandenabeele, J. and Wildermeersch, D. (1998) "Learning for Sustainable Development: Examining Life-World Transformation Among Farmers", in D. Wildermeersch, M. Finger and T. Jansen, eds., *Adult Education and Social Responsibility,* Frankfurt am Main: Peter Land Verlag, pp. 115-132.

Vilela, M. and Corcoran, P.B. (2005) "Building Consensus on Shared Values", in P.B. Corcoran, ed., *The Earth Charter in Action: Toward a Sustainable World,* Amsterdam: Royal Tropical Institute (KIT) Publishers, pp. 17-22.

Wals, A.E.J. and Heymann, F. (2004) "Learning on the Edge: Exploring the Change Potential of Conflict in Social Learning for Sustainable Living", in A. Wenden, ed., *Educating for a Culture of Social and Ecological Peace,* New York: State University of New York Press, pp. 123-144.

Yap, K.P. (2005) "Using the Earth Charter in Local Campaigns Against Genetically Modified Organisms", in P.B. Corcoran, ed., *The Earth Charter in Action: Toward a Sustainable World,* Amsterdam: Royal Tropical Institute (KIT) Publishers, pp. 69-70.

Appendix 27.1

The Earth Charter Preamble and main principles

http://www.earthcharter.org

The full Earth Charter consists of a Preamble, sixteen main principles with sixty-one subprinciples, and "A way forward". This appendix includes the Preamble and the main principles only.

Preamble

We stand at a critical moment in Earth's history, a time when humanity must choose its future. As the world becomes increasingly interdependent and fragile, the future at once holds great peril and great promise. To move forward we must recognize that in the midst of a magnificent diversity of cultures and life forms we are one human family and one Earth community with a common destiny. We must join together to bring forth a sustainable global society founded on respect for nature, universal human rights, economic justice, and a culture of peace. Towards this end, it is imperative that we, the peoples of Earth, declare our responsibility to one another, to the greater community of life, and to future generations.

Earth, our home

Humanity is part of a vast evolving universe. Earth, our home, is alive with a unique community of life. The forces of nature make existence a demanding and uncertain adventure, but Earth has provided the conditions essential to life's evolution. The resilience of the community of life and the well-being of humanity depend upon preserving a healthy biosphere with all its ecological systems, a rich variety of plants and animals, fertile soils, pure waters, and clean air. The global environment with its finite resources is a common concern of all peoples. The protection of Earth's vitality, diversity, and beauty is a sacred trust.

The global situation

The dominant patterns of production and consumption are causing environmental devastation, the depletion of resources, and a massive extinction of species. Communities are being undermined. The benefits of development are not shared equitably and the gap between rich and poor is widening. Injustice, poverty, ignorance, and violent conflict are widespread and the cause of great suffering. An unprecedented rise in human population has overburdened ecological and social systems. The foundations of global security are threatened. These trends are perilous – but not inevitable.

The challenges ahead

The choice is ours: form a global partnership to care for Earth and one another or risk the destruction of ourselves and the diversity of life. Fundamental changes are needed in our values, institutions, and ways of living. We must realize that when basic needs have been met, human development is primarily about being more, not having more. We have the knowledge and technology to provide for all and to reduce our impacts on the environment. The emergence of a global civil society is creating new opportunities to build a democratic and humane world. Our environmental, economic, political, social, and spiritual challenges are interconnected, and together we can forge inclusive solutions.

Universal responsibility

To realize these aspirations, we must decide to live with a sense of universal responsibility, identifying ourselves with the whole Earth community as well as our local communities. We are at once citizens of different nations and of one world in which the local and global are linked. Everyone shares responsibility for the present and future well-being of the human family and the larger living world. The spirit of human solidarity and kinship with all life is strengthened when we live with reverence for the mystery of being, gratitude for the gift of life, and humility regarding the human place in nature.

We urgently need a shared vision of basic values to provide an ethical foundation for the emerging world community. Therefore, together in hope we affirm the following interdependent principles for a sustainable way of life as a common standard by which the conduct of all individuals, organizations, businesses, governments, and transnational institutions is to be guided and assessed.

Principles

I. Respect and care for the community of life

1. Respect Earth and life in all its diversity.
2. Care for the community of life with understanding, compassion, and love.
3. Build democratic societies that are just, participatory, sustainable, and peaceful.
4. Secure Earth's bounty and beauty for present and future generations.

In order to fulfill these four broad commitments, it is necessary to:

II. Ecological integrity

5. Protect and restore the integrity of Earth's ecological systems, with special concern for biological diversity and the natural processes that sustain life.
6. Prevent harm as the best method of environmental protection and, when knowledge is limited, apply a precautionary approach.
7. Adopt patterns of production, consumption, and reproduction that safeguard Earth's regenerative capacities, human rights, and community well-being.
8. Advance the study of ecological sustainability and promote the open exchange and wide application of the knowledge acquired.

III. Social and economic justice

9. Eradicate poverty as an ethical, social, and environmental imperative.
10. Ensure that economic activities and institutions at all levels promote human development in an equitable and sustainable manner.
11. Affirm gender equality and equity as prerequisites to sustainable development and ensure universal access to education, health care, and economic opportunity.
12 Uphold the right of all, without discrimination, to a natural and social environment supportive of human dignity, bodily health, and spiritual well-being, with special attention to the rights of indigenous peoples and minorities.

IV. Democracy, nonviolence, and peace

13. Strengthen democratic institutions at all levels, and provide transparency and accountability in governance, inclusive participation in decision making, and access to justice.
14. Integrate into formal education and life-long learning the knowledge, values, and skills needed for a sustainable way of life.
15. Treat all living beings with respect and consideration.
16. Promote a culture of tolerance, nonviolence, and peace.

Epilogue

Creating networks of conversations

Arjen E.J. Wals

So where are we now? Having travelled through the various parts of this book, are there any enlightening emerging patterns? Does social learning have any value in co-creating pathways towards a world that is more sustainable than the world today? Or is it just a new hype or at best a new label for many of its predecessors, some of which are still on-going: action research, community problem-solving, grassroots learning, collaborative learning, action learning, and so on? Depending on the route you have taken to get to this point you may find different answers to these questions, none of them being definitive in all likelihood. Still there are some enlightening patterns emerging from the principles, perspectives and praxis of social learning as presented and discussed in the 27 chapters of this volume. Indeed, social learning appears to be more than just a hype and helps in reconceptualising learning.

If there is any agreement among the contributors to this book, it is that the interactions between people are viewed as providing possibilities or opportunities for meaningful learning. But there is more of course. Many chapters highlight the value of 'difference' and 'diversity' in energizing people, creating dissonance and unleashing creativity. The importance of both reflection and reflexivity in social learning has been repeatedly emphasized in all three sections of the book. Many contributors speak of the power of 'social cohesion' and 'social capital' in creating change in complex situations characterised by varying degrees of uncertainty. And, finally, the power of collaborative action that preserves the (unique) qualities of each individual should not be underestimated.

The success of social learning depends a great deal on the collective goals and/or visions shared by those engaged in the process. Whether such collective goals and/or visions can actually be achieved depends, to a degree, on the amount of space for possible conflicts, oppositions and contradictions. Moving towards sustainability or sustainable living, inevitably involves diverging norms, values, interests and constructions of reality. If there is one guiding principle to be distilled from the chapters in this book, it is that such differences need to be explicated rather than concealed. By explicating and deconstructing the oftentimes diverging norms, values, interests and constructions of reality people bring to a sustainability challenge, it not only becomes possible to analyze and understand their roots and their persistence, but also to begin a collaborative change process in which shared meanings and joint actions emerge.

Learning often results from a critical analysis of one's own norms, values, interests and constructions of reality (deconstruction), exposure to alternative ones (confrontation) and the construction of new ones (reconstruction). Such a change process is greatly enhanced when the learner is mindful and respectful of other perspectives. Obviously, not all participants in a social learning process, as we have seen in this book, display the same amount of initial openness and respect, but as they develop social relationships and mutual respect (social capital), they not only become more open towards ideas alternative to their own, they, as a group, also become more resilient and responsive to challenges both from within and from outside.

Given the importance of conflict and dissonance in social learning, it is important to be mindful of people's comfort zones or dissonance thresholds. Some people are quite comfortable with dissonance and are challenged and energized by radically different views, while others have a much lower tolerance with regards to ideas conflicting with their own. The trick is to learn on the edge of peoples' individual comfort zones with regards to dissonance: if the process takes place too far outside of this zone, dissonance will not be constructive and will block learning. However, if the process takes place well within peoples' comfort zones – as is the case when homogenous groups of like-minded people come together – learning is likely to be blocked as well. Put simply: there is no learning without dissonance, and there is no learning with too much dissonance! Ideally facilitators of social learning become skilful in reading peoples' comfort zones, and when needed, expanding them little by little. An important role of facilitators of social learning is to create space for alternative views that lead to the various levels of dissonance needed to trigger learning both at the individual and at the collective level.

Frame awareness, frame deconstruction and reframing (Kaufman and Smith 1999) can be viewed as central steps in transformative social learning. People can become so stuck in their own frames – ideas, ways of seeing things, ways of looking at the world, ways of interpreting reality – that they may fail to see how those frames colour their judgment and interaction. Perhaps the essence and success of social learning lies in people's ability to transcend their individual frames, so that they can reach a plane where they are able find each other and create enough 'chemistry' to feel empowered to work jointly on the challenges they come to share. An important first step in social learning is becoming aware of one's own frames. Only then can deconstruction (sometimes referred to as deframing) begin (Wals and Heymann 2004). Deconstruction is then seen as a process of untangling relationships, becoming aware of one's own hidden assumptions, their ideological underpinnings and the resulting blinding insights they provide. When this is done in a collaborative setting, where dissonance is properly managed, cultivated and utilised, participants become exposed to the deconstructed frames of others, begin to rethink their old ideas and are challenged to jointly create new ones (co-creation).

It is hard to capture social learning in a neat process or cycle, and virtually all contributors steer clear of doing so, but there are some 'sequential moments' or activities that might be helpful when trying to design and monitor social learning (see also Wals and Heymann 2004):

- *Orientation and exploration*: Identifying key actors and, with them, key issues of concern or key challenges to address in a way that connects with their own prior experiences and background thereby increasing their motivation and sense of purpose.
- *(Self)awareness raising*: Eliciting one's own frames relevant to the issues or challenges identified.
- *Deframing or deconstructing*: Articulating and challenging one's own and each other's frames through a process of clarification and exposure to conflicting or alternative frames.
- *Co-creating*: Joint (re)constructing of ideas, prompted by the discomfort with one's own de-constructed frames and inspired by alternative ideas provided by others.
- *Applying / experimenting*: Translating emergent ideas into collaborative actions based on the newly co-created frames, and testing them in an attempt to meet the challenges identified.
- *Reviewing*: Assessing the degree to which the self-determined issues or challenges have been addressed, but also a review of the changes that have occurred in the way the issues/challenges were originally framed, through a reflective and evaluative process.

Arguably, a preliminary phase is needed, before entering this cycle of activities. In this phase the initiators of the change process, reflect on the nature of the change process by asking questions such as: "Is the kind of change that is desired of a more emancipatory or of a more instrumental nature?" And, "Is there sufficient political and organisational space available for engaging people in a participatory process characterised by high levels of self-determination and autonomy?" need to be asked in order to be able to confidently introduce and enhance social learning as an important vehicle for realising change.

It should be noted that although these activities can be distinguished, they are hard to separate in reality as they interrelate and overlap. They also suggest a linearity one seldom finds in social learning processes since social learning, as pointed out by virtually all the contributors, is more of an on-going, cyclical and emergent process. Furthermore, having an evaluation moment at the end suggests that this is a one-off activity which it obviously is not: social learning requires reflection and reflexivity throughout the entire process, if only to improve the quality of the process itself and to monitor change and progress throughout. Interestingly enough the sequence of activities as presented here resembles the conceptual change process as described by Driver and Oldham (1986) in the context of children's learning in science.

But what about sustainability? After all, the title of this book is *'Social learning towards a sustainable world'.* People around the world, scientists and policy-makers alike, are working on identifying 'indicators of sustainable development' (128,000 Google hits on November 17, 2006) or 'sustainability indicators' (411,000 Google hits on November 17, 2006). Many scientists working on sustainability are doing so at the request of international organisations like UNESCO, UNECE, UNEP and the World Bank, or at the request of national governments. Sustainability and sustainable development – but also 'Education for Sustainable Development' as a means to 'realise' sustainability – have deeply penetrated the world of policy. There is a strong need to translate these policies into concrete actions with measurable outcomes by creating benchmarks and standards that heavily rely on Specific, Measurable, Acceptable, Realistic, Time-specified (SMART) goals. To have an exhaustive list of sustainability indicators seems very handy for becoming SMART in working towards a more sustainable world. Interestingly enough none of the contributions focus on sustainability as a measurable outcome. Instead they focus on the *processes* and the *conditions* needed to engage people in issues related to sustainability.

Although all authors will probably agree that our current way of living on this planet is unsustainable and something needs to, indeed, radically change in the way we live, interact, do business, use resources, and so on, most authors stop short of defining sustainability. They suggest – some more explicitly than others – that it would be pretentious to declare what 'sustainability' is exactly, let alone how it should be implemented. In fact, they suggest that doing so would take the learning out of creating a world that is more sustainable than the one we currently live in. A red thread running through this book is that the key to creating a more sustainable world lies precisely in *learning*, and not just any learning, but rather in transformative learning that leads to a new kind of thinking, alternative values and co-created, creative solutions, co-owned by more reflexive citizens, living in a more reflexive and resilient society.

Hence, sustainable living requires more than consensus in the present about what sustainability is or even might be. While there is a constellation of ideas as to what a sustainable world might entail, the lack of consensus about the implications of an exact meaning – if this were at all possible – in variable contexts, prevents global prescriptions. Instead contextual solutions are required that are, at least partly, co-created and co-owned by those who are to (want to?) live sustainably. Forcing consensus on how people should live their lives is undesirable from a deep democracy perspective, and from an emancipatory education perspective it is essentially 'mis-educative' (Dewey 1916, Wals and Jickling 2002). This is not to say that having indicators for sustainability is necessarily a bad thing, but the questions then become: For whom are these indicators? How have they been created? By whom? Are they carved in stone or subject to change and even abolition? The process of identifying indicators can in and by itself be a very useful part of social learning, but when indicators are then authoritatively generated and prescribed, the transformative learning disappears and is replaced by conditioning and training.

Social learning – albeit as a spontaneously emerging property of people interacting together or as an intentionally introduced and facilitated process of change – not only allows for commonalities and social capital to form, it also provides space for discord and 'dissensus'. From this perspective democracy and participation, much like social learning, depend on this space for difference, dissonance, conflict, and antagonism. This also suggests that deliberation is radically indeterminate (Goodman and Saltman 2002). The conflicts that emerge in the exploration of sustainable living become prerequisites *for* rather than barriers *to* learning. Sustainable living requires dialogue to continuously shape and re-shape ever changing situations and conditions. A dialogue here requires that stakeholders involved *can* and *want* to negotiate as *equals* in an *open* communication process which *celebrates diversity* and *conflict* as the driving forces for development and social learning (Wals and Bawden 2000). Such dialogue can indeed spontaneously emerge, but can be enhanced and up-scaled with careful designing and planning, as some of the chapters in this book have shown.

To what extent does this book itself have the potential to trigger social learning? This book indeed consists of, in referring to Fritjof Capra's preface, a network of conversations. Some authors refer to the same key thinkers as inspirational or influential to their own thinking. Most authors have never met each other but appear to be bound by common interests and common sources –albeit not always interpreted in the same way. Figure E.1 illustrates how the various chapters are linked by some key thinkers. Included are those key thinkers to whom four or more chapters refer. Of course, this is not to suggest that that there aren't any other influential sources in this field. Inevitably there are lesser known, but to some, very influential works that did not make the figure purely because of an arbitrary threshold. Nonetheless the figure does show, at a glance, which names appear to be influential to many who have contributed to this book: Wenger, Senge, Argyris, Schön, Leeuwis, Giddens, Habermas and Röling, but also the recent work of Keen,

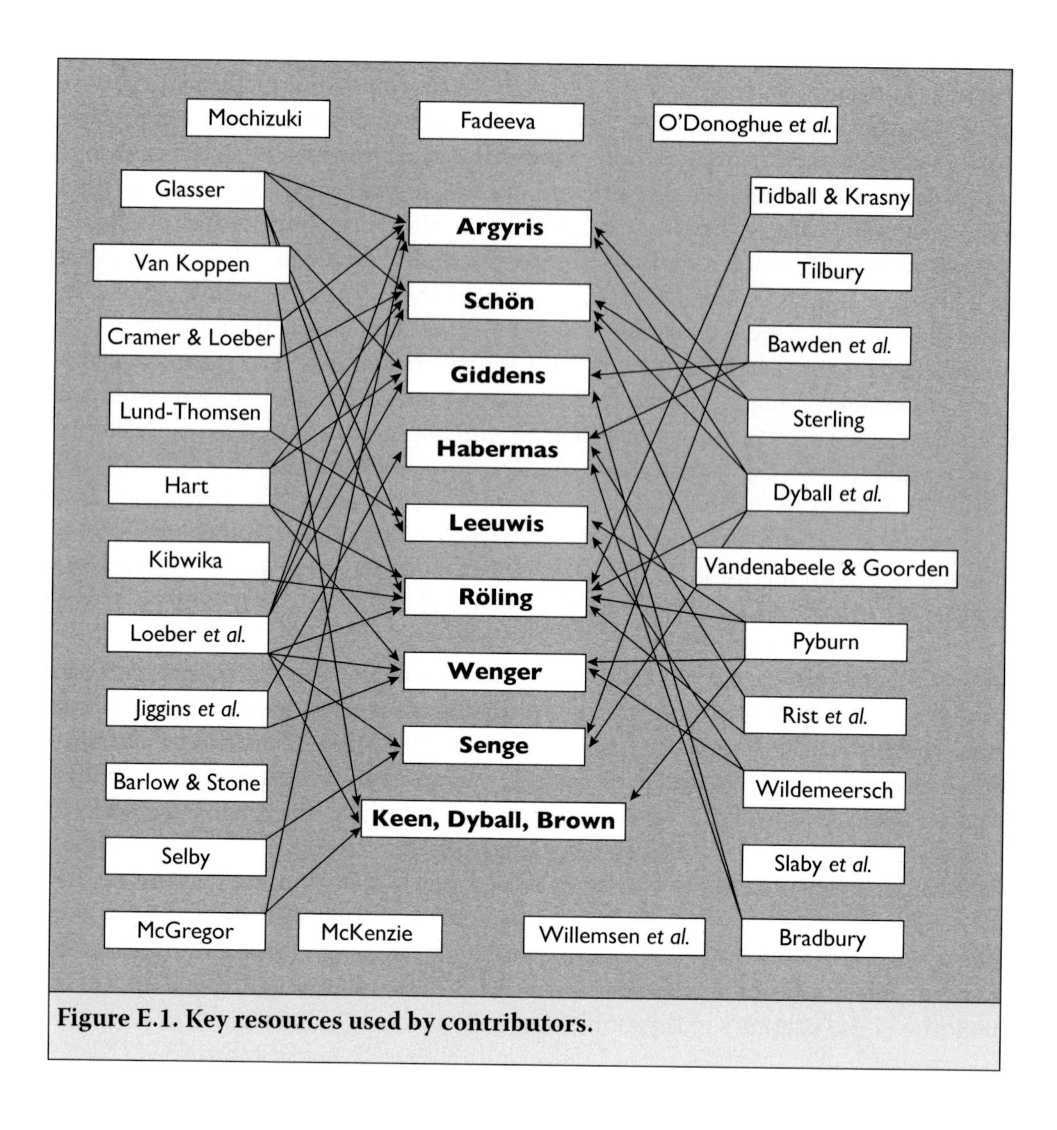

Figure E.1. Key resources used by contributors.

Dyball and Brown on social learning in the context of environmental management (Keen *et al.* 2005). It also shows that the work of Bandura, seen by some as the founding father of social learning, is not referred to by many of the contributors to this book.

To what extent then is this book a network of *diverse* conversations? One could argue that most contributors are people who by and large have the luxury to reflect on and write about things like social learning in the context of sustainability. Some indeed do know each other or know of each other's work. Figure E.2 illustrates

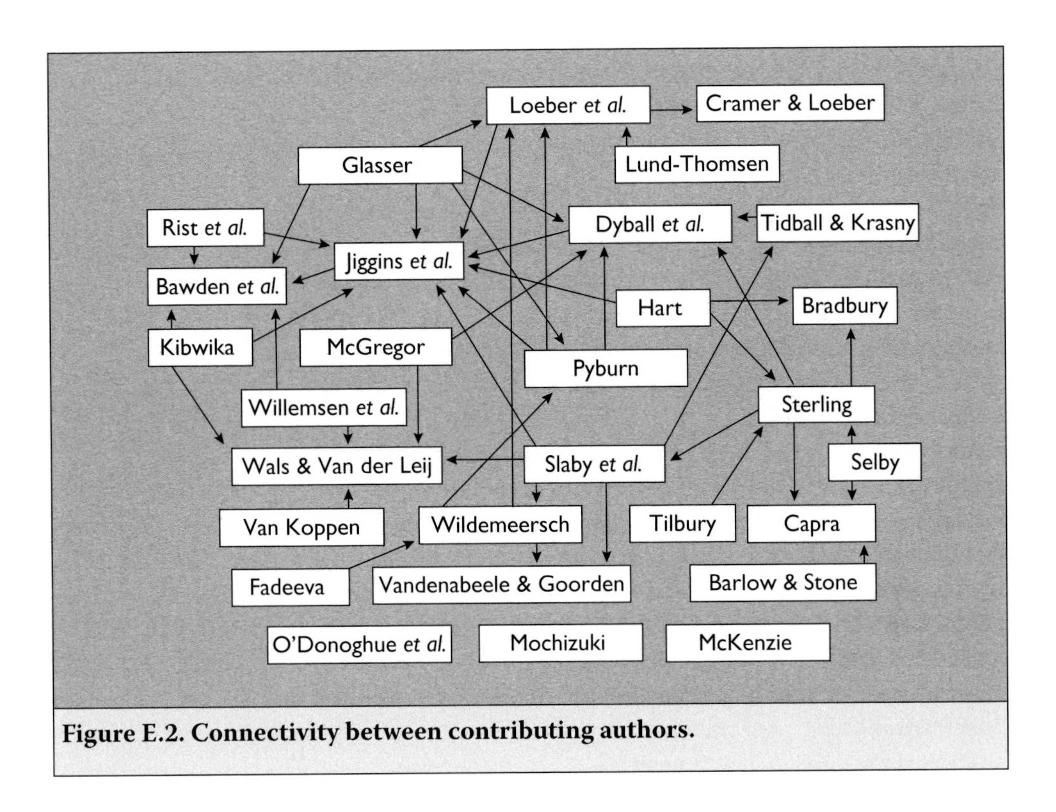

Figure E.2. Connectivity between contributing authors.

the extent to which the various contributing authors refer to one another.[57] Some authors are cited by a number of other contributors; others are not and tend to come from different disciplinary perspectives.

Nonetheless, one could argue that most of the contributors have an academic background, although some would rather describe themselves as reflective practitioners or change agents. In this sense the book is somewhat one-dimensional. But the glass is half-full as well: there appears to be a balance between men and women contributing to the book, different generations are a part of this book (varying from the low twenties to the low seventies), many regions of the world are represented in some way (Table E.1), and some have never heard of the other authors contributing and are rooted in other networks of conversations.

[57] Note that when an arrow points to a chapter written by multiple authors it could be that a chapter refers to the work of just one of them.

Table E.1. Geographical distribution of contributor's workplaces.

Name	Country of workplace	Continent
Fritjof Capra	United States of America	North America
Harold Glasser	United States of America	North America
Stephen Sterling	United Kingdom	Europe
Anne Loeber	The Netherlands	Europe
Barbara van Mierlo	The Netherlands	Europe
John Grin	The Netherlands	Europe
Cees Leeuwis	The Netherlands	Europe
Danny Wildemeersch	Belgium	Europe
Daniella Tilbury	Australia	Australia
Richard Bawden	United States of America	North America
Irene Guijt	The Netherlands	Europe
Jim Woodhill	The Netherlands	Europe
Keith Tidball	United States of America	North America
Marianne Krasny	United States of America	North America
David Selby	United Kingdom	Europe
Robert Dyball	Australia	Australia
Valerie A. Brown	Australia	Australia
Meg Keen	Australia	Australia
Joke Vandenabeele	Belgium	Europe
Lieve Goorden	Belgium	Europe
Rhiannon Pyburn	The Netherlands	Europe
Stephan Rist	Switserland	Europe
Freddy Delgado	Bolivia	South America
Urs Wiesmann	Switserland	Europe
Zinaida Fadeeva	Japan	Asia
Jacqueline Cramer	The Netherlands	Europe
Hilary Bradbury	United States of America	North America
Peter Lund-Thomsen	Denmark	Europe
Paul Hart	Canada	North America
Marcia McKenzie	Canada	North America
Sue McGregor	Canada	North America
Kris van Koppen	The Netherlands	Europe
Yoko Mochizuki	Japan	Asia
Michael K. Stone	United States of America	North America
Zenobia Barlow	United States of America	North America
Janice Jiggins	The Netherlands	Europe
Niels Röling	The Netherlands	Europe

Table E.1. Continued.		
Erik van Slobbe	The Netherlands	Europe
Rob O'Donoghue	South Africa	Africa
Heila Lotz-Sisitka	South Africa	Africa
Robert Asafo-Adjei	South Africa	Africa
Lutho Kota	South Africa	Africa
Nosipho Hanisi	South Africa	Africa
Paul Kibwika	Uganda	Africa
Marleen Willemsen	The Netherlands	Europe
Julio Beingolea	Ecuador	South America
Conny Almekinders	The Netherlands	Europe
Michael C. Slaby	Germany	Europe
Brandon P. Hollingshead	United States of Amercica	North America
Peter Blaze Corcoran	United States of Amercica	North America
Arjen Wals	The Netherlands	Europe
Tore van der Leij	The Netherlands	Europe

So indeed, this book can be seen as a network of somewhat diverse conversations, but whether these conversations, and the stories they draw from, lead to meaningful new conversations and new networks of conversations will depend on whether they travel further, beyond this book. This will depend on what the reader makes of them. Do they, or at least some of them, resonate and create a healthy dissonance that promotes reflexivity? How are they mirrored against one's own ideas, experiences and stories? How are they interpreted and re-interpreted against these ideas? How might they influence theory and practice elsewhere? The answers to these questions determine to a large extent the value of this book. That is: the value of this book for those outside the community that created it. For those who contributed to this network of conversations, by writing, editing and in some cases reading each others work and re-reading or re-interpreting one's own work, in all likelihood, will have benefited from the book before it even went to print.

References

Dewey, J. (1916) *Education and democracy*. New York: The Free Press.

Driver, R. and Oldham, V.A. (1986) Constructivist Approach to Curriculum Development in Science Education. *Studies in Science Education,* 13, 105-122.

Goodman, R.T. and Saltman, K.J. (2002) *Strange Love or how we learn to stop worrying and love the market*. Oxford: Rowan and Littlefield.

Kaufman, S. and Smith, J. (1999) "Framing and Reframing in Land Use Change Conflicts", *Journal of Architectural and Planning Research*, 16(2): 164-180.

Keen, M., Brown, V.A. and Dyball, R. (Eds.) (2005) *Social Learning in Environmental Management: Towards a Sustainable Future.* London: Earthscan,

Wals, A.E.J. and Heymann, F.V. (2004) "Learning on the edge: exploring the change potential of conflict in social learning for sustainable living", in A. Wenden, ed., *Educating for a Culture of Social and Ecological Peace.* New York: SUNY Press, pp. 123-145.

Wals, A.E.J. and Jickling, B. (2002) "Sustainability in Higher Education from Doublethink and Newspeak to Critical Thinking and Meaningful Learning", *Higher Education Policy,* 15: 121-131.

Wals, A.E.J. and Bawden, R. (2000) *Integrating sustainability into agricultural education: dealing with complexity, uncertainty and diverging worldviews.* Gent: ICA.

Afterword

Michael W. Apple

As this book so clearly demonstrates, we are living in a period of very real crises. They are ones that have infected all of our economic, political, and cultural institutions and have challenged the very notion of sustainability. The effects of these crises on our environment, our daily lives, and our future are numerous and extraordinarily dangerous. There are many questions one could ask about how we should respond to the issues we confront. But a crucial issue involves the following: What is the place of education in responding to the crises we are experiencing? We are told by neo-liberals that only by turning our societies over to the competitive market will we find a solution. Thus, education should be about and for 'the market'. This belief in market certainties is something of a religious commitment it seems, since we have a large amount of evidence that neoliberal policies both locally and internationally have helped create, not lessen, the immense economic, cultural, and environmental problems we are facing (Apple 2006, Davis 2006). But, if neoliberal 'solutions' and the possessive individualism associated with markets are not the answer, where should we turn? Are there models of deliberation and examples of engaging in critical reflection and action on alternatives that are more responsive?

Serious and sustained discussions of the implications of all this for education and learning are unfortunately few and far between or are scattered. Examples of how education and learning can be organized to interrupt dominant assumptions, policies, and practices are less apt to be found in official circles than are those that seem to ratify dominance. This is where the book you have read enters in thoughtful and important ways. It documents a considerable number of the conceptual, political, educational, and practical ways in which a renovated vision of social learning can be employed productively to generate imaginative possibilities.

Imaginative possibilities grow best when they are collective. During one of the times I was working in Brazil, I remember Paulo Freire repeatedly saying to me that education must begin in critical dialogue. Both of these last two words were crucial to him. Education must hold our dominant institutions in education and the larger society up to rigorous questioning and at the same time this questioning must deeply involve those who benefit least from the ways these institutions now function. Both conditions were necessary, since the first without the second was simply insufficient to the task of creating a critically democratic education and an engaged citizenry. This book is within the expanding tradition of work that enlarges the collective dialogue.

The idea of an expanding tradition is important, since this volume does not stand alone. After reading this book, it would be worthwhile to compare it to and build further upon other powerful models of critical education and critical learning. How can we enable all of our people, including youth, to understand and act on their world in critical ways? Work in the democratic schools movement in the United States provides some examples for what can be done in schools, for instance (see, e.g. Apple and Beane 2007).

But there are other sources to which we can turn to go further. The impressive and ongoing efforts to build 'Citizen Schools' and to employ 'participatory budgeting' in Porto Alegre, Brazil also documents how an education that is grounded in processes of social learning can lead to not only greater understanding of one's social and natural environments, but also to a willingness and ability to act to transform them (see Apple *et al.* 2003, Apple and Buras 2006). What is happening in Porto Alegre is connected to a larger project of using models of critical education to teach the state, to activate the voices of those who are often silenced in public dialogue, to build popular and effective alliances, and to challenge the ways in which accepted policies and practices contribute to the economic, cultural, and environmental crises that are so pervasive.

Joining what is so clear and refreshing in this book with the larger movements toward a critically democratic and activist education that is worthy of its name, is but one step in the struggle for sustainability. But it is an essential step if we are to use the insights that are included in this book.

References

Apple, M.W. (2006). *Educating the 'Right' Way: Markets, Standards, God, and Inequality*. New York: Routledge.

Apple, M.W. and Beane, J.A. (2007). *Democratic Schools*. Portsmouth, NH: Heinemann.

Apple, M.W. and Buras, K.L., eds. (2006). *The Subaltern Speak*. New York: Routledge.

Apple, M.W., Aasen, P., Kim Cho, M., Gandin, L.A., Oliver, A., Sung, Y.-K., Tavares, H. and Wong, T.H. (2003). *The State and the Politics of Knowledge*. New York: Routledge.

Davis, M. (2006). *Planet of Slums*. New York: Verso.

About the contributors

Conny Almekinders is a research fellow at the Chair Group Technology and Agrarian Development, Social Science Department of Wageningen University. She has worked on seed and crop genetic diversity issues in various settings and with a range of different organisations. Her involvement in the PhD programme on Participatory Approaches and Up-scaling reflects the opening up of her work to a wider field of agricultural technology and related issues of institutional change for more effective agricultural research. http://www.tad.wur.nl/UK/People/faculty/Conny/

Michael W. Apple is John Bascom Professor of Curriculum and Instruction and Educational Policy Studies at the University of Wisconsin, Madison and Professor of Educational Policy Studies at the University of London Institute of Education. Among his many books are *Ideology and Curriculum, Education and Power, Cultural Politics and Education, Official Knowledge*, and *Educating the 'Right' Way.*

Robert Asafo-Adjei, is a high school teacher in the uplands of the Eastern Cape near Queenstown in South Africa. He has been involved in teacher training and as an examiner in the district as well as teaching the Agricultural Sciences curriculum in the Further Education and Training Band (FET). Bob has recently co-authored a textbook in his field and the relating of classroom knowledge to local realities has been a key concern in his work.

Zenobia Barlow is a cofounder of the Center for Ecoliteracy. She has been the Center's executive director since its founding, providing leadership in applying theory to practice in a philanthropic strategy committed to education for sustainability. She previously served as executive director of The Elmwood Institute, an ecological think tank and international network of independent scholars and activists, after serving as executive editor of an international publishing company and as an academic administrator at Sonoma State University in California. She is an accomplished still photographer. She is coeditor, with Michael K. Stone, of *Ecological Literacy*. Email: Zenobia@ecoliteracy.org

Richard Bawden has been a Distinguished Visiting University Professor at Michigan State University since 2000. From 1978 to 1994 he was the Dean of Agriculture and Rural Development at Hawkesbury Agricultural College which was incorporated into the University of Western Sydney in 1989, at which time he was concurrently appointed Professor of Systemic Development. The Central thrust of his work relates systemic rural development to the intellectual and moral development of all who are engaged in the process. He is a joint author (with Frank Fear, Cheryl Rosaen and Penny Foster Fishman), of the recently published Coming to Public Engagement.

Julio Beingolea Ochoa is the National Coordinator of the NGO World Neighbours in Ecuador and Peru. He impels the programs of community development, the integrated programmes (population-environment), programmes involving participatory research with farmers using experiments, and programmes centring on capacity building and the strengthening of local – and farmers' organisations. He was professor at the National University of San Cristóbal of Huamanga in Peru, Director of the Agronomy School, and Coordinator of the Research Programme of Nutritious and Andean Cultivars, realising different types of research in quínoa, maize, potato, vegetables and pasture.

Hilary Bradbury (PhD) is Director of Sustainable Business Programmes at the University of Southern California, Center for Sustainable Cities. She brings her expertise in action research – a community design approach – to work with businesses around sustainability. Prior to this she was Associate Professor of Organizational Behavior at Case University, Weatherhead School of Management in Cleveland, Ohio. She has published widely in journals including *Organization Science* and *Academy of Management Executive*. She is co-editor of *Action Research* and the bestselling *Handbook of Action Research*. She grew up in Ireland and lived/ worked in Germany, Switzerland and Japan. She lives in LA with her family.

Valerie A. Brown AO (BSc, Med, PhD) is Director of a Local Sustainability Project and a Visiting Fellow at the School of Resources, Environment and Society of the Australian National University. She sits on national and international advisory committees on environmental sustainability and on public health. Her collaborative action research seeks community-based solutions to local and global sustainability, linking community, specialist, and administrative decision-makers. Her most recent books, alone and with co-authors, *Towards whole-of-community engagement: a practical toolkit*, 2004; *Sustainability and Health: supporting global ecological integrity in public health,* 2005 and *Social learning and environmental management: towards a sustainable future,* 2005; *Leonardo's vision: a guide to collective thinking and action, 2007.*

Fritjof Capra is a cofounder and chair of the board of directors of the Center for Ecoliteracy in Berkeley, California. In addition to research in physics and systems theory, he has been engaged in a systematic examination of the philosophical and social implications of contemporary science for the past 30 years. His books include *The Tao of Physics, The Turning Point, The Web of Life,* and *The Hidden Connections.* He serves on the faculty of Schumacher College in the United Kingdom, and lectures widely to lay and professional audiences in Europe, Asia, and North and South America.

Peter Blaze Corcoran is Professor of Environmental Studies and Environmental Education and Director of the Center for Environmental and Sustainability Education at Florida Gulf Coast University. He has written widely on environmental education, education for sustainable development, Earth Charter education, progressive education, and nature study. He is a Senior Fellow at University Leaders for a Sustainable Future and a Senior Advisor to the Earth Charter Initiative. With Arjen Wals, he was co-editor of *Higher Education and the Challenge of Sustainability: Problematics, Promise, and Practice* (Kluwer, 2004). He served as editor-in-chief of *The Earth Charter in Action: Toward a Sustainable World* (Royal Tropical Institute (KIT) Publishers, 2005).

Jacqueline Cramer is director of Cramer Environmental Consultancy after working as an associate professor at the University of Amsterdam (1976- 1989) and as senior researcher at the Centre 'Strategy, Technology and Policy' of TNO (1989-1999). During the last ten years she has worked with many companies on the implementation of sustainable entrepreneurship. Presently she is affiliated as well with the University of Utrecht as professor in sustainable entrepreneurship. Moreover, she is member of various (inter) national advisory boards of the government, industry and non-profit organisations. Email: jmcramer@xs4all.nl

Freddy Delgado first studied agronomic sciences at the University of Cochabamba in Bolivia and subsequently completed a PhD in rural sociology at the Institute of Peasant Studies and History (ISEC) at the University of Cordoba in Spain. Since 2000 he is the director of the Agroecology Programme of the University of Cochabamba (AGRUCO) in Bolivia (www.agruco.org). In research and lecturing he is focussing on bio-cultural diversity, agro-ecology and the relationship between ethno-history and sustainable development. Since 2005 he is director of the regional programme BioAndes financed by Swiss Development Corporation (SDC), which is an collaborative programme between Bolivia, Ecuador and Peru focussing on transdisciplinary research, capacity building and education in regard to the development of a concept and implementing strategy for a new development approach aiming to integrate biological and cultural diversity.

Robert Dyball lectures in the Human Ecology Programme at the Australian National University, which offers undergraduate courses and graduate research degrees in Human Ecology. Undergraduate courses are strongly focussed on experiential learning, and are designed to help students to understand their own agency in the dynamics of the systems about which they are concerned. Robert's current research is on furthering the application of dynamic systems thinking as a powerful means of understanding the characteristic changes in human-ecological situations. He is particularly interested in urban-rural interdependencies.

Zinaida Fadeeva is Research Associate for the Education for Sustainable Development programme at United Nations University Institute for Advanced Studies UNU-IAS). Her main responsibilities include the undertaking of research projects and organising development activities within the framework of the programme. Before joining UNU-IAS, Zinaida worked at the International Institute for Industrial Environmental Economics at Lund University, Sweden. Her research activity was related to inter-organizational environmental management. Her work focused on building an understanding of structural, dynamic and contextual characteristics affecting the cross-sectoral networking processes.

Harold Glasser is an Associate Professor in Environmental Studies at Western Michigan University. His research explores how individuals, organizations, and societies – from pre-agricultural to contemporary – make choices that affect the environment, which, in turn, affects them. An overarching theme is an effort to understand – and help bridge – the gap between stated, widespread ecocultural concern and present-day unsustainable actions, lifestyles, and policies. Glasser is Series Editor of the ten-volume *Selected Works of Arne Naess* (Springer, 2005). He has written on the philosophy of Arne Naess, deep ecology, environmental values and policy, multicriteria analysis, green accounting, education for ecocultural sustainability, and campus sustainability assessment. Email: harold.glasser@wmich.edu

Lieve Goorden received training in political and social sciences (University of Antwerp) and in labour and economics (Ph.D. University of Antwerp). She has been working for several years at the Flemish Foundation for Technology Assessment. Her research and advisory task focused on technology policy and technology assessment. At the TNO-centre in the Netherlands (Strategy, Technology, Policy-centre, University of Delft) she did research in the domain of information and communication technology. From October 1999 she became the director of the research cluster 'technology assessment' within STEM (Research Centre on Technology, Energy and Environment, University of Antwerp). Lieve.goorden@ua.ac.be. STEM-University of Antwerp, Kleine Kauwenberg 12, B-2000 Antwerpen, Belgium.

John Grin (MSc in physics, 1986; PhD in technology assessment, 1990) is a full professor in policy science at the University of Amsterdam. Throughout his career, his research interests have focused on issues of science, technology, society and policy. His recent work focuses on system innovations, such as the transformation of intensive agriculture to a more sustainable agriculture, or of current health care systems to a more demand driven, prevention oriented system. He is scientific director of the Amsterdam School of Social science Research (ASSR; www.assr.nl)

and co-director of the Dutch Knowledge Network on System Innovations (www.ksinetwork.nl). He is responsible for a post-graduate course on these issues.

Irene Guijt is an independent advisor and researcher focusing on learning processes and systems (including monitoring and evaluation) in rural development and natural resource management, particularly where this involves collective action. She is completing her PhD on the contribution of monitoring to trigger learning. Key publications include *Participatory Learning and Action: A Trainer's Guide* (co-author) and *The Myth of Community: Gender Issues in Participatory Development* (with M.K. Shah).

Nosipho Hanisi is a science teacher in Grahamstown. She has been active in the development of Eco-School and Health Promoting School Schools where school gardens are developed as sites of learning science and for the food production in response to poverty in the local community. Here the relating of indigenous knowledge practices and scientific ideas has become a key focus of her work.

Paul Hart is Professor of Science and Environmental Education at the University of Regina, Canada. He is an Executive Editor for the *Journal of Environmental Education* and consulting editor for several other journals. He has contributed to publishing in environmental education for many years, served on the NAAEE Board and has received several research and service awards. His current research interest is in the genealogical dimensions of learning as an interactional process.

Brandon P. Hollingshead is a recent graduate of Florida Gulf Coast University, where he studied environmental communication and literature. He is a research associate with University Leaders for a Sustainable Future and the Center for Environmental and Sustainability Education at FGCU. Hollingshead is a member of the Earth Charter Youth Initiative Core Group and actively participated in the editing of the book *The Earth Charter in Action: Toward a Sustainable World* (Royal Tropical Institute (KIT) Publishers, 2005). He plans to continue his interdisciplinary education at the graduate level, studying the role the humanities play in environmental and sustainability discourse.

Janice Jiggins is former Foundation Chair in Human Ecology, Department of Rural Development Studies, Swedish University of Agricultural Sciences, Uppsala, Sweden, 1998-2001. A social scientist who trained in the UK and Sri Lanka, she has worked for leading bilateral and multilateral international development agencies in Africa and South Asia for more than twenty-five years, in the fields of farming systems research, extension, participatory development, integrated pest management and gender issues. She continues to travel extensively and write in support of smallholder systems development, and adult learning-based approaches to sustainable resource management. From March 1, 2002, she has worked as a quest researcher at the Communication Science Group of the Wageningen University in the Netherlands and from December 2006 as a visiting fellow at IIED, London. Current research topics include sustainable water management and innovation in agriculture research systems.

Meg Keen is a visiting fellow at the Asia Pacific School of Economics and Government, Australian National University. She has worked on a range of issues related to social learning in environmental management including participatory resource management, local environmental management systems (EMS), and the use of EMS to improve aid effectiveness. She has conducted research throughout the Asia Pacific, but has recently concentrated on the South Pacific region. Her most recent book (with V. Brown and R. Dyball), *Social Learning in Environmental Management*, develops and applies social learning theory to environmental management in the global context.

Paul Kibwika is a lecturer in the Department of Agricultural Extension/Education, Makerere University in Uganda. He is the author of a book entitled *Learning to make change: developing innovation competence for recreating the African university for the 21st century*, also a PhD thesis obtained from Wageningen University, The Netherlands. He is also engaged in facilitating processes for transformation of organisations and people with research interests in: learning for change; innovations for rural/agricultural development; local organisations development; and knowledge management. Email: pkibwika@agric.mak.ac.ug

Lutho Kota teaches Consumer Studies in a High School near King Williamstown in the Eastern Cape in South Africa. She has been active in teacher professional development related to the implementation of the new Outcomes-Based Education System in the High School and the Further Education and Training Band (FET). A practical linking of classroom learning on issues of health and nutrition with local community needs has been one of her main interests.

Marianne E. Krasny is Professor and Chair in the Department of Natural Resources at Cornell University, and Director of the Cornell Initiative for Civic Ecology. Her scholarship focuses on the interface between science education, civic participation, cultural diversity, and the environment. She has provided leadership for programs focusing on urban environmental science and community education, graduate student outreach to high schools, and K-12 curriculum development.

Cees Leeuwis is professor of Communication and Innovation Studies at Wageningen University, the Netherlands. He regards innovations as a balanced whole of technical devices, mental models and institutional arrangements. His research focuses on (1) the value of cross-disciplinary approaches to bringing about coherent innovations, (2) the analysis of social learning and conflict management in networks, (3) changing dynamics and arrangements in knowledge networks due to among other things privatization of research and extension, and (4) the reflexive monitoring and evaluation of innovation support strategies and trajectories. Email: cees.leeuwis@wur.nl

Anne Loeber is a senior researcher and lecturer in the Department of Political Science at the University of Amsterdam, the Netherlands. She is also a member of the Technology Assessment steering committee that advises the Dutch Ministry of Agriculture. She obtained her PhD with a thesis on the role of interpretive analysis in inducing processes of learning for sustainable development, which was awarded a second place in 2005 in the Dutch Political Science Association's yearly selection of the best academic thesis in the field. As a participant in an EU-funded project on participatory governance and institutional innovation (www.paganini-project.net), her current research focuses on novel participatory arrangements as a component of the European polity for governing normatively and technically complex issues such as food safety. Email: a.m.c.loeber@uva.nl

Heila Lotz-Sisitka joined the environmental education unit at Rhodes University in 1995 and was appointed to The Murray and Roberts Chair of Environmental Education in 2001. She co-ordinates a National Research Foundation programme alongside an MEd coursework and research course and has worked extensively in support of the development of environmental education processes in South and southern Africa.

Peter Lund-Thomsen is Assistant Professor of Corporate Social Responsibility and Aid Management at the Centre for Business and Development Studies, Copenhagen Business School, Denmark. His current research focuses on corporate social and environmental responsibility (CSER) in developing countries with a regional focus on Southern Africa and South Asia. His main concern is the elaboration of a critical research agenda on CSER in the South while special interests include donor funding and the sustainability of CSER interventions in the South, public-private partnerships, and the role of community-based accountability strategies. Email: plt.ikl@cbs.dk

Sue McGregor is a Professor in the Faculty of Education, Mount Saint Vincent University and Graduate Coordinator, Educational Foundations Program (formerly Coordinator of the undergraduate Peace and Conflict Studies Program) . Her recent entry into the peace education field provides exciting synergy between 30 years of consumer, family and home economics education and peace, citizenship and human rights/responsibility education. She is Principal Consultant for the McGregor Consulting Group http://www.consultmcgregor.com. She recently completed her book on *Transformative Practice: New pathways to Leadership.* http://www.kon.org/news.html

Marcia McKenzie is a Social Sciences and Humanities Research Council of Canada Postdoctoral Scholar. Her research interests include education as socio-cultural practice, teacher education, epistemological issues of educational research, digital ethnography, student agency and activism, globalization, social justice, and the ecological. Funded by a SSHRC Standard Grant (2005-2008), Marcia's current research centres on a collaborative web-based hypermedia project entitled, *Discursive Approaches to Teaching and Learning.* Marcia has published in a range of journals and is co-editor of an in-progress book, *Fields of Green: Restorying Education* (Hampton Press). Marcia is also involved with a number of community-based non profit organizations. For more information, visit www.otherwise-ed.ca.

Yoko Mochizuki is a Postdoctoral Fellow for the Education for Sustainable Development Programme at the United Nations University-Institute of Advanced Studies (UNU-IAS). Prior to joining UNU-IAS, she was an Adjunct Assistant Professor at the Department of Human Development, Teachers College, Columbia University, and taught courses in Comparative Sociology of Education. Her current work explores the interactions of the global and the local in shaping education through a case study of *Regional Centres of Expertise on Education for Sustainable Development* (RCE).

Rob O'Donoghue co-ordinated environmental education in Ezemvelo KZN Wildlife for 18 years before being appointed Director of the Gold Fields Environmental Education Service Centre at Rhodes University in 2002. His research interests are culture, learning and social change.

Rhiannon Pyburn graduated from the University of Toronto in 1995 with a BSc in *International Development Studies – Resource Management.* For several years in between academic degrees she supervised youth programmes across rural Canada and Indonesia facilitating development and environmental education in inter-cultural contexts. Since 2000 Rhiannon has been based at Wageningen

University in the Netherlands where she obtained an MSc in the *Management of Agro-Ecological Knowledge and Socio-technical Change (MAKS)* programme and is currently writing her PhD dissertation. Both her MSc and doctoral research explore social learning and group certification for small farmers in developing countries. In 2002 she co-edited the academic book, *Wheelbarrows full of frogs – social learning in rural resource management.*

Niels Röling is Emeritus Professor of Communication and Innovation Studies, Wageningen University, The Netherlands. Trained in the Netherlands and the USA, he is a leading teacher, thinker and writer on interactive learning-based approaches to innovation. His career spans more than forty years' as a university-based teacher, professor and consultant. His work initially focused on extension and smallholder development, but gradually shifted to knowledge systems and social learning for sustainable natural resource management. His books include *Extension Science* (1988), and *Facilitating Sustainable Agriculture* (1998). Current research and PhD supervision includes collaboration in an EU-funded project on Social Learning for Integrated Water for Integrated Water Management and Sustainable Use of Water at Catchment Scale. Niels is a Board member of IIED (International Institute for Environment and Development, London). He is married to Janice Jiggins.

Stephan Rist studied agronomic sciences at the Swiss Federal Institute of Technology (ETHZ) and complemented his knowledge by a PhD in the field of rural sociology carried out at the Technical University of Munich in Germany. During 9 years he was co-director of the Agroecology Programme of the University of Cochabamba (AGRUCO) in Bolivia. Since 1999 he works at the Centre for Development and Environment (CDE) as a senior research scientist and lecturer focussing on social learning processes, transdisciplinarity and ethnoecology in the context of sustainable development in Europe, Latin America and India. Since 2005 he acts as scientific coordinator of the integrative component, called 'Transversal Package' consisting of 10 major inter- and transdisciplinary research partnership projects related to the mitigation of the syndromes of global change of the Swiss National Centre of Competence in Research (NCCR) North-South (http://www.nccr-north-south.unibe.ch).

David Selby David Selby is Professor for Education for Sustainability at the University of Plymouth where he directs the Centre for Sustainable Futures. He was previously (1992-2003) Professor of Education at the Ontario Institute for Studies in Education of the University of Toronto. His books include *Global Teacher, Global Learner* (1988), *Earthkind: A Teachers' Handbook on Humane Education* (1995), *In the Global Classroom, Books One and Two* (1999, 2000) *Weaving Connections:*

Educating for Peace, Social and Environmental Justice (2000). His immediately forthcoming book (with Jamie Gray-Donald) is *Green Frontiers: Environmental Educators Dancing Away from Mechanism* (Rotterdam, Sense Publishers, 2007).

Michael C. Slaby is Inter-Faith Coordinator of Earth Charter International, which is the coordinating secretariat for the world-wide Earth Charter Initiative. In this function, Slaby is currently operating in the Earth Charter Center for Strategy and Communication in Stockholm, Sweden. Having been actively involved in youth-led initiatives since 1996, Michael can look back on volunteer work focussing on sustainable development, human rights, and refugee aid. Inter alia, Slaby volunteered as the International Coordinator of the Earth Charter Youth Initiative for more than three years. Slaby holds a Master degree in comparative religion, international law and politics at Heidelberg University, Germany, and has written his master thesis on the discourses on religious diversity at the World's Parliament of Religions of 1893 and its Centenary.

Stephen Sterling is Schumacher Reader in Education for Sustainability at the Centre for Sustainable Futures, University of Plymouth, UK, and Senior Adcisor on ESD to the UK Higher Education Academy. He is also a Visiting Research Fellow at London South Bank University, and at the Centre for Research in Education and the Environment (CREE) at the University of Bath. His research interest is in the interrelationships between systemic change, learning, ecological thinking and sustainability, and his publications include the Schumacher Briefing *Sustainable Education – Re-visioning Learning and Change*, (Green Books/Schumacher Society, 2001). Email: stephen.sterling@plymouth.ac.uk

Michael K. Stone is senior editor at the Center for Ecoliteracy, a public foundation located in Berkeley, California. He was previously managing editor of *Whole Earth* magazine and the *Millennium Whole Earth Catalogue.* He has written for several publications, including *The New York Times* and the *Toronto Star,* and served on the staffs of the Lt. Governor of Illinois and the Illinois Arts Council. He was a founding faculty member at World College West, an innovative undergraduate institution in northern California, where he served as academic vice president. He is coeditor, with Zenobia Barlow, of *Ecological Literacy.* Email: mkstone@ ecoliteracy.org

Keith G. Tidball is an Extension Associate and Associate Director of the Initiative for Civic Ecology in the Department of Natural Resources at Cornell University, where he works to connect people with plants in urban contexts for purposes of education, community restoration and regeneration, and biodiversity conservation. He previously served as an International Affairs Specialist with the US Department of Agriculture's Foreign Service, where he worked on international development and sustainability in urban contexts, primarily in Asia and Africa. His research focuses on Environmental Security and Peacemaking, Community Based Approaches to Urban Natural Resource Management and Civic Ecology.

Daniella Tilbury is an Associate Professor in Environmental Education and Sustainable Development at Macquarie University, Sydney and Director of the Australian Research Institute in Education for Sustainability (ARIES). Daniella's doctoral study undertaken at the University of Cambridge in the early 1990s developed a framework for Education for Sustainability. Since then she has lectured at the postgraduate and undergraduate level in many universities across the globe and facilitated programmes for NGOs, corporate and government agencies in this area of learning. Daniella holds a number of international responsibilities: she is the Australian research fellow on the OECD's international research programme ENSI and is currently working with UNESCO to develop ESD indicators to monitor progress during the UN Decade of Education for Sustainable Development.

Joke Vandenabeele is a senior lecturer and part of the Centre for Research on Lifelong Learning and Participation at the University of Leuven in Belgium. She is responsible for the education and research on nonformal education and citizenship. The key topics of her research are: democratic practices, social learning, public debate, policy participation. Contact information: joke.vandenabeele@ped.kuleuven.be. Centre for Research on Adult and Continuing Education (CRACE, Department of Educational Sciences), Catholic University of Leuven, Vesaliusstraat 2, 3000 Leuven, Belgium.

Tore van der Leij is a graduate from Wageningen University in The Netherlands. He worked at the Yukon College, Canada, on a thesis on the role of environmental ethics in environmental education. Once back in The Netherlands, he co-authored an articles on the dangers of standardization in environmental education for the *Canadian Journal of Environmental Education*. After a number of years he chose to become a teacher in the Dutch high school system. For three years he combined the teaching job with working on organic farms. Presently Tore lives on a farm in a Dutch polder landscape.

C.S.A. (Kris) van Koppen is senior lecturer at the Environmental Policy Group of Wageningen University, and professor in Environmental Education, by special appointment, at the University of Utrecht. At both universities, he is involved in research and education in the field of management and policy of nature and environment, with a special interest in social learning. He is also a member of the Wageningen University Board of Education, and coordinator of the Environment and Society Research Network of the European Sociological Association. Email: kris.vankoppen@wur.nl

Barbara van Mierlo studied sociology at the University of Amsterdam. She works as Assistant Professor in 'Management of Innovations, Environmental Studies and Ecology' at the sub-department of Communication Science of the Wageningen University in the Netherlands. For more than ten years she was a senior consultant and researcher at an environmental research center connected to the University of Amsterdam. In 2002, she finished her dissertation on the way in which pilot projects can contribute to the diffusion of new sustainable technologies. Her current, main interest is in processes of (system) innovation towards a sustainable development by means of innovation projects. In her work she focusses on communicative processes like network formation, negotiating and learning in relation to change of practices and institutional change. Email: barbara.vanmierlo@wur.nl

Erik van Slobbe is senior consultant in the field of policy analysis and planning in river basin management. He is an employee of ARCADIS; a major Dutch consultancy firm. His clients are governments in the Netherlands and other European countries. Currently he is involved in the implementation process of the EU Water Frame Work Directive in the international river basins of the Rhine and the Maas. He is leader of the research programme "Performance indicators of interactive methods in water management in the context of local and regionally oriented planning and execution of works". In this role he is connected to the Communication and Innovation Studies Department of Wageningen University. His research is focused on social learning processes in water management.

Arjen Wals is an Associate Professor within the Education & Competence Studies Group of the Department of Social Sciences of the Wageningen University in the Netherlands. His PhD – obtained in 1991 from the University of Michigan in Ann Arbor, U.S.A., under the guidance of UNESO's first Director for Environmental Education, the late Professor William B. Stapp – explored the crossroads between environmental education and environmental psychology. He is the (co)author and (co)editor of over 120 publications on environmental education and learning in the context of sustainability. Email: arjen.wals@wur.nl

Urs Wiesmann is co-founder and deputy of the Centre for Development and Environment (CDE) at the University of Bern (www.cde.unibe.ch). He is a Professor of Human Geography specialised on peasant studies, integrative approaches to nature conservation, sustainable regional development and the development of transdisciplinary approaches of research focussing on North-South relationships. Since 2001 he is deputy director of the Swiss National Centre of Competence in Research (NCCR) North-South (http://www.nccr-north-south.unibe.ch) which is an international research partnership programme of seven Swiss research institutes and over 80 partner organisations representing South and Central America, East-, Southeast- and Central Asia, Africa, and Europe.

Danny Wildemeersch is a full professor of 'Social and Intercultural Education' at the University of Leuven in Belgium. Earlier, he was a full professor of 'Social Pedagogy and Andragogy' at the University of Nijmegen in the Netherlands. He is head of the research centre on Social Pedagogy and coordinates a research group on Intercultural and International Education. His research focuses on a variety of themes such as intercultural learning, learning and social participation, intercultural dialogue, learning and citizenship, environmental learning, transitions from school to work, and participation in development cooperation. He has published widely and in various languages, in books and papers on subjects like 'Experiential Learning', 'Learning for Social Responsibility', 'Learning for Inclusion', 'Learning Citizenship' and 'Social Learning'.

Marleen Willemsen works as a junior Programme Officer at the Sustainable Economic Development bureau of HIVOS, a Dutch NGO (http://www.hivos.nl/). She finished her Masters in International Development Studies at Wageningen University in 2005. She did field work for her two theses in the highlands of Mexico on empowerment processes (2003/2004), and in Ecuador about the design of a new project with farmers, Local Organisations and NGO's (2005). Her thesis in Ecuador about 'multiple level learning in participatory project design' formed the basis for the chapter she co-authored for this book. In 2006 she finished a postgraduate master course in advanced development studies at the Center for International Development Issues Nijmegen (CIDIN).

Jim Woodhill is Director of the Programme for Capacity Development and Institutional Change at Wageningen University and Research Centre in the Netherlands. He has worked for 20 years as a manager, researcher, educator and consultant in the fields of agricultural, rural development, natural resource management and international development assistance. His particular expertise is in the application of systems thinking and participatory learning approaches to complex development issues. Over recent years he has worked in particular on methodologies for the facilitation of multi-stakeholder processes and innovative approaches to monitoring and evaluation.

Index

A

action
- communicative – 289
- competence – 19
- concerted – 426
- dimension – 108
- inquiry – 313
- learning – 121, 497
- networks – 286
- research – 27, 121, 497

actionability – 280, 287
active social learning – 49, 51, 52
activism – 28
activist education – 508
adaptive
- capacity – 462
- confrontation – 461
- learning – 36, 150
- management – 47

advanced liberalism – 103, 115
advocacy – 285
agency – 28
- class-based sense of – 343
- contingent – 331, 332, 346, 348
- human – 446
- human reflexive – 436
- moral – 364
- reflexive – 437
- sense of – 341

agri-culture – 465
Agricultural Age – 35
Agricultural Science curriculum – 438
Alm, Per Uno – 283
alternative rice farming – 388
American Community Gardening Association – 157
Amin, Idi – 451
Amnesty International – 334
Andean
- cosmovision – 229
- land-use – 25

anthropocentrism – 39, 167
apartheid – 441- 443
Appalachian Trail – 52
asset-based approaches – 150
atmosphere of trust – 95, 270, 271
attention – 49
attentiveness – 170
authentic – 364
- participation – 491

authenticity – 331
autonomy – 499

B

Bandura, Albert – 48-50
Beck, Ulrich – 436
behavioural
- potential – 359

behaviourial
- capability – 356

biodiversity – 29, 198, 199, 203, 386, 389-391, 394
- loss – 42, 43

biological diversity – 36
biomimicry – 53
Bolivia – 235
Bosnia-Herzegovina – 155
Boston Metropolitan Park System – 52
bottom-up approach – 302
Brazil – 507
brokerage – 462
Brower, Dave – 44
Bush, George – 53
business management literature – 298, 309

C

California freshwater shrimp – 409
Canada World Youth – 334
capabilities
  development – 136
  intellectual – 136
  moral – 136
capacity
  building – 117, 121
  carrying – 43
  for learning and adaptation – 156
  to forget – 361
capital
  financial – 153
  human – 153
  natural – 153
  physical – 153
  social – 153
Case Western Reserve University – 288
CAT (Confrontation, Accusation, Threat) – 413
catchments – 421
certification – 209
change
  agents – 293
  conceptual – 500
  ecological – 47
  human-cultural – 279
  in participation – 481
  on system level – 268
  planned – 47
  social – 47, 361
  systemic – 119, 280, 292
  technical – 279, 292
  towards sustainability – 467
chaos theory – 29
children's play – 378
Christian religiosity – 239
Ciskei Homeland – 442
citizen involvement – 421
citizen participation
  in advocacy and structural change – 142
  in economic life – 142
  in local development and service delivery – 142
Citizen Schools – 508
citizenship – 103
  strengthening – 141
city forest dialogue – 106
civic
  activism – 23
  ecology – 23, 151, 158
civil society – 133, 372
  activism – 22, 134
  domains of participation – 135
  organisations – *See:* CSO
  participation – 134
classifications of participation – 466
clearing house of information – 273
co-creation – 499
co-creation of knowledge – 429
co-evolution – 68
co-learning – 51
co-management – 111, 112
co-operation – 181
co-produce knowledge for sustainable development – 229
codes of conduct – 210
cognition – 140
  collective – 213, 454
  distributed – 213, 454
  epistemic – 140
  meta – 140
  multiple – 213, 454
cognitive
  competence – 25, 240
  convergence – 209
  individual development – 135
  social development – 135
coherence – 213
collaboration – 188
collaborative – 53

action – 497
learning – 497
reframing – 483
collective – 241
dialogue – 507
problem solving – 20, 100, 101
colonial intrusion – 439
comfort zones – 498
communal capital – 30
Communal Nature Development Plans – 104
communication
multilateral – 109, 113
skills – 175
unilateral – 109, 113
communion skills – 175
community
education – 120
greeners – 156
of life – 494
problem-solving – 497
sustainability indicators – 52
workshops – 473, 477
competences – 102
cognitive – 231
emotional – 231
social – 231
complex human systems – 287
conflict – 453, 498
environmentalists-farmers – 402
management – 187
peaceful – 489
transformation – 489
confrontation – 498
congruency – 90, 94, 95
connectedness – 364
consensus – 290
building – 90, 285
seeking – 290
consequences – 352
consultation – 199
consumer
accountability – 29
action – 119
behaviour – 358
education – 28, 351
empowerment – 363
lifestyle – 443
type 4 education – 364
types of education – 364
consumer-citizens – 364
consumerism – 29, 342, 369
political – 373
sociological theories – 369
Consumer Studies curriculum – 444
consumption – 369, 493
contextualization – 53
contextual praxis – 437
continuous experimentation – 460
conventionalism – 114
conversing – 286
cooperative – 222
coordination mechanisms – 422
Copenhagen Business School – 305
Corporate Social Responsibility – *See:* CSR
correspondence – 213
corruption – 124
creative
balance – 102
tension – 102
creativity – 174, 175, 497
crisis – 76
critical
education – 508
learning – 508
level of consciousness – 480
meaning-making – 436
networking considerations – 253
pedagogy – 364
reflective thinking – 125
research agenda – 300, 301, 306, 308
thinking – 120, 125
thinking skills – 53

CSO – 22, 134
governance, programming, monitoring, and accountability' – 142
CSR – 26, 27, 89, 122, 210, 265, 267, 269, 270, 274
and development – 297, 298
business case approach – 299
critical approach to – 309
critical research agenda – 298
impact assessment – 304
impact of initiatives – 300
implementing – 270, 274
power and participation in – 300, 301, 304
target audiences – 303
voices in the debate – 302
cultural
capital – 30
diversity – 56
maladaptation – 50
narratives – 333, 341
pillar – 362
pluralism – 447
culture
of nonviolence – 495
of peace – 493, 495
of tolerance – 495
shock – 177
curriculum changes – 440
cyberspace – 490

D

debate – 169
deconstruction – 498
deconstructive postmodernism – 75
defensive routines development – 88
deforestation – 42
deframing – 498
deliberation – 445, 501
deliberative
dialogue – 289
mediation – 447
democracy – 135
anticipatory – 173
democratic
governance – 133
institutions – 495
participation – 134
pedagogy – 365
schools movement – 508
Denmark – 41
depletion of resources – 493
DESD – 117, 126, 129, 246, 248
vision – 247
design of a project – 472, 476
determinism – 175
Det Naturliga Steget – 281
development empowerment – 363
Dewey – 51
dialectic processes – 445
dialogical
learning circle – 171
social learning – 23
dialogue – 138, 170, 285, 290
dichotomies – 166, 206
dis-integrative practice – 69
disaster in cities – 151
disconnective thinking – 69
discourse
analysis – 332
anthropocentric – 334
of contingent knowing – 343
of critique – 344
of educational bias – 344
of lifestyle activism – 341
of neutrality – 334, 341
of subjective knowing – 344
of subjectivity – 335
unexamined – 341, 345
discursive
constitution – 332
frame – 348

framing – 348
understandings – 28
discussion – 169
disequilibrium – 176
dissensus – 501
dissipation – 176
dissipative structures – 175
dissonance – 19, 497, 498
distributed cognition – 90, 485
diversity – 149
dualism – 66, 166
Dutch National Initiative for Sustainable Development – 26, 265

E

Earth Charter – 31, 483, 493
Youth Initiative – 31, 483
Earth Council – 484
eco-tourism – 392
ecocultural
degradation – 46
sustainability – 20, 36, 39, 44, 46, 51, 52, 54
sustainable behaviors – 56
unsustainability – 53, 55
ecofootprints – 279
ecological
consciousness – 21, 64
epistemology – 21
integrity – 495
rationality – 375
realism – 75
sustainability, definition – 407
worldview – 21, 67
ecologists – 430
economic
justice – 489, 493, 495
pillar – 362
Ecosystem Wellbeing Index – 41
Eco Trekkers
Earth Charter Youth Group – 487
Society, Inc. – 486
Ecuador – 31
educated incapacity – 55
education
adult – 99
comparative – 99
continuing – 99
for All movement – 246, 247
for sustainability – 13, 122
for sustainable development – *See:* ESD
intercultural – 99
educational neutrality – 339
educative deliberations – 439, 441, 444
eidos – 66
Einstein – 17
Elkington – 265
emancipation – 102
emergence – 14, 407
facilitation of – 14
emergent curriculum – 170
emotional
competences – 25
intelligence – 283
employee motivation – 299
empowerment – 289, 351
enabling environment – 158
engagement – 189
England – 43
enlightenment – 102
entrepreneurial
development – 31
skills – 461, 462
environmental
action – 378
action networks – 379
attitudes – 41
degradation – 41, 124
devastation – 493
education – 377, 437
management – 39, 334
pillar – 362

policy – 39
project-based learning – 408
sociology – 373
environmentalism – 120
Environment and School Initiatives – 27
envisioning – 124
episteme – 315
epistemological underpinnings – 314
equilibrium – 175
Eriksson, Karl Erik – 281
ESD – 22, 26, 246, 249, 400, 500
content – 247
strategic perspectives – 247
ethical – 361
ethnocentric approaches – 39
ethos – 66
European Union's Water Framework Directive – 30
exchange visits – 457
expectations – 355
experts – 401
and non-experts – 254
experience-based – 255, 256
non-certified – 401
explicate order – 167
extension – 450, 452, 459, 461
external locus of control – 360
extinction of species – 493

F

facilitation – 121, 426
facilitation-team level learning – 480
facilitator – 173
factors
behaviour – 362
environment – 362
personal – 362
failed cities – 149
Fair-trade – 210
Labelling Organizations International – 25, 210
farmers empowerment – 402
farmer support – 393
Finland – 41
Flanders – 104, 105
flow – 168
food security – 31
foresightful behaviour – 358
forest extension – 105
Foucauldian perspective – 99
Foucault – 331
frame – 356
awareness – 498
deconstruction – 498
framing – 112, 332
freedom – 138
Freire, Paulo – 507
future-oriented thinking – 362
future activities – 177
fuzzy logic – 292

G

gardening – 153
Gardner, John – 54
geese
greater white-fronted – 385-387
wild – 385-387, 389, 392-394, 402
gender – 458
generative order – 167
genetic erosion – 468
genuine participation – 128
global
climate change – 56
education – 331
education pedagogy – 29
security – 494
warming – 401
Global Education – 333, 341
globalization – 242, 400
globalized economic development – 334
Global Learning Space – 251, 252, 260
GMOs – 488

Google-growth pattern – 32
Google-hits – 32
Gorbachev, Mikhail – 484
governance – 103
for sustainability – 230
mechanism for ESD – 250
governmentality – 100, 102, 103
greening of progress – 44-46
Green Wall – 268
group certification – 210

H

Habermas – 20, 230
habitat
destruction – 410
restoration – 406
habituation – 167
Hanoi – 107
hierarchy – 422
higher scales – 431
Hillview Global Education – 334, 335
Hillview Secondary School – 333
Hindu – 485
historical necessity – 77
Hitler – 53
HIV-AIDS – 441, 443
holarchies – 68
holism – 124
holistic thinking – 119
Holmberg, John – 281, 295
hologram – 168
holomovement – 29, 167
human
diversity – 154
ignorance – 46
services provision – 39
solidarity – 494
human-induced species extinction – 42
human/nature divide – 166
hurricane Katrina – 157

I

I-Thou – 68
identity – 371
Imaginal exercises – 177
imifino – 439, 440
implicate order – 167
inclusive participation – 495
inconvenient truths – 17
indigenous
fermented foods – 441
food preparation – 445
heritage – 444
knowledge – 398, 435, 436
land-use system – 229
peoples – 495
propositions – 445
vegetable plants – 439
ways of knowing – 30
Industrial Revolution – 35
innovation diffusion – 20
inquiry – 285
instability – 175
institutional
self-renewal – 44
structures – 293
transformations and sustainable development – 93
institutions – 419
of higher education – 255
instruments, systemic – 92
integrated
curriculum – 414
understanding – 187
integrated-systems design – 53
integration
horizontal – 186
vertical – 186
integrative thinking – 124
intelligence – 169
interactive dialogue – 460
intercultural

interdiscursivity – 348
interdependence – 419
interdiscursivity – 335, 348
interface between social institutions – 363
intergenerational
exchange – 458, 459
platforms – 458
practice – 445
sustainability – 458
ways of knowing – 435
Intergovernmental Panel on Climate Change – 41
internal
control systems – 210
locus of control – 360
reinforcement – 356
International Affairs – 306, 307
International Federation of Organic Agriculture Movements – 25, 210, 211
International Labor Organization – 212
International Research Network on Business, Development and Society – 305, 307, 308
interpersonal
connections – 457
dialogue – 280
interrelatedness – 363
Inuit
– 485
Inuit Circumpolar Conference – 485
invasive species – 42
investment in social learning – 431
ISEAL Alliance (International Social and Environmental Accrediting and Labelling Alliance) – 224
Israel – 154
issue of
allocation – 204, 205
location – 204, 205
regime – 204, 205
IUCN – 123

J

Japan – 29, 385
JAWPG (Japanese Association for Wild Geese Protection) – 29, 386-389, 395
joined-up thinking – 124

K

Kabukuri-numa – 29, 385-387
Kirkwood Secondary School – 338
knowledge
base – 304
co-creation – 250
context-specific – 127
dimensions – 242
expert – 206, 255
local – 127
loss of – 469
matrix – 191
production and consumption – 254
systems – 446
transfer – 435
Kramer, Samuel Noah – 40

L

labelling – 210
landscape heterogeneity – 153
Lao Tzu – 54
Latour – 230
Lawson College – 343
learning – 71
action-oriented processes – 85
and action – 84
anticipatory – 47
arrangements – 380
as an essentially social practice – 88
as the process of reviewing the 'theories-in-use' – 94
at company level – 271

at group level – 269
at system level – 273
based change – 117, 119, 120, 122
by default – 73
by design – 73
by doing – 103
by groups and organizations – 88
by individuals – 85, 266
collective – 230, 471, 478
community-based – 13
community level – 479
conditions for – 86, 88, 89
congruency as the outcome of – 90
cross-cultural – 177
cycle of Kolb – 481
cycles – 467, 470
double loop – 87, 354
ecological – 174
endogenous – 174
essential for sustainable development – 83
experiential – 100
first order – 71, 87, 267
grassroots – 497
higher order – 72
in social interaction – 94
levels – 470
life-long – 254
multiple level – 479
on company level – 268
organization – 119, 212
process – 469
reflexive – 435
second order – 72, 87, 89, 267, 271
second order, conditions for – 268
single loop – 87, 111, 354
societies – 23, 63, 133
theory – 313
third order – 72
transformative – 174
transposed – 354
trial-and-error – 48
triple loop – 189
vis-a-vis exchange of information – 275
levels of awareness – 480
liberation – 174
license to operate – 299
lifestyle
activism – 338
choices – 30, 437, 440
lifeworld – 288, 289, 293
Limits to Growth – 42
livelihoods – 30
living systems – 67
LLINC (Limestone Landscape Improvement and Nature Conservation) – 107, 110
Local Agenda 21 – 122, 126
locus of control – 360

M

Madagascar – 455
mainstream business settings – 297
maladaptation – 71
maladaptive behaviors – 50
management
system – 212
theory – 39
mapping – 291
market – 422
dynamics – 459
matters of concern – 203
Meadows, Donella – 17
meaning
coherence of – 173
incoherence of – 172
meaning-making interaction – 435
mechanistic worldview – 23, 165
mediated interactions – 445
memory – 168
Mencius – 43
mentoring – 121

metaphysical assumptions – 165
methodological perspectives – 313
Millennium Development Goals – 246
Millennium Ecosystem Assessment – 41
modernisation – 238
modernity
  institutions – 136
Montessori – 331, 338, 341
Montreal Protocol – 42
Motivation – 49
multi-stakeholder
  constituencies – 136
  dialogues – 460, 462
  learning – 472, 478
  social learning – 406
  workshop – 478
multilateral communication – 105
multiple discourses – 332
Mumford, Lewis – 52

N

nature
  conflicts – 203
  conservation – 24, 110
negotiation – 24, 109, 187, 188
neoliberal – 507
nested systems – 412
The Netherlands – 53
network building – 285
networking – 257
networks – 422
  informal – 453
neurobiology – 39, 45
neutral technologies – 115
New York – 152, 156
New Zealand – 53
NGO-level learning – 480
NGOs – 134
non-tillage farming method – 388, 390, 393, 395, 401
non-violence – 489
nonlinear relationships – 414
nonviolence – 495
normative methods – 285
norming – 339
Norway – 41

O

observational learning and modeling – 352
online communication – 31
onto-political proceedings – 206
  Plato's allegory of the cave – 205
  process criteria – 205
  separation of facts and values – 205
organic rice – 396
  farming – 388, 390, 392, 393, 395
organizational
  change – 39, 122
  learning – 212
Organizational Development – 288, 289
Our Common Future (1987) – 38
out-of-the-box thinking – 53
outcome – 360
Outcomes Based Education – 441
ownership – 109, 114, 398

P

Palestine – 154
panexperientialist – 68
paradigm
  change – 75
  shift – 270
Parks, Rosa – 332
participation – 127, 188, 466
  content-based arguments – 198, 202
  moral arguments – 197
  pragmatic arguments – 197
participative
  inquiry – 121
  reality – 63

participatory
- approaches – 424
- democracy – 178
- learning processes – 22
- planning – 24, 102, 103, 113
- processes – 22
- thought – 171

Participatory Rural Development – 291
partnerships – 182, 402
- cross-sectoral – 126
- environmentalists-farmers – 385, 399, 401
- North-South – 491
- school-community – 436

passive social learning – 49, 51
patterns of interpretation – 237
peacemaking – 489
pedagogy – 361
Perellona – 396, 397
performativity – 316
personal engagement – 280
phronesis – 315
The Philippines – 486
platform – 424, 456
Plato – 43
plenary dialogue – 291
plurality of ways of knowing – 435
policy
- amnesia – 182
- planning – 115

pooling of resources – 460
Porto Alegre – 508
portraits of resistance – 28, 331, 332
positive
- feedback loops – 150
- reinforcement – 354

possessive individualism – 507
post-apartheid – 30
post critical – 325
poverty – 124, 449
- relief – 445

power
- cube – 143
- dynamics – 104
- inequality – 112
- relational – 133
- relations – 107, 181, 461

practices – 371
practitioners
- methodical principles for – 84
- principles for – 92

praxis – 66
price mechanisms – 424
principles of dialogue – 187
privatisation – 238
problem-solving capacity – 100
problem of losing seeds – 475
professional ignorance – 449, 452
proprioception of thought – 170
prototyping ideas – 291
public participation – 379
public relations benefits – 299

Q

quantum
- drama – 177
- leap – 23, 176
- learning – 17, 23, 177
- physics – 29, 173
- society – 178

R

radical value change – 45
Rainforest Alliance – 25
Ramsar – 385-387, 393, 395-397, 399, 402
RCE – 25, 246, 248, 249, 385
- administrative structure – 258
- assessment of a role – 260
- challenges – 253
- collaborative networks – 262
- concept – 253
- contributions – 260

diversity – 258
expectations and assessments – 258
functions – 250
geographical range – 249
goal – 249, 258
governance mechanisms – 262
heterogeneous actors – 257
homogeneous actors – 257
interactive structure – 258
membership – 256, 257, 258
model – 250
network – 249
new governance structure – 250
power – 258
principles of action selection and assessment – 259
research and development activities – 250, 251
scale of activities – 260
scientific community – 255
status of implementation – 251
strategy – 249
transformative education – 251
value of collaboration – 263
re-cognition – 70
re-perception – 70
reciprocal determinism – 352
reconstruction – 498
recursiveness of practices – 93
reductionism – 165
reflection – 183, 497
dimension – 108
reflection-in-action – 86
reflective case studies – 20
reflexive
engagement – 437
modernization – 136
perspective – 84
praxis – 435
reflexivity – 28, 104, 183, 332, 341, 347, 348, 372, 436, 497
reframing – 412, 417, 498
Regional Centre of Expertise – *See:* RCE
regulation – 423
relational thinking – 124
reproduction – 49
resilience – 23, 78
theory – 23, 151
Resilience Alliance – 149
resistance – 348
resolution – 489
resource
dilemmas – 419
management – 39
respectful interaction – 287
response-ability – 70
retention – 49
rice-eating pest – 386
Rice Paddy
Fauna Survey – 390
rice paddy – 29, 385, 390
conservation value – 391
ecosystem – 391
in Monsoon Asian – 398
multi-functionality – 395
reassessment – 394
Rio Summit – 126
Rio United Nations Conference on Environment and Development – 484
Risk Society – 436
Robèrt, Karl-Henrik – 281
Roeselare – 106
roles of officials – 202
Royal Institute of International Affairs – 306
rural way of life – 443
Rwanda – 53

S

San Francisco Bay Area – 30
Satoyama – 394, 400
Schön – 24

school-community interactions – 437
school-in-community
  interface – 437
  research – 446
scientific
  expertise – 256
  propositions – 445
seed – 474
  as a public good – 479
  system conservation – 31
self
  as agent – 360
  discrete – 171
  extended – 171
  liberation from – 174
  liberation of – 174
self-determination – 241, 332, 499
self-efficacy – 356
self-esteem – 358
self-evident assumptions – 105
self-organization – 150
  capacity – 155
self-reflective capability – 357
self-regulation – 357
self-worth – 358
separation of facts and values – 206
shape the environment – 158
Sierra Leone – 486, 488-490
SLIM – 420
slow food – 399
small-scale farmers – 450
smallholder
  farmers – 31
  producers – 209
SMART – 500
social
  action – 436
  auditing – 210
  capital – 25, 231, 361, 497
  certification – 209
  cohesion – 497
  constructivist – 436
  dialogue – 303, 305-309
  energy – 457
  innovations – 50
  justice – 495
  maladaptation – 50
  narratives – 333
  practices – 374
  relations – 361
  spaces for interaction – 425
  technologies – 115
  transformation – 112, 113, 361
social-ecological systems – 149
Social Accountability in Sustainable Agriculture – 25, 210
Social Accountability International – 25, 210
socialisation – 237
societal learning – 23, 134
society pillar – 362
socio-ecological
  activism – 28, 341
  change – 346
  realities – 446
  risk – 435
socio-technical objects – 427
socio-technological development – 21
soft systems thinking – 47
South Africa – 30, 437
Southern-centred perspectives – 298, 300, 302, 304
Soweto – 152
species extinction – 124
sphere
  of concern – 74
  of influence – 74
spiritual well-being – 495
stakeholder
  management – 301
  theory – 301
standards
  environmental – 210
  human rights – 210

labor – 210
social – 210
trade – 210
status competition – 370
Stone Age – 35
strands for social learning – 183
strategic problem orientation – 95
Strong, Maurice F. – 484
Students and Teachers Restoring a Watershed – 30
subjectivity – 332, 345
Sumerian civilization – 40
sustainability – 229
as a network property – 415
as a property of networks – 407
thinking – 292
transition – 65
sustainable
agriculture – 29, 209, 386, 397, 398, 400, 401
community – 13, 407
consumer empowerment – 351
consumption – 355, 374
education – 78
food choices – 444
human livelihoods – 30
land-use system – 229
lifestyle choices – 446
livelihood – 461, 462
water supply system – 109
Sustainable Agriculture Network – 25, 210
sustainable development – 133, 297
and learning – 83
contestable concept – 84
definitions of – 83
human and social – 359
normative concept – 84
revolutionary concept – 84
three pillars – 362
Sweden – 281
SWOT – 474, 476
symbolic modeling – 353
synthesis – 186
system
behaviour – 185
conditions – 295
imperfections – 91, 92
innovation – 91
learning – 20, 94, 95
theory – 20
thinking – 24, 89, 185
systemic
instruments – 95
socio-ecological activism – 348
thinking – 120, 123

T

Tajiri – 385, 393
Taoism – 54
techne – 315
technology
of power – 331
of the self – 331
terrorism – 124
the economy – 363
The Natural Step – 26, 279, 280
theories-in-use – 52, 87
theory
of change – 290
of living systems – 14
of organizational flexibility – 284
of practice – 290
Theory of Communicative Action – 20
thought – 168
tolerance of diversity – 445
traditional crops – 449
transformative
change – 365
learning – 23, 183
transition
real – 176
virtual – 176

Triple-P (People, Planet, Profit) – 275
trust/identity – 141

U

Uganda – 31, 449-451, 461
umfuno – 438, 440
umqombothi – 441, 442
unbroken wholeness – 168
UN Decade in Education for Sustainable Development – *See:* DESD
UNECE – 500
UNEP – 500
UNESCO – 22, 37, 38, 42, 129, 246, 400, 500
  Decade of Education for Sustainable Development – 35
UNICEF – 108
unilateral communication – 105
United Nations University – 246, 248
United States – 43
universal human rights – 493
UN Literacy Decade – 246, 247
unprofessional experts – 452
unsustainability – 122
unsustainable – 17
urban-socio-ecological systems – 23
urban community
  greeners – 23
  greening – 151-153

V

vanilla – 31, 450
  farmers – 449, 454
viable alternatives – 363
VIBEKAP (Vietnamese Belgian Karst Project) – 107
Vietnam – 107, 110
virtual reality – 377
visualization – 291
visualizing – 291
Vitrivius – 43
volunteer conservation action – 119

W

Wals, Harry – 5
waterfowl habitat – 389
Water Framework Directive – 420
wave/particle duality – 173
wave function – 175
Ways of Working for Sustainable Development – 303
weir – 427
wet-rice agriculture – 390, 396
wetlands
  wise use – 388
WFRF – 388-395, 399
wicked problems – 181
win-win situations – 299
Winter-Flooded Rice Fields – *See:* WFRF
World
  Bank – 110, 287, 288, 500
  Commission on Environment and Development – 38, 83
  Trade Centers – 157
world
  problematique – 64
  view – 361
WTO – 400

X

xenophanes – 42

Z

zeitgeist – 79